VOLUME FOUR HUNDRED AND FORTY-SEVEN

METHODS IN
ENZYMOLOGY

RNA Turnover in
Prokaryotes,
Archaea and Organelles

METHODS IN ENZYMOLOGY

Editors-in-Chief

JOHN N. ABELSON AND MELVIN I. SIMON

Division of Biology
California Institute of Technology
Pasadena, California, USA

Founding Editors

SIDNEY P. COLOWICK AND NATHAN O. KAPLAN

VOLUME FOUR HUNDRED AND FORTY-SEVEN

METHODS IN
ENZYMOLOGY

RNA Turnover in Prokaryotes, Archaea and Organelles

EDITED BY

LYNNE E. MAQUAT
Department ot Biochemistry and Biophysics
University of Rochester
School of Medicine and Dentistry
Rochester, NY, USA

CECILIA M. ARRAIANO
ITQB-Instituto de Tecnologia Química e Biológica
Universidade Nova de Lisboa
Oeiras, Portugal

AMSTERDAM • BOSTON • HEIDELBERG • LONDON
NEW YORK • OXFORD • PARIS • SAN DIEGO
SAN FRANCISCO • SINGAPORE • SYDNEY • TOKYO

Academic Press is an imprint of Elsevier

ELSEVIER

Academic Press is an imprint of Elsevier
525 B Street, Suite 1900, San Diego, California 92101-4495, USA
84 Theobald's Road, London WC1X 8RR, UK

For information on all Elsevier Academic Press publications
visit our Web site at elsevierdirect.com

ISBN-13: 978-0-12-374377-0

PRINTED IN THE UNITED STATES OF AMERICA
08 09 10 11   9 8 7 6 5 4 3 2 1

CONTENTS

Contributors

Hiroji Aiba
Division of Biological Science, Graduate School of Science, Nagoya University, Chikusa, Nagoya, Japan

Soraya Aït-Bara
Laboratoire de Microbiologie et Génétique Moléculaire, UMR 5100, Centre National de la Recherche Scientifique et Université Paul Sabatier, Toulouse, France

Mónica Amblar
Unidad de Investigación Biomédica, Instituto de Salud Carlos III (Campus de Majadahonda), Madrid, Spain

Cecília Maria Arraiano
Instituto de Tecnologia Química e Biológica, Universidade Nova de Lisboa, Oeiras, Portugal

Ana Barbas
Instituto de Tecnologia Química e Biológica, Universidade Nova de Lisboa, Oeiras, Portugal

David H. Bechhofer
Mount Sinai School of Medicine of New York University, Department of Pharmacology and Systems Therapeutics, New York, USA

Joel G. Belasco
Kimmel Center for Biology and Medicine at the Skirball Institute and Department of Microbiology, New York University School of Medicine, New York, USA

Lionel Bénard
CNRS UPR 9073 (affiliated with Université de Paris 7; Denis Diderot), Institut de Biologie Physico-Chimique, Paris, France

Jonathan A. Bernstein
Department of Pediatrics, Stanford University, Stanford, CA 94305, USA

Sandrine Boisset
INSERM U851, Centre National de Référence des Staphylocoques, Université de Lyon I, Lyon, France

Douglas J. Briant
Department of Biochemistry and Molecular Biology, University of British Columbia, Vancouver, British Columbia, Canada

Agamemnon J. Carpousis
Laboratoire de Microbiologie et Génétique Moléculaire (LMGM), Centre National de la Recherche Scientifique (CNRS) and Université Paul Sabatier, Toulouse, France

Helena Celesnik
Kimmel Center for Biology and Medicine at the Skirball Institute and Department of Microbiology, New York University School of Medicine, New York, USA

Clément Chevalier
Architecture et Réactivité de l'ARN, CNRS, Université de Strasbourg, Institut de Biologie Moléculaire et Cellulaire, Strasbourg, France

Mikkel Christensen-Dalsgaard
Institute for Cell and Molecular Biosciences, Medical School, University of Newcastle, Newcastle, United Kingdom

Ciarán Condon
CNRS UPR 9073 (affiliated with Université de Paris 7; Denis Diderot), Institut de Biologie Physico-Chimique, Paris, France

Elena Conti
Max-Planck Institute of Biochemistry, Martinsried, Germany

Glen A. Coburn
Department of Biochemistry and Molecular Biology, University of British Columbia, Vancouver, British Columbia, Canada

Atilio Deana
Kimmel Center for Biology and Medicine at the Skirball Institute and Department of Microbiology, New York University School of Medicine, New York, USA

Hili Giladi
Department of Genetics, University of Georgia, Athens, Georgia, USA

Gianni Dehò
Dipartimento di Scienze biomolecolari e Biotecnologie, Università degli Studi di Milano, Milan, Italy

Gintaras Deikus
Mount Sinai School of Medicine of New York University, Department of Pharmacology and Systems Therapeutics, New York, USA

Murray P. Deutscher
Department of Biochemistry and Molecular Biology, University of Miami School of Medicine, Miami, Florida, USA

André Dietrich
Institut de Biologie Moléculaire des Plantes, Centre National de la Recherche Scientifique, Unité Propre de Recherche 2357, Université de Strasbourg, France

Marc Dreyfus
École Normale Supérieure, Laboratoire de Génétique Moléculaire, Paris, France, and Centre National de la Recherche Scientifique (CNRS), Paris, France

Daniel Karzai Dulebohn
Department of Biochemistry and Cell Biology and Center for Infectious Diseases of Stony Brook University, Stony Brook, New York, USA

Elena Evguenieva-Hackenberg
Institut für Mikrobiologie und Molekularbiologie, University of Giessen, Giessen, Germany

Pierre Fechter
Architecture et Réactivité de l'ARN, Université de Strasbourg, CNRS, IBMC, Strasbourg, France

Dominique Gagliardi
Institut de Biologie Moléculaire des Plantes, Centre National de la Recherche Scientifique, Unité Propre de Recherche 2357, Université de Strasbourg, France

Thomas Geissmann
Architecture et Réactivité de l'ARN, Université de Strasbourg, CNRS, IBMC, Strasbourg, France

Zhiyun Ge
Department of Biochemistry and Cell Biology and Center for Infectious Diseases of Stony Brook University, Stony Brook, New York, USA

Kenn Gerdes
Institute for Cell and Molecular Biosciences, Medical School, University of Newcastle, Newcastle, United Kingdom

Pawel Golik
Department of Genetics and Biotechnology, Warsaw University, Warsaw, Poland, and Institute of Biochemistry and Biophysics PAS, Warsaw, Poland

Eliane Hajnsdorf
CNRS UPR9073; Institut de Biologie Physico-Chimique; Université Paris Diderot, Paris7, Paris, France

Janet S. Hankins
Department of Biochemistry and Molecular Biology, University of British Columbia, Vancouver, British Columbia, Canada

Narumi Hino-Shigi
Department of Chemistry and Biotechnology, Graduate School of Engineering, University of Tokyo, Tokyo, Japan

Sarah Holec
Institut de Biologie Moléculaire des Plantes, Centre National de la Recherche Scientifique, Unité Propre de Recherche 2357, Université de Strasbourg, France

Eric Huntzinger
Max Planck Institute for Developmental Biology, Tübingen, Germany

Robert Jedrzejczak
Synchrotron Radiation Research Section, MCL, National Cancer Institute, Argonne National Puchta Laboratory, Argonne, Illinois, USA

Stefanie S. Jourdan
Astbury Centre for Structural Molecular Biology, University of Leeds, Leeds, LS2 9JT, United Kingdom

A. Wali Karzai
Department of Biochemistry and Cell Biology and Center for Infectious Diseases of Stony Brook University, Stony Brook, New York, USA

Vanessa Khemici
Laboratoire de Microbiologie et Génétique Moléculaire (LMGM), Centre National de la Recherche Scientifique (CNRS) and Université Paul Sabatier, Toulouse, France

Louise Kime
Astbury Centre for Structural Molecular Biology, University of Leeds, Leeds, LS2 9JT, United Kingdom

Gabriele Klug
Institut für Mikrobiologie und Molekularbiologie, University of Giessen, Giessen, Germany

Sidney R. Kushner
Department of Genetics, University of Georgia, Athens, Georgia, USA

Heike Lange
Institut de Biologie Moléculaire des Plantes, Centre National de la Recherche Scientifique, Unité Propre de Recherche 2357, Université de Strasbourg, France

Pei-Hsun Lin
Institute of Molecular Biology, Academia Sinica, Taipei 115, Taiwan

Sue Lin-Chao
Institute of Molecular Biology, Academia Sinica, Taipei 115, Taiwan

Zhongwei Li
Department of Biomedical Science, Florida Atlantic University, Boca Raton, Florida, USA

Esben Lorentzen
Birkbeck College London, Institute of Structural Molecular Biology, London, United Kingdom

George A. Mackie
Department of Biochemistry and Molecular Biology, University of British Columbia, Vancouver, British Columbia, Canada

Kimika Maki
Division of Biological Science, Graduate School of Science, Nagoya University, Chikusa, Nagoya, Japan

Michal Malecki
Department of Genetics and Biotechnology, Warsaw University, Warsaw, Poland

Valerie F. Maples
Department of Genetics, University of Georgia, Athens, Georgia , USA

Nathalie Mathy
CNRS, Institut de Biologie Physico-Chimique, Paris, France

Pierluigi Mauri
Istituto di Tecnologie Biomediche, Consiglio Nazionale delle Ricerche, Segrate (Milan), Italy

Kenneth J. McDowall
Astbury Centre for Structural Molecular Biology, University of Leeds, Leeds, LS2 9JT, United Kingdom

Wenzhao Meng
Departments of Chemistry, Temple University, Philadelphia, Pennsylvania, USA

Xin Miao
Department of Biochemistry and Molecular Biology, University of British Columbia, Vancouver, British Columbia, Canada

Bijoy K. Mohanty
Department of Genetics, University of Georgia, Athens, Georgia, USA

Teppei Morita
Division of Biological Science, Graduate School of Science, Nagoya University, Chikusa, Nagoya, Japan

Asuteka Nagao
Department of Chemistry and Biotechnology, Graduate School of Engineering, University of Tokyo, Tokyo, Japan

Lilian Nathania
Departments of Chemistry, Temple University, Philadelphia, Pennsylvania, USA

Rhonda H. Nicholson
Departments of Biology, Temple University, Philadelphia, Pennsylvania, USA

Allen W. Nicholson
Departments of Biology and Chemistry, Temple University, Philadelphia, Pennsylvania, USA

Irina A. Oussenko
Mount Sinai School of Medicine of New York University, Department of Pharmacology and Systems Therapeutics, New York, USA

Martin Overgaard
Department of Biochemistry and Molecular Biology, University of Southern Denmark, Odense, Denmark

Olivier Pellegrini
CNRS UPR 9073 (affiliated with Université de Paris 7; Denis Diderot), Institut de Biologie Physico-Chimique, Paris, France

Alexandre V. Pertzev
Departments of Chemistry, Temple University, Philadelphia, Pennsylvania, USA

Leonora Poljak
Laboratoire de Microbiologie et Génétique Moléculaire (LMGM), Centre National de la Recherche Scientifique (CNRS) and Université Paul Sabatier, Toulouse, France

Victoria Portnoy
Department of Biology Technion, Israel Institute of Technology, Haifa, Israel

Florence Proux
École Normale Supérieure, Laboratoire de Génétique Moléculaire, Paris, France, and Centre National de la Recherche Scientifigure (CNRS), Paris, France

Annie Prud'homme-Genereux
Department of Biochemistry and Molecular Biology, University of British Columbia, Vancouver, British Columbia, Canada

Olga Puchta
Department of Genetics and Biotechnology, Warsaw University, Warsaw, Poland, and Institute of Biochemistry and Biophysics PAS, Warsaw, Poland

Yulia Redko
CNRS UPR 9073 (affiliated with Université de Paris 7; Denis Diderot), Institut de Biologie Physico-Chimique, Paris, France

Philippe Régnier
CNRS UPR9073; Institut de Biologie Physico-Chimique; Université Paris Diderot, Paris7, Paris, France

Jamie Richards
Department of Biochemistry and Cell Biology and Center for Infectious Diseases of Stony Brook University, Stony Brook, New York, USA

Pascale Romby
Architecture et Réactivité de l'ARN, Université de Strasbourg, CNRS, IBMC, Strasbourg, France

Gadi Schuster
Department of Biology Technion, Israel Institute of Technology, Haifa, Israel

Dharam Singh
Institute of Molecular Biology, Academia Sinica, Taipei 115, Taiwan

Shimyn Slomovic
Department of Biology Technion, Israel Institute of Technology, Haifa, Israel

Piotr P. Stepien
Department of Genetics and Biotechnology, Warsaw University, Warsaw, Poland, and Institute of Biochemistry and Biophysics PAS, Warsaw, Poland

Leigh M. Stickney
Department of Biochemistry and Molecular Biology, University of British Columbia, Vancouver, British Columbia, Canada

Thomas Sundermeier
Department of Biochemistry and Cell Biology and Center for Infectious Diseases of Stony Brook University, Stony Brook, New York, USA

Tsutomu Suzuki
Department of Chemistry and Biotechnology, Graduate School of Engineering, University of Tokyo, Tokyo, Japan

François Vandenesch
INSERM U851, Centre National de Référence des Staphylocoques, Université de Lyon I, Lyon, France

Steffen Wagner
Institut für Mikrobiologie und Molekularbiologie, University of Giessen, Giessen, Germany

Kristoffer Skovbo Winther
Institute for Cell and Molecular Biosciences, The Medical School, University of Newcastle, Newcastle, United Kingdom

Mieko Yagi

Division of Biological Science, Graduate School of Science, Nagoya University, Chikusa, Nagoya, Japan

Shiyi Yao

Mount Sinai School of Medicine of New York University, Department of Pharmacology and Systems Therapeutics, New York, USA

PREFACE

The cellular cleavage of RNA has remarkably diverse biologic consequences. With growing recognition that RNA turnover has a profound impact on gene expression and that rates of RNA decay can be modulated in response to environmental and developmental signals, significant progress has been made toward explaining cleavage mechanisms and how such mechanisms are regulated.

In view of research developments over the past 20 years, it is surprising that a *Methods in Enzymology* on RNA turnover is not a well-worn tome on our bookshelves. To fill the void, this volume, which addresses RNA turnover in bacteria, archaea, and organelles, and two companion volumes (Volumes 448 and 449), which focus on RNA turnover in eukaryotes, are hoped to serve both as useful references for specialists in the field and as a helpful guides for the broader community of research scientists wishing to initiate studies of RNA decay. Fundamental principles govern RNA decay and how to study it. By incorporating this information for a diverse array of organisms, we hope to stimulate the cross-fertilization of concepts and techniques.

A wide range of methods and reagents are presented, often in the context of answering important biologic questions. The authors have offered detailed rationalizations for and descriptions of their work, endeavoring to ensure that important technical points are made clearly. We wish to express our appreciation to the authors for their thoughtful contributions and willingness to share expertise.

Most chapters provide not only methodologic recipes but also short reviews that place the methods in a proper biologic framework, taking into account the interplay between RNA turnover and other cellular processes. Some degree of overlap between chapters on related topics is unavoidable. This allows each chapter to be read and understood as an independent unit while permitting the expression of diverse viewpoints on alternate methods.

Chapters are written to emphasize the characterization of RNA elements, protein factors, and cellular processes that mediate and/or regulate RNA turnover. RNA elements, including primary sequences and, frequently, higher-order structures, can influence the decay functions of a variety of protein factors such as exoribonucleases, endoribonucleases, helicases, and other types of RNA-binding proteins, all of which often exist in multiprotein complexes. Cellular processes having the potential to affect RNA turnover include transcription, translation, and polyadenylation. Many experimental approaches for analyzing RNA decay are presented, each generally focusing on a specific organism and a particular class of RNA such as mRNA, a stable RNA, or a type of small noncoding RNA. Techniques presented include isolating total-cell or organelle RNA by different methods, determining RNA half-lives, establishing *in vitro* and *in vivo* RNA degradation assays, mapping full-length RNAs and decay intermediates, and purifying and characterizing ribonucleases and RNA binding proteins. Global approaches are also discussed, such as genomic analyses of RNA decay with DNA microarrays and proteomic analyses of the RNA degradosome with two-dimensional chromatography coupled to tandem mass spectrometry.

The knowledge collected makes clear how far our understanding of RNA degradation has come in the past few years and how much about this important regulatory process remains to be discovered. This volume should be of lasting value in providing techniques and tools for studying RNA turnover mechanisms. Applications of this knowledge to medicine and biotechnology are underway.

It is our hope that this volume will reflect the excitement currently held by RNA turnover specialists and will serve as a source of inspiration for scientists entering this rapidly moving and important field.

Lynne E. Maquat and Cecilia M. Arraiano

METHODS IN ENZYMOLOGY

VOLUME 71. Lipids (Part C)
Edited by JOHN M. LOWENSTEIN

VOLUME 72. Lipids (Part D)
Edited by JOHN M. LOWENSTEIN

VOLUME 73. Immunochemical Techniques (Part B)
Edited by JOHN J. LANGONE AND HELEN VAN VUNAKIS

VOLUME 74. Immunochemical Techniques (Part C)
Edited by JOHN J. LANGONE AND HELEN VAN VUNAKIS

VOLUME 75. Cumulative Subject Index Volumes XXXI, XXXII, XXXIV–LX
Edited by EDWARD A. DENNIS AND MARTHA G. DENNIS

VOLUME 76. Hemoglobins
Edited by ERALDO ANTONINI, LUIGI ROSSI-BERNARDI, AND EMILIA CHIANCONE

VOLUME 77. Detoxication and Drug Metabolism
Edited by WILLIAM B. JAKOBY

VOLUME 78. Interferons (Part A)
Edited by SIDNEY PESTKA

VOLUME 79. Interferons (Part B)
Edited by SIDNEY PESTKA

VOLUME 80. Proteolytic Enzymes (Part C)
Edited by LASZLO LORAND

VOLUME 81. Biomembranes (Part H: Visual Pigments and Purple Membranes, I)
Edited by LESTER PACKER

VOLUME 82. Structural and Contractile Proteins (Part A: Extracellular Matrix)
Edited by LEON W. CUNNINGHAM AND DIXIE W. FREDERIKSEN

VOLUME 83. Complex Carbohydrates (Part D)
Edited by VICTOR GINSBURG

VOLUME 84. Immunochemical Techniques (Part D: Selected Immunoassays)
Edited by JOHN J. LANGONE AND HELEN VAN VUNAKIS

VOLUME 85. Structural and Contractile Proteins (Part B: The Contractile Apparatus and the Cytoskeleton)
Edited by DIXIE W. FREDERIKSEN AND LEON W. CUNNINGHAM

VOLUME 86. Prostaglandins and Arachidonate Metabolites
Edited by WILLIAM E. M. LANDS AND WILLIAM L. SMITH

VOLUME 87. Enzyme Kinetics and Mechanism (Part C: Intermediates, Stereo-chemistry, and Rate Studies)
Edited by DANIEL L. PURICH

VOLUME 88. Biomembranes (Part I: Visual Pigments and Purple Membranes, II)
Edited by LESTER PACKER

VOLUME 435. Oxygen Biology and Hypoxia
Edited by HELMUT SIES AND BERNHARD BRÜNE

VOLUME 436. Globins and Other Nitric Oxide-Reactive Protiens (Part A)
Edited by ROBERT K. POOLE

VOLUME 437. Globins and Other Nitric Oxide-Reactive Protiens (Part B)
Edited by ROBERT K. POOLE

VOLUME 438. Small GTPases in Disease (Part A)
Edited by WILLIAM E. BALCH, CHANNING J. DER, AND ALAN HALL

VOLUME 439. Small GTPases in Disease (Part B)
Edited by WILLIAM E. BALCH, CHANNING J. DER, AND ALAN HALL

VOLUME 440. Nitric Oxide, Part F Oxidative and Nitrosative Stress in Redox
Regulation of Cell Signaling
Edited by ENRIQUE CADENAS AND LESTER PACKER

VOLUME 441. Nitric Oxide, Part G Oxidative and Nitrosative Stress in Redox
Regulation of Cell Signaling
Edited by ENRIQUE CADENAS AND LESTER PACKER

VOLUME 442. Programmed Cell Death, General Principles for Studying Cell
Death (Part A)
Edited by ROYA KHOSRAVI-FAR, ZAHRA ZAKERI, RICHARD A. LOCKSHIN,
AND MAURO PIACENTINI

VOLUME 443. Angiogenesis: *In Vitro* Systems
Edited by DAVID A. CHERESH

VOLUME 444. Angiogenesis: *In Vivo* Systems (Part A)
Edited by DAVID A. CHERESH

VOLUME 445. Angiogenesis: *In Vivo* Systems (Part B)
Edited by DAVID A. CHERESH

VOLUME 446. Programmed Cell Death, The Biology and Therapeutic
Implications of Cell Death (Part B)
Edited by ROYA KHOSRAVI-FAR, ZAHRA ZAKERI, RICHARD A. LOCKSHIN,
AND MAURO PIACENTINI

BACTERIA

ANALYSIS OF RNA DECAY, PROCESSING, AND POLYADENYLATION IN *ESCHERICHIA COLI* AND OTHER PROKARYOTES

Bijoy K. Mohanty, Hili Giladi, Valerie F. Maples, *and* Sidney R. Kushner

Contents

Department of Genetics, University of Georgia, Athens, Georgia, USA

Methods in Enzymology, Volume 447

ISSN 0076-6879, DOI: 10.1016/S0076-6879(08)02201-5

Abstract

This chapter provides detailed methodologies for isolating total RNA and poly-adenylated RNA from *E. coli* and other prokaryotes, along with the procedures necessary to analyze the processing and decay of specific transcripts and determine their 3′- and 5′-ends. The RNA isolation methods described here facilitate isolating good-quality RNA in a very cost-effective way compared to the commercially available RNA isolation kits, without employing phenol and/or alcohol precipitation. We also discuss the limits associated with polyacrylamide and agarose gels for the separation of small and large RNAs. Methods useful for the analysis of post-transcriptionally modified transcripts and the processing of very large polycistronic transcripts are also presented.

1. INTRODUCTION

RNA turnover and processing have now been demonstrated to be important steps that directly affect protein synthesis and the cell's ability to survive in nature. However, the analysis of mRNA decay and polyadenylation in *Escherichia coli* has long been considered technically difficult. The development over the past 15 years of methods for the isolation and characterization of both mRNA and polyadenylated species has made the study of these important pathways of RNA metabolism more straightforward (Arraiano *et al.*, 1988; O'Hara *et al.*, 1995; Mohanty and Kushner, 1999).

By employing these techniques and using isogenic strains carrying mutations in a variety of genes encoding ribonucleases, poly(A) polymerases, and accessory proteins, it has been possible to develop a better understanding of the multiple pathways by which mRNA decay and polyadenylation take place (Arraiano *et al.*, 1988, 1993, 1997; O'Hara *et al.*, 1995; Hajnsdorf and Régnier, 2000; Mohanty and Kushner, 2000; Ow *et al.*, 2000, 2003; Mohanty *et al.*, 2004; Perwez and Kushner, 2006). Furthermore, the ability to examine the *E. coli* transcriptome in a variety of mutants deficient in the enzymes involved in mRNA decay, processing, and polyadenylation (Lee *et al.*, 2002; Li *et al.*, 2003; Mohanty and Kushner, 2003, 2006; Bernstein *et al.*, 2004) has provided a valuable new approach for identifying targets of specific enzymes.

The use of reliable and reproducible methodologies is paramount to successfully investigate RNA metabolism in any organism. In the sections that follow, methods are described to allow any laboratory interested in studying RNA metabolism in any prokaryote to successfully carry out such experiments. These procedures are the results of many years of experimentation.

2. General Considerations When Working with RNA

Because RNA is much less stable than DNA and is prone to hydrolysis by ubiquitous RNases, extra precautions are necessary. In particular, always wear a lab coat and powder-free gloves. Periodically change gloves and never touch an RNA sample with bare hands. All glassware for RNA work should be kept separate from everything else in the laboratory and never be used for anything else. This is particularly important, since RNase A is extremely stable and tends to stick to glassware. Glassware should be baked at 180 °C for 8 h prior to use. All plastic ware, gel apparatus, and other materials that cannot be baked and may come in contact with RNA should be treated with a commercially available RNase inactivating solution (RNAaseZap, Ambion, TX, USA) or equivalent solutions from other manufacturers and thoroughly washed with RNase-free water (devoid of any RNase contamination). Never autoclave any of the material used for RNA analysis. Pipette tips and Eppendorf tubes are always used new, as these are packaged RNase- and DNase-free.

Water is a major source of RNase contamination. We generate RNase-free water by purifying through a Millipore water-filtration system (Durapore CVDI Cartridge Filter) that has been retrofitted to the deionized water source in the laboratory. By using this system, we have eliminated our RNase problem and thus do not treat water with DEPC (diethyl pyrocarbonate), which was formerly our standard process for making water RNase free. In addition, RNase-free water is now available through many commercial suppliers (Ambion, TX, USA; Promega, WI, USA; and Invitrogen, CA, USA). All reagents and buffers are prepared using RNase-free water only. All appropriate safety procedures should be followed when working with toxic and carcinogenic materials, such as phenol, ethidium bromide, formaldehyde, polyacrylamide, and UV transilluminators (Sambrook *et al.*, 1989).

2.1. Importance of using isogenic strains

Although it might seem obvious, it is critical to carry out RNA experiments in isogenic strains. It is not clear at this point if all the enzymes involved in RNA metabolism have been identified in *E. coli* or, for that matter, in any prokaryote.

2.2. Inhibition of new transcription

The level of each transcript in an exponentially growing cell is maintained at steady state by a combination of synthesis and degradation. The half-lives of *E. coli* transcripts vary between less than 30 sec to more than 20 min (Kushner, 2007). Although the steady-state level of a transcript can be measured by isolating RNA at any given time, its half-life is measured after new transcription has been inhibited. In *E. coli*, inhibition of new transcription is accomplished by adding rifampicin to a growing culture. The first time point (time zero) is taken 75 sec after the addition of the drug to allow for equilibration within the cell.

2.3. Kits versus detergent methods for RNA isolation

We have previously taken advantage of a cationic surfactant called Catrimox-14 (tetradecyltrimethylammonium oxalate), which has the ability to lyse cells and precipitate nucleic acids (Dahle and Macfarlane, 1993; Macfarlane and Dahle, 1993), to isolate RNA from *E. coli* (O'Hara *et al.*, 1995; Mohanty and Kushner, 1999). This detergent was initially developed and marketed by Iowa Biotechnology Corp. (Coralville, IA). However, shortly after Qiagen acquired Iowa Biotechnology, Catrimox-14 was taken off the market. Because the detergent-based RNA-isolation method was simple and inexpensive, we subsequently determined that a similar cationic surfactant, trimethyl(tetradecyl)ammonium bromide (Sigma) was equally effective in the isolation of RNA. For simplicity, we have called this compound Catrimide.

Although there are many commercially available RNA isolation kits, they can be expensive ($5-8/sample), particularly if multiple samples have to be isolated. In contrast, the detergent-based method described here is very low cost (~$0.20/sample), avoids the use of phenol and/or chloroform along with alcohol precipitation, and provides many specific advantages relative to the kits. For example, the RNA yields are consistently higher with the detergent method and the samples retain low molecular weight RNAs such as tRNAs and regulatory RNAs. In contrast, most of the column-based kits do not retain these small RNAs (Table 1.1). In addition, the detergent-LiCl-based method removes >98% of the DNA contamination from the RNA (Macfarlane and Dahle, 1997), such that for many

Table 1.1 Comparison of commonly used RNA isolation methods/kits

RNA isolation method/kit	Average yield (μg)/ sample (*E. coli*)	Exclusion of small RNAs
Catrimide-LiCl	200*	No
Ribopure-Bacteria Kit (Ambion)	90#	Yes
RNeasy Protect Bacteria Mini Kit (Qiagen)	70#	Yes
Purelink Micro-to-Midi Total Purification System (Invitrogen)	50#	Partial
SV Total RNA Isolation System (Promega)	36#	Unknown

* Typical actual yield.
# As claimed by the manufacturer.

experiments no further DNase I treatment is necessary. Furthermore, most of the commercially available kits do not provide a mechanism to stop further metabolic activity after initial sampling of the cells. This is a significant limitation when dealing with multiple samples, particularly those used for half-life determinations.

3. ISOLATION OF TOTAL RNA FROM EXPONENTIALLY GROWING CELLS

The RNA isolation procedure from exponentially growing cells is a detergent-based procedure using Catrimide that has been modified from a previously described method (Macfarlane and Dahle, 1993).

3.1. Materials

Centrifuge tubes (Corex or equivalent, 15 ml), screw-cap tubes (Pyrex, 13 × 100 mm), 1.5-ml Eppendorf tubes, refrigerated centrifuge (Beckman Coulter, Avanti-J25 or equivalent), tabletop microcentrifuge (with at least 13,000-rpm rating), vortex mixer, water bath (37 °C), cold room (optional), Savant Speed Vac concentrator (optional).

3.2. Reagents and buffers

Tris-MgCl$_2$ (TM) buffer (5×): 50 mM Tris base, 25 mM MgCl$_2$, pH 7.2. This solution is stable at room temperature (RT) for 3 months and at 4 °C for 6 months.

Sodium azide (1 *M*): The solution is stored at 4 °C.

Stop buffer (SB): Chloramphenicol, 50 mg (dissolved in 12.5 ml of 95% ethanol); TM buffer (5×), 20 ml; RNase-free water, 80 ml; sodium azide (1 *M*), 2.5 ml. The volume of SB prepared should be proportionally increased or decreased according to the number of samples in a given experiment. Seven ml of the SB are dispensed into baked prelabeled centrifuge tubes and placed in a −20 °C freezer at least 1 hour before use. This provides enough time for the buffer to freeze.

Rifampicin (prepared fresh): 33.3 mg/ml in DMSO (Sigma).

Nalidixic acid (prepared fresh): 2 mg/ml in sterile water (Sigma).

Note: Rifampcin and nalidixic acid are only required if the RNA samples will be used for half-life determinations.

Lysis buffer (prepared fresh): 303 μl of freshly prepared lysozyme (Sigma) solution (10 mg/ml in 1× TM buffer), 33 μl of RNase-free DNase I (Roche, 10 μg/μl), 10 ml of 1× TM buffer. Store the lysis buffer in ice before use.

Catrimide: A 10% solution (w/v) of Catrimide stored at RT.

Other solutions: Acetic acid (20 m*M*), 2 *M* LiCl in 35% EtOH, 2 M LiCl in water, ethanol (EtOH) (70%, 95%, and 100%).

Dry-ice slurry: 1–2 lbs of dry ice is crushed in an ice bucket and mixed with 100% EtOH to make the slurry.

3.3. Procedure

1. An overnight inoculum from a single colony of each bacterial strain to be studied is prepared in 5 ml of Luria broth (LB) containing appropriate antibiotics (if applicable) at the desired temperature. Faster-growing strains such as wild-type controls should be grown as standing cultures whereas slower growing strains should be shaken.

2. The next day, the overnight inoculum is diluted (2-5%) into a desired volume of LB, which is grown with shaking (always grow a few ml more than you need) (with appropriate antibiotic, if applicable) to Klett 50 (No. 42 green filter or OD_{600} of 0.4 or ~10^8 cells/ml). The cultures should undergo at least two doublings when growing at either 30 °C or 37 °C. For temperature-sensitive mutants (e.g., *rne-1* and *mpA49*), the cultures are initially grown at 30 °C until they reach Klett 40 and then are shifted to 44 °C for up to 2 h. The cultures should be periodically diluted with prewarmed (44 °C) LB to maintain their exponential growth.

3. When a culture is ready for sampling, remove a centrifuge tube containing the frozen SB from the freezer and crush the frozen buffer using a glass rod to make an icy slurry. To maintain accurate sampling times, in half-life experiments multiple tubes can be removed from the freezer if they will be used within 5 min.

4. If the RNA will be used for half-life measurements, appropriate volumes of rifampicin (500 μg/ml final concentration) and nalidixic

acid (40 μg/ml final concentration) are added to the culture to stop new transcription. The first (0 min) sample is removed after 75 sec to permit complete blockage of new transcription by the rifampicin. Typical time points for half-life determinations of most *E. coli* mRNAs are 0,1,2,4,8, and 16 minutes.

5. At each desired time point 7 ml of cells are removed and mixed thoroughly with the crushed SB using a vortex mixer. This step blocks all further *in vivo* metabolic activity including RNA degradation.

6. Each sample is held in ice until the time course has been completed. At this time the lysis buffer is prepared, the dry ice–EtOH slurry is made in an ice bucket, and all the screw-cap tubes are labeled with corresponding sample numbers and are put in dry ice–EtOH slurry for prechilling.

7. The cells are harvested by centrifuging the samples at 5000 rpm for 5 min at 4 °C in a refrigerated centrifuge. The supernatants are poured off and the centrifuge tubes are inverted on a couple of layers of Kimwipes to drain the tubes thoroughly (10–20 sec). The outsides of the tubes are wiped to remove excess liquid and are then placed in ice.

8. The cell pellets are resuspended in 400 μl of lysis buffer and transferred to the corresponding prechilled screw-cap tubes in dry ice–EtOH.

9. The cells are lysed by repeatedly (four times) thawing the frozen cells in a 37 °C water bath and refreezing the cells in the dry ice–EtOH. After the last freezing step, the cells can be stored at −70 °C until ready for RNA extraction.

3.4. RNA extraction

1. To further reduce any possible RNase activity after lysis, 67 μl of 20 mM acetic acid is added to the frozen lysed cells and each sample is placed in a 37 °C water bath.

2. After the lysed cell suspensions have completely melted, 400 μl of Catrimide is added and the mixture is incubated for another 2 min in the water bath. The suspension, which is cloudy at this step, is then transferred to a 1.5-ml Eppendorf tube and centrifuged for 1 min at 5000 rpm in a microcentrifuge in the cold room.

3. The supernatants are removed either with a suction apparatus or with a pipette, being careful not to suck up the pellets (containing RNA).

4. The pellets are resuspended in 1 ml of 2 M LiCl in 35% ethanol and incubated at RT for 5 min.

5. The suspensions are then centrifuged at 13,000 rpm for 6 min in the cold room. The supernatants are removed as in step 3.

6. The pellets are then resuspended in 1 ml of 2 M LiCl in water and incubated at RT for 2 min.

7. The suspensions are then centrifuged at 13,000 rpm as described in step 5. The supernatants are removed as in step 3.

8. The pellets are subsequently washed by adding 200 μl of 70% EtOH and centrifuging at 13,000 rpm for 5 min, followed by removal of the supernatants as described in step 3. The pellets are then dried in a Speed Vac concentrator for 5 min (or air-dried for 15 min).

   Note: The centrifugations in steps 5, 6, and 8 can be carried out at RT without any significant adverse effect.

9. The pellets are finally resuspended in 200 μl of RNase-free water by incubating them for 30 min at room temperature. If time permits, overnight incubation at 4 °C yields maximum RNA levels. After resuspension, the samples are centrifuged for 1 min in the microcentrifuge at full speed to pellet any cell debris. The supernatants, containing the RNA, are transferred to new Eppendorf tubes and should be stored at 4 °C if they are to be used immediately. Long-term storage should be at −20 °C. Repeated freezing and thawing should be avoided to prevent degradation.

4. Isolation of Total RNA from Stationary Phase Cells

Isolation of RNA from stationary phase cells has been found to be difficult, because of poor cell lysis, using either Catrimox-14 or Catrimide. Thus, the preceding method has been modified as follows.

4.1. Reagents and buffers

TM buffer (5×), sodium azide (1 M), SB, lysis buffer, acetic acid (20 mM) are as described in the section on isolation of total RNA from exponentially growing cells. SDS (sodium dodecyl sulfate, 20%), sodium acetate (3 M, pH 5.2), acid phenol equilibrated with Tris-HCl (pH 4.5-5.2) (Sambrook *et al.*, 1989), phenol:chloroform (5:1 solution, pH 4.5, Ambion, TX, USA), and EtOH (70% and 100%).

4.2. Procedure

1. Cells are grown overnight with shaking ($\sim$16–18 h) from a single colony. No more than 2 ml of culture (faster-growing strains need only 1 ml of culture) are mixed with 7 ml of crushed ice SB in a centrifuge tube and held on ice until all samples are collected.

2. The cells are collected by centrifugation at 5000 rpm for 5 min at 4 °C in a refrigerated centrifuge. Each pellet is resuspended in 600 μl of ice-cold lysis buffer.

3. The suspension is transferred to a chilled (in dry ice–EtOH), baked, screw-cap tube for four cycles of freeze and thaw treatment in dry ice–EtOH and 37 °C water bath.

4. After the fourth cycle of freezing, 100 μl of 20 mM acetic acid is added to the frozen cells and each sample is transferred to a 37 °C water bath. After 1 min, 30 μl of SDS (20%) is added and the suspension is incubated further for 2 min. Clearing of the suspension is indicative of good cell lysis.

5. The cell lysate is carefully transferred to a new 1.5-ml Eppendorf tube containing 600 μl of acid phenol (pH 4.5–5.2) and mixed thoroughly by vortexing for 1 min. The mixture is centrifuged at 13,000 rpm in a microcentrifuge for 5 min at RT.

6. The top aqueous layer is transferred to a new Eppendorf tube and extracted with an equal volume of acid phenol. The final extraction is carried out an additional two times using an equal volume of phenol:chloroform (5:1 solution, pH 4.5).

7. After the fourth extraction, 0.11 vol of 3 M NaOAc (pH 5.2) and 2.5 vol of 100% EtOH are added to the supernatant.

8. The RNA is precipitated at −20 °C for at least 2 hr (can be left overnight). The RNA is pelleted by centrifuging for 15 min at 13,000 rpm in a microcentrifuge in the cold room. The RNA pellet is then washed with 1 ml of cold 70% ethanol and dried either at RT (15 min) or in a Speed Vac Concentrator (5 min maximum). The RNA pellets are dissolved in 200 μl of RNase-free water and stored as described previously.

5. REMOVAL OF DNA CONTAMINATION

DNA contamination can be a major problem, particularly if the RNA will be used for either dot-blot analysis or cDNA synthesis. RNA isolated from exponentially growing cells using the Catrimide method is relatively free from DNA contamination (containing <1.5%, Fig. 1.1, lanes 1–3;

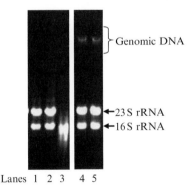

Lanes 1 2 3   4 5

Figure 1.1 Rapid analysis of integrity and purity of RNA samples. Total RNA (500 ng/lane) isolated from exponentially growing (lanes 1–3) and stationary phase (lanes 4–5) *E. coli* were separated in 1% agarose in TAE buffer. The positions of genomic DNA, 23S and 16S rRNA in the gel are indicated to the right. The RNA in lane 3 is partially degraded.

Macfarlane and Dahle, 1997), thus requiring no further DNase treatment when used for Northern analysis. However, RNA isolated from stationary phase cells by the SDS-lysis method contains significant amounts (>5%) of residual DNA contamination (Fig. 1.1, lanes 4–5) despite a DNase I step during the initial isolation. This contamination is due primarily to a higher cell concentration in stationary phase cultures compared to the log phase cultures and can be minimized by using smaller cell volumes. We routinely use the DNA-free kit (Ambion, Austin, TX) as per manufacturer's instructions to remove DNA from total RNA. This is a quick process and avoids use of phenol and chloroform to remove the DNase I from the reaction mixtures.

6. RNA YIELD AND PURITY ASSESSMENT

The RNA samples are diluted 100-fold in RNase-free water and the optical density (OD) at 260 and 280 nm is measured in duplicate. The amount of RNA ($\mu g/ml$) in the original sample is calculated by using the formula: average OD at 260 nm × 40 × dilution factor (100). The ratio of $A_{260/280}$ is obtained to determine if RNA is contaminated with protein. A typical RNA isolation from a wild-type strain using preceding methods yields ∼200 μg/sample (Table 1.1) with an $A_{260/280}$ ratio of ∼1.8–2.0. Although any spectrophotometer is acceptable for these measurements, extremely accurate measurements of 1-μl samples are obtained using an ND-1000 instrument (Nanodrop, Thermo Fisher Scientific).

7. RNA INTEGRITY ASSESSMENT

It is very important to check the integrity of each RNA sample before carrying out any further experiments, as OD measurements do not indicate whether a sample has been degraded. We routinely check the quality of the RNA by running the samples in agarose minigels. Five hundred ng of total RNA is mixed with 3–4 μl of 1× RNA loading dye (Table 1.2) and separated in a 1% agarose minigel (5 × 10 cm) in 1× TAE buffer (Tris-acetate/EDTA, [Sambrook *et al.*, 1989]) at 100 V. The gel is stained with ethidium bromide and viewed with a UV transilluminator. Visualization of intact rRNAs with a ratio of 23S to 16S rRNA of approximately 2/1 (Fig. 1.1, lanes 1–2) ensures that the RNA has not been significantly degraded (Fig. 1.1, lane 3).

In many cases, we see that the amount of RNA observed in the ethidium bromide stained gel varies among the samples even though all lanes have been loaded with 500 ng of RNA based on OD measurements. These discrepancies in quantification may be due to either pipetting errors or

Table 1.2 Polyacrylamide gel solution and loading dye for Northern analysis

Gel solution (6%) with 7 M urea	Gel
Urea	14.7 g
TBE (5×)	7.0 ml
Acrylagel (30%)	7.0 ml
Bis–acrylagel (2%)	5.6 ml
Water	4.3 ml
Ammonium persulfate (APS)(10%)	280 μl
TEMED	28 μl
Acrylagel and Bis-acrylagel solutions are obtained from National Diagnostics, GA, USA. APS is made fresh. To obtain 8.3-M urea concentration, 17.4 g of urea is added to the solution.	
RNA loading dye (2×)	
Xylene cyanol	30 mg
Bromophenol blue	30 mg
$Na_2EDTA.2H_2O$	37 mg
Deionized formamide	10 ml
Tris-acetate-EDTA (TAE) (10×)	
Tris base	100 mM
Sodium acetate	50 mM
EDTA	5 mM
Adjust the pH to 7.8 with glacial acetic acid	

incorrect OD measurements. To distinguish these possibilities, we use a second method to normalize samples in a particular experiment. The amount of RNA loaded onto the minigel is reduced from 500 ng to less than 100 ng and stained with Vistra Green (GE Healthcare). The gel is then scanned using a PhosphorImager (GE Healthcare, Storm 480) and the rRNA bands (both 16S and 23S) are quantified using Imagequant software. The difference in the amount of rRNA bands in each sample compared to the wild-type strain is then used to normalize all the samples.

8. Isolation of Polyadenylated RNA

Isolation of polyadenylated RNA from *E. coli* is a challenge as only <1% to 10% of the total RNA is polyadenylated at any given time, the majority of the poly(A) tails are less than 10 nt in length, and many post-transcriptionally added tails are not added by poly(A) polymerase I (PAP I), but rather by polynucleotide phosphorylase (PNPase) resulting in heteropolymeric tails (Mohanty and Kushner, 1999, 2000, 2006; Mohanty *et al.*, 2004). However, the level of polyadenylated transcripts in *E. coli* can be significantly increased by overexpressing PAP I (Mohanty and Kushner, 1999). We have successfully isolated polyadenylated transcripts from both the wild-type and PAP I–overexpressing strains (Mohanty and Kushner, 2006) using Dynabeads, oligo(dT)$_{25}$ coupled magnetic beads (Dynabeads mRNA direct kit, Dynal, NY, USA).

The amount of Dynabeads needed may vary depending on the source of RNA and percentage of RNA polyadenylated. However, use of more Dynabeads than needed will result in significant backgrounds. Similarly, use of less Dynabeads will result in poor recovery of polyadenylated RNA. From *E. coli* RNA (wild-type or PAP I–induced), 10 μl (5 mg/ml) of Dynabeads is sufficient to process 20 μg of total RNA.

8.1. Procedure

1. Dynabeads are thoroughly mixed in a vortex mixer. Ten μl of the beads are pipetted into a new Eppendorf tube and transferred into a magnetic separator (12-Tube Magnet, Qiagen, CA, USA) for 30 sec. The supernatant is removed carefully by pipetting.
2. The beads are washed once with 50 μl of 2× binding buffer (20 mM Tris HCl, pH 7.6; 1 M LiCl; 2 mM EDTA, pH 8.0 and 0.1% SDS) and resuspended in 50 μl of 2× binding buffer.
3. RNA samples (20 μg) are prepared by adding RNase-free water (if necessary) to make the total volume 50 μl. The mixtures are heated at 65 °C for 2 min and added to 50 μl Dynabeads in 2× binding buffer (step 2). The samples are mixed thoroughly and allowed to anneal for 15 min by rotating in a hybridizing oven at room temperature.
4. Each tube is placed in the magnetic stand for 30 sec and the supernatant is removed by pipetting. The beads are washed three times with 100 μl of washing buffer-A, and two times with washing buffer-B (both are supplied in the kit).
5. The polyadenylated RNA is then eluted by adding 10 μl of RNase-free water and heating at 65 °C for 3 min. The sample is briefly centrifuged and placed in the magnetic stand for 30 sec. The supernatant containing the poly(A$^+$) RNA is removed to a new Eppendorf tube.

9. NORTHERN ANALYSIS

Northern analysis is a powerful technique that is used to analyze both a full-length transcript and its processing and/or breakdown intermediates. It is carried out by separating total RNA in either a polyacrylamide or agarose gels, followed by transferring the RNA to a nylon membrane that is subsequently hybridized to various ^{32}P-labeled probes to identify the transcripts of interest. The choice of gel matrix is crucial for successful separation of different length transcripts. In general, polyacrylamide is the choice of matrix for separation of transcripts below 900 nt and provides well-defined sharp bands. A transcript larger than 900 nt is more suitable for agarose gels. In either case, RNA molecular-weight size standards (Invitrogen or equivalent) are run in the same gel to help determine the size of the transcripts.

9.1. Separation of RNA in polyacrylamide gels

9.1.1. Sample preparation
For each sample, 5–12 μg of total RNA are dried in a Speed Vac Concentrator and resuspended in 1× RNA loading dye (Table 1.1). Each sample is then heated at 65 °C in a water bath for 15 min and quickly chilled in ice. Subsequently, each sample is briefly centrifuged in the cold room and held in ice until loaded onto the gel.

9.1.2. Polyacrylamide gel electrophoresis
We generally use a 5% acrylamide gel for separation of 700–900-nt transcripts and 6% acrylamide gel (Table 1.2) for separation of 200–700-nt transcripts. Transcripts less than 200 nt should be run in 8–10% gels. Although 7 M urea in the gel is sufficient to provide excellent denaturation for most of the transcripts, we have obtained better resolution for highly structured molecules such as tRNAs when 8.3 M urea was used in the gel (Mohanty and Kushner, 2007, 2008). The gel (18 cm length and 18.5 cm wide) is run in 1× TBE (Tris–Borate/EDTA buffer) (Sambrook *et al.*, 1989) using a custom-made vertical gel apparatus. The gel is prerun at a constant current of 60 mA for 20–30 min to raise the buffer temperature to ~55 °C and to leach out excess urea. This prevents the formation of RNA secondary structures during running. Before loading the samples, all the wells are flushed with a syringe to remove the leached urea. The denatured samples on ice should be loaded while the gel temperature is greater than 50 °C. The gel is run with a constant current of 50 mA, which helps maintain the temperature of the gel at ~50 °C. For most mRNAs the gel is run until both dyes run out. For tRNAs, the gel is run until the bromophenol blue reaches ~14 cm from the loading well.

9.2. Separation of RNA in agarose gels

Two types of agarose Northern gels are generally used for analysis of high-molecular-weight RNA (Sambrook *et al.*, 1989) based on the choice of either formaldehyde or glyoxal as a denaturant. Because formaldehyde is toxic, its use in gel electrophoresis requires a special setup and often does not provide good results. In contrast, the glyoxal method described here does not require a special setup, is carried out on a regular gel bench in a commercially available horizontal gel apparatus, and provides excellent resolution. The method has been described in detail (Burnett, 1997). We use either 0.8% or 1% agarose for analysis of most transcripts.

9.3. Transferring RNA to membrane

If possible, it is always a good idea to check the quality of RNA after it has been separated in a gel and before transferring it to a membrane. This is accomplished for agarose gels by using an UV transilluminator, as the RNA sample is prestained with ethidium bromide. The presence of intact rRNA bands (both the 16S and 23S) indicates that the RNA was not degraded during the run and is ready for transferring.

Electroblotting is used to transfer RNA from both polyacrylamide and agarose gels to membranes using a custom-made electroblotting apparatus that can hold two 18 × 18 cm gels. The RNA is transferred in 1× TAE (Table 1.2) buffer for either 3 hr (at 15 V constant for the first hour followed by 40 V constant for the next 2 h) or overnight (at 15 V constant) to a membrane. Charged nylon membranes (Magnacharge nylon membrane, GE Water & Processing Technologies or Biotrans(+) nylon membranes, ICN Biomedicals) are strongly recommended, as they provide low backgrounds and are highly durable for multiple probings and strippings.

After the transfer, a polyacrylamide gel should be stained with ethidium bromide and the efficiency of transfer verified using an UV transilluminator. Agarose gels can be checked directly using a UV transilluminator, as the RNA samples are prestained with ethidium bromide. The large rRNAs (16S and 23S) should be still visible as a single bands in a polyacrylamide gel, while no RNA should be visible in the agarose gel when there is an efficient transfer. If the transfer has been successful, the membrane is then baked at 80 °C for 30 min followed by a 10-sec exposure to UV light (optional) to fix the RNA to the membrane. At this point the membrane can be stored indefinitely at RT by wrapping it in Saran^{TM} Wrap.

9.4. Choice of probes

Three types of probes can be used to hybridize to a Northern membrane: an oligonucleotide complementary to the gene of interest, an antisense runoff transcript, or a DNA fragment of the gene of interest. Because we routinely only use either an oligonucleotide or a DNA fragment for probing the Northern blots, we will restrict our discussion to these types of probes.

The choice of probe depends on the nature of the target transcript. An oligonucleotide probe is particularly useful when probing for small transcripts such as tRNAs or regulatory RNAs. When designing an oligonucleotide probe the following criteria should be considered. The length of an oligonucleotide probe is usually between 18 and 25 nt, though a longer length can be selected. A 50% or higher G/C content is preferred because this will provide a T_m (melting temperature) of $\sim$60 °C or more, resulting in less background hybridization. G as the first nucleotide of the oligonucleotide is avoided wherever possible because it labels poorly with T4 polynucleotide kinase (PNK) and γ-[^{32}P]-ATP.

There are several advantages to using oligonucleotide probes. They are extremely useful in differentiating various processing intermediates of tRNA operon transcripts (Ow and Kushner, 2002; Mohanty and Kushner, 2007, 2008) and specific breakdown products of a larger transcript. In addition, an oligonucleotide probe can be stripped easily (see below), so a membrane can be reused multiple times (up to 10 times), providing an opportunity for comparison of several transcripts in the same blot and thereby reducing the experimental errors due to pipetting, loading, and transferring. However, there are a few drawbacks associated with oligonucleotide probes. First, not all oligonucleotides label efficiently, and the specific activities of oligonucleotide probes are generally lower compared to DNA probes. In our experience, the labeling efficiency of oligonucleotide probes ranges from <20% to >70%. Furthermore, oligonucleotides may not be suitable in determining all the processing or breakdown products of larger transcripts, as they only hybridize to very specific regions of individual transcripts.

In contrast, a DNA fragment probe provides a higher specific activity allowing quick detection of both the full-length and breakdown products of a transcript. Fragments are easily generated by PCR from genomic DNA using two gene-specific primers that cover either the full or partial length of the gene of interest. However, it is relatively difficult to strip a DNA fragment probe compared to an oligonucleotide probe, thus limiting the ability to reuse a Northern blot to probe for more than one transcript.

9.5. Probe labeling

Although nonradioactive probe labeling for Northern blot analysis has been available for quite some time, we have not used this system because of its incompatibility with many imaging and quantification systems and its overall lower sensitivity. Accordingly, we routinely use ^{32}P for labeling all probes. Oligonucleotide probes are labeled at the 5′-end with γ-(^{32}P)-ATP in the presence of PNK according to the manufacturer's (NEB, MA, USA) instructions. DNA fragment probes are uniformly labeled with (α-^{32}P)-dATP using random primers, the Klenow fragment of DNA polymerase I, and the Strip-EZ DNA kit (Ambion). This labeling kit has an advantage over the traditional random primer labeling because it generates a probe that is stable under the conditions for hybridization and washing but can be stripped relatively easily using the reagent provided in the kit.

The labeling efficiency of each probe is determined prior to hybridization with the membrane. To do this, 1 μl of the sample is applied to a
DE-81 filter paper (2.4-cm circle, Whatman) and dried under a halogen
lamp for 10 min. The total radioactivity (A) on the filter is measured in a
liquid scintillation system (Beckman LS 6000) by Cherenkov counting
(without scintillation fluid). The filter is washed two times (5 min each)
with 0.5 M Na_2HPO_4 one time (1 min) with distilled water and finally
rinsed with 100% EtOH. The filter is dried under the halogen lamp for
10 min and counted again for the radioactivity remaining (B). The percentage of the labeling efficiency is calculated by the formula $(B/A) \times 100$. In our
experience, oligonucleotides with labeling efficiencies of 20–70% and DNA
fragments with labeling efficiencies of 30–90% work well during hybridization. Fragments or oligonucleotides with lower labeling efficiencies should
not be used.

9.6. Hybridization, washing, and analysis

9.6.1. Reagents and solutions

SSC (20×), SDS (20%, w/v), EDTA (10 mM) (Sambrook *et al.*, 1989),
nonfat dry milk. The hybridizing solution for oligonucleotides (HSO) is
made up of 0.25% nonfat dry milk, 2× SSC, and 0.2% SDS (SDS is added
last to avoid precipitation). The solution is autoclaved and stored at RT.
PerfectHyb Plus (Sigma) is used as the hybridizing solution for DNA-
fragment probes. Washing buffer I (WB-I) is made up of 0.3% SDS and
2× SSC. Washing buffer II (WB-II) is made up of 0.1% SDS and 2× SSC.

9.6.2. Procedure

Before hybridization, the membrane is presoaked in 2× SSC for 5 min and
subsequently hybridized either in a large or small hybridization bottle
(depending on the size) rotating in a hybridizing oven. The hybridization
is carried out in the presence of either 5 ml (small hybridization bottle) or
7 ml (large hybridization bottle) of hybridizing solution at a temperature
appropriate for either the oligonucleotide probe (HSO) or DNA-fragment
probe (PerfectHyb Plus). The hybridization temperature should be set at
10 °C below the T_m of the oligonucleotide probe or 65 °C for the DNA
fragment probe. The membrane is prehybridized for 4–5 h for the oligonucleotide probes and for minimum of 30 min for a DNA-fragment probe.
The labeled probe is denatured (the DNA-fragment probe is diluted 10-fold
with 10 mM EDTA) at 100 °C in a sand bath for 5–10 min and quickly
chilled on ice. The probe is briefly centrifuged to collect it at the bottom of
the Eppendorf tube and added directly to the hybridizing solution in the
hybridization bottle. For best results, the blot should be hybridized
overnight.

Blots hybridized with oligonucleotide probes are washed two times (15 min each) with WB-I and one time (1 min) with the same buffer that has been heated to the hybridization temperature. A blot hybridized with DNA fragment should be washed two times (15 min each) with WB-I and two times (15 min each) with WB-II in a 50 °C water bath. The washed membrane is wrapped in Saran^{TM} Wrap and exposed to a PhosphorImager screen for scanning and analysis.

9.6.3. Stripping of hybridized probes

The ability to strip a probe from the Northern blots provides the major advantage of reusing the blot for more than one probe. The benefits are multifold. First of all, use of the same blot for the analysis of more than one transcript helps avoid inconsistencies associated with handling of multiple RNA samples during pipetting, loading, and transferring steps, leading to experimental variations. Second, it limits the amount of RNA needed for an experiment. Finally, it reduces the cost by using fewer reagents and membranes. We have successfully stripped and reprobed a Northern blot with different DNA fragments up to six times and with oligonucleotide probes up to ten times.

A Northern blot hybridized with an oligonucleotide probe is stripped by submerging it in 100–200 ml of 0.5% SDS in a large Tupperware container. The solution is brought to a boil in a microwave oven and kept boiling for 2–3 min. The radioactive SDS solution is disposed of following standard laboratory safety procedures. More than 90% of the radioactivity should be removed in the first washing. The radioactivity remaining on the membrane is checked with a Geiger counter. The washing step is repeated two to three times with fresh 0.5% SDS until little or no radioactivity is detected with the Geiger counter. A Northern blot hybridized with a DNA fragment labeled with modified dCTP can be stripped by using Strip-EZ DNA kit (Ambion, Tx, USA) according to the manufacturer's protocol.

Although a blot that will be stripped and reprobed should not be allowed to dry and should be stripped as soon as the first analysis is over, it can be briefly stored at −20 °C without any adverse effect on stripping. Due to *in vivo* differences in the amount of each transcript and the G content of the sequences being analyzed (Strip EZ kit uses modified dCTP to facilitate the probe degradation and removal), some probes strip less efficiently than others. However, often the hybridization and analysis of the second probe are not seriously affected, as the residual level of the first probe is significantly reduced after stripping (requiring longer exposure time) and may be in a different location in the blot than the second probe.

10. ANALYSIS OF RNA POLYADENYLATION IN *E. COLI*

Posttranscriptional addition of poly(A) tails to RNA plays an important role during RNA metabolism in both prokaryotes and eukaryotes. Although the regulation of poly(A) tail length has been implicated in translation control and decay of mRNA in eukaryotes (Baker, 1993), addition of 3′-poly(A) tails onto prokaryotic RNAs is generally regarded as a quality-control mechanism for initiating RNA decay (O'Hara *et al.*, 1995; Mohanty and Kushner, 1999; Li *et al.*, 2002). Techniques to study eukaryotic polyadenylation have been well established over the years, facilitating its in-depth characterization. In contrast, understanding of the nature and role of prokaryotic polyadenylation is still in its infancy in part due to the lack of appropriate techniques to analyze shorter poly(A) tails and the heteropolymeric tails generated by polynucleotide phosphorylase (Mohanty and Kushner, 2000; Mohanty *et al.*, 2004) that are associated with prokaryotic RNAs. Here we describe four important techniques that we use most to analyze polyadenylated transcripts in *E. coli*.

10.1. Dot-blot analysis

Dot-blot analysis is the quickest way to compare the total poly(A) level in strains based on the hybridization of oligo(dT)$_{20}$ to total RNA.

10.1.1. Materials and reagents

Dot-blot apparatus (Schleicher and Schuell Minifold), water bath (65 °C), SSC (20×), formaldehyde (37% solution).

10.1.2. Procedure

1. Total RNA (10–20 μg) is pipetted into new 1.5-ml Eppendorf tube.
2. The total volume of all the RNA samples is adjusted to 55 μl by adding diluted SSC (0.133×) followed by 165 μl of a mixture of 20× SSC/formaldehyde solution (1:1).
3. The samples are incubated in a 65 °C water bath for 20 min and quickly chilled in ice for 1 min.
4. The dot-blot apparatus should be assembled following the manufacturer's instructions and connected to a vacuum source. A charged nylon membrane (Magnacharge nylon membrane, GE Water & Processing Technologies) is placed inside the dot-blot apparatus to collect the RNA.
5. Each sample is mixed thoroughly by vortexing just before loading into a dot-blot well. Samples (100 μl) are pipetted in duplicate to the respective

wells. Once all samples are loaded, the vacuum is turned on slowly to draw the samples onto the membrane.

6. All the wells are washed once with 100 μl of 10× SSC. The membrane is removed after the vacuum is turned off and the dot-blot apparatus is disassembled.

7. The membrane is baked at 80 °C for 30 min followed by 10 sec of UV irradiation (optional) to fix the RNA.

8. The membrane is hybridized to ^{32}P-labeled oligo(dT)$_{20}$ at 30 °C as described in the previous section. A typical dot-blot is shown in Fig. 1.2.

9. The amount of ^{32}P-labeled oligo(dT)$_{20}$ that hybridizes to total RNA is quantified using a PhosphorImager and directly represents the amount of poly(A) within each sample. Because *E. coli* only contains a low level of oligo(A) sequences greater than 7 nt in length, the background one obtains is relatively low. If background is a problem, an oligo(dT)$_{30}$ can be used.

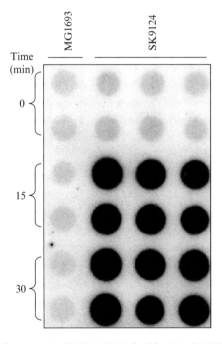

Figure 1.2 Results from a typical RNA-DNA dot blot. Total RNA isolated from *E. coli* (MG1693, wild-type; SK9124, wild-type/pBMK11[*pcnB*$^+$]) after IPTG (350 μmol) induction (0, 15, and 30 min) was immobilized onto a charged nylon membrane using S&S Minifold and probed with ^{32}P-labeled oligo(dT)$_{20}$. The plasmid pBMK11 contains *pcnB* coding sequence under the control of *lacZ* promoter and thus is responsible for the increase in poly(A) level after IPTG induction with increase in time (Mohanty and Kushner, 1999).

10.2. Poly(A) sizing assay

Dot-blot analysis is useful for comparing the relative level of polyadenylation among various strains. However, it does not provide any information about the length of the poly(A) tails. Accordingly, the poly(A) sizing assay is used to compare the relative length of poly(A) tails among various strains. In this procedure total RNA is first labeled at the 3′-end using ^{32}pCp and RNA ligase. The labeled RNA is then digested with RNase A, which cleaves after C and U residues, and RNase T1, which cleaves after G residues, leaving polymer of A residues labeled at the 3′-end. The internally encoded stretches of A residues are not labeled by this procedure and thus do not contribute to the analysis of post-transcriptionally added poly(A) tails at the 3′-ends. The labeled poly(A) tails are resolved in a polyacrylamide gel to determine their sizes. The detailed protocol described below is modified from Sachs and Davis (1989).

10.2.1. Reagents and solutions

End labeling mix: RNase-free water, 13 μl; RNA ligase buffer (10×), 3 μl; DMSO, 3 μl; RNasin [diluted to 10 U/μl, Promega], 4 μl; ^{32}pCp [Cytidine 3′,5′-bis(phosphate), 5′-(^{32}P)], 3000 Ci/mmol, NEN, Perkin Elmer), 5 μl; T4 RNA ligase (20 U/μl, NEB), 2 μl.

Stop solution: NaCl, 500 mM; EDTA, pH 8, 10 mM.

Digestion buffer (2×): Tris–HCl, pH 8, 20 mM; NaCl, 600 mM; MgCl$_2$, 2 mM; RNase A, 0.5 μg/μl; RNase T1, 8 U/μl.

Phenol:chloroform solution: Acid phenol (pH 4.5–5.2) mixed with equal volume of chloroform.

Sodium citrate loading dye (2×): Sodium citrate, pH 5, 40 mM; urea, 10 M; EDTA, pH 8, 2 mM; yeast tRNA, 0.5 mg/ml; xylene cyanol, 0.05%; bromophenol blue, 0.05%.

High-pH TBE (Tris–Borate–EDTA), pH 8.8 (5×): Tris base, 670 mM; boric acid, 222 mM; EDTA, 12.8mM. The pH is adjusted to 8.8 before final volume makeup.

Acrylamide solution (40%): Acrylamide, 38 g; bis-acrylamide, 2 g.

Acrylamide gel solution (14%): Urea, 43 g; high-pH TBE (5×), 20 ml; acrylamide solution (40%), 35 ml; water, 20 ml. The gel is polymerized by adding 400 μl of ammonium persulfate (APS, 10%, freshly made) and 30 μl of TEMED to the 50 ml of gel solution. The unpolymerized gel solution can be stored at 4 °C for later use.

Fixing solution: 20% EtOH, 10% acetic acid.

10.2.2. Methods

1. Each total RNA sample (1–20 μg) is dried in a Speed Vac concentrator and resuspended in 2 μl of RNase-free water.
2. 6 μl of the end-labeling mix is added to the sample and incubated overnight (~14 h) at RT.
3. 50 μl of the stop solution is added to stop the reaction.

4. The labeling efficiency (%/μl of sample) is calculated as described above using DE-81 filter paper.

5. Equal amounts (μl) of end-labeled RNAs, based on the percentage of labeling efficiency, are pipetted into new Eppendorf tubes.

6. The RNAs are digested by adding 2× Digestion buffer equal to the volume of end-labeled RNA to each sample and incubated at 37 °C for 30 min (an additional incubation for 30 min at 50 °C may be helpful for highly structured RNAs).

7. The total reaction mixture is extracted once with an equal volume of phenol:chloroform (1:1, pH 4.5–5.2).

8. The supernatant is transferred to a new Eppendorf tube. The RNA is precipitated overnight by incubating at −20 °C after adding 1 μl of Glycoblue (15 μg/μl, Ambion, TX, USA) and 2.5 vol of 100% ethanol to the supernatant. The sample is centrifuged at 13,000 rpm for 30 min in the cold room and the blue pellet is washed once with 200 μl of 70% ethanol. The pellet is dried in a Speed Vac concentrator for less than 5 min and dissolved in 2.5 μl of 1× sodium citrate dye.

9. Molecular-size standards of 10, 20 and 30 nt dA are mixed (5 pmol each) and 5′-end-labeled with γ-^{32}P ATP and T4 PNK.

10. A high-pH (8.8) polyacrylamide sequencing gel (12–20%) with 7 M urea is used to separate the poly(A) sequences. To detect the smaller poly(A) species, a 20% PAGE is used.

11. The gel (19.5 × 38 cm) is prerun using a custom-made vertical gel apparatus at 70 W (constant power) in 1× high-pH (8.8) TBE buffer until the buffer temperature reaches ∼55 °C. The samples and the molecular-size standards in the loading dye are heated in a 100 °C sand bath for 2 min, chilled in ice for 1 min, centrifuged briefly in the cold room, and loaded onto the gel.

12. The gel is run at 50 W (constant power) until the bromophenol blue dye travels ∼30 cm from the wells.

13. The gel is fixed for 3–5 min in the fixing solution, washed with deionized water 2-3 times to remove the acetic acid, transferred to a Whatman paper (3 mm), and dried in a gel dryer at 70 °C for 30 min.

14. The dried gel is initially analyzed with the PhosphorImager and subsequently exposed to ×-ray film. A typical poly(A) sizing gel is shown in Fig. 1.3.

11. REVERSE TRANSCRIPTION PCR (RT-PCR) TO ANALYZE 3′-ENDS OF SPECIFIC TRANSCRIPTS

The preceding two methods for analysis of polyadenylated transcripts are helpful in understanding the global polyadenylation profile of a strain. However, they do not provide any information on the nature of poly(A)

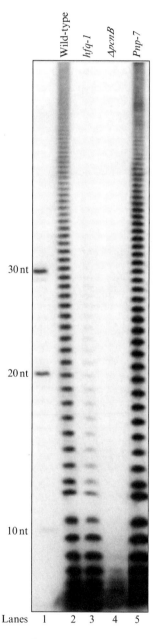

Lanes   1   2   3   4   5

Figure 1.3 Results from a typical poly(A) tail sizing assay comparing length of poly (A) tails in various *E. coli* genetic backgrounds. Total RNA samples (18 μg per lane) from exponentially growing cultures were processed for the poly(A) sizing assay on a 14% polyacrylamide gel. Lane 1, ^{32}P-labeled oligo d(A) size standards (nt) as marked. The genotype of each strain is as noted. Mutations in the *hfq* (*hfq-1*) and *pcnB* (*ΔpcnB*) genes reduce the poly(A) level and length of poly(A) tails whereas a mutation in *pnp* (*pnp-7*) increases poly(A) levels and average tail length (O'Hara *et al.*, 1995; Mohanty and Kushner, 2000; Mohanty *et al.*, 2004).

tails associated with a specific transcript in *E. coli*. Furthermore, many tails are synthesized by PNPase and are thus heteropolymeric (Mohanty and Kushner, 2000; Mohanty *et al.*, 2004). RT-PCR cloning and sequencing of 3′-ends of specific transcripts are alternative ways to analyze the nature and composition of individual poly(A) tails. The method involves reverse transcription of polyadenylated transcripts using an oligo(dT)$_{17}$ adapter primer (AP) followed by amplification of the cDNAs using a gene-specific primer (GSP) and abridged adapter primer (AAP) homologous to the 5′-end of AP. The PCR products are cloned into a suitable vector and sequenced (Fig. 1.4A).

11.1. Primer design for RT and PCR

AP: Adapter primer contains 17 dT residues at the 3′-end and multiple cloning sites (MCS) at the 5′-end [5′-GATGGTACCTCTAGAGCTC (T)$_{17}$].

AAP: Abridged adapter primer homologous to the MCS of AP (5′-GATGGTACCTCTAGAGCTC).

GSP: For effective PCR the gene-specific primer should be designed at $\sim$300–500 nt upstream of the 3′-end of the gene of interest. A suitable restriction site is engineered into the 5′-sequences of the GSP for directed cloning of PCR products.

11.2. Procedure

Reverse transcription of the polyadenylated RNAs is carried out in a 20-μl reaction volume using Superscript III reverse transcriptase (Invitrogen) essentially as suggested by the manufacturer with the following modifications. Total RNA (2–3 μg) containing 50 pmol of AP (total volume of 12.5 μl) is denatured for 5 min at 65 °C and quickly chilled in ice for 1 min. The mixture is centrifuged briefly and kept in ice. First strand buffer (5×, 4 μl), DTT (0.1 M, 1μl), dNTP mix (10 mM each of dATP, dCTP, dGTP, and dTTP, 1 μl), RNasin (Promega, 40 U/μl, 0.5 μl), and Superscript III RT (200 U/μl, 1 μl) are added to the reaction. Reverse transcription is carried out at 43 °C for 1 hr. The reaction is terminated by heating the sample at 70 °C for 15 min. The 3′-ends of the specific gene of interest are amplified in a 50-μl reaction volume containing 2.5 μl of the cDNA using JumpStart REDTaq DNA polymerase (Sigma) in the presence of a 5′-GSP and AAP.

11.3. Cloning and sequencing of PCR products

The expected size of the PCR product is confirmed by running 5 μl of the PCR reaction on a 1.5% agarose minigel and are subsequently purified using a Qiagen PCR purification kit. The PCR products are cloned by

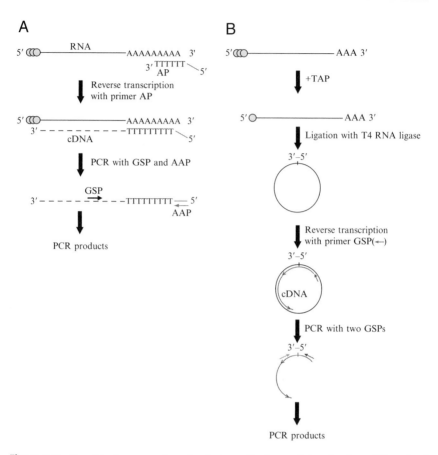

Figure 1.4 Graphical presentation of techniques for determining the 5′- and 3′-ends of specific transcripts employing RT-PCR. (A) RT-PCR cloning to analyze 3′-ends containing poly(A) tails of specific transcripts. The polyadenylated RNA is reverse transcribed using an oligo(dT)$_{17}$ adapter primer (AP) followed by PCR amplification of the cDNA using a gene specific primer (GSP) and an abridged adapter primer (AAP). (B) RNA self-ligation method for determining both the 3′- and 5′-ends of a transcript. The 5′-triphosphate of the transcript is converted to a monophosphate by treating with tobacco acid phosphatase (TAP). The 5′- and 3′-end of the transcript is ligated with t4 RNA ligase. The resulting 5′ to 3′-junction of the transcript is reverse transcribed using a GSP followed by PCR amplification of the cDNA by two GSPs.

using the restriction sites present in the 5′- and 3′-primers (GSP and AAP) and inserting them into the MCS of the cloning vector (pWSK29; Wang and Kushner, 1991) employing the Fast-Link DNA ligation kit (Epicenter Technologies, WI, USA). The ligated PCR products are transformed into DH5α by electroporation (Gene Pulsar, Biorad). White clones are selected for further analysis. The recombinant plasmids are isolated using a Qiagen (TX, USA) mini–prep kit and sequenced using either the *fmol* DNA cycle sequencing kit (Promega) or an automated sequencer (Applied Biosystems 3730xl DNA analyzer).

The major limitation of this technique is that it does not provide the actual length of the tails, as an oligo(dT) primer can prime anywhere along a long tail. In addition, it will not prime transcripts with tails shorter than 10 nt and thus underestimates the size and population of polyadenylated transcripts. Nevertheless, it is a powerful technique and an important tool in the analysis of polyadenylated transcripts (Mohanty and Kushner, 2000; Mohanty *et al.*, 2004; Mohanty and Kushner, 2006).

11.4. RNA self-ligation for determining 3′- and 5′-ends

As described in the previous section, one of the drawbacks of the RT-PCR cloning analysis of polyadenylated transcripts is that it cannot identify transcripts with very short (<10 nt) poly(A) tails. To analyze such transcripts we use an RT-PCR technique employing 5′ to 3′-end-ligated transcripts (Fig. 1.4B) (Yokobori and Pääbo, 1995) to clone and sequence both the 5′- and 3′-termini of the transcripts simultaneously (Mohanty & Kushner, 2008).

Total steady-state RNA (3 μg) is self-ligated in the presence of T4 RNA ligase (NEB) according to manufacturer's instructions. For analysis of primary transcripts, such as full-length mRNAs that contain 5′-triphosphate, the total RNA is initially treated with TAP (tobacco acid pyrophosphatase) according to the manufacturer's instructions (Epicenter Technologies, WI, USA) to convert the 5′-triphosphate termini to phosphomonoesters prior to self-ligation. The circular transcripts are then reverse transcribed using a gene-specific primer close to the 5′ to 3′-junction in the presence of Superscript III reverse transcriptase (Invitrogen) as described previously. The 5′ to 3′-junctions of the resulting cDNAs are amplified using either a pair of gene-specific primers (for monocistronic transcripts) or one upstream and one downstream gene-specific primers (for polycistronic transcripts) in the presence of JumpStart REDTaq DNA polymerase (Sigma). All the primers used in the cDNA amplifications are engineered to contain a suitable restriction site for directed cloning into pWSK29.

This method also has its limitations. In particular, unlike the oligo(dT)-dependent RT-PCR cloning procedure (Fig. 1.4A), there is no enrichment of polyadenylated transcripts during the self-ligation of transcripts using total RNA. Thus, the chances of obtaining polyadenylated transcripts are significantly reduced, requiring a lot more cloning and sequencing of cDNA clones. Furthermore, we have not been able to identify poly(A) tails longer than 7 nt using this method, which raises the question of whether transcripts with longer poly(A) tails are good substrates for efficient self-ligation.

12. Conclusion

Posttranscriptional modifications of RNA molecules, as well as the maturation and decay of specific transcripts in prokaryotes, have not been studied in great detail in part because of the lack of reliable and reproducible techniques. Understanding the mechanisms and regulation of RNA modifications in prokaryotes will be significant in answering a wide range of important questions in biomedical research and diagnostics. The general approaches and the methods described here should be useful for the study of RNA processing and posttranscriptional modifications in all organisms.

ACKNOWLEDGMENTS

This work was supported in part by a grant from the National Institutes of Health (GM57220) to SRK.

REFERENCES

Arraiano, C. M., Yancey, S. D., and Kushner, S. R. (1988). Stabilization of discrete mRNA breakdown products in *ams pnp rnb* multiple mutants of *Escherichia coli* K-12. *J. Bacteriol.* **170**, 4625–4633.

Arraiano, C. M., Yancey, S. D., and Kushner, S. R. (1993). Identification of endonucleolytic cleavage sites involved in the decay of the *Escherichia coli trxA* mRNA. *J. Bacteriol.*, **175**, 1043-1052.

Arraiano, C. M., Cruz, A. A., and Kushner, S. R. (1997). Analysis of the *in vivo* decay of the *Escherichia coli* dicistronic *pyrF-orfF* transcript: Evidence for multiple degradation pathways. *J. Mole. Biol.* **268**, 261-272.

Baker, E. J. (1993). Control of poly(A) length. *In* "Control of messenger RNA stability" (J. Belasco and G. Brawerman, eds.), pp. 367–417. Academic Press, New York.

Bernstein, J. A., Lin, P.-H., Cohen, S. N., and Lin-Chao, S. (2004). Global analysis of *Escherichia coli* RNA degradosome function using DNA microarrays. *Proc. Natl. Acad. Sci. USA* **101**, 2748–2763.

Burnett, W. V. (1997). Northern blotting of RNA denatured in glyoxal without buffer recirculation. *Bio. Techn.* **22**, 668–671.

Dahle, C. E., and Macfarlane, D. E. (1993). Isolation of RNA from cells in culture using Catrimox-14^{TM} cationic surfactant. *Bio. Techn.* **15**, 1102–1105.

Hajnsdorf, E., and Régnier, P. (2000). Host factor Hfq of *Escherichia coli* stimulates elongation of poly(A) tails by poly(A) polymerase I. *Proc. Natl. Acad. Sci. USA* **97**, 1501–1505.

Kushner, S. R. (2007). Messenger RNA decay. *In* "*Escherichia coli* and *Salmonella*: Cellular and molecular biology" (A. Böck *et al.*, eds.). American Society for Microbiology Press, Washington, DC (available online at http://www.ecosal.org).

Lee, K., Bernstein, J. S., and Cohen, S. N. (2002). RNase G complementation of *rne* null mutation identified functional interrelationships with RNase E in *Escherichia coli*. *Mol. Microbiol.* **43**, 1445–1456.

Li, Y., Cole, K., and Altman, S. (2003). The effect of a single, temperature-sensitive mutation on global gene expression in *Escherichia coli*. *RNA* **9**, 518–532.

Li, Z., Reimers, S., Pandit, S., and Deutscher, M. P. (2002). RNA quality control: Degradation of defective transfer RNA. *EMBO J.* **21,** 1132–1138.

Macfarlane, D. E., and Dahle, C. E. (1993). Isolating RNA from whole blood: The dawn of RNA-based diagnostics. *Nature* **362,** 186–188.

Macfarlane, D. E., and Dahle, C. E. (1997). Isolating RNA from clinical samples with Catrimox-12TM and lithium chloride. *J. Clin. Lab. Anal.* **11,** 132–139.

Mohanty, B. K., and Kushner, S. R. (1999). Analysis of the function of *Escherichia coli* poly(A) polymerase I in RNA metabolism. *Mol. Microbiol.* **34,** 1094–1108.

Mohanty, B. K., and Kushner, S. R. (2000). Polynucleotide phosphorylase functions both as a 3′ to 5′-exonuclease and a poly(A) polymerase in *Escherichia coli*. *Proc. Natl. Acad. Sci. USA* **97,** 11966–11971.

Mohanty, B. K., and Kushner, S. R. (2003). Genomic analysis in *Escherichia coli* demonstrates differential roles for polynucleotide phosphorylase and RNase II in mRNA abundance and decay. *Mol. Microbiol.* **50,** 645–658.

Mohanty, B. K., and Kushner, S. R. (2006). The majority of *E. coli* mRNAs undergo post-transcriptional modification in exponentially growing cells. *Nuc. Acids Res.* **34,** 5695–5704.

Mohanty, B. K., and Kushner, S. R. (2007). Ribonuclease P processes polycistronic tRNA transcripts in *Escherichia coli* independent of ribonuclease E. *Nuc. Acids Res.* **35,** 7614–7625.

Mohanty, B. K., and Kushner, S. R. (2008). Rho-independent transcription terminators inhibit RNase P processing of the *secG leuU* and *metT* tRNA polycistronic transcripts in *Escherichia coli*. *Nuc. Acids Res.* **36,** 364–375.

Mohanty, B. K., Maples, V. F., and Kushner, S. R. (2004). The Sm-like protein Hfq regulates polyadenylation dependent mRNA decay in *Escherichia coli*. *Mol. Microbiol.* **54,** 905–920.

O'Hara, E. B., Chekanova, J. A., Ingle, C. A., Kushner, Z. R., Peters, E., and Kushner, S. R. (1995). Polyadenylation helps regulate mRNA decay in *Escherichia coli*. *Proc. Natl. Acad. Sci. USA* **92,** 1807–1811.

Ow, M. C., and Kushner, S. R. (2002). Initiation of tRNA maturation by RNase E is essential for cell viability in *Escherichia coli*. *Genes Dev.* **16,** 1102–1115.

Ow, M. C., Liu, Q., and Kushner, S. R. (2000). Analysis of mRNA decay and rRNA processing in *Escherichia coli* in the absence of RNase E-based degradosome assembly. *Mol. Microbiol.* **38,** 854–866.

Ow, M. C., Perwez, T., and Kushner, S. R. (2003). RNase G of *Escherichia coli* exhibits only limited functional overlap with its essential homologue, RNase E. *Mol. Microbiol.* **49,** 607–622.

Perwez, T., and Kushner, S. R. (2006). RNase Z in *Escherichia coli* plays a significant role in mRNA decay. *Mol. Microbiol.* **60,** 723–737.

Sachs, A. B., and Davis, R. W. (1989). The poly(A) binding protein is required for poly(A) shortening and 60S ribosomal subunit-dependent translation initiation. *Cell* **58,** 857–868.

Sambrook, J., Fritsch, E. F., and Maniatis, T. (1989). "Molecular cloning: A laboratory manual." Cold Spring Harbor, NY: Cold Spring Harbor, Laboratory Press.

Wang, R. F., and Kushner, S. R. (1991). Construction of versatile low-copy-number vectors for cloning, sequencing and expression in *Escherichia coli*. *Gene* **100,** 195–199.

Yokobori, S.-I., and Pääbo, S. (1995). Transfer RNA editing in land snail mitochondria. *Proc. Natl. Acad. Sci. USA* **92,** 10432–10435.

CHAPTER TWO

ANALYZING THE DECAY OF STABLE RNAs IN *E. COLI*

Zhongwei Li* *and* Murray P. Deutscher[†]

Contents

Abstract

Stable RNA, mainly comprised of rRNA and tRNA, accounts for the majority of cellular RNA. Although normally stable under favorable growth conditions in the laboratory, these RNA species undergo extensive degradation responding to many environmental changes and stress conditions. Multiple ribonucleases and other enzymes may be involved in the decay of stable RNA. The onset and rate of degradation are probably determined by the status of the RNA as well as the availability of the degrading activities. The elucidation of pathways for stable RNA decay has been benefited by many biochemical and genetic approaches. These include purification of the enzymes and characterization of their substrate specificity *in vitro*, and studies of stable RNA decay by inactivating and over-expressing the degradation activities *in vivo*. Furthermore, RNA degradation intermediates have been characterized in detail, such as determining the sizes, the sequences, the 5'- and 3'-termini, etc. In this work, we describe the

* Department of Biomedical Science, Florida Atlantic University, Boca Raton, Florida, USA
† Department of Biochemistry and Molecular Biology, University of Miami School of Medicine, Miami, Florida, USA

Methods in Enzymology, Volume 447

ISSN 0076-6879, DOI: 10.1016/S0076-6879(08)02202-7

methods that are most commonly used in the study of the degradation and processing of stable RNA in E. coli. Most of them should be also useful in studies of other RNA species or RNA from other organisms.

1. INTRODUCTION

Stable RNAs constitute >95% of total RNA in an exponentially growing *E. coli* cell, of which approximately 80% are rRNAs and approximately 15% are tRNAs. In contrast to unstable mRNAs, stable RNAs are resistant to the cellular activities that degrade RNA under normal growth conditions. Their half-lives are usually longer than the doubling time of cells in culture, allowing their function to be maintained. Stable RNAs are protected from degradation through formation of highly stable secondary and tertiary structures, by complexing with proteins such as in the ribosome, and by amino acylation in the case of tRNA.

Under certain circumstances, stable RNAs may also undergo efficient decay (Deutscher, 2003). For example, ribosomes are extensively degraded to as much as 95% during starvation. This massive degradation seems to involve primarily ribosomal RNAs, not ribosomal proteins. Ribosome degradation begins with the conversion of polysomes to monosomes and then to ribosome subunits. Interestingly, once ribosome breakdown starts, degradation is quickly completed. The remaining ribosomes are intact and presumably functional. Similar ribosome decay may happen under stationary phase and slow growth conditions. This process seems to play a major role in bacterial adaptation to changing environments under natural conditions. Bacterial cells are able to survive under nutrient-limiting conditions solely from nutrients provided by the degradation of ribosomes and rRNAs. The residual ribosomes may keep cells viable and allow them to recover from unfavorable growth conditions. Despite its importance, our understanding of the activities and regulation of rRNA degradation during starvation, stationary phase, and slow growth conditions is still in its infancy.

When *E. coli* cultures are treated with certain chemical agents or antibiotics, cellular RNAs also undergo extensive breakdown (Deutscher, 2003). Many of these agents alter permeability of the cell membrane, resulting in loss of ions (including Mg^{++}) from the cytoplasm. The reduction in $[Mg^{++}]$ may affect ribosome structure and render the rRNAs more accessible to RNases. RNase I, a nonspecific endoribonuclease present primarily in the periplasmic space, is largely responsible for RNA degradation in these cases, resulting in formation of $3'$-mononucleotides. Presumably, membrane damage allows RNase I to enter the cytoplasm and destroy RNA. Loss of Mg^{++}, an inhibitor of RNase I, and increased accessibility of rRNAs all contribute to the extensive RNA decay. It should be noted that certain toxin–antitoxin pairs, such as the hok/sok system encoded by R1 plasmids, work by a similar mechanism (Gerdes and Wagner, 2007). The activated toxin causes depolarization of the cellular membrane, resulting in RNase I influx and subsequent massive RNA degradation and cell death (Nielsen *et al.*, 1991).

Another important aspect of stable RNA degradation was revealed by observations that stable RNA species with abnormal sequence or structure may be degraded. Such quality control of the RNA pool has recently attracted considerable attention. In one example, it was shown that a mutant tRNA^{Trp} is unstable and degraded at the precursor level (Li *et al.*, 2002). Degradation involves the exoribonucleases, polynucleotide phosphorylase (PNPase), and RNase R and is promoted by polyadenylation by poly(A) polymerase (Deutscher, 2006; Li *et al.*, 2002). In the absence of these enzymes, a precursor to the mutant tRNA^{Trp} accumulates to high levels. In many respects, degradation of the mutant tRNA^{Trp} is very similar to the pathway for mRNA decay (Deutscher, 2006). An additional example of stable RNA degradation occurs during misassembly of ribosomes. rRNA fragments generated in this process are normally degraded by PNPase and RNase R. When both enzymes are inactive, the RNA fragments accumulate to high levels, and the cells lose viability (Cheng and Deutscher, 2003). Apparently, the fragments need to be removed rapidly, because their presence interferes with ribosome assembly and leads to even more rRNA fragmentation. In contrast to the aforementioned RNA breakdown catalyzed by RNase I, this degradation is confined to misassembled ribosomes and is carried out by RNase activities that produce 5′-mononucleotides.

Study of stable RNA decay has benefited tremendously from mutants that display altered degradation. These include mutants that lead to increased membrane permeability, altered ribosome assembly, misfolded RNA structure, and inactive RNases. At least 17 ribonucleases have been identified in *E. coli*. Most of them have been extensively studied for their role in RNA degradation with genetic and biochemical approaches (Li and Deutscher, 2004). Although the function of many RNases in mRNA decay is well understood (Kushner, 2002), relatively little is known about how these enzymes might degrade stable RNAs (Deutscher, 2003). Early work suggested that stable RNA degradation, similar to mRNA decay, is initiated by endonucleolytic cleavages followed by 3′ to 5′-exonucleolytic action on the resulting RNA fragments (Kaplan and Apirion, 1975). As discussed previously, RNase I also participates in nonspecific RNA decay under certain unusual conditions, resulting in extensive loss of cellular RNA and possibly cell death. However, it is unlikely that RNase I plays any role in stable RNA decay under physiologic conditions of starvation, stationary phase, and slow growth, or in the surveillance of defective RNA. In fact, breakdown of stable RNAs under these conditions seems to be more controlled by use of many of the same enzymes that are involved in mRNA turnover (Deutscher, 2006). It is likely that poly(A) tails are added to the $3'$-ends of both mRNAs and stable RNAs, or their fragments, to facilitate binding of the processive exoribonucleases that are needed for degradation of highly structured RNAs (Cohen, 1995; Hajnsdorf *et al.*, 1995; Li *et al.*, 2002; Vincent and Deutscher, 2006).

Many ribonucleases participate in stable RNA degradation. RNase E is an essential endoribonuclease involved in the processing of rRNA and tRNA and degradation of mRNA. Numerous studies have shown that RNase E cleavages are responsible for initiating degradation of many mRNAs. RNase E is also the central component of the degradosome, a multienzyme complex also containing PNPase, RNA helicase B, and enolase. It is believed that such organization ensures that mRNA degradation will proceed to completion efficiently. However, it remains to be determined whether RNase E and/or the degradosome is also responsible for breakdown of stable RNAs. Other endoribonucleases, such as RNase III, RNase P, and the RNase E homolog, RNase G, are also likely candidates for this process. Some protein toxins, such as MazF and RelE, have been shown to possess endoribonuclease activities and to cleave and inactivate mRNA in response to stress conditions (Gerdes *et al.*, 2005). Whether any of the RNase toxins are involved in breakdown of stable RNA is not known. The exoribonucleases PNPase, RNase II, and RNase R progressively degrade RNA in the $3'$ to $5'$-direction. Two of them, PNPase and RNase R, have been shown to participate in degradation of rRNA fragments and unstable tRNA precursors, RNAs that contain a large amount of structured regions. Polyadenylation and RNA helicase activities also are probably required for efficient digestion (Cheng and Deutscher, 2005; Deutscher, 2006; Khemici and Carpousis, 2004).

Study of stable RNA decay requires quantification and characterization of the stable RNA species, their degradation intermediates, and their end products. The amount of intact RNAs and of their decay products provides a measure of the extent of degradation, whereas the decay intermediates provide important information on how degradation proceeds. Depending on the RNase(s) responsible for degradation, RNA products may contain a $3'$-hydroxyl or a $3'$-phosphate, whereas the released nucleotides may be $3'$- or $5'$-monophosphates or $5'$-diphosphates. The length of RNA decay intermediates is determined by the positions of the endonucleolytic cleavages and by stop sites for exonucleolytic trimming. Modified nucleotides are abundant in rRNAs and tRNAs and may play a role in the degradation of these RNAs (see later).

Many of the methods used to determine the level, half-life, and end products during mRNA decay are also applicable to the study of stable RNA degradation. Methods specifically useful for detection and characterization of stable RNAs and their degradation intermediates are discussed in the following.

2. Preparation of Stable RNA Substrates for *In Vitro* Degradation Assays

Various stable RNA substrates have been used extensively to identify and characterize degradation activities of purified RNases or in cell extracts. Sometimes, extracts from cells lacking specific enzymes (Li and Deutscher, 1994) or overexpressing certain activities (Li *et al.*, 1999) are used to study the function of relevant activities. Because of the high abundance of rRNAs and tRNAs in preparations of total RNA, it is relatively easy to isolate these RNA species to sufficient purity for *in vitro* assays. Frequently, the RNA is labeled by radioisotopes to simplify detection of products. Cellular RNA can be labeled with ^{32}P by the addition of carrier-free ^{32}P orthophosphate to a culture growing in low-phosphate medium. Low-phosphate TB medium (Gegenheimer *et al.*, 1977) is easy to prepare and provides good labeling of RNA. Typically, incubation of an exponentially growing culture for 1 h in the presence of 20 μCi/ml ^{32}Pi results in a specific radioactivity of $\sim$50,000 cpm per μg RNA. Alternatively, RNA can be labeled *in vivo* with ^{3}H-uridine by including the radioactive nucleoside in a rich medium (Li *et al.*, 1999). Finally, purified RNA can also be labeled *in vitro* (see the following).

rRNAs are usually prepared from isolated ribosomes. For this purpose, cell extracts should be made under conditions that preserve ribosome structure such as by grinding or by French press treatment. After removing cell debris by centrifugation at 30,000g, ribosomes in the supernatant fraction (S30) are pelleted at 100,000g. Ribosomes in the pellet P100 are often pure enough for most purposes and sometimes are used directly as substrates in RNA processing or degradation reactions. In these cases, ribosomes may be washed with high concentrations of KCl or NH$_4$Cl (0.5 or 1 M) and repelleted.

To isolate rRNAs from ribosome preparations, ribosomal proteins are denatured by the addition of sodium dodecyl sulfate (SDS, 0.5%) followed by extraction with phenol and chloroform. Protein-free rRNAs recovered in the aqueous phase are precipitated by ethanol. The resulting mixture of rRNAs can be used directly. Alternatively, the three rRNA components (23S, 16S, 5S) can be separated in agarose gels and purified by extraction (Cheng and Deutscher, 2002; Li *et al.*, 1999). Up to 10 mg of rRNA can be obtained from 1 g of exponentially growing cells.

A rapid procedure may be used to obtain small amounts of rRNA directly from cell lysates (Li *et al.*, 1999). This can be important when large-scale isolation is not feasible or necessary. Cells are pelleted from several milliliters of culture and resuspended in a lysis buffer (10 mM Tris-Cl, pH 7.4, 10 mM Na$_2$EDTA, pH 7.4, 1% SDS, 40% glycerol, 0.1% DEPC, and 0.1% bromphenol blue, as described in Gegenheimer *et al.*, [1977]). After boiling to release RNA, the lysate is directly applied to an agarose gel for electrophoretic separation. The 23S and 16S rRNAs are then individually extracted from the gel (Li et al., 1999b). A few micrograms of 23S and 16S rRNAs can be obtained from 1 ml of an exponential phase culture. This method is not suitable for isolation of 5S rRNA, because it is not separated well from tRNAs on an agarose gel.

tRNA can be prepared in large amounts by direct phenol extraction of cells followed by isopropanol fractionation (Deutscher and Hilderman, 1974). Cells are first suspended in a small volume of saline solution and are extracted with an equal volume of 88% phenol. Because the extraction procedure does not result in complete disruption of cells, large RNAs are only partially released. The RNA in the aqueous phase is then recovered by ethanol precipitation. RNAs are further purified by isopropanol fractionation, which separates on the basis of their sizes. RNA is dissolved in 0.3 M sodium acetate, pH 7.0, at room temperature. Larger RNAs are precipitated with 0.54 volumes of isopropanol and small RNAs with 0.98 volumes. The tRNA preparations are normally devoid of any high-molecular-weight material but can be contaminated with a small amount of 5S RNA (Deutscher and Hilderman, 1974). Two to three milligrams of tRNA can be isolated from 1 g of log-phase cells (Deutscher and Hilderman, 1974).

As noted previously, rRNA and tRNA preparations also can be labeled after they are isolated. Labeling of the 5'-end of a dephosphorylated RNA can be accomplished with polynucleotide kinase and γ-^{32}P-ATP. The 3'-end can be labeled by addition of [^{32}P]pCp with RNA ligase. In the case of tRNA, the 3'-terminal A residue can be removed by periodate treatment and β-elimination. This process causes oxidation of the terminal nucleoside which contains a vicinal 2',3'-hydroxyl, and in the presence of an amine, eliminates the oxidized residue from the phosphate. The resulting phosphate at the 3'-end of the tRNA is subsequently removed with alkaline phosphatase. This process can be repeated to remove the C residues at the second and third positions from the 3'-end. The resulting truncated tRNAs are treated with tRNA nucleotidyltransferase (the CCA enzyme) and CTP or ATP to add back radioactive C or A residues (Deutscher and Ghosh, 1978). Such substrates are useful in tRNA degradation experiments, especially for study of degradation of the 3'-termini.

Specific rRNA and tRNA species also can be easily prepared by in vitro transcription and have been widely used for RNA processing and degradation assays (e.g., Li and Deutscher [1994] and Tuohy et al. [1994]). RNA transcripts can be uniformly labeled by including one or more radioactive ribonucleoside triphosphates in the reaction, or they can be end labeled as described previously. If a radioactive nucleotide is used in the in vitro transcription reaction, the same nonradioactive nucleotide sometimes is added at a low concentration to reach a sufficient concentration for efficient labeling. Approximately 1 μg of RNA transcript can be produced in a 25-μl reaction mixture containing 60 μM of the radioactive ribonucleoside triphosphate.

If each ribonucleoside triphosphate is provided at 0.5 mM in a 25-μl reaction mixture, approximately 5 μg of RNA is normally produced. The RNA is often gel-purified before enzyme treatment (Li and Deutscher, 1999). It should be noted that tRNA or rRNA prepared by *in vitro* transcription lacks the modifications present in the corresponding cellular RNAs.

3. DETECTION OF DEGRADATION PRODUCTS *IN VITRO*

All eight exoribonucleases identified in *E. coli* degrade RNA in the 3' to 5'-direction, generating 5'-mononucleotides. The endoribonuclease RNase I cleaves RNA nonspecifically, producing short RNAs and 3'-mononucleotides. These products are soluble in cold trichloroacetate (TCA), whereas longer RNAs are precipitated. Quantification of acid-soluble products from radioactively labeled RNA substrates has been widely used to study the activity of these RNases *in vitro*. To study the rate of stable RNA decay, radioactively labeled stable RNAs are treated with RNase preparations, followed by addition of carrier yeast RNA to 0.5% and cold TCA to 10% final concentration (Cheng and Deutscher, 2002). Long RNAs precipitate quickly in 10 min on ice and are present in the pellet after centrifugation. The supernatant fractions are counted in a scintillation counter to quantify the release of acid-soluble material. By controlling the amount of RNA substrate, the amount of enzyme preparation, and the length of incubation, conversion to acid-soluble products can be kept low (<20%), ensuring a nearly linear rate of enzyme action.

Degradation intermediates of stable RNA frequently accumulate because of their resistance to RNases. This resistance may be conferred by highly stable structures, nucleotide modifications, and, in many cases, association with RNA-binding proteins. These intermediates are readily detectable by gel electrophoresis. Depending on the size of the RNAs to be detected, one may use an agarose gel (for RNAs from hundreds to thousands of nucleotides) (Lalonde *et al.*, 2007) (Fig. 2.1) or a denaturing polyacrylamide gel (for RNAs of 1 to hundreds of nucleotides) (Li and Deutscher, 1994, 1995, 1996; Li *et al.*, 1998a, 1998b). A polyacrylamide gel containing 0.2% SDS and 3% glycerol is also useful in separating large rRNAs (Gegenheimer *et al.*, 1977; Li *et al.*, 1999). RNA labeled with ^{32}P is detected by autoradiography. Nonlabeled RNAs can be stained by fluorescent dyes. The dye SYBR Gold (Invitrogen, Carlsbad, CA) stains single-stranded nucleic acids with at least 10 times the sensitivity of the double-strand specific dye, ethidium bromide.

When nonlabeled RNA is used as substrate, Northern blotting is useful for detection of degradation intermediates derived from specific RNAs. Because of the high abundance of stable RNAs and their decay intermediates, a very sensitive probe is usually not necessary. For instance,

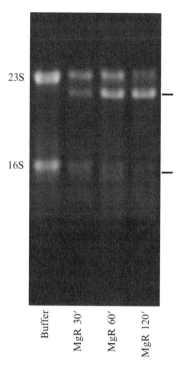

Figure 2.1   Degradation of *E. coli* 23S and 16S rRNAs by RNase R from *Mycoplasma genitalium* (MgR). rRNAs isolated from ribosomes (see text) were incubated with buffer alone for 120 min, or with 0.5 μg of MgR in a 10-μl reaction for 30, 60, or 120 min under conditions described previously (Lalonde *et al.*, 2007). The resulting products were separated in a 1.5% agarose gel, stained with SYBR Gold, and visualized under UV light. The starting 23S and 16S RNAs are labeled on the left. Prominent degradation intermediates are labeled with bars on the right.

oligonucleotide DNA probes labeled at their 5′-end with ^{32}P are usually sufficient to detect the degradation products of tRNA (Li *et al.*, 2002) or rRNA (Cheng and Deutscher, 2003). A carefully performed Northern blot can achieve single-nucleotide resolution for products of stable RNAs in the range from tens to more than 300 nucleotides in length (Li and Deutscher, 1995, 1996; Li *et al.*, 1998a,b).

4. Examination of Stable RNA Decay *In Vivo*

As discussed previously, stable RNA molecules are broken down by complicated processes involving both endoribonucleases and exoribonucleases, and possibly regulatory proteins as well (Deutscher, 2003, 2006). Mutants affecting various RNases have proven extremely useful for determining pathways of stable RNA decay *in vivo*. Such mutants may harbor interrupted and/or deleted genes or may contain temperature-sensitive alleles. For studies of essential enzymes, temperature-sensitive mutants are most useful. In addition, introduction of cloned genes that overexpress a particular activity or that rescue a missing activity in a mutant strain provide additional important information regarding the function of the genes and their products.

In the absence of a particular RNase, specific degradation intermediates may accumulate to high levels, which would suggest a key role of that RNase in the removal of the intermediate. Roles for other proteins such as poly(A) polymerase and RNA helicases may be assessed in a similar manner (Deutscher, 2006). Certain steps of stable RNA degradation are often carried out by multiple activities with overlapping functions. In these situations, degradation intermediates may only accumulate when most or all of the overlapping activities are inactivated. For instance, mutant *E. coli* cells devoid of RNase R or PNPase show little defect in rRNA degradation or in cell growth. In contrast, mutant cells lacking both activities accumulate degradation intermediates of 23S and 16S rRNAs to high levels and lose viability (Cheng and Deutscher, 2003), and these could be identified by Northern blotting.

Mutant forms of stable RNA species have been proven equally important in studying stable RNA decay. A mutant tRNA^{Trp} containing a single-nucleotide change in the acceptor stem renders this tRNA temperature-sensitive and exceedingly sensitive to degradation such that it is present at only ~15% the level of wild-type tRNA^{Trp} (Li et al., 2002). Degradation of this mutant tRNA is impeded in cells lacking PNPase and/ or poly(A) polymerase. In the absence of both enzymes, the amount of mutant tRNA^{Trp} increases to 60% of the wild-type level, and much of it is present as the tRNA precursor (Li et al., 2002). These data suggest that both enzymes and RNase R as well (Deutscher, 2006) are responsible for the degradation of this mutant tRNA. Moreover, mutations affecting the stability of an RNA greatly facilitate elucidation of RNA stability determinants. For example, it was noted that most stable RNA species in *E. coli* undergo exonucleolytic trimming to form their mature 3′-ends (Li and Deutscher, 2004; Li et al., 1998a). Remarkably, a base-paired structure, formed between the 3′ and 5′-terminal sequences, is present in each of these stable RNAs. Disruption of this structure may result in a marked decrease in the RNA's stability as shown in the case of the tRNA^{Trp}.

5. DETERMINATION OF THE 3′- AND 5′-TERMINI OF INTERMEDIATES DURING THE PROCESSING AND DEGRADATION OF STABLE RNAs

The processing of stable RNA precursors and the degradation of stable RNAs produce intermediates and breakdown products that often contain defined 5′- and 3′-termini. Because a 5′ to 3′-exonucleolytic activity has not yet been identified in *E. coli*, the 5′-ends of these intermediates are most likely generated by endonucleolytic cleavages (Li and Deutscher, 2004). The 3′-end, however, can be formed by the action of either an endoribonuclease or exoribonuclease. Knowledge of the 5′- and 3′-termini of intermediates often helps to explain RNA sequence and structural features that govern endonucleolytic cleavage sites or that impede exonucleolytic digestion. The methods discussed in the following generally apply to determination of the termini of a specific stable RNA present in a total RNA preparation. When other RNA species might interfere with detection of specific RNA products, isolation of the specific RNA under study may be required before the analysis (Li *et al.*, 1999).

Primer extension has been widely used to determine the 5′-termini of stable RNA products (Li and Deutscher, 1995, 1996; Li *et al.*, 1998a). A cDNA is synthesized from a labeled oligonucleotide primer that is annealed to a site on the target RNA. From the length of the cDNA product and the site of the primer, it is possible to determine exactly the 5′-end of the RNA. Mapping with S1 nuclease is often used to determine 3′-ends of RNAs. S1 nuclease digests single-stranded nucleic acids much more efficiently than double-stranded molecules. A single-stranded DNA probe hybridized to the RNA of interest is protected from S1 digestion in the region in which it is complementary to the stable RNA, whereas the rest is digested. This results in a shortened, labeled probe whose size indicates the 3′-terminus of the stable RNA. Although used routinely, S1 nuclease also often cleaves at weak double-stranded regions of the hybrid. Moreover, the "breathing" at the end of the double-stranded region leads S1 to generate staggered ends. As a consequence, S1 mapping may not be an ideal method for precise determination of 3′-termini and of their relative abundance.

Site-directed RNase H cleavage is another method that has proven useful for determining both the 5′- and 3′-ends of stable RNA products (Li *et al.*, 1999a,b). The endoribonuclease RNase H nonspecifically cleaves the RNA in an RNA/DNA duplex. For this method, a chimeric complementary oligonucleotide containing a short stretch of DNA and flanking 2-O-methyl RNA regions is used to force RNase H cleavage to occur at only a single position on the RNA (Fig. 2.2A). The number, position(s), and efficiency of RNase H cleavages depend on the length and sequence of the DNA in the chimera and on the source of the RNase H (Lapham *et al.*, 1997). A DNA stretch of 3 to 4 nt usually results in a single cleavage. Sometimes different RNase H enzymes need to be tested with a particular chimeric oligonucleotide to find the best match that cleaves efficiently at a single position on target RNA (Zhongwei Li, Shilpa Pandit, and Murray P. Deutscher, unpublished observation). The complementary chimera is usually designed to anneal to the target RNA at a position close to its 5′ or 3′-end. After RNase H treatment, the resulting shortened RNA fragments are separated to single-nucleotide resolution in a denaturing polyacrylamide gel and are detected

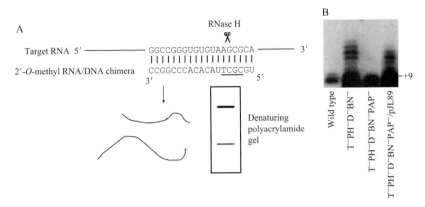

Figure 2.2 Analysis of RNA termini by site-directed RNase H cleavage. (A) Illustration of the method. A chimeric oligonucleotide containing a stretch of DNA (underlined sequence) is used to direct RNase H cleavage at a position close to the 3'-end of 23S rRNA. (B) The 3'-ends of precursor to 23S RNA revealed by site-directed RNase H cleavage (adapted from Li *et al.* [1999] with permission). The products were separated in a denaturing polyacrylamide gel and were detected by Northern blotting with a probe specific for the 3'-terminal sequence of 23S RNA. A small amount of precursor with eight extra nucleotides at the 3'-end is present in 23S RNA from wild-type cells. In a mutant strain deficient in the exoribonucleases, RNase T, PH, D, and BN (T⁻PH⁻D⁻BN⁻), a precursor containing 9 or more 3'-extra residues was found. These longer products are formed by polyadenylation. Removal of poly(A) polymerase from the multi-RNase–deficient strain (T⁻PH⁻D⁻BN⁻PAP⁻) restored the precursor to +8nt size. Introduction of the plasmid pJL89 harboring the poly(A) polymerase gene into the PAP⁻ background (T⁻PH⁻D⁻BN⁻PAP⁻/pJL89) resulted in formation of the polyadenylated species.

either by autoradiography for ³²P-labeled RNAs (Li *et al.*, 1999) or by Northern blotting for nonlabeled RNA (Li *et al.*, 1999). On the basis of the sizes of the RNA fragments, the position of the 5' or 3'-terminus can be determined. Site-directed RNase H cleavage works better than primer extension or S1 mapping to eliminate nonspecific products and to determine the relative abundance of different products. Figure 2.2B shows the heterogeneous 3'-termini of 23S rRNA revealed by a site-directed RNase H cleavage experiments. These 3'-ends were generated by polyadenylation in a mutant strain deficient in multiple exoribonucleases (Li *et al.*, 1999).

The 3'-end of a stable RNA often is sufficiently homogeneous that its identity can be determined by a modification of rapid amplification of cDNA 3'-ends (3'-RACE). This method requires ligation of a short oligonucleotide linker to the 3'-end of the RNA, followed by reverse transcription from an oligonucleotide primer complementary to the linker. The cDNA for the RNA of interest is then specifically amplified by polymerase chain reactions (PCR) with the same primer originally used for cDNA synthesis and a second primer specific for the RNA of interest. The PCR product can then be directly sequenced with the RNA-specific primer (Fig. 2.3) and the 3′-end of the RNA determined.

By use of this method, degradation of *E. coli* 23S rRNA by RNase R from *Mycoplasma genitalium* was shown to stop 1-nt downstream of two closeby ribose-methylation sites (shorter 23S products in Fig. 2.1), demonstrating the sensitivity of this particular RNase R to such RNA modifications (Lalonde *et al.*, 2007). Although RNA linkers are often used (Li *et al.*, 1998b, 2002), DNA linkers work equally well (Lalonde *et al.*, 2007). Linkers should be phosphorylated at their 5′-ends, and to prevent self-ligation, linkers should be blocked at their 3′-end by phosphorylation or other modifications. The sequence of the linker may include useful restriction sites, enabling subsequent digestion of the PCR products to remove the linkers, to eliminate unwanted cDNA products, or to generate cloning sites (see following). In addition to overcome occasional difficulties of the site-directed RNase H cleavage method, 3′-RACE can provide sequence information that may reveal posttranscriptional changes in the RNA sequence (see following).

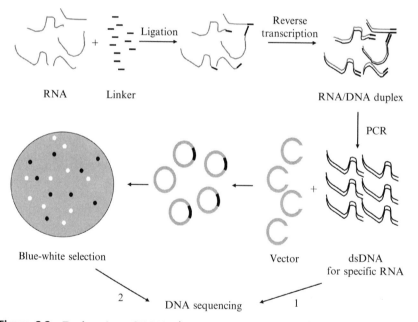

Figure 2.3 Explanation of RNA 3′-terminal sequences by 3′-RACE and cloning. The sequence of RNA is converted to that of cDNA by ligation of a linker to the 3′-end of the RNA and reverse transcription. Specific RNA sequences are amplified by PCR. The resulting PCR product can be directly sequenced (option 1) if a homogeneous 3′-end is expected (Lalonde *et al.*, 2007). Alternately, multiple individual clones can be constructed and sequenced (option 2) if heterogeneous 3′-ends are analyzed (Li *et al.*, 1998b).

6. DETECTION OF POLY(A) TAILS ON PRODUCTS GENERATED BY PROCESSING OR DEGRADATION OF STABLE RNA

In a mutant strain lacking multiple exoribonucleases, unexpected elongated 3′-sequences were found on precursors of some RNAs. This observation led to the discovery of polyadenylation of these RNA species (Li *et al.*, 1998b). Poly(A) tails were found on all stable RNA precursors examined in this mutant strain, and to a much less extent, also in wild-type cells. In the case of mutant tRNA^{Trp}, poly(A) tails were detected when PNPase was inactivated (Li *et al.*, 2002). Polyadenylation seems to promote degradation of stable structures in RNA and participates in mRNA turnover and stable RNA quality control (Deutscher, 2006). The poly(A) tails on stable RNAs were identified by cloning the 3′-RACE products (see earlier) followed by DNA sequencing of individual clones (Fig. 2.3) (Li *et al.*, 1998b, 2002). Because of the heterogeneous nature of the 3′-ends of these RNAs, multiple clones were sequenced for each stable RNA species. The results revealed RNA molecules containing all possible 3′-ends, including the normal mature end, encoded sequences shorter or longer than mature, and polyadenylated ends. For example, in a mutant lacking multiple RNases, approximately half of the precursors to 23S rRNA contain 1 to 5 adenylates at their 3′-ends (Fig. 2.2B) (Li *et al.*, 1998b). Because of the laborious clone selection and sequencing, such an approach is only used when the population of termini needs to be clearly identified.

7. CONCLUDING REMARKS

Degradation of stable RNAs is of fundamental importance for bacteria to adapt to environmental changes and to maintain a functional RNA reservoir. Various biochemical and genetic approaches have been applied to study stable RNA decay in *E. coli*. The methods described here have been used to study processing and degradation of a variety of stable RNA species. Many of the methods can characterize RNA at single-nucleotide resolution or can provide well-defined sequence information. The use of mutants and of overexpressed and purified enzymes has enabled identification of activities responsible for specific steps of degradation pathways. These powerful approaches will eventually lead to the explanation of detailed molecular mechanisms of stable RNA decay. It should be noted that many features contribute to the normally high stability of the stable RNAs. Changes in these RNA features and alteration of the cellular environment may affect accessibility of degradation activities to RNA, triggering rapid breakdown of stable RNAs (Deutscher, 2003). Moreover, stable RNA decay may be regulated by cell metabolism and stress. Therefore, processes such as signal transduction and multiple molecular interactions may regulate stable RNA decay. If so, other methods that address these additional features may also be useful in future studies.

ACKNOWLEDGMENTS

We thank Maureen S. Lalonde and Xin Gong for providing Fig. 2.1 for this work. This work was supported by NIH Grant GM16317 to M. P. D. and NIH grant S06 GM073621 and Florida Atlantic University Startup Fund to Z. L.

REFERENCES

Cheng, Z. F., and Deutscher, M. P. (2002). Purification and characterization of the *Escherichia coli* exoribonuclease RNase R. Comparison with RNase II. *J. Biol. Chem.* **277**, 21624–21629.

Cheng, Z. F., and Deutscher, M. P. (2003). Quality control of ribosomal RNA mediated by polynucleotide phosphorylase and RNase R. *Proc. Natl. Acad. Sci. USA* **100**, 6388–6393.

Cheng, Z. F., and Deutscher, M. P. (2005). An important role for RNase R in mRNA decay. *Mol. Cell.* **17**, 313–318.

Cohen, S. N. (1995). Surprises at the 3′-end of prokaryotic RNA. *Cell* **80**, 829–832.

Deutscher, M. P. (2003). Degradation of stable RNA in bacteria. *J. Biol. Chem.* **278**, 45041–45044.

Deutscher, M. P. (2006). Degradation of RNA in bacteria: Comparison of mRNA and stable RNA. *Nucleic Acids Res.* **34**, 659–666.

Deutscher, M. P., and Ghosh, R. K. (1978). Preparation of synthetic tRNA precursors with tRNA nucleotidyltransferase. *Nucleic Acids Res.* **5**, 3821–3829.

Deutscher, M. P., and Hilderman, R. H. (1974). Isolation and partial characterization of *Escherichia coli* mutants with low levels of transfer ribonucleic acid nucleotidyltransferase. *J. Bacteriol.* **118**, 621–627.

Gegenheimer, P., Watson, N., and Apirion, D. (1977). Multiple pathways for primary processing of ribosomal RNA in *Escherichia coli*. *J. Biol. Chem.* **252**, 3064–3073.

Gerdes, K., Christensen, S. K., and Løbner-Olesen, A. (2005). Prokaryotic toxin-antitoxin stress response loci. *Nat. Rev. Microbiol.* **3**, 371–382.

Gerdes, K., and Wagner, E. G. (2007). RNA antitoxins. *Curr. Opin. Microbiol.* **10**, 117–124.

Hajnsdorf, E., Braun, F., Haugel-Nielsen, J., and Régnier, P. (1995). Polyadenylylation destabilizes the rpsO mRNA of *Escherichia coli*. *Proc. Natl. Acad. Sci. USA* **92**, 3973–3977.

Kaplan, R., and Apirion, D. (1975). Decay of ribosomal ribonucleic acid in *Escherichia coli* cells starved for various nutrients. *J. Biol. Chem.* **250**, 3174–3178.

Khemici, V., and Carpousis, A. J. (2004). The RNA degradosome and poly(A) polymerase of *Escherichia coli* are required *in vivo* for the degradation of small mRNA decay intermediates containing REP-stabilizers. *Mol. Microbiol.* **51**, 777–790.

Kushner, S. R. (2002). mRNA decay in *Escherichia coli* comes of age. *J. Bacteriol.* **184**, 4658–4665.

Lalonde, M. S., Zuo, Y., Zhang, J., Gong, X., Wu, S., Malhotra, A., and Li, Z. (2007). Exoribonuclease R in *Mycoplasma genitalium* can carry out both RNA processing and degradative functions and is sensitive to RNA ribose methylation. *RNA* **13**, 1957–1968.

Lapham, J., Yu, Y. T., Shu, M. D., Steitz, J. A., and Crothers, D. M. (1997). The position of site-directed cleavage of RNA using RNase H and 29-O-methyl oligonucleotides is dependent on the enzyme source. *RNA* **3,** 950–951.

Li, Z., and Deutscher, M. P. (1994). The role of individual exoribonucleases in processing at the 3′-end of *Escherichia coli* tRNA precursors. *J. Biol. Chem.* **269,** 6064–6071.

Li, Z., and Deutscher, M. P. (1995). The tRNA processing enzyme RNase T is essential for maturation of 5S RNA. *Proc. Natl. Acad. Sci. USA* **92,** 6883–6886.

Li, Z., and Deutscher, M. P. (1996). Maturation pathways for *E. coli* tRNA precursors: A random multienzyme process *in vivo*. *Cell* **86,** 503–512.

Li, Z., and Deutscher, M. P. (2004). Exoribonucleases and endoribonucleases. *In EcoSal-Escherichia coli and Salmonella: Cellular and Molecular Biology*. (R. Curtiss, III, ed.). Chapter 4.6.3, ASM Press, Washington, DC; ([online] http://www.ecosal.org).

Li, Z., Pandit, S., and Deutscher, M. P. (1998a). 3′-Exoribonucleolytic trimming is a common feature of the maturation of small, stable RNAs in *Escherichia coli*. *Proc. Natl. Acad. Sci. USA* **95,** 2856–2861.

Li, Z., Pandit, S., and Deutscher, M. P. (1998b). Polyadenylation of stable RNA precursors *in vivo*. *Proc. Natl. Acad. Sci. USA* **95,** 12158–12162.

Li, Z., Pandit, S., and Deutscher, M. P. (1999a). Maturation of 23S ribosomal RNA requires the exoribonuclease RNase T. *RNA* **5,** 139–146.

Li, Z., Pandit, S., and Deutscher, M. P. (1999b). RNase G (CafA protein) and RNase E are both required for the 5′-maturation of 16S ribosomal RNA. *EMBO J.* **18,** 2878–2885.

Li, Z., Reimers, S., Pandit, S., and Deutscher, M. P. (2002). RNA quality control: Degradation of defective transfer RNA. *EMBO J.* **21,** 1132–1138.

Nielsen, A. K., Thorsted, P., Thisted, T., Wagner, E. G., and Gerdes, K. (1991). The rifampicin-inducible genes srnB from F and pnd from R483 are regulated by antisense RNAs and mediate plasmid maintenance by killing of plasmid-free segregants. *Mol. Microbiol.* **5,** 1961–1973.

Tuohy, T. M. F., Li, Z., Atkins, J. F., and Deutscher, M. P. (1994). A functional mutant of tRNA$^{Arg}_2$ with 10 extra nucleotides in its TΨC arm. *J. Mol. Biol.* **235,** 1369–1376.

Vincent, H. A., and Deutscher, M. P. (2006). Substrate recognition and catalysis by the exoribonuclease RNase R. *J. Biol. Chem.* **281,** 29769–29775.

CHAPTER THREE

GENOMIC ANALYSIS OF mRNA DECAY IN *E. COLI* WITH DNA MICROARRAYS

Pei-Hsun Lin,* Dharam Singh,* Jonathan A. Bernstein,[†] *and* Sue Lin-Chao*

Contents

Abstract

The decay of mRNA plays an important role in the regulation of gene expression. Although relatively ignored for many years and regarded as a simple ribonucleotide salvage pathway, mRNA decay has been established in recent years as a well-defined cellular process that plays an integral role in determining gene expression. The recent application of microarray methods to the study of diverse organisms will help us to better understand these gene regulatory circuits and the influence of transcript stability on gene expression. DNA microarray technology is the method of choice to study individual mRNA half-lives on

* Institute of Molecular Biology, Academia Sinica, Taipei 115, Taiwan
† Department of Pediatrics, Stanford University, Stanford, CA 94305, USA

Methods in Enzymology, Volume 447
ISSN 0076-6879, DOI: 10.1016/S0076-6879(08)02203-9

a global scale. It is important to standardize these methods to generate reproducible and reliable results. In this chapter, we describe experimental designs for the analysis of mRNA decay on a genome-wide scale and provide detailed protocols for each experimental step. We also present an analysis of the decay of chromosomally encoded mRNAs in *E. coli*.

1. INTRODUCTION

Prokaryotic mRNAs are generally unstable relative to those of eukaryotes. Their decay rates can differ considerably within a single cell, and their stability depends on growth conditions, environmental signals, and the efficiency of translation. The degradation of mRNA is an important step in the controlled expression of genes at the posttranscriptional level. It also defines the rate at which nascent transcripts can adopt a new steady-state level when there is a change in transcription in response to a change in the environmental conditions. The level of protein-encoding gene expression is primarily determined by three factors: the efficiency of transcription of DNA into mRNA, the stability of mRNA, and the frequency of its translation into proteins (Régnier and Arraiano, 2000). Transcriptional and translational factors have been well studied, but research into mRNA stability and factors that contribute to mRNA stability is only now gaining momentum. At one time, mRNA decay in prokaryotes was regarded as a nonspecific and unorganized activity, but the discovery of RNA degradation machinery known as the RNA degradosome (Miczak *et al.*, 1996; Py *et al.*, 1996) has generated more interest in understanding mRNA decay.

Recent studies have revealed multiple pathways for the degradation of specific mRNAs, including RNase E–dependent endonucleolytic cleavage, 3′ to 5′-exonucleolytic decay, and decay initiated at internal positions within mRNAs (reviewed by Steege [2000]). It is also evident that mRNA decay and the maturation of rRNA and tRNA share common features such as initial endonucleolytic cleavage followed by 3′ to 5′-exoribonucleolytic digestion (reviewed by Deutscher [2006] and Condon [2007]). In *E. coli*, mRNA decay generally involves endonucleolytic cleavage by RNase E. RNase E has recently been demonstrated to preferentially cleave RNA with a 5′-monophosphate (Celesnik *et al.*, 2007). This work also revealed that the first committed step in mRNA decay is the conversion of the triphosphorylated 5′-end to a monophosphorylated 5′-end by an unknown enzyme in a step that precedes RNase E cleavage and thus influences the half-lives of primary transcripts (Celesnik *et al.*, 2007). Recently Deana *et al.* (2008) showed that the *E. coli* protein RppH (formerly NudH/YgdP) is the RNA pyrophosphohydrolase that initiates mRNA decay by a 5′-end dependent pathway. Together with the finding that poly(A) tails play a role in the control of mRNA decay in prokaryotes, a new and unexpected link to eukaryotic mRNA decay was revealed (Bernstein and Ross, 1989; Ross, 1996), raising the possibility that bacterial mRNA decay, like eukaryotic mRNA decay, is a tightly regulated process (Schoenberg, 2007). The recent identification of an endoribonuclease RNase J1 having 5′ to 3′-exoribonuclease activity in *Bacillus subtilis* suggests there is much still to be learned about mRNA decay in prokaryotes (Mathy *et al.*, 2007).

In *E. coli*, mRNA half-life can range from a fraction of a minute to half an hour (Belasco, 1993; Coburn and Mackie, 1999; Nierlich and Murakawa, 1996). The most common method of mRNA half-life analysis is measurement of RNA stability using Northern blotting after an experimentally induced block in transcription, which for practical reasons cannot be used to analyze large numbers of transcripts. However, the emergence of microarray technology has facilitated the study of mRNA decay on a genome-wide scale. Only a few reports of large-scale RNA degradation analyses are found in the literature: in human (Lam *et al.*, 2001), yeast (Wang *et al.*, 2002), two species of the hyperthermophilic crenarchaeon *Sulfolobus* (Andersson *et al.*, 2006), the extremely halophilic euryarchaeon *H. salinarum* (Hundt *et al.*, 2007), *B. subtilis* (Hambraeus *et al.*, 2003), and *E. coli* (Bernstein *et al.*, 2002, 2004; Selinger *et al.*, 2003).

Here, we describe one commonly used approach: microarray analysis that uses PCR products spotted on a poly-L-lysine–coated glass microscope slide (Schena *et al.*, 1995; Shalon *et al.*, 1996). The degree of mRNA hybridization is quantified by comparing signals generated during concurrent hybridization of two differentially labeled cDNA pools with target DNAs arrayed on the surface of the slide. This method has been widely used to study the steady-state level of RNAs, but there has only been limited application of DNA microarrays to the study of kinetic events such as mRNA decay (Holstege *et al.*, 1998; Lam *et al.*, 2001; Wang *et al.*, 2002). We routinely investigate the individual half-lives of transcripts encoded by the *E. coli* genome with DNA microarrays and will now describe the method and discuss technical aspects of the study of mRNA decay on a genome-wide basis.

2. EXPERIMENTAL STRATEGY FOR mRNA HALF-LIFE ANALYSIS

2.1. Experimental design

Experimental designs for studying mRNA half-life on a global scale require careful selection of bacterial strains and culture conditions to best address hypotheses under investigation. The experimental process involves cell harvesting, RNA extraction, cDNA probe synthesis, and hybridization

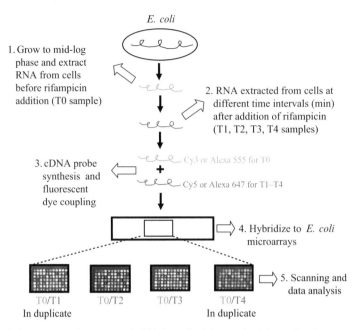

Figure 3.1   Strategy for mRNA half-life analysis in *E. coli* with DNA microarrays. Bacterial cells grown to mid-log phase were collected before and after the addition of rifampicin (500 μg/ml) as T0 to T4 samples. T1, T2, T3, and T4 represent samples at different time points of 1.5 min, 3.0 min, 4.5 min, and 6.0 min after rifampicin addition, respectively, followed by RNA extraction from all samples. RNA from the T0 sample was reverse transcribed into cDNA and coupled with Cy3 or Alexa 555 dye, whereas cDNA obtained after reverse transcribing RNA from T1 to T4 samples was coupled with Cy5 or Alexa 647 dye. cDNAs labeled with Alexa 555 and Alexa 647 dyes were mixed together and then hybridized to *E. coli* DNA microarrays at 65 °C for 5 to 6 h. The first and last time points were performed in duplicate. Finally, slides were scanned with a GenePix 4000B scanner, and data were used for mRNA half-life analyses. (See Color Insert.)

steps before data analysis (Fig. 3.1). The most important factors here are the quality of the isolated RNA and the sets of control spots used during normalization to correct for experimental variations. In the following sections, we will discuss the important technical aspects of the RNA half-life determination with microarrays.

2.2. Control spots for normalization

Microarray data must be normalized with control spots after scanning to account for a number of variables, including differences in the amounts of starting RNA, labeling efficiencies of samples, hybridization and scanning procedures. Normalization of microarray data is a key step in data processing that can affect later evaluations, so it is advisable to include a large number of controls in every microarray analysis. The use of multiple controls can help determine and improve the limit of detection on the array. For example, control spots that serve as negative controls should not hybridize to any sequence present in the experimental sample. Hybridization to a negative control indicates that significant nonspecific hybridization is occurring, and the stringency of the washing conditions should be increased. Among these control spots, transcripts that are assumed to be present in samples before and after inducing a block in transcription at unchanging ratios can be used as normalization controls. These normalization controls are housekeeping transcripts that cover a range of abundances. For example, total yeast genomic transcripts can be used to normalize the signal between the two samples for yeast gene expression experiments (Wang *et al.*, 2002).

Here, we use ribosomal DNA (rDNA) as our control spots for normalizing *E. coli* microarrays. In each array, we print different concentrations of rDNAs several times in a random pattern throughout the array. For example, we print 64 rDNA spots, including seven large and small rDNAs (RRLA, RRLB, RRLC, RRLD, RRLE, RRLG, RRLH, RRSA, RRSB, RRSC, RRSC, RRSD, RRSE, RRSG, RRSH) individually and different concentrations of all rDNA mixtures in different locations on each array. After hybridization, these normalization controls show high fluorescence signals and ratio of red and green intensities for each spot during scanning is close to 1.0. A normalization coefficient is then calculated by averaging the log base 2 ratio of 64 rDNA spots on each array. Finally, Normalization of all expressed genes is calculated by using the normalization coefficient from each successive time point. All half-life and normalization calculations are carried out with Microsoft Excel. Details of the calculations are described in data acquisition and analysis.

3. METHODS

3.1. Cell harvest

1. Inoculate 2 ml of medium with bacterial culture and grow overnight at 37 °C.
2. Dilute overnight culture into a 1-L flask containing 250 ml of culture medium and incubate at 30 °C with shaking. The use of lower culture temperatures for large-scale total-RNA preparations results in slower mRNA decay, facilitating the measurement of half-life. The dilution (1:100 or 1:50) needed depends on the turbidity of the overnight culture.
3. Follow the growth curve and take OD readings to obtain doubling time. The OD wavelength depends on which medium is used: OD_{600} is used for rich media such as LB, and OD_{540} is used for minimal media such as M9 medium.

4. When OD reaches mid-log phase ($OD_{600} = 0.4$ to 0.5; $OD_{540} = 0.6$ to 0.7), remove 30 ml of culture into 50-ml screw top tubes containing 5 ml of ice-cold stop solution (5% buffer saturated phenol, pH 7 to 8, in 95% ethanol) and mix well, then place on ice. Mark this as the T0 sample, and prepare in duplicate to increase the yield of RNA.

5. Add rifampicin (Sigma R3050, dissolved in DMSO) to the bacterial culture at a final concentration of 500 μg/ml to stop further transcription and be careful to reduce the speed of the bacterial incubator (normally waterbath shaker) to half the original speed. At regular intervals (for example, T1 = 1.5 min, T2 = 3 min, T3 = 4.5 min, T4 = 6 min), remove 30 ml of culture into 50-ml screw top tubes containing 5 ml of ice-cold stop solution and mix well, then place on ice until finished collecting all samples.

6. Spin down all culture samples at 3000g for 10 min. After centrifugation, carefully remove supernatant, and completely remove any residual medium (it will interfere during lysozyme digestion by changing pH) by blotting with clean Kim wipes with forceps. Samples can be stored at $-20\,^{\circ}$C or you can directly extract total-cell RNA with the Qiagen Midi RNA kit (cat. 75142).

3.2. RNA extraction and quantification

It is extremely important that all procedures involving RNA are performed in a ribonuclease-free environment. The use of DEPC-treated water, clean gloves, and "clean" equipment is essential to prevent RNA degradation, and, as a consequence, poor hybridization results. We recommend the use of commercially available total-RNA isolation kits, because the reagents have been quality controlled for the absence of ribonucleases and they are also very convenient. Typically, we use Qiagen RNeasy Midi Kit for our RNA preparation according to the manufacturer's instructions with some modifications as follows:

1. Loosen the bacterial pellets by flicking the bottom of the tube at $-20\,^{\circ}$C.
2. Resuspend the bacterial cells completely in 1 ml of lysozyme-TE buffer (10 mg/TE buffer, pH 8.0, RNase-free). Incubate at room temperature or 30 $^{\circ}$C for 20 to 30 min, and check for complete digestion of cell wall (gives a milky appearance).
3. The subsequent RNA extraction steps are performed at room temperature and are similar to manufacturer's instructions except that we apply 250 to 260 μl of RNase-free DNase I working reagent (RNase-free DNase I, Qiagen, cat. 79254) to the RNeasy Midi spin column and incubate at 37 $^{\circ}$C for 0.5 to 1 h.
4. Finally, elute RNA by adding 200 to 250 μl RNase-free water to RNeasy Midi spin column, let stand 5 min, and then centrifuge at 4000 rpm for 10 min. Transfer RNA sample into a 1.5-ml RNase-free Eppendorf tube (Axygen, CA) and keep at $-20\,^{\circ}$C.

5. Dilute RNA sample 1:100 with 10 mM Tris buffer (pH 7.5) and check the OD_{260}/OD_{280} ratio before analysis of RNA quantity and quality. For microarray experiments to be successful, RNA quality and quantity are crucially important. We use the Agilent 2100 Bioanalyzer LabChip (Agilent Technoliges Ltd., UK) to analyze RNA quality and quantity. This chip assigns an RNA integrity number (RIN) to each sample, and the RIN software algorithm then ranks the integrity of total-cell RNA on a scale of 1 to 10 (most to least degraded) on the basis of the most informative features of the ribosomal RNA genes (Schroeder *et al.*, 2006).

6. Apply 1 μl of RNA sample (1:10 to 20 dilution) for RNA LabChip analysis with the BioAnalyzer 2100 according to the manufacturer's instructions. The results can easily be shown as the peaks of 23S rRNA, 16S rRNA, and chromosomal DNA. Normally, high-quality RNA has a 23S/16S rRNA ratio of 1.6 to 2.0, and the ribosomal RNA peak should be very sharp with no degraded pattern. Degraded RNA or samples contaminated with chromosomal DNA should not be used as a template for cDNA synthesis.

3.3. Probe preparation

A labeled cDNA probe can be obtained from RNA either by direct labeling through incorporation of Cy3 or Cy5-dUTP during reverse transcription or by indirect labeling with amino-allyl dUTP followed by chemical coupling of Cy3 or Cy5 dyes by the amino-allyl group. Indirect labeling offers several advantages over direct incorporation, including (1) Cy3 and Cy5 are not incorporated equally into cDNA during direct incorporation, and (2) it is less expensive and more consistent. In addition, alternative dyes such as Alexa 555 and Alexa 647 can be used without major changes to the protocol. Fluorescent cDNA probes from 20 μg of total-cell RNA have previously been produced by amino allyl reverse transcription (Brumbaugh *et al.*, 1988; Hughes *et al.*, 2001; Randolph *et al.*, 1997). We use a modification of these methods as described in the following.

3.3.1. Prepare first strand amino-allyl cDNA(aa-cDNA) with amino-allyl-dUTP

1. Vacuum dry 20 μg of total-cell RNA but do not overdry, or it becomes difficult to dissolve RNA pellets later.
2. Make random primer stock by adding 550 μl of RNase-free water to pd $(N)_6$ random hexamer powder stock (50 A_{260} units = 1325 μg) (Amersham Pharmacia, cat. 8982) and keep at $-20\,^\circ$C.
3. Mix 20 μg of total-cell RNA and 10 μg of random primers ($\sim$4 μl) completely in 15 μl of RNase-free water to a total volume of 19 μl (primer-RNA tube). Incubate at 70 $^\circ$C for 10 min and then keep on ice.

4. Add 1.5 μl of 20× aa–dUTP/dNTP stock (final concentration 500 μM of each of dATP, dCTP, dGTP, 400 μM amino-allyl dUTP, 100 μM dTTP), 6 μl of 5× RT buffer, 1.5 μl of 0.1 M DTT, and 2 μl SSIII (reverse transcriptase is added last; SuperScript plus indirect cDNA labeling kit, cat. 466340, Invitrogen) to the 19 μl of primer-RNA tube to a final volume of 30 μl.

5. Incubate at room temperature for 5 to 10 min and then 46 °C for 2 to 3 h.

6. Degrade template RNA by adding 10 μl of 1 N NaOH, incubate at 70 °C for 15 min, and then add 25 μl of 1 M HEPES, pH 7.4; mix completely for neutralization.

7. Purify the first strand aa–cDNA with QIAquick PCR purification Kit (Qiagen, cat. 28106).

8. Add 500 μl of PB buffer to column and mix the labeling reaction (aa-cDNA, earlier) with PB buffer above the column, then spin at room temperature for 30 to 60 sec at 13,000 rpm.

9. Add 600 μl of 80% ethanol to column and spin at room temperature for 30 to 60 sec at 13,000 rpm. Repeat ethanol wash. Spin another 5 min to remove any residual ethanol. Avoid the use of the PE wash buffer provided by the kit, because the buffer PE contains Tris, which has a free amine that may interfere with the subsequent fluorescent dye coupling reaction.

10. Place column in a new 1.5-ml Eppendorf tube and apply 60 μl of 0.1 M NaHCO$_3$, pH 9.0, to the column. Incubate at room temperature for 3 min.

11. Spin at room temperature for 2 min at 13,000 rpm to elute aa-cDNA. Adjust volume to 60 μl with 0.1 M NaHCO$_3$, pH 9.0, then store aa-cDNA at −20 °C until needed or use directly for fluorescent dye labeling.

3.3.2. Fluorescent dye labeling—Alexa dye coupling

1. Resuspend one aliquot of Alexa 555 or Alexa 647 dye (SuperScript plus indirect cDNA labeling kit, Invitrogen) in 60 μl of aa–cDNA labeled probe. All steps involving Alexa dyes should be performed in the dark.

2. Incubate in the dark at room temperature for 60 to 90 min. Mix completely every 15 min.

3. Add 15 μl of 4 M NH$_2$OH.HCl to each reaction to quench unreacted Alexa dye to prevent cross coupling. Mix well and incubate in the dark at room temperature for 15 min. These samples are referred to as Alexa Dye-aa-cDNA.

3.3.3. Purify Alexa Dye-aa-cDNA with Qiagen QIAquick PCR purification kit

1. Add 500 μl of PB buffer to column, mix the labeling reaction (Alexa Dye-aa-cDNA) with PB buffer above the column, and spin at room temperature for 30 to 60 sec at 13,000 rpm.

2. Add 600 μl of buffer PE to column and spin at room temperature for 30 to 60 sec at 13,000 rpm. Repeat wash. Spin another 3 min to remove residual ethanol.

3. Place column in a new 1.5-ml Eppendorf tube and apply 30 μl EB buffer (prewarmed to 50 °C) to the column. Incubate at room temperature for 3 min.

4. Spin at room temperature for 2 min at 13,000 rpm to elute Alexa Dye-aa-cDNA. If not used immediately, then store at 4 °C in the dark. It is better to use Alexa Dye-aa-cDNA as soon as possible.

3.4. Microarray slide preparation

3.4.1. PCR amplification of *E. coli* genes

1. Genomic DNA extracted from the MG1655 strain of *E. coli* with the Blood and Cell Culture DNA maxi kit (Qiagen, cat. 13362) provides the template for PCR amplification.

2. Commercial primers available from Sigma-Genosys cover 4298 primer pairs (Richmond *et al.*, 1999). Resuspend the primer master powder stock and dilute to final working concentration of 2.5 μM (approximately 2.5 pmol/μl).

3. HotstarTaq (Qiagen, cat. 203205) DNA polymerase is used for PCR amplification of 4298 *E. coli* ORFs. Typically, each PCR reaction contains 0.25 μM primers, 1× PCR buffer with Mg^{2+}, 200 μM dNTPs, 10 ng *E. coli* genomic DNA, and 2.5 units of HotstarTaq DNA polymerase in 100-μl reaction mixtures.

4. The required PCR conditions are predenaturation at 95 °C for 15 min, 35 cycles of denaturing at 94 °C for 30 sec, annealing at 59 °C for 45 sec, and extension at 72 °C for 3 min, with a final cycle at 72 °C for 10 min.

3.4.2. PCR product verification

1. Confirm the size of the PCR products (2 μl of the total reaction volume) by electrophoresis in a 1.4% (w/v) agarose gel in 0.5× TBE buffer and staining with ethidium bromide. Standardize PCR conditions to get the optimum amount of the desired PCR products.

2. Precipitate the amplified DNAs in 96-well plates with ethanol. Fill 96-well V-bottom plates with 210 μl per well of ethanol/acetate solution (200 μl 95% ethanol + 10 μl 3 M sodium acetate, pH 5.2 to 6.0).

3. Transfer PCR products to 96-well V-bottom plates containing ethanol/acetate solution and mix by pipetting up and down four times. Seal with Nunc silicone tape (Nunc, cat. 236366) and place the plates overnight at −20 °C. The next day, plates are centrifuged at 3000g for 60 min at 4 °C.

4. Discard the supernatant by inversion and replace with 100 μl of ice-cold 70% ethanol and spin for an additional 30 min at 3000g at 4 °C. It is important that the 95% and 70% ethanol solutions not be made from 100% ethanol, because 100% ethanol may contain traces of fluorescent material that can interfere with array hybridization. It can be alternatively made or purchased industrial ethanol (95 %).

5. Discard the supernatant again and remove drops with paper towel or Kim wipes. Dry pellets overnight on bench top covered with paper tissue. Seal with silicone sealing tape (Nunc, cat. 236366) and store plates at 4 °C or −20 °C until transfer to 384-well plates.

6. Determine DNA concentration with a microplate spectrophotometer (Spectramax Plus[384], Molecular Devices, Sunnyvle, CA, USA). We use 100 ng/μl of DNA for array printing.

7. Transfer PCR products from 96-well plate to 384-well plate with a robotic transfer machine (Tecan Freedom EvoR Liquid Handling System, Tecan Trading AG, Switzerland).

8. Pick five PCR products for pGEM-T vector system T-A cloning (Promega, A3600) and sequencing. DNA sequence verification is obtained to confirm the correct identity.

3.4.3. cDNA microarray fabrication
Dissolve DNA pellets in 5 μl of spotting buffer (3× SSC) and print onto glass slides coated with poly-L-lysine (Eisen and Brown, 1999; Shalon *et al.*, 1996) or commercial amino-saline (Corning GAPSII, cat. 40004, Corning, NY) by OmniGrid arrayer (GeneMachine, Genomic Solutions, Ann Arbor, MI, USA) with Telechem SMP4 split pins. The diameter of each spot is approximately 120 μm, and the spot center–center distance is approximately 225 μm.

3.4.4. Slide postprocessing
A critical step in the construction of microarrays is postprocessing before hybridization. After the DNA is spotted onto the microarrays, the unbound lysine on the slide surface must be chemically blocked to prevent nonspecific binding of probe DNA to the array.

1. The microarrays must then be rehydrated to evenly distribute the mass of DNA within each spot. Place the array DNA facedown in a slide hydrating chamber (Humid Chamber, Sigma H-6644) over a room-temperature solution of 1× SSC and incubate until the spots on the array are visibly hydrated and glistening (20 to 30 min).

2. The DNA is then cross-linked to the slides with a UV light at 3000 mJ.

3. Before postprocessing, mark the array boundaries with a diamond pen on the back of the slide, because the array will disappear after postprocessing.

4. The free lysine groups on the slide surface are postprocessed (Eisen and Brown, 1999) by soaking slides in 317 ml of m-methylpyrrilidinone

(Aldrich, cat. 328634) with 6.18 g of succinate anhydrate (Aldrich, cat. 239690) and 35 ml sodium borate solution (0.2 M, pH 8.0) for 20 min.

5. Wash slides with 95 °C water for 2 min, and then dip rinse in 95% ethanol and dry by centrifugation. Store slides at room temperature. It is better to use slides as soon as possible, because prolonged storage can result in decreased performance.

3.5. Hybridization

1. The Alexa dye-labeled cDNA is combined with 1× SSC, 0.25% SDS, and 40 μg yeast tRNA in a final volume of 20 μl (probe mixture).
2. Mix completely, centrifuge briefly, and incubate at 95 °C for 10 min.
3. Spin down for 1 min and place a microarray slide in a hybridization chamber (GeneMachine, Genomic Solutions, Ann Arbor, MI, USA).
4. Place a TaKaRa spaced cover glass (TAKARA cat. TX702) on a microarray slide. Pipette probe mixture to the junction between the microarray slide and the cover glass, and allow it to spread by capillary action between cover and slide. The TaKaRa Spaced Cover Glass has printed protuberances (20 μl thick) at the four corners to make a space for hybridization. This allows the probe mixture to spread easily over the microarray without any bubbles. See Fig. 3.2 for the method of applying the probe onto the DNA microarray.

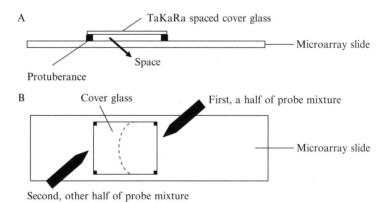

Figure 3.2 Applying labeled cDNA probes to DNA microarray slides with TaKaRa spaced cover glass. (A) A TaKaRa spaced cover glass is printed with protuberances at the four corners which create a space between the cover glass and microarray slide. (B) Half of the probe mixture was added to one side of the cover glass and allowed to spread across the slide by capillary action, followed by adding the rest of the mixture from the other side of the cover glass.

5. Moisten small pieces of 3MM paper with sterilized water in the wells in both sides of the hybridization chamber to maintain humidity during hybridization.

6. Close hybridization chamber and place in a water bath preheated to 65 °C. Incubate for 5 to 6 h.

3.6. Washing

1. Remove slides from the hybridization chamber and quickly put on a slide rack that has been submerged in wash buffer I (1× SSC, 0.03% SDS).

2. The cover slip can easily be dislodged during submerging process. Shake the slides at room temperature for 2 min.

3. Transfer the slides to wash buffer II (0.05× SSC) and shake at room temperature for 5 min.

4. Dry the slides by centrifuging at 1000 rpm for 10 min at room temperature. Store the dried slides in a dark box until scanning.

4. DATA ACQUISITION AND ANALYSIS

4.1. Microarray scanning

To detect the fluorescent signal of the DNA microarray after hybridization, the slides should be scanned with a microarray scanner. Among the most popular of the scanning laser devices available is the GenePix scanner from Axon Instruments (Union City, CA). Dual lasers simultaneously excite the different dyes (Cy3 and Cy5 or Alexa 555 and Alexa 647) at 532 nm and 635 nm, respectively, and the emitted fluorescence of each dye is measured independently. The scanner parameters can be adjusted during scanning, as described in the GenePix manual. Here, we use the GenePix 4000B scanner in the half-life experiments.

1. First, scan the T0/T1 slide initially at a low resolution of 50 μm to obtain a quick display image and the appropriate PMT, and then scan again at 10 μm resolution. The ribosomal DNA (rDNA) spots are used as our control for slide scanning. The Cy5/Cy3 or Alexa 647/Alexa 555 ratio of rRNAs should be equal to 1.0 during scanning, because they are housekeeping genes with long half-lives and, therefore, should not change in abundance after rifampicin-mediated transcriptional arrest. When the appropriate PMT is selected to keep the ratio of rDNA signal intensity close to 1.0, the other slides (T0/T2, T0/T3, and T0/T4) can then be scanned with the same PMT parameters.

2. After scanning, data are extracted through a procedure known as "gridding." In this step, the location of a spot is defined. This step is very important, because spots might be dislocated from the expected position or have different sizes because of variations during the printing process.
3. Save the scanned images as a 16-bit TIFF file and quantify the intensity of each spot with GenePix software.
4. After gridding, save as .gps files and export the raw data as .gpr files for data submission and half-life calculations.

4.2. Creating a data set for half-life calculation

Before half-life calculation, all the TIFF images and raw data are analyzed with Stanford Microarray database. Spots that are physically damaged or give signal within 150 units of background in the reference channel should be excluded from further analysis. The raw data are then extracted, and all half-life calculations carried out with Microsoft Excel. Figure 3.3 shows a flowchart of half-life calculation, and Fig. 3.4 is an example of half-life calculation. After half-life calculations, the data can also be further analyzed by using other software such as GeneSpring (Silicon Genetics) or GABRIEL (Pan *et al.*, 2002).

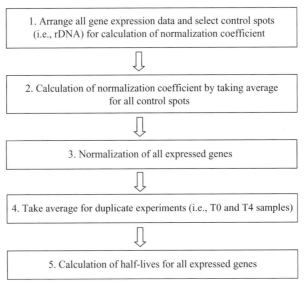

Figure 3.3 Flowchart for microarray half-life calculations. All half-lives and normalization calculations were carried out with Microsoft Excel. A normalization coefficient was calculated by averaging rDNA spots on each array. Duplicated samples were analyzed for the first and last time points. The average of normalized fluorescence ratios from the duplicate hybridizations was used in subsequent calculations. Detailed calculations are described in data acquisition and analysis.

A

| D4 | ▼ | *fx* | =IF(B4<>"", B4-C4,"") |

	A	B	C	D	E	F	G	H
1			1.5 min after treatment with Rif		3.0 min after treatment with Rif			4.5 min after
2	Gene	Raw data	Norm. coeff.	Normalization	Raw data	Norm. coeff.	Normalization	Raw data
3								
4	a	−0.992	0.141	−1.133	−2.409	0.08	−2.489	−2.234
5	b	−0.34	0.141	−0.481	−0.12	0.08	−0.2	−0.382

B

| F5 | ▼ | *fx* | =−1/INDEX(LINEST(B5:E5, $B$3:$E$3,TRUE, TRUE), 1) |

	A	B	C	D	E	F	G	H
1		1.5 min after treatment with Rif	3.0 min after treatment with Rif	4.5 min after treatment with Rif	6.0 min after treatment with Rif	half-life	R^2	
2						(min)		
3	Time	1.5	3	4.5	6			
4	Gene							
5	a	−1.133	−2.489	−2.493	−3.715	1.935	0.900	
6	b	−0.571	−0.2	−0.641	−1.757	3.751	0.591	

C

| G5 | ▼ | *fx* | =INDEX (LINEST (B5:E5, $B$3:$E$3, TRUE, TRUE), 3) |

	A	B	C	D	E	F	G	H
1		1.5 min after treatment with Rif	3.0 min after treatment with Rif	4.5 min after treatment with Rif	6.0 min after treatment with Rif	half-life	R^2	
2						(min)		
3	Time	1.5	3	4.5	6			
4	Gene							
5	a	−1.133	−2.489	−2.493	−3.715	1.935	0.900	
6	b	−0.571	−0.2	−0.641	−1.757	3.751	0.591	

Figure 3.4 Examples of half-life calculations from microarray experiments. (A) Microsoft Excel was used to normalize raw data obtained with the GenePix software package for different time points after rifampicin addition. The values of raw data were the log base2 of Cy5/Cy3 or Alexa 647/Alexa 555 ratios. The normalization coefficient (Norm. Coeff.) was computed by averaging in log space the fluorescence hybridization ratios for rDNA spots on each array (different time points). Normalized ratios were calculated by subtracting the value of normalization coefficient from the value of raw data. The D4 cell shows the equation (fx) of IF function for calculating the normalization coefficient. The result after calculation is shown in red. (B and C) Calculation of half-life for different time points after rifampicin addition. Microsoft Excel was used to calculate mRNA half-life from a least-squares linear fit to semilog (log base 2) of mRNA abundance (normalized raw data) versus time. Regression coefficients (R^2) are a measure of linear fits. The F5 and G5 show the half-life and R^2 values based on the INDEX and LINEST functions, respectively. The results of the calculation are shown in red. *fx* refers to the formula/function used in normalization (A), calculating the half-life (B) and Regression coefficient (C). (See Color Insert.)

1. Arrange all gene expression data to easily select all control spots (rDNAs) for normalization coefficient calculation with the data sorting function of Microsoft Excel.
2. The normalization coefficient is computed by averaging the log base 2 ratio of rDNAs.

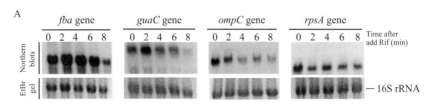

	Half-life determination		
Gene	Microarrays	Northern blots	RT-PCR
Strain	MG1655	MG1655	
fba	7.2 ± 1.9	6.9 ± 1.7	
guaC	8.4 ± 2.2	7.4 ± 1.0	
ompC	9.7 ± −1.2	8.4 ± 0.8	
rpsA	4.2 ± 0.2	4.8 ± 0.2	

Strain Gene	Wild type	Mutant	Wild type	Mutant
Enolase *cysN*	1.0	3.1	1.0 ± 0.1	2.0 ± 0.8
RhlB *accC*	1.6 ± 0.4	3.3 ± 0.4	2.5 ± 0.1	3.5 ± 1.0
PNPase *pntA*	2.9 ± 0.1	6.1	3.8 ± 0.1	5.8 ± 0.1
Rne C-ter truction *sdhA*	2.5	6.6	2.0 ± 0.1	4.6 ± 0.8

Figure 3.5 mRNA half-lives as determined by DNA microarray, Northern blotting, and qRT-PCR. (A) Northern blots of *fba*, *guaC*, *ompC*, and *rpsA* transcripts from *E. coli* strain MG1655. The number indicates the time (min) after rifampicin addition. The 16S rRNA abundance was determined by scanning densitometry of ethidium bromide–stained gels (shown as EtBr gel). Northern blot signals were quantified with either a Molecular Dynamics PhosphorImager or an FLA-5000 imager (Fuji). RNA half-lives were calculated from normalized hybridization intensity data by use of linear fits to a semilog plot. (B) Comparison of RNA half-lives determined by DNA microarray, Northern blot, and qRT-PCR methods. These data were previously published (Bernstein *et al.*, 2002, 2004). Half-life measurements are shown as mean ± standard error or mean alone where the standard error was not estimable.

3. Normalization of data is conducted by use of the raw data value minus the value of the normalization coefficient. Figure 3.4A shows how to use Microsoft Excel to normalize the raw data. The IF function used here is to avoid applying the normalization coefficient to empty cells.
4. After normalization of all raw data in each array, take the average of the duplicated experiments in first and last time points.
5. Half-lives are determined from a least-squares linear fit to a log base 2 of mRNA abundances (normalized data) versus time. Regression coefficients (R^2) should also be calculated to assess the quality of the linear fits used in half-life estimation. The INDEX and LINEST function for half-life and regression coefficient calculations may be used. The LINEST function is used to calculate the half-life and regression coefficient. The INDEX function is used to display the data. In Fig. 3.4B and C, sample equations for half-life and regression coefficient calculation with Microsoft Excel are shown.

4.3. Data verification

Microarray methods provide a global view of gene expression in response to genetic or environmental changes. Before drawing conclusions about the decay of specific transcripts, it is appropriate to independently verify the data obtained. Several techniques and strategies are available, including the use of previously published literature values, hybridization-based techniques (e.g., Northern blotting or RNase protection), or real-time RT-PCR. Several studies also have demonstrated agreement between microarray and RT-PCR or primer extension results (VanDyk *et al.*, 2001; Wei *et al.*, 2001; Zheng *et al.*, 2001). As shown in Fig. 3.5, we used Northern blots or real-time RT-PCR to check half-life data obtained in the microarray experiments; these data are taken from our previously published papers (Bernstein *et al.*, 2002, 2004).

ACKNOWLEDGMENTS

We thank Dr. S. Y. Tung, IMB microarray core facility, for excellent technical support and Dr. H. Wilson for critical reading and comments on the manuscript. This work was supported by grants from the National Science Council, Taiwan (NSC 94/95-2311-B-001-034; NSC 97-2321-B-001-014) and by an intramural fund from Academia Sinica to S. L.-C., and D. S. received a postdoctoral fellowship from the National Science Council, Taiwan.

REFERENCES

Andersson, A. F., Lundgren, M., Eriksson, S., Rosenlund, M., Bernander, R., and Nilsson, P. (2006). Global analysis of mRNA stability in the archaeon *Sulfolobus*. *Genome Biol.* **7**, R99.

Belasco, J. G. (1993). mRNA degradation in prokaryotic cells: An overview. *In* "Control of Messenger RNA Stability." (J. G. Belasco and G Brawerman, eds.), pp. 3–12. Academic Press, San Diego.

Bernstein, J. A., Khodursky, A. B., Lin, P.-H., Lin-Chao, S., and Cohen, S. N. (2002). Global analysis of mRNA decay and abundances in *Escherichia coli* at single-gene resolution using two-color fluorescent DNA microarrays. *Proc. Natl. Acad. Sci. USA* **99,** 9697–9702.

Bernstein, J. A., Lin, P. H., Cohen, S. N., and Lin-Chao, S. (2004). Global analysis of *Escherichia coli* RNA degradosome function using DNA microarrays. *Proc. Natl. Acad. Sci. USA* **101,** 2758–2763.

Bernstein, P., and Ross, J. (1989). Poly(A), poly(A) binding protein and the regulation of mRNA stability. *Trends Biochem. Sci.* **14,** 373–377.

Brumbaugh, J. A., Middendorf, L. R., Grone, D. L., and Ruth, J. L. (1988). Continuous, online DNA sequencing using oligodeoxynucleotide primers with multiple fluorophores. *Proc. Natl. Acad. Sci. USA* **85,** 5610–5614.

Celesnik, H., Deana, A., and Belasco, J. G. (2007). Initiation of RNA decay in *Escherichia coli* by 5′-pyrophosphate removal. *Mol. Cell* **27,** 79–90.

Coburn, G. A., and Mackie, G. A. (1999). Degradation of mRNA in *Escherichia coli*: An old problem with some new twists. *Prog. Nucleic Acid Res. Mol. Biol.* **62,** 55–108.

Condon, C. (2007). Maturation and degradation of RNA in bacteria. *Curr. Opin. Microbiol.* **10,** 271–276.

Deana, A., Celesnik, H., and Belasco, J. G. (2008). The bacterial enzyme RppH triggers messenger RNA degradation by 5′-pyrophosphate removal. *Nature* **451,** 355–358.

Deutscher, M. P. (2006). Degradation of RNA in bacteria: Comparison of mRNA and stable RNA. *Nucleic Acid Res.* **34,** 659–666.

Eisen, M. B., and Brown, P. O. (1999). DNA arrays for analysis of gene expression. *Methods Enzymol.* **303,** 179–205.

Hambraeus, G., von Wachenfeldt, C., and Hederstedt, L. (2003). Genome-wide survey of mRNA half-lives in *Bacillus subtilis* identifies extremely stable mRNAs. *Mol. Genet. Genomics* **269,** 706–714.

Holstege, F. C., Jennings, E. G., Wyrick, J. J., Lee, T. I., Hengartner, C. J., Green, M. R., Golub, T. R., Lander, E. S., and Young, R. A. (1998). Dissecting the regulatory circuitry of a eukaryotic genome. *Cell* **95,** 717–728.

Hughes, T. R., Mao, M., Jones, A. R., Burchard, J., Marton, M. J., Shannon, K. W., Lefkowitz, S. M., Ziman, M., Schelter, J. M., Meyer, M. R., Kobayashi, S., and Davis, C. (2001). Expression profiling using microarrays fabricated by an ink-jet oligonucleotide synthesizer. *Nat. Biotechnol.* **19,** 342–347.

Hundt, S., Zaigler, A., Lange, C., Soppa, J., and Klug, G. (2007). Global analysis of mRNA decay in *Halobacterium salinarum* NRC-1 at single-gene resolution using DNA microarrays. *J. Bacteriol.* **189,** 6936–6944.

Lam, L. T., Pickeral, O. K., Peng, A. C., Rosenwald, A., Hurt, E. M., Giltnane, J. M., Averett, L. M., Zhao, H., Davis, R. E., Sathyamoorthy, M., Wahl, L. M., and Harris, E. D. (2001). Genomic-scale measurement of mRNA turnover and the mechanisms of action of the anti-cancer drug flavopiridol. *Genome Biol.* **2,** 1–11.

Mathy, N., Bénard, L., Pellegrini, O., Daou, R., Wen, T., and Condon, C. (2007). 5′ to 3′-exoribonuclease activity in bacteria: Role of RNase J1 in rRNA maturation and 5′-stability of mRNA. *Cell* **129,** 681–692.

Miczak, A., Kaberdin, V. R., Wei, C.-L., and Lin-Chao, S. (1996). Proteins associated with RNase E in a multicomponent ribonucleolytic complex. *Proc. Natl. Acad. Sci. USA* **93,** 3865–3869.

Nierlich, D. P., and Murakawa, G. J. (1996). The decay of bacterial messenger RNA. *Prog. Nucleic Acid Res. Mol. Biol.* **52,** 153–216.

Pan, K.-H., Lih, C.-J., and Cohen, S. N. (2002). Analysis of DNA microarrays using algorithms that employ rule-based expert knowledge. *Proc. Natl. Acad. Sci. USA* **99,** 2118–2123.

Py, B., Higgins, C. F., Krisch, H. M., and Carpousis, A. J. (1996). A DEAD-box RNA helicase in the *Escherichia coli* RNA degradosome. *Nature* **381,** 169–172.

Randolph, J. B., and Waggoner, A. S. (1997). Stability, specificity and fluorescence brightness of multiply-labeled fluorescent DNA probes. *Nucleic Acids Res.* **25,** 2923–2929.

Régnier, P., and Arraiano, C. M. (2000). Degradation of mRNA in bacteria: Emergence of ubiquitous features. *BioEssays* **22,** 235–244.

Richmond, C. S., Glasner, J. D., Mau, R., Jin, H., and Blatter, F. R. (1999). Genome-wide expression profiling in *Escherichia coli* K-12. *Nucleic Acids Res.* **27,** 3821–3835.

Ross, J. (1996). Control of messenger RNA stability in higher eukaryotes. *Trends Genet.* **12,** 171–175.

Schena, M., Shalon, D., Davis, R. W., and Brown, P. O. (1995). Quantitative monitoring of gene expression patterns with a complementary DNA microarray. *Science* **270,** 467–470.

Schoenberg, D. R. (2007). The end defines the means in bacterial mRNA decay. *Nat. Chem. Biol.* **3,** 535–536.

Schroeder, A., Mueller, O., Stocker, S., Salowsky, R., Leiber, M., Gassmann, M., Lightfoot, S., Menzel, W., Granzow, M., and Ragg, T. (2006). The RIN: An RNA integrity number for assigning values to RNA measurements. *BMC Mol. Biol.* **7,** 3.

Selinger, D. W., Saxena, R. M., Cheung, K. J., Church, G. M., and Rosenow, C. (2003). Global RNA half-life analysis in *Escherichia coli* reveals positional patterns of transcripts degradation. *Genome Res.* **13,** 216–223.

Shalon, D., Smith, S. J., and Brown, P. O. (1996). A DNA micro-array system for analyzing complex DNA samples using two-color fluorescent probe hybridization. *Genome Res.* **6,** 639–645.

Steege, D. A. (2000). Emerging features of mRNA decay in bacteria. *RNA* **6,** 1079–1090.

VanDyk, T. K., Wei, Y., Hanafey, M. K., Dolan, M., Reeve, M. G., Rafalski, J. A., Rothman-Denes, L. B., and LaRossa, R. A. (2001). A genomic approach to gene fusion technology. *Proc. Natl. Acad. Sci. USA* **98,** 2555–2560.

Wang, Y., Liu, C. L., Storey, J. D., Tibshirani, R. J., Herschlag, D., and Brown, P. O. (2002). Precision and functional specificity in mRNA decay. *Proc. Natl. Acad. Sci. USA* **99,** 5860–5865.

Wei, Y., Lee, J.-M., Smulski, D. R., and LaRossa, R. A. (2001). Global impact of sdiA amplification revealed by comprehensive gene expression profiling of *Escherichia coli*. *J. Bacteriol.* **183,** 2265–2272.

Zheng, M., Wang, X., Templeton, L. J., Smulski, D. R., LaRossa, R. A., and Storz, G. (2001). Microarray-mediated transcriptional profiling of the *Escherichia coli* response to hydrogen peroxide. *J. Bacteriol.* **183,** 4562–4570.

CO-IMMUNOPURIFICATION OF MULTIPROTEIN COMPLEXES CONTAINING RNA-DEGRADING ENZYMES

Agamemnon J. Carpousis, Vanessa Khemici, Soraya Aït-Bara, *and* Leonora Poljak

Contents

Abstract

Co-immunopurification is a classical technique in which antiserum raised against a specific protein is used to purify a multiprotein complex. We describe work from our laboratory in which co-immunopurification was used to characterize the RNA degradosome of *Escherichia coli*, a multiprotein complex involved in RNA processing and mRNA degradation. Polyclonal rabbit antibodies raised against either RNase E or PNPase, two RNA degrading enzymes in the RNA degradosome, were used in co-immunopurification experiments aimed at studying the assembly of the RNA degradosome and mapping protein–protein interactions within the complex. In *E. coli*, this method has been largely supplanted by approaches in which proteins are engineered to contain tags that interact with

Laboratoire de Microbiologie et Génétique Moléculaire (LMGM), Centre National de la Recherche Scientifique (CNRS) and Université Paul Sabatier, Toulouse, France

Methods in Enzymology, Volume 447
ISSN 0076-6879, DOI: 10.1016/S0076-6879(08)02204-0

commercially available antibodies. Nevertheless, we believe that the method described here is valid for the study of bacteria in which the genetic engineering needed to introduce tagged proteins is difficult or nonexistent. As an example, we briefly discuss ongoing work in our laboratory on the characterization of RNase E in the psychrotolerant bacterium *Pseudoalteromonas haloplanktis*.

1. INTRODUCTION

RNase E is a single-strand specific endoribonuclease that has a general role in RNA processing and mRNA degradation in *Escherichia coli*. Homologs of RNase E are found in many gram-negative bacteria. *E. coli* also encodes a paralog of RNase E named RNase G. Together, RNase E and RNase G belong to a large family of bacterial ribonucleases known as the RNase E/G family (Carpousis, 2002; Condon and Putzer, 2002; Lee and Cohen, 2003). Originally discovered as a ribonuclease necessary for the maturation of 5S ribosomal RNA, RNase E was subsequently identified as an enzyme important in the degradation of mRNA and the maturation of tRNA (Babitzke and Kushner, 1991; Mudd *et al.*, 1990; Ow and Kushner, 2002; Taraseviciene *et al.*, 1991). In many cases, RNase E is the key endoribonuclease in initiating mRNA degradation, although recent evidence suggests that an initial cleavage by RNase E might be preceded by the conversion of the 5′-triphosphate of a primary transcript to a 5′-monophosphate (Celesnik *et al.*, 2007). RNase E is also involved in the degradation of mRNAs that are targeted by noncoding regulatory RNAs (Carpousis, 2003; Masse *et al.*, 2003; Morita *et al.*, 2005). Most work on the role of RNase E in RNA metabolism has been reviewed previously, and we direct the reader to a selection of reviews (Carpousis, 2007; Carpousis *et al.*, 1999; Coburn and Mackie, 1999; Grunberg-Manago, 1999; Kushner, 2002).

RNase E is an unusually large protein composed of a 1061-residue protomer that associates to form a tetrameric enzyme with a molecular weight of 472 kDa (Callaghan *et al.*, 2005a,b). RNase E interacts with three other enzymes by means of protein–protein contacts, RhlB (RNA helicase B), enolase, and PNPase (polynucleotide phosphorylase), to form a complex now known as the RNA degradosome (Carpousis *et al.*, 1994; Py *et al.*, 1994, 1996; Vanzo *et al.*, 1998). It is believed that RNase E normally exists in the cell complexed with these proteins (Liou *et al.*, 2001; Taghbalout and Rothfield, 2007), although evidence suggests that, under certain conditions, the RNA degradosome can be remodeled to form different RNase E–based complexes (Gao *et al.* 2006; Khemici *et al.*, 2004; Lee *et al.*, 2003; Morita *et al.*, 2005; Prud'homme-Genereux *et al.*, 2004). The 1061-residue protomer of RNase E is organized into two distinct regions. The N-terminal half of RNase E (529 residues) forms a structure that assembles into the tetrameric catalytic domain. The C-terminal half of RNase E (532 residues) has no known catalytic activity. RNase G and its homologs lack the large C-terminal extension that distinguishes RNase E from RNase G. The noncatalytic region of RNase E is a natively unstructured protein (Callaghan et al., 2004). That is, under physiologic conditions, the noncatalytic region has little, if any, propensity to fold into secondary and higher order structures. Nevertheless, this region is the scaffold for the interaction of RNase E with RhlB, enolase, and PNPase. These interactions involve relatively short motifs within the noncatalytic region of RNase E. Different proteins interact with different residues in the C-terminal of RNase E: RhlB and RNase E involve residues 698 to 762 of RNase E (Chandran et al., 2007), enolase and RNase E involve residues 833 to 850 of RNase E (Chandran and Luisi, 2006), and the interaction between PNPase and RNase E involves residues 1023 to 1061 of RNase E (Callaghan et al., 2004). Indeed, the crystal structure of enolase complexed with a polypeptide derived from residues 833 to 850 has been solved to high resolution, revealing an elaborate network of hydrogen-bonding interactions between the polypeptide and enolase (Chandran and Luisi, 2006).

Homologs of RNase E can be identified by sequence comparisons of the N-terminal catalytic domain, which is highly conserved, and the presence of a large noncatalytic region. However, the sequence of the non-catalytic regions of these homologs is highly variable (Marcaida et al., 2006). It should be noted that despite the lack of primary sequence conservation, a noncatalytic region of 400 to 600 residues is a hallmark of RNase E homologs (Carpousis, 2002; Condon and Putzer, 2002; Lee and Cohen, 2003; Marcaida et al., 2006). Among the RNase E homologs of the enterobacteria, the γ-Proteobacteria most closely related to E. coli, conserved sequence motifs of 20 to 50 residues can be identified within the noncatalytic region. Three of these motifs correspond to the sites for protein–protein interactions of E. coli RNase E with RhlB, enolase, and PNPase. For this reason, we believe that the enterobacteria contain an RNA degradosome similar to the complex in E. coli, although this prediction has not been tested experimentally (Marcaida et al., 2006). RNase E–based degradosomes from distantly related gram-negative bacteria have also been identified, but their composition differs from that of E. coli. A complex composed of RNase E, a DEAD-box protein, and RNase R has been identified in Pseudomonas syringae (Purusharth et al., 2005); RNase E, two DEAD-box proteins, and the Rho transcription termination factor in Rhodobacter capsulatus (Jager et al., 2001). We believe that the variable composition of the different RNase E–based complexes is related to the sequence plasticity of the noncatalytic region of RNase E (Marcaida et al., 2006). On the basis of these considerations, it has been suggested that plasticity in RNA degradosome composition might confer adaptive advantages during evolution by permitting modulation of the posttranscriptional control of gene expression (Marcaida *et al.*, 2006).

In previous work on the identification and characterization of the RNA degradosome in *E. coli*, co-immunopurification experiments had an important role in demonstrating that RNase E is part of an multiprotein complex and in identifying the other components of the complex (Carpousis *et al.*, 1994; Py *et al.*, 1994; Vanzo *et al.*, 1998). Antibodies against RNase E were used in co-immunopurification experiments with partially purified proteins. Initially, three major polypeptides of 85, 50, and 48 kDa were found to co-immunopurify with RNase E (Carpousis *et al.*, 1994). The 85-kDa protein was identified as PNPase on the basis of enzymatic activity, protein fingerprinting, and cross-reactivity with antibodies raised against purified PNPase. The 50- and 48-kDa proteins were identified as RhlB and enolase by N-terminal protein sequencing (Py *et al.*, 1996). In subsequent work, antibodies against RNase E and PNPase were used in co-immuno-purifications with mutant strains in which various regions within the noncatalytic region of RNase E were deleted (Vanzo *et al.*, 1998). This work showed that the noncatalytic region of RNase E contains elements that are essential for the interaction with RhlB, enolase, and PNPase. More recently, co-immunopurification experiments were used to demonstrate that the presence of RhlB in the degradosome is not necessary for the interaction of enolase and PNPase with RNase E and that catalytically inactive RhlB still associates with RNase E as part of the RNA degradosome (Khemici *et al.*, 2005). Here we describe the method that was used in the co-immunopurification work on the RNA degradosome of *E. coli* and discuss the possible application of this method to the characterization of the RNA degradosome in other bacteria.

2. GENERAL CONSIDERATIONS

Immunopurification covers a large variety of techniques in which antibodies specific to a particular antigen are used to purify that antigen from a complex mixture. The term *co-immunopurification* as used here refers to a technique in which antibodies against a specific protein are used to identify and characterize other proteins (or nucleic acids) that form a stable complex with the protein to which the antibodies were raised. A general discussion about techniques that use antibodies is largely beyond the scope of this chapter. We have routinely used *Antibodies* by E. Harlow and D. Lane (Cold Spring Harbor Laboratories, 1988) as a general reference, and we highly recommend it as a comprehensive resource on the raising and handling of antibodies.

Antibodies for co-immunopurification can be either polyclonal or monoclonal. Each has its advantages and disadvantages. In our work, we used polyclonal antibodies raised in rabbits, and this choice will be discussed here. In principal, monoclonal antibodies are directed against a single site (epitope) in a protein, whereas polyclonal antibodies are a mixture of IgGs that interact with many different epitopes in a specific protein. It is also possible to raise polyclonal antibodies against a synthetic peptide. In this case, the polyclonal antibody will react against one or a small number of epitopes in the polypeptide. The advantage of a carefully selected mono-clonal antibody or polyclonal antibody raised against a peptide is their high affinity and specificity against one or a few epitopes. Nevertheless, because co-immunopurification experiments are often performed under native con-ditions, it is difficult to know in advance which epitopes in a multiprotein complex will be available on the surface for interaction with an antibody. Furthermore, the mammalian immune system does not produce antibodies against all possible epitopes. The advantage of polyclonal antibodies raised against a whole protein is that they will interact with many different epitopes, and at least a few of them will, hopefully, be on a surface available for interaction with an antibody. For this reason, we prefer to perform co-immunopurification experiments with polyclonal antibodies. Nevertheless, it should be noted that rabbit antisera raised against whole proteins tend to cross-react with other proteins in whole-cell extracts. Solutions to this problem will be discussed in a subsequent section. Finally, it should be mentioned that purified proteins used as antigens for immunization can be either native or denatured. We have routinely used denatured proteins to raise high-titer antisera that are active in both Western blotting and the co-immunopurification of multiprotein complexes.

Another consideration in co-immunopurification experiments is the preparation of cell extracts. We published previously in *Methods in Enzy-mology* protocols for the large-scale purification of the RNA degradosome from *E. coli* and for the small-scale preparation of extracts that can be used in co-immunopurification experiments (Carpousis *et al.*, 2001). Here we will only make a few general comments about the preparation of cell extracts. Extracts or partially purified fractions for co-immunopurification experi-ments should consist of soluble protein that is free of cell debris, ribosomes, and nucleic acids. In the case of the RNase E of *E. coli*, a number of points also apply to other multiprotein complexes containing RNA-degrading enzymes. Numerous trials showed that RNase E requires a nonionic deter-gent such as Triton X-100 and high ionic strength (0.5 to 1.0 M NaCl) for efficient solubilization during the preparation of cell extracts (Carpousis *et al.*, 1994; Ehretsmann *et al.*, 1992). At low ionic strength, RNase E avidly binds RNA and ribosomes. For these reasons, we prepared high-speed supernatants at high ionic strength to remove ribosomes and then precipitated RNase E with ammonium sulfate to remove soluble RNA. As an alter-native, the high-speed centrifugation step can be replaced by precipitating ribosomes and nucleic acids with polyethylenimine (Carpousis *et al.*, 2001).

Because a multiprotein complex like the RNA degradosome contains several enzymes that bind RNA avidly, there is always the question of whether an interaction detected by co-immunopurification is a direct protein–protein interaction or an RNA-mediated interaction between two RNA binding proteins. It is, in principle, possible to address this issue by treating protein extracts with RNase A or micrococcal nuclease to degrade RNA (Carpousis *et al.*, 1994; Morita *et al.*, 2005). Nevertheless, adequately addressing this question requires additional approaches such as the direct detection of a protein–protein interaction with purified proteins or an analysis with the yeast two-hybrid system to validate the interaction (Vanzo *et al.*, 1998). It should be noted that the method described here could easily be adapted to identify and characterize RNA that is tightly bound to a multiprotein complex (MPC).

A final general consideration is the choice of resin used to immobilize antibodies for co-immunopurification work. In the protocol described in the following, protein A covalently attached to a Sepharose bead is used to immobilize IgG (immunoglobulin G) molecules from rabbit antiserum. Protein A is a bacterial protein that makes a strong noncovalent interaction with the constant region of an IgG molecule. Protein A Sepharose beads are sold commercially for the affinity purification of IgG from antiserum. In our hands, commercially available protein A Sepharose beads have been a reliable support for immobilizing IgGs from rabbit antisera.

Figure 4.1 shows a schematic diagram of the overall strategy used with protein A Sepharose beads. In step 1, the beads are incubated with antiserum and washed. In step 2, the IgG molecule is chemically cross-linked to protein A. In step 3, the beads are incubated with a cell extract and washed. In step 4, the multiprotein complex (MPC) is eluted under denaturing conditions, and protein A and IgG, which remain coupled to the Sepharose bead, are removed by sedimentation. In step 5, the MPC is analyzed by SDS-PAGE (sodium dodecylsulfate polyacrylamide gel electrophoresis).

Two covalent interactions hold together the protein A–IgG complex (Fig. 4.1). The heavy and light chains of each IgG molecule are linked by disulfide bonds. The glutaraldehyde treatment covalently links protein A to IgG. The protein A–IgG complex linked to the Sepharose bead can be maintained intact under denaturing conditions if precautions are taken to avoid breaking these covalent links. Reducing agents such as β-mercaptoethanol should be avoided to preserve the disulfide bridges, and heating above 55 °C should be avoided to preserve the covalent links formed by glutaraldehyde. Two points should be noted here: (1) Polyclonal antisera contain a complex mixture of IgGs and only a small proportion of these molecules (probably less than 1%) are specific to the antigen used to immunize the rabbit. (2) The Sepharose beads used here are very stable. They can be boiled in SDS-loading mix or briefly treated with strong acid or base without harming the bead.

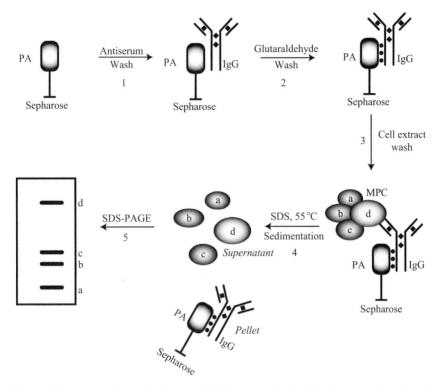

Figure 4.1 Preparation of IgG cross-linked to protein A (PA) Sepharose beads and immunopurification of a multiprotein complex (MPC). See text for details.

3. MATERIALS

RNase E was overexpressed with the pET15 expression vector and BL21(DE3) as the host strain (Novagen). The N-terminal 6× His-tag encoded by pET15 was used to purify RNase E under denaturing conditions (8 M urea). Purification was performed with an Ni-NTA Spin Kit (Qiagen) following the manufacturer's protocol. The yield from four spin columns, typically 200 to 400 μg of protein, is sufficient to immunize two rabbits. Rabbit antisera were raised commercially (Eurogentec) following the standard inoculation schedule. Normal rabbit serum (Sigma) can be used as a generic preimmune serum if authentic preimmune serum is not available. Protein A Sepharose CL-4B beads were purchased from GE Health Care (formerly Pharmacia). The beads were rehydrated in 1× PBS and washed according to the manufacturer's instructions. All other reagents were the highest grade available.

4. PROTOCOLS

4.1. Buffers and stock solutions

$10\times$ PBS: 1.37 M NaCl, 27 mM KCl, 100 mM Na$_2$HPO$_4$, 20 mM KH$_2$PO$_4$. This stock is autoclaved and stored at room temperature.

$5\times$ CLB: 100 mM Na$_2$HPO$_4$, 25 mM NaH$_2$PO$_4$, 1 M NaCl, 2.5 mM EDTA. This stock is autoclaved and stored at room temperature.

BG300: 10 mM Tris-HCl (pH 7.5), 300 mM NaCl, 5% glycerol, 0.5% Genapol X-080, 1 mM EDTA, 1 mM DTT, 1 mM PMSF, aprotinin (2 μg/ml), pepstatin A (0.8 μg/ml), leupeptin (0.8 μg/ml). PMSF, aprotinin, pepstatin A, and leupeptin can be replaced by complete protease inhibitor cocktail tablets, EDTA-free (Roche Diagnostics). This buffer, which has been optimized to stabilize the RNA degradosome and protect it from proteolysis, is made fresh from concentrated stock solutions. For further details see Carpousis *et al.* (2001).

$2.5\times$ SDS-PAGE loading buffer: 160 mM Tris-HCl (pH 6.8), 5% SDS, 25% glycerol, 0.1% BPB (bromophenol blue). This stock is stored at room temperature. If β-mercaptoethanol is required, the buffer is diluted to $1\times$, and β-mercaptoethanol is added to a final concentration of 5%.

4.2. Cross-linking IgG to protein A Sepharose beads

This protocol involves the absorption of IgG from the rabbit antiserum by the protein A bead and chemical cross-linking of the protein A–IgG complex with glutaraldehyde (Fig. 4.1). As a rule of thumb, the concentration of IgG in rabbit antiserum is approximately 10 mg/ml, and the rehydrated protein A beads can bind at least 10 mg of IgG/ml. To be sure that the beads are saturated, we generally use 2 to 3 volumes of antiserum per volume of bead. Here, the volume of bead refers to the volume of the packed pellet after low-speed centrifugation. Because high-titer antisera are precious, we recommend initially testing this protocol with normal rabbit serum. For all steps, we use disposable plastic culture tubes, preferably with a conical bottom. Agitation is performed by gentle rocking on a horizontal platform. The volume of the tube should be at least 10 times the volume of the antiserum, permitting the slurry of antiserum and beads to move from end to end of the tube during the rocking motion. Sedimentation of the beads can be performed by gravity or at low speed in a tabletop centrifuge. Precautions should be taken to avoid breaking or crushing the beads: avoid pipetting with small orifice pipette tips; avoid high-speed centrifugation.

All steps are at 4 °C unless indicated otherwise. Typically, 0.4 ml of bead is mixed with 1.0 ml of antiserum. The beads and antiserum are agitated for 1 to 3 h. After 1 min of low-speed centrifugation, the supernatant is removed and discarded; 4 ml of cold 1× PBS is added, the beads are agitated for 15 min, and then pelleted by low speed centrifugation for 1 min. This step, hereafter referred to as a wash, is repeated twice (3 washes total). The beads are then washed in 4 ml of 1× CLB three times. The beads are suspended in 4 ml of 1× CLB containing 1% glutaraldehyde (freshly diluted) and agitated for 1 h. The beads are washed three times with 4 ml of 1× CLB and twice with 4 ml of BG300.

The efficiency of binding and cross-linking to protein A Sepharose can be checked by a simple test with SDS-PAGE. Rabbit IgGs are composed of two 55-kDa heavy chains and two 25-kDa light chains. The chains can be visualized on Coomassie blue–stained gels, although the bands are diffuse because of substantial heterogeneity among various classes of IgG heavy and light chains. Prepare the beads as described previously except that after absorbing the IgG and washing, split the beads into two aliquots. Cross-link one aliquot and perform a mock cross-link on the other aliquot by omitting the glutaraldehyde. Suspend 50 μl of beads from each aliquot in 100 μl of 1× SDS-PAGE loading buffer without β-mercaptoethanol, heat to 55 °C for 15 min, and cool to room temperature. Pellet the beads by brief (1 min) sedimentation in a microfuge at room temperature. Transfer the supernatant to a fresh tube, add β-mercaptoethanol (5% final) and heat at 95 °C for 3 min. Cool to room temperature and run the samples side by side on SDS-PAGE. The mock cross-linking permits quantification of how much IgG was absorbed on the protein A bead. The cross-linked sample permits quantification of the efficiency of cross-linking with glutaraldehyde. IgG from the beads that were not cross-linked should be clearly visible on a Coomassie blue–stained gel. In contrast, little, if any, IgG should be detected with the glutaraldehyde cross-linked beads.

4.3. Co-immunopurification with antibodies coupled to protein A beads

All steps are at 4 °C unless indicated otherwise. Agitation and washing are performed as described in the previous section. Cell extracts, prepared as described previously (Carpousis *et al.*, 2001), usually have concentrations ranging from 2 to 4 mg/ml of protein. The protein concentration can be determined by UV absorbance if the extract is free of nucleic acid or by Lowry assay. To 100 μl of bead, add 2 mg of protein (the volume depends on the concentration). The beads are agitated for 2 to 3 h and then washed three times with 1 ml of GB300. At room temperature, suspend the beads in 200 μl of 1× SDS-PAGE loading buffer without β-mercaptoethanol, heat to 55 °C for 15 min, and cool to room temperature. Pellet the beads by brief (1 min) sedimentation in a microfuge at room temperature. Transfer the supernatant to a fresh tube, add β-mercaptoethanol, and heat at 95 °C for 3 min. Cool to room temperature and analyze by SDS-PAGE.

Figure 4.2 shows examples where this protocol has been used in previously published work. Panel A shows lanes from a silver-stained SDS gel of RNA degradosome co-immunopurified with antisera raised against RNase E or PNPase. In this analysis, authentic preimmune serum was used as a control. These co-immunopurifications were part of a larger study in which the composition of the RNA degradosome was analyzed in strains in which different parts of the noncatalytic region of RNase E were deleted. For

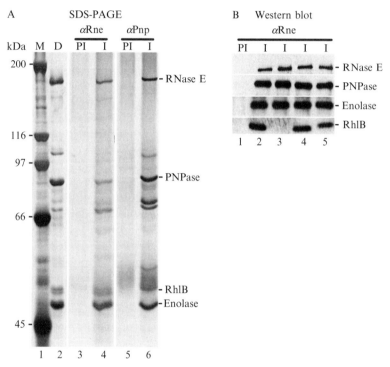

Figure 4.2 (A) Silver-stained SDS-PAGE of the co-immunopurification of the RNA degradosome from *E. coli*. Lanes 1 and 2 are controls showing protein size standards (M) and a preparation of the RNA degradosome purified by a large-scale, classical biochemical approach (Carpousis *et al.*, 1994) (D). Lanes 3 and 4, co-immunopurification from protein extracts with preimmune (PI) control or antiserum (I) raised against RNase E (αRne). Lanes 5 and 6, same as lanes 3 and 4, except that the antiserum was raised against PNPase (αPnp). The molecular weight of the markers is indicated to the left of the panel; the positions of RNase E, PNPase, RhlB, and enolase are indicated to the right of the panel. (B) Western blot analysis of immunopurified RNA degradosome. Lane 1, control with preimmune serum (PI); lanes 2 to 5, co-immunopurification with antiserum (I) raised against RNase E (αRne). Lanes 1 and 2 are the analysis of cell extracts from the wild-type control; lane 3, a strain in which the gene encoding RhlB was deleted; lanes 4 and 5, strains producing catalytically inactive RhlB because of point mutants targeting residues important for enzymatic activity. This figure was adapted from previously published work (Khemici *et al.*, 2005; Vanzo *et al.*, 1998).

further discussion of these results, see Vanzo *et al.* (1998). Panel B shows Western blots that were performed after co-immunopurification with anti-sera against RNase E. In the analysis, normal rabbit serum was used as a control, because the authentic preimmune serum had been exhausted in previous work. RNA degradosome co-immunopurified with antiserum raised against RNase E was separated on SDS-PAGE, blotted, and then analyzed with antiserum raised against RNase E, PNPase, enolase, or RhlB. This analysis allowed us to conclude that catalytically inactive variants of RhlB were assembled into the RNA degradosome. For further discussion of these results see Khemici *et al.* (2005).

4.4. Small-scale affinity purification of polyclonal antibodies

We have occasionally found it useful to affinity-purify antibodies on a small scale for Western blotting. Here, we describe a simple protocol that can be performed at the same time that a blot is prepared for Western analysis. In this technique, purified RhlB is run on SDS-PAGE and transferred to a nitrocellulose filter following any standard Western blotting procedure. We usually run the RhlB in two or three lanes of the gel (1 μg per lane). The remaining lanes of the gel can be used for the samples that are to be analyzed by Western blotting. After coloration of the filter with Ponceau red, the region containing RhlB is marked with a soft pencil. After blocking the filter, the strip of nitrocellulose containing the RhlB is excised with a sharp scissors or a scalpel. The following steps are performed at 4 °C. The strip is placed on the side of a plastic disposable tube, and 1 ml of antiserum is layered on top of the filter. The tube is agitated on rocking platform as described previously for 2 to 3 h to bind the RhlB-specific antibodies to the Rh1B on the filter. The antiserum is removed, and the filter is washed twice for 15 min with 4 ml of 1× PBS containing 0.1 mg/ml BSA (bovine serum albumin) and 0.02% Tween 20. The antibody is then eluted from the filter by brief treatments with acid and base. The filter is washed for 4 min with 1 ml of 1× PBS containing 0.02 *N* HCl and 0.02% Tween 20 and then for 4 min with 1 ml of 1× PBS containing 0.02 *N* NaOH and 0.02% Tween 20. The washes are combined immediately.

The following should be noted regarding this protocol. In principle, the elution is nondenaturing. The acid and base treatments disrupt the antigen–antibody complex by altering the charge of acidic and basic amino acid residues at the interface between the antigen and antibody; 1× PBS buffer has a pH of approximately 7.5. The addition of 0.02 *N* HCl forms an acid phosphate buffer with a pH of approximately 2.5; the addition of 0.02 NaOH forms a basic phosphate buffer with a pH of approximately 12.5. Long–term exposure to either of these extremes of pH can lead to irrevers-ible damage because of denaturation and hydrolysis of protein. When the acid and base washes are mixed, the pH is restored to neutrality, minimizing possible damage. We perform the Western analysis the same day that the antibody is affinity purified. We use half of the affinity-purified antibody, holding the other half on ice as a reserve in case there is a problem with the initial Western blot. This protocol was developed to affinity-purify antiserum raised against *E. coli* RhlB, because it is a rather poor antigen-producing antisera with low titers and extensive cross-reactivity to other *E. coli* proteins. For high-titer antisera, it should be possible to use significantly less antiserum (diluted into $1 \times$ PBS if necessary).

5. PERSPECTIVE

Here we have described techniques for use of polyclonal rabbit antibodies in the identification and characterization of multiprotein complexes containing RNA degrading enzymes. This method was used over the past decade to help characterize the RNA degradosome of *E. coli*. When a model organism such as *E. coli* is used, technology that is more recent has largely supplanted the approach described here. RNase E tagged with FLAG or other epitopes can be immunopurified with commercially available high-specificity antibodies coupled to beads (Miczak *et al.*, 1996; Morita *et al.*, 2005). This technology can be used with any protein that can be engineered to present an accessible epitope. Indeed, an *E. coli* protein "interactome" has recently been determined on the basis of the systematic analysis of nearly 1000 protein-coding genes that were engineered to add a C-terminal SPA tag (Butland *et al.*, 2005). The composition of SPA-tagged MPCs was determined by mass spectrometry, and a network of protein–protein interactions was mapped. All of the protein–protein interactions described previously in the identification and characterization of the RNA degradosome were detected in the systematic analysis of the *E. coli* interactome.

Nevertheless, systematic studies of protein–protein interactions by a proteomics approach are available only in a few well-characterized model organisms. Indeed, although genomic sequences of prokaryotic microorganisms are now being produced at the rate of several hundred per year, most of these organisms are refractory to the genetic engineering needed to introduce epitope-tagged proteins. This method requires the capacity to introduce DNA vectors that either self-replicate or integrate into the chromosome. Although certain bacteria are naturally transformable and others can be transformed by conjugation, optimizing these processes for a specific bacterium is difficult and time consuming. Thus, there are ample reasons to consider the use of the method described here for the analysis of MPCs in microorganisms where the possibilities for genetic engineering are limited or nonexistent. In this context, we will briefly describe unpublished work from our laboratory aimed at identifying and characterizing the RNA degradosome from *Pseudoaltermonas haloplanktis*.

5.1. Characterization of the RNase E of psychrotolerant bacteria

Pseudoalteromonas haloplanktis is a marine bacterium isolated in Antarctica whose genome has recently been sequenced (Medigue *et al.*, 2005). One interest in studying this bacterium is that it grows well at temperatures ranging from 10 to 30 °C, with generation times that are not very different from *E. coli* growing in the range of 20 to 40 °C. Bacteria like *P. haloplanktis* are known as *psychrotolerant* or *psychrophilic*, meaning that they tolerate or grow well in the cold. Like *E. coli*, *P. haloplanktis* is a γ-proteobacterium, although it is not an enterobacterium. *E. coli* and *P. haloplanktis* have many homologs in common, and conservation at the level of protein sequence ranges from 70 to 90%. *P. haloplanktis* has clearly identifiable homologs for all of the components of the RNA degradosome (see, for instance, Iost and Dreyfus, 2006). The RNase E of *P. haloplanktis* is 1071 residues in length compared with 1061 for *E. coli*. The first 530 residues of each protein make up the catalytic domain whose sequence is conserved at the level of 85%. However, there is little, if any, conservation of sequence in the noncatalytic region. Furthermore, there are no clear signatures in the RNase E noncatalytic region of *P. haloplanktis* for the motifs identified in the enterobacteria as necessary for protein–protein interactions between RNase E and RhlB, enolase, and PNPase. Two scenarios that are not necessarily mutually exclusive could explain these observations. Either the sequences of the motifs for protein–protein interactions have diverged significantly or the RNase E of *P. haloplanktis* interacts with a different set of proteins.

Our long-term interest is to identify and characterize the RNA degradosome of *P. haloplanktis* to understand how it has been modified to work efficiently at lower temperatures. To initiate this project, we raised polyclonal antibodies against the RNase E of *P. haloplanktis* by overexpressing it in *E. coli*, purifying it under denaturing conditions, and immunizing rabbits. Figure 4.3 shows the characterization of this antiserum. Panel A shows a Coomassie blue–stained SDS gel loaded with whole cell extracts used to perform the Western blot in panel B. The Western blot in panel B was performed with the antiserum against the RNase E of *P. haloplanktis* diluted 1/4000. Lane 1 in each panel is a whole cell extract from *P. haloplanktis*, whereas lanes 2 to 5 are extracts from *E. coli*. Note that to get approximately equal signals in the Western blot, 4-fold less *P. haloplanktis* protein than *E. coli* protein was loaded. Thus, the Coomassie blue–stained protein in lane 1 of panel A is barely visible in this image. In panel B, lane 1, the antiserum reacts with two proteins from *P. haloplanktis*. The larger protein running at approximately 180 kDa is RNase E. The smaller protein running at

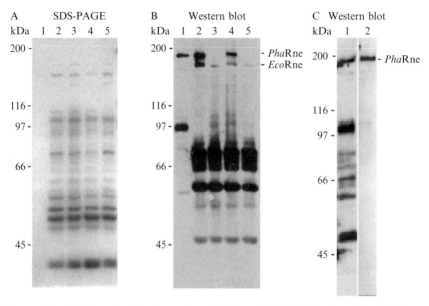

Figure 4.3 Polyclonal antibodies against the RNase E of *P. haloplanktis*. (A and B) Lane 1, *P. haloplanktis* whole cell extract; lanes 2 to 5, *E. coli* whole cell extracts. In lanes 2 and 4, the DH5α strain of *E. coli* was transformed with a low copy number plasmid containing a 5-kbp insert of *P. haloplanktis* genomic DNA encoding RNase E. Lanes 3 and 5 are the empty vector controls, pLN135.1 and pAM238, respectively. Note that the protein loaded on lane 1, which is fourfold less than in lanes 2 to 5, is barely visible in the Coomassie-stained SDS-PAGE (panel A). The antiserum, raised against the RNase E of *P. haloplanktis*, was diluted 1/4000, the secondary antibody was anti-rabbit IgG conjugated to peroxidase (Sigma A8275), and the blot was developed with an ECL kit (GE Healthcare/Amersham). (C) Western blot before (lane 1) and after (lane 2) affinity purification of the antiserum raised against the *P. haloplanktis* RNase E. The whole cell extract analyzed in this Western blot was from *P. haloplanktis*. The position of protein size markers is indicated to the left of each panel. *Pha*Rne and *Eco*Rne indicate the position of the *P. haloplanktis* RNase E and the *E. coli* RNase E, respectively.

approximately 97 kDa could be either a proteolysis product of RNase E or a different protein that cross-reacts with the antiserum. To distinguish between these two possibilities, we performed a small-scale affinity purification of the antiserum against full-length *P. haloplanktis* RNase E following the protocol described previously. Panel C shows lanes from Western blots of a whole cell extract of *P. haloplanktis* with antiserum (lane 1) or affinity-purified antibody (lane 2). Note that, as expected, the affinity-purified antibody interacts with the full-length RNase E. Because the 97-kDa protein and other smaller polypeptides were not detected after affinity purification, the possibility that they are proteolysis products of RNase E can be excluded. Finally, in other Western blots, the antiserum has been diluted to 1/20,000 with little loss of the specific signal against full-length RNase E and a significant reduction of the cross-reaction with the protein at

97 kDa (data not shown). From these results, we conclude that the antiserum has a high titer of IgGs specific to the RNase E of *P. haloplanktis,* and, on the basis of our previous experience with antisera raised against the RNase E and PNPase of *E. coli*, this antiserum can be used in co-immunopurification experiments without further purification.

Lanes 2 and 4 in panel B (Fig. 4.3) are *E. coli* whole cell extracts prepared from cells harboring low copy number plasmids containing *P. haloplanktis* genomic DNA encoding RNase E. Lanes 3 and 5 are the empty vector controls. In this Western blot, the antiserum was diluted 1/4000 to permit visualization of the *E. coli* RNase E. This interspecies cross-reactivity is due to antibodies against epitopes that are conserved between the RNase E of these bacteria. Note that expression of the RNase E of *P. haloplanktis* is clearly detected in lanes 2 and 4, showing that it can be expressed in *E. coli* with endogenous transcription and translation signals from the chromosome of *P. haloplanktis*. We are currently testing whether these plasmids can complement *E. coli* strains lacking RNase E. Lanes 2 to 5 show that, at this dilution of antiserum, there is a strong cross-reaction with *E. coli* proteins in the range of 40 to 80 kDa. This could be due to either the presence of reactive antibodies before the rabbit was immunized (rabbits are susceptible to infections by pathogenic enterobacteria) or to contaminants in the RNase E used for immunization (the *P. haloplanktis* RNase E for immunization was overexpressed in *E. coli*). Regardless of these considerations, the Western blot in panel C demonstrates that, if necessary, the antiserum can be affinity purified for analysis of the expression of *P. haloplanktis* RNase E in *E. coli*.

We have adapted protocols used previously with *E. coli* (Carpousis *et al.,* 2001) to make protein extracts from *P. haloplanktis* for co-immunopurification experiments. Although preliminary results are encouraging, we have encountered a problem that needs to be resolved. The RNase E of *E. coli* is known to be sensitive to proteolysis in crude protein extracts. This problem is even more severe in protein extracts from *P. haloplanktis*. On the basis of experiments in which we have mixed extracts from *E. coli* and *P. haloplanktis* and analyzed the degradation of the endogenous RNase E, we believe that *P. haloplanktis* contains psychrophilic proteases that are active on ice even in the presence of the battery of protease inhibitors that we normally use. With the possible exception of EDTA, increasing the concentration of the protease inhibitors does not seem to help. Ongoing work, which will not be detailed here, suggests that it should be possible to overcome the proteolysis problem by changing the procedure that we use to lyse the cells.

ACKNOWLEDGMENTS

Our research is supported by the Centre National de la Recherche Scientifique (CNRS) with additional funding from the Agence Nationale de la Recherche (ANR, grant NT05_1-44659).

REFERENCES

Babitzke, P., and Kushner, S. R. (1991). The AMS (altered mRNA stability) protein and ribonuclease E are encoded by the same structural gene of *Escherichia coli*. *Proc. Natl. Acad. Sci. USA* **88,** 1–5.

Butland, G., Peregrin-Alvarez, J. M., Li, J., Yang, W., Yang, X., Canadien, V., Starostine, A., Richards, D., Beattie, B., Krogan, N., Davey, M., Parkinson, J., Greenblatt, J., and Emili, A. (2005). Interaction network containing conserved and essential protein complexes in *Escherichia coli*. *Nature* **433,** 531–537.

Callaghan, A. J., Aurikko, J. P., Ilag, L. L., Gunter Grossmann, J., Chandran, V., Kuhnel, K., Poljak, L., Carpousis, A. J., Robinson, C. V., Symmons, M. F., and Luisi, B. F. (2004). Studies of the RNA degradosome-organizing domain of the *Escherichia coli* ribonuclease RNase E. *J. Mol. Biol.* **340,** 965–979.

Callaghan, A. J., Marcaida, M. J., Stead, J. A., McDowall, K. J., Scott, W. G., and Luisi, B. F. (2005a). Structure of *Escherichia coli* RNase E catalytic domain and implications for RNA turnover. *Nature* **437,** 1187–1191.

Callaghan, A. J., Redko, Y., Murphy, L. M., Grossmann, J. G., Yates, D., Garman, E., Ilag, L. L., Robinson, C. V., Symmons, M. F., McDowall, K. J., and Luisi, B. F. (2005b). "Zn-link": A metal-sharing interface that organizes the quaternary structure and catalytic site of the endoribonuclease, RNase E. *Biochemistry* **44,** 4667–4675.

Carpousis, A. J. (2002). The *Escherichia coli* RNA degradosome: Structure, function and relationship to other ribonucleolytic multienzyme complexes. *Biochem. Soc. Trans.* **30,** 150–155.

Carpousis, A. J. (2003). Degradation of targeted mRNAs in *Escherichia coli*: Regulation by a small antisense RNA. *Genes Dev.* **17,** 2351–2355.

Carpousis, A. J. (2007). The RNA degradosome of *Escherichia coli*: An mRNA-degrading machine assembled on RNase E. *Annu. Rev. Microbiol.* **61,** 71–87.

Carpousis, A. J., Leroy, A., Vanzo, N., and Khemici, V. (2001). *Escherichia coli* RNA degradosome. *Methods Enzymol.* **342,** 333–345.

Carpousis, A. J., Van Houwe, G., Ehretsmann, C., and Krisch, H. M. (1994). Copurification of *E. coli* RNAase E and PNPase: Evidence for a specific association between two enzymes important in RNA processing and degradation. *Cell* **76,** 889–900.

Carpousis, A. J., Vanzo, N. F., and Raynal, L. C. (1999). mRNA degradation: A tale of poly(A) and multiprotein machines. *Trends Genet.* **15,** 24–28.

Celesnik, H., Deana, A., and Belasco, J. G. (2007). Initiation of RNA decay in *Escherichia coli* by 5′-pyrophosphate removal. *Mol. Cell* **27,** 79–90.

Chandran, V., and Luisi, B. F. (2006). Recognition of enolase in the *Escherichia coli* RNA degradosome. *J. Mol. Biol.* **358,** 8–15.

Chandran, V., Poljak, L., Vanzo, N. F., Leroy, A., Miguel, R. N., Fernandez-Recio, J., Parkinson, J., Burns, C., Carpousis, A. J., and Luisi, B. F. (2007). Recognition and cooperation between the ATP-dependent RNA helicase RhlB and ribonuclease RNase E. *J. Mol. Biol.* **367,** 113–132.

Coburn, G. A., and Mackie, G. A. (1999). Degradation of mRNA in *Escherichia coli*: An old problem with some new twists. *Prog. Nucleic Acid Res. Mol. Biol.* **62,** 55–108.

Condon, C., and Putzer, H. (2002). The phylogenetic distribution of bacterial ribonucleases. *Nucleic Acids Res.* **30,** 5339–5346.

Ehretsmann, C. P., Carpousis, A. J., and Krisch, H. M. (1992). Specificity of *Escherichia coli* endoribonuclease RNase E: *In vivo* and *in vitro* analysis of mutants in a bacteriophage T4 mRNA processing site. *Genes Dev.* **6,** 149–159.

Gao, J., Lee, K., Zhao, M., Qiu, J., Zhan, X., Saxena, A., Moore, C. J., Cohen, S. N., and Georgiou, G. (2006). Differential modulation of *E. coli* mRNA abundance by inhibitory proteins that alter the composition of the degradosome. *Mol. Microbiol.* **61,** 394–406.

Grunberg-Manago, M. (1999). Messenger RNA stability and its role in control of gene expression in bacteria and phages. *Annu. Rev. Genet.* **33**, 193–227.

Iost, I., and Dreyfus, M. (2006). DEAD-box RNA helicases in *Escherichia coli*. *Nucleic Acids Res.* **34**, 4189–4197.

Jager, S., Fuhrmann, O., Heck, C., Hebermehl, M., Schiltz, E., Rauhut, R., and Klug, G. (2001). An mRNA degrading complex in *Rhodobacter capsulatus*. *Nucleic Acids Res.* **29**, 4581–4588.

Khemici, V., Poljak, L., Toesca, I., and Carpousis, A. J. (2005). Evidence *in vivo* that the DEAD-box RNA helicase RhlB facilitates the degradation of ribosome-free mRNA by RNase E. *Proc. Natl. Acad. Sci. USA* **102**, 6913–6918.

Khemici, V., Toesca, I., Poljak, L., Vanzo, N. F., and Carpousis, A. J. (2004). The RNase E of *Escherichia coli* has at least two binding sites for DEAD-box RNA helicases: Functional replacement of RhlB by RhlE. *Mol. Microbiol.* **54**, 1422–1430.

Kushner, S. R. (2002). mRNA decay in *Escherichia coli* comes of age. *J. Bacteriol.* **184**, 4658–4665.

Lee, K., and Cohen, S. N. (2003). A *Streptomyces coelicolor* functional orthologue of *Escherichia coli* RNase E shows shuffling of catalytic and PNPase-binding domains. *Mol. Microbiol.* **48**, 349–360.

Lee, K., Zhan, X., Gao, J., Qiu, J., Feng, Y., Meganathan, R., Cohen, S. N., and Georgiou, G. (2003). RraA. a protein inhibitor of RNase E activity that globally modulates RNA abundance in *E. coli*. *Cell* **114**, 623–634.

Liou, G. G., Jane, W. N., Cohen, S. N., Lin, N. S., and Lin-Chao, S. (2001). RNA degradosomes exist *in vivo* in *Escherichia coli* as multicomponent complexes associated with the cytoplasmic membrane via the N-terminal region of ribonuclease E. *Proc. Natl. Acad. Sci. USA* **98**, 63–68.

Marcaida, M. J., DePristo, M. A., Chandran, V., Carpousis, A. J., and Luisi, B. F. (2006). The RNA degradosome: Life in the fast lane of adaptive molecular evolution. *Trends Biochem. Sci.* **31**, 359–365.

Masse, E., Escorcia, F. E., and Gottesman, S. (2003). Coupled degradation of a small regulatory RNA and its mRNA targets in *Escherichia coli*. *Genes Dev.* **17**, 2374–2383.

Medigue, C., Krin, E., Pascal, G., Barbe, V., Bernsel, A., Bertin, P. N., Cheung, F., Cruveiller, S., D'Amico, S., Duilio, A., Fang, G., Feller, G., *et al.* (2005). Coping with cold: The genome of the versatile marine Antarctic bacterium *Pseudoalteromonas haloplanktis* TAC125. *Genome. Res.* **15**, 1325–1335.

Miczak, A., Kaberdin, V. R., Wei, C. L., and Lin-Chao, S. (1996). Proteins associated with RNase E in a multicomponent ribonucleolytic complex. *Proc. Natl. Acad. Sci. USA* **93**, 3865–3869.

Morita, T., Maki, K., and Aiba, H. (2005). RNase E-based ribonucleoprotein complexes: Mechanical basis of mRNA destabilization mediated by bacterial noncoding RNAs. *Genes Dev.* **19**, 2176–2186.

Mudd, E. A., Krisch, H. M., and Higgins, C. F. (1990). RNase E, an endoribonuclease, has a general role in the chemical decay of *Escherichia coli* mRNA: Evidence that rne and ams are the same genetic locus. *Mol. Microbiol.* **4**, 2127–2135.

Ow, M. C., and Kushner, S. R. (2002). Initiation of tRNA maturation by RNase E is essential for cell viability in *E. coli*. *Genes Dev.* **16**, 1102–1115.

Prud'homme-Genereux, A., Beran, R. K., Iost, I., Ramey, C. S., Mackie, G. A., and Simons, R. W. (2004). Physical and functional interactions among RNase E, polynucleotide phosphorylase and the cold-shock protein, CsdA: Evidence for a "cold shock degradosome." *Mol. Microbiol.* **54**, 1409–1421.

Purusharth, R. I., Klein, F., Sulthana, S., Jager, S., Jagannadham, M. V., Evguenieva-Hackenberg, E., Ray, M. K., and Klug, G. (2005). Exoribonuclease R interacts with endoribonuclease E and an RNA helicase in the psychrotrophic bacterium *Pseudomonas syringae* Lz4W. *J. Biol. Chem.* **280,** 14572–14578.

Py, B., Causton, H., Mudd, E. A., and Higgins, C. F. (1994). A protein complex mediating mRNA degradation in *Escherichia coli*. *Mol. Microbiol.* **14,** 717–729.

Py, B., Higgins, C. F., Krisch, H. M., and Carpousis, A. J. (1996). A DEAD-box RNA helicase in the *Escherichia coli* RNA degradosome. *Nature* **381,** 169–172.

Taghbalout, A., and Rothfield, L. (2007). RNaseE and the other constituents of the RNA degradosome are components of the bacterial cytoskeleton. *Proc. Natl. Acad. Sci. USA* **104,** 1667–1672.

Taraseviciene, L., Miczak, A., and Apirion, D. (1991). The gene specifying RNase E (rne) and a gene affecting mRNA stability (ams) are the same gene. *Mol. Microbiol.* **5,** 851–855.

Vanzo, N. F., Li, Y. S., Py, B., Blum, E., Higgins, C. F., Raynal, L. C., Krisch, H. M., and Carpousis, A. J. (1998). Ribonuclease E organizes the protein interactions in the *Escherichia coli* RNA degradosome. *Genes Dev.* **12,** 2770–2781.

CHAPTER FIVE

PABLO ANALYSIS OF RNA: 5′-PHOSPHORYLATION STATE AND 5′-END MAPPING

Helena Celesnik, Atilio Deana, *and* Joel G. Belasco

Contents

Abstract

Recent studies have revealed that 5′-end-dependent RNA degradation in pro-karyotes is triggered by pyrophosphate removal from the 5′-terminus to gener-ate a monophosphorylated intermediate that is readily degraded. This chapter describes how to examine the 5′-phosphorylation state of any specific bacterial RNA by PABLO analysis. The method is based on the ability of monophos-phorylated, but not triphosphorylated, RNA 5′-ends to undergo splinted ligation to a DNA oligonucleotide when juxtaposed by base pairing to a bridging

Kimmel Center for Biology and Medicine at the Skirball Institute and Department of Microbiology, New York University School of Medicine, New York, USA

Methods in Enzymology, Volume 447
ISSN 0076-6879, DOI: 10.1016/S0076-6879(08)02205-2

oligonucleotide. PABLO analysis not only makes it possible to quantify the proportion of a particular RNA that is monophosphorylated in bacterial cells but also provides a more reliable method than primer extension for high-resolution mapping of RNA 5′-termini.

1. INTRODUCTION

Together with transcription and translation, RNA turnover plays a key role in the regulation of gene expression. In bacteria, where messenger RNAs typically survive for only minutes after their synthesis, it had long been assumed that the degradation of primary transcripts begins with ribonuclease cleavage. However, this view has recently been challenged by the discovery of a prior event that marks many transcripts for rapid decay: the conversion of the 5′-terminal triphosphate to a monophosphate by a cellular RNA pyrophosphohydrolase (Celesnik *et al.*, 2007; Deana *et al.*, 2008). This subtle modification at the 5′-end can trigger the destruction of an entire transcript by facilitating attack by ribonucleases that preferentially degrade 5′-monophosphorylated RNAs, such as the *Escherichia coli* endonucleases RNase E and RNase G (Jiang *et al.*, 2000; Mackie, 1998; Tock *et al.*, 2000). In principle, it might also play a role in species, such as *Bacillus subtilis*, that produce a 5′-exonuclease (RNase J) sensitive to the phosphorylation state of RNA (Mathy *et al.*, 2007).

The mechanism of this 5′-end–dependent degradation pathway came to light with the development of an assay for detecting and quantifying monophosphorylated RNAs within preparations of total RNA extracted from bacterial cells (Celesnik *et al.*, 2007). This analytical method—*phosphorylation assay by ligation of oligonucleotides* (PABLO)—makes it possible to determine what percentage of an RNA is monophosphorylated *in vivo*. In addition, PABLO has significant potential as a high-resolution method for mapping RNA 5′-ends.

2. CONCEPT

PABLO analysis of the 5′-phosphorylation state of RNA is based on the ability of T4 DNA ligase to covalently join the 3′-end of a DNA oligonucleotide (oligo X) to the 5′-end of RNA when the RNA is monophosphorylated and the two termini are juxtaposed by a bridging DNA oligonucleotide (oligo Y) to which both are base paired (Figs. 5.1 and 5.2A) (Celesnik *et al.*, 2007; Fareed *et al.*, 1971; Kleppe *et al.*, 1970; Moore and Sharp, 1972; Nath and Hurwitz, 1974). Because of the specificity of that enzyme, only monophosphorylated 5′-ends can undergo ligation; those that

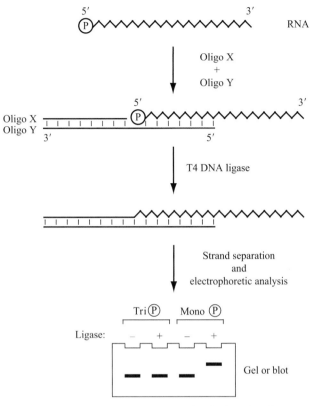

Figure 5.1 The PABLO method for determining the 5'-phosphorylation state of RNA. Total cellular RNA is combined with DNA oligonucleotides X and Y, which form a duplex with a particular transcript of interest. T4 DNA ligase is added to join oligo X specifically to 5'-monophosphorylated (but not 5'-triphosphorylated) RNA, thereby producing an extended DNA-RNA chimera that can be detected by Northern blotting because of its reduced electrophoretic mobility. Alternatively, the ligation products of radiolabeled monophosphorylated transcripts synthesized *in vitro* can be visualized directly by gel electrophoresis and autoradiography. (Reprinted from Celesnik *et al.*, 2007, with permission from Cell Press.)

are triphosphorylated are completely unreactive (Fig. 5.2B). The greater length and reduced electrophoretic mobility of the ligation products makes them readily distinguishable from unligated RNA. By use of this procedure, it is possible to determine the phosphorylation state of a cellular transcript by customizing oligos X and Y so as to target that particular transcript in total cellular RNA and by subjecting the ligation products to Northern blot analysis with a suitable radiolabeled probe. Figure 5.2C illustrates the results of such an analysis for cellular RNA I.613, a derivative of the ColE1 plasmid replication inhibitor RNA I.

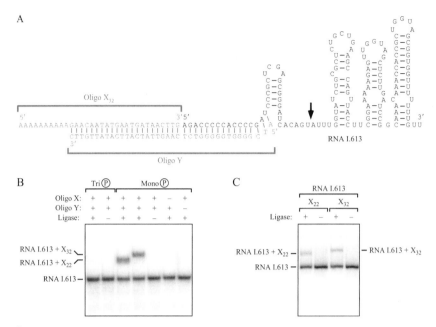

Figure 5.2 PABLO analysis of RNA I.613. (A) Diagrammatic representation of RNA I.613 (black), a derivative of the ColE1 plasmid replication inhibitor RNA I, complexed with oligo X_{32} (grey) and an RNA I.613-specific oligo Y (grey). The principal site at which RNase E cleaves RNA I.613 is marked with an arrow. (B) PABLO analysis of RNA I.613 bearing a 5′-monophosphate or a 5′-triphosphate. Internally radiolabeled RNAs that were fully monophosphorylated or fully triphosphorylated were prepared by *in vitro* transcription in the presence or absence of excess AMP (see section 7.3), and the importance of various reaction components for PABLO ligation was determined. The two different oligonucleotides X that were used, X_{22} (lanes 1, 2, 3, 5, and 7) and X_{32} (lane 4), were 22 and 32 nucleotides long, respectively. An intrinsic ligation efficiency of ~50% was determined by averaging ligation yields from several PABLO experiments with fully monophosphorylated RNA I.613. This efficiency was not limited by incomplete monophosphorylation, because the same *in vitro* transcript could be fully degraded by treatment with a 5′-monophosphate–dependent exoribonuclease (Terminator; see section 8.1), whereas triphosphorylated RNA I.613 was entirely resistant to digestion. (Reprinted from Celesnik *et al.*, 2007, with permission from Cell Press.) (C) 5′-Phosphorylation state of RNA I.613 in *E. coli*. Total RNA extracted from *E. coli* cells containing plasmid-encoded RNA I.613 was examined by PABLO analysis to determine the 5′-phosphorylation state of that transcript *in vivo*. The ligation yield of ~15% indicates that ~30% of RNA I.613 is monophosphorylated. (Reprinted from Celesnik *et al.*, 2007, with permission from Cell Press.)

3. Detection Method

The visualization of PABLO ligation products by Northern blotting has distinct advantages over other detection methods, such as RT-PCR. By allowing simultaneous detection of both the ligation product and the unligated transcript, Northern hybridization makes it possible to quantify their relative abundance and thus to calculate the percentage of the transcript that is monophosphorylated. It also makes it possible to compare the phosphorylation state of a full-length transcript to that of a degradation intermediate truncated at the 3'-end.

4. Efficiency of Ligation

The PABLO ligation yield, defined as the percentage of a particular transcript that undergoes covalent joining to oligo X, reflects not only the fraction of that RNA that is monophosphorylated but also its intrinsic ligation efficiency (i.e., the maximum ligation yield that can be achieved when the RNA is fully monophosphorylated). The ligation efficiency of a monophosphorylated RNA is influenced both by its 5'-terminal sequence and structure and by the design of PABLO oligos X and Y. For instance, stable structure in the 5'-portion of an RNA can diminish the efficiency of PABLO ligation by impeding duplex formation with oligo Y.

The intrinsic ligation efficiency of an RNA of interest can be determined empirically by measuring the PABLO ligation yield when the transcript is produced in a fully monophosphorylated state by *in vitro* synthesis. By then comparing that intrinsic efficiency to the PABLO ligation yield measured when the same RNA has instead been extracted with total RNA from bacterial cells, it is possible to calculate how much of the transcript is monophosphorylated *in vivo* (monophosphorylated fraction = PABLO yield/ligation efficiency). For example, a maximum ligation efficiency of ~50% has been measured for monophosphorylated RNA I.613 synthesized *in vitro* (Fig. 5.2B). From that value and the ~15% ligation yield observed when this transcript is isolated from *E. coli* (Fig. 5.2C), one can estimate that ~30% of RNA I.613 bears a 5'-monophosphate *in vivo*.

5. Determining the 5'-Phosphorylation State of Long Transcripts

The accurate measurement of PABLO ligation yields requires the electrophoretic separation of PABLO ligation products from unligated transcripts. This is relatively straightforward for RNAs less than 650 nucleotides (nt) in length. However, resolving these bands becomes increasingly difficult for longer RNAs and would be impossible for many polycistronic transcripts. One way to circumvent this problem is to fragment the RNA of interest by *in vitro* cleavage at a defined site with a complementary DNA-zyme (an RNA-cleaving deoxyribozyme) (Santoro and Joyce, 1997) before

PABLO analysis. Doing so generates a 5'-terminal fragment that can more readily be resolved from its PABLO ligation product by gel electrophoresis (Fig. 5.3). Because the cleavage site specificity of DNAzymes is determined by sequence complementarity, they can be made to cut RNA almost anywhere that is desired. To maximize their usefulness for PABLO analysis, they should be designed to cleave 100 to 400 nt from the 5'-end, preferably in a single-stranded region so as to facilitate DNA-RNA base pairing. In addition, it is important for DNAzyme cleavage to occur at a site well downstream of the 3'-end of any 5'-terminal decay intermediate(s) that may accumulate to a significant cellular concentration; otherwise, the PABLO measurement will reflect not only the 5'-phosphorylation state of the full-length RNA but also that of the degradation intermediate(s), which may be different.

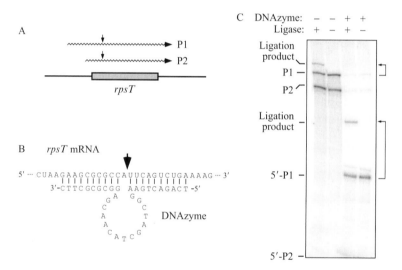

Figure 5.3 The use of a DNAzyme to improve the electrophoretic resolution of a PABLO ligation product. (A) The *E. coli rpsT* transcriptional unit. The *rpsT* gene, which encodes ribosomal protein S20, is transcribed from two promoters (P1 and P2) to generate a pair of transcripts, 0.45 kb and 0.36 kb in length, that can be cleaved *in vitro* at a chosen site (arrows) by treatment with a DNAzyme. (B) RNA-DNA duplex formed by *rpsT* mRNA and a 10-23 DNAzyme (Santoro and Joyce, 1997). Cleavage by this DNAzyme occurs at a unique site (arrow), generating 5'-terminal P1 and P2 fragments 163 nt and 74 nt long, respectively. (C) PABLO analysis of *rpsT* P1 transcripts cleaved *in vitro* by a DNAzyme. In the presence or absence of the DNAzyme shown in (B), total RNA from *E. coli* was subjected to PABLO analysis with a 90-nt oligo X and a P1-specific oligo Y. Detection by Northern blotting was performed with a radiolabeled probe complementary to both *rpsT* transcripts. Arrows identify PABLO ligation products.

6. ASCERTAINING WHETHER THE DECAY OF A TRANSCRIPT BEGINS WITH PYROPHOSPHATE REMOVAL

Although suggestive, the percentage of a particular full-length RNA that is monophosphorylated is not alone sufficient to prove whether the degradation of that transcript occurs primarily by means of a 5′-end–dependent pathway involving pyrophosphate removal as the initial step. Because the size of that monophosphorylated subpopulation is determined by the ratio of the rate constants for pyrophosphate removal and subsequent ribonuclease attack, it can be substantial not only when pyrophosphate removal is rapid but also when the resulting monophosphorylated intermediate is resistant to further degradation. Conversely, a failure to observe significant PABLO ligation does not rule out an important contribution from a decay pathway involving pyrophosphate removal, as some monophosphorylated RNAs (e.g., those with structured 5′-ends) may be poor PABLO substrates, while others may be too labile to detect because they are swiftly degraded by a 5′-monophosphate–dependent ribonuclease as soon as they form.

Additional evidence for the importance of pyrophosphate removal in the decay of a transcript can be obtained in a number of ways. One such test is to determine whether the lifetime of the transcript can be prolonged by adding a 5′-terminal stem loop (Arnold et al., 1998; Baker and Mackie, 2003; Bouvet and Belasco, 1992; Bricker and Belasco, 1999; Celesnik et al., 2007; Deana et al., 2008; Emory et al., 1992; Hambraeus et al., 2002; Sharp and Bechhofer, 2005), as the enzymes that participate in 5′-end–dependent decay prefer RNA substrates with single-stranded 5′-termini (Deana et al., 2008; Mackie, 1998; Mathy et al., 2007). In addition, the importance of pyrophosphate removal can be corroborated more directly in E. coli by showing that the decay rate of a transcript decreases when its 5′-triphosphate is replaced with a hydroxyl group so as to prevent the RNA from ever attaining a monophosphorylated state (Celesnik et al., 2007).

PABLO analysis of endonucleolytic cleavage products can provide another means to determine whether the degradation of a particular E. coli transcript proceeds principally by a pathway involving 5′-pyrophosphate removal before internal cleavage by RNase E. In that case, the 5′-terminal fragment that results from such cleavage should be more highly monophosphorylated than its full-length precursor because of the marked preference of RNase E for substrates bearing a 5′-monophosphate. On the other hand, those two RNAs should instead be monophosphorylated to a similar degree if decay occurs by means of a 5′-end–independent mechanism. For example, in the case of RNA I.613, cleavage by RNase E in E. coli generates a 5′-fragment that undergoes PABLO ligation with a yield of 53%, significantly higher than the yield for its full-length precursor (15%) and comparable to that observed for both when they are synthesized *in vitro* in a fully mono-phosphorylated state (Fig. 5.4). That the 5′-fragment, unlike the full-length transcript, is almost completely monophosphorylated in *E. coli* suggests that pyrophosphate removal from RNA I.613 followed by RNase E cleavage is considerably faster than RNase E cleavage of the triphosphorylated transcript.

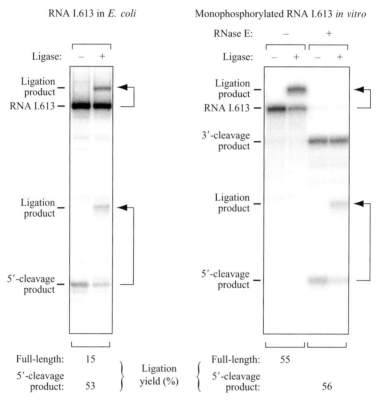

Figure 5.4 Phosphorylation state of a 5′-terminal decay intermediate produced by RNase E cleavage of RNA I.613 in *E. coli*. Left, Comparison of PABLO ligation yields for RNA I.613 and its 5′-terminal decay intermediate produced in *E. coli* by RNase E cleavage 38 nt from the 5′-end (see diagram in Fig. 5.2A). Right, PABLO ligation yields for internally radiolabeled 5′-monophosphorylated RNA I.613 synthesized *in vitro* and its 38-nt 5′-fragment produced by RNase E cleavage. To allow direct comparison of the ligation yields, the same oligos X and Y were used in each experiment. (Reprinted from Celesnik *et al.*, 2007, with permission from Cell Press.)

7. PABLO PROTOCOLS

7.1. Oligonucleotide design

The choice of oligo X for use in PABLO analysis is fairly flexible. Its sequence is relatively unimportant as long as it does not fold into a stably base-paired structure. On the other hand, its length must be carefully chosen to provide a shift in electrophoretic mobility sufficient to resolve the ligation product from unligated RNA. For transcripts up to 400 nt in length, an oligo X that is 20 to 45 nt long is generally adequate (e.g., X_{22}: 5′-GAACAATATGAATGATAACTTG-3′ or X_{32}: 5′-AAAAAAAAAA-GAACAATATGAATGATAACTTG-3′); for RNAs more than 400 nt in length, an oligo X that is at least 90 nt long is more appropriate (e.g., X_{90}: 5′-CCCCCCCCCCCCCCCCCCCCCCCCCCCCCCCCCCCCCCCCCCC-CCCCCCCCCCCCCCCCCCCCCCCCCCCCCCCGAACAATATG-AATGATAACTTG-3′), although the ligation yield may decrease in extreme cases. Oligo X should be gel-purified to eliminate any heterogeneity in length that may arise during oligonucleotide synthesis.

Proper design of the bridging oligonucleotide (oligo Y) is crucial, because it must precisely juxtapose the 3′-end of oligo X and 5′-end of the transcript of interest by base pairing with both. Typically, oligo Y is complementary to the 3′-terminal ∼22 nt of oligo X and the 5′-terminal 16 to 20 nt of the RNA (Fig. 5.2A). A bridging oligonucleotide with more extensive complementarity to the RNA (e.g., 30 nt or more) can sometimes result in poor electrophoretic resolution, possibly because of incomplete dissociation of the nucleic acid strands. Before designing oligo Y, the 5′-terminus of the transcript must be mapped at high resolution by a reliable method (see section 9). It is also important to bear in mind that secondary structure at or near the RNA 5′-end can interfere with PABLO analysis by making it difficult for the bridging oligonucleotide to anneal.

7.2. PABLO ligation and analysis

When performing PABLO analysis, measures should be taken to ensure that all of the reaction components are nuclease free. For example, oligos X and Y and any DNAzymes should first be treated with phenol-chloroform and ethanol precipitated, and T4 DNA ligase should be purchased from a manufacturer that guarantees it to be free of nuclease activity (e.g., Fermentas).

Total cellular RNA can be extracted by the hot-phenol procedure or another method, as PABLO ligation yields are not affected by the extraction procedure (Celesnik et al., 2007).

7.2.1. Extraction of bacterial RNA by the hot-phenol procedure

E. coli cultures are grown to log phase, and 10-ml aliquots are chilled rapidly by pipetting them into 50-ml centrifuge tubes containing crushed ice. The bacteria are pelleted by centrifugation at 5000 rpm for 8 min at 4 °C, resuspended on ice in 125 μl of cold buffer A (0.3 M sucrose, 0.01 M sodium acetate [pH 4.5], 50 mM EDTA), transferred to chilled microfuge tubes, and combined with 125 μl of room temperature buffer B (2% SDS, 0.01 M sodium acetate [pH 4.5]). The mixtures are vortexed and placed in a 75 °C heat block for 3 min; 250 μl of preheated water-saturated phenol (prepared without pH neutralization) is added, the samples are vortexed, incubated at 75 °C for another 3 min, vortexed again, and quickly cooled for 15 sec in a dry ice/ethanol bath. After centrifugation at 14,000 rpm for 5 min, the aqueous phase is collected into a new microfuge tube. The extraction with hot water–saturated phenol is repeated two more times. The RNA is precipitated from the final aqueous phase by adding 30 μl of 3 M sodium acetate (pH 4.8) and 900 μl of ethanol. The pellets are washed with 70% ethanol and air-dried. The RNA is redissolved in 90 μl of water and stored at -20 °C.

The three phenol extractions typically result in RNA preparations that are free of DNA. However, if desired, any remaining DNA contaminants can be removed by DNase I treatment. 90 μl of RNA (100 to 200 μg) is combined with 10 μl of 10× DNase I buffer (100 mM Tris Cl [pH 7.6], 25 mM MgCl$_2$, 5 mM CaCl$_2$) and 2 μl of DNase I (4 U, New England Biolabs). The mixture is incubated at 37 °C for 30 min. The reaction is stopped by adding 1 μl of 0.5 M EDTA (pH 8.0) and heating at 75 °C for 10 min. The RNA is then extracted twice with phenol/chloroform, ethanol precipitated, and dissolved in 100 μl of water.

7.2.2. Annealing of DNA oligonucleotides to RNA

Total cellular RNA (20 μg) (or ~1 pmol of an internally radiolabeled *in vitro* transcript, when measuring an intrinsic PABLO ligation efficiency) is combined with 2 μl of 10 μM oligo X and 4 μl of 1 μM oligo Y. To improve electrophoretic resolution of the eventual ligation product, 4 μl of a 100-μM solution of a DNAzyme oligonucleotide (Santoro and Joyce, 1997) can be included as well. Water is added to bring the final volume to 52.5 μl. The samples are heated at 75 °C for 5 min and then cooled gradually to 30 °C before being placed on ice for at least 1 min.

7.2.3. Ligation

A premixture (27.5 μl) containing the following components is added to each sample of RNA complexed with oligos X and Y: 2.5 μl of T4 DNA ligase (12.5 Weiss units; Fermentas), 1 μl of RNasin (40 U; Promega), 8 μl of 10× ligation buffer (400 mM Tris Cl [pH 7.8], 100 mM MgCl$_2$, 100 mM

dithiothreitol, 5 mM ATP), and 16 μl of 1 mM ATP. The resulting mixtures are incubated at 37 °C for 4 h. The ligation reactions are quenched by adding 120 μl of 4 mM EDTA, and the products are phenol extracted and ethanol precipitated in the presence of glycogen. The pellets are washed with 70% ethanol and air-dried.

7.2.4. Gel electrophoresis and Northern blotting

The pellets containing the ligation products are redissolved in 5 μl of water, combined with 15 μl of RNA loading buffer (95% v/v formamide, 20 mM EDTA [pH 8.0], 0.025% w/v bromophenol blue, 0.025% w/v xylene cyanol), and heated at 95 °C for 5 min. Electrophoresis is performed in a polyacrylamide gel (5 to 6% polyacrylamide for RNAs 100 to 750 nt in length or 11% polyacrylamide for RNAs <100 nt long) containing 8 M urea. The gel is electroblotted onto a Hybond-XL membrane (Amersham), and after UV crosslinking, the membrane is probed with radiolabeled DNA complementary to the transcript of interest. Radioactive bands corresponding to ligated and unligated RNA are visualized with a PhosphorImager, and ligation yields are calculated from the measured band intensities (yield = ligated/[unligated + ligated]).

7.3. *In vitro* synthesis of internally radiolabeled RNA bearing a 5'-monophosphate

To measure an intrinsic ligation efficiency, internally radiolabeled RNA bearing a monophosphorylated guanosine nucleotide at the 5'-terminus can be synthesized by *in vitro* transcription in the presence of a 60-fold molar excess of GMP over GTP. A typical reaction mixture (40 μl) contains 4 μl of 10× transcription buffer (400 mM Tris Cl [pH 7.9], 60 mM MgCl$_2$, 100 mM dithiothreitol, and 20 mM spermidine), 6 μl of 100 mM GMP, 1 μl of 10 mM GTP, 4 μl of 10 mM CTP, 4 μl of 10 mM UTP, 4 μl of 10 mM ATP, 5 μl of [α-^{32}P]CTP (50 μCi, 3000 Ci/mmol), 0.5 μl of RNasin (20 U; Promega), 5 μl of 0.1 μg/μl linearized plasmid DNA template, 5.7 μl of water, and 0.8 μl of T7 RNA polymerase (40 U; New England Biolabs). (*In vitro* synthesis of monophosphorylated RNAs that begin with A, C, or U is performed in a reaction mixture that contains 6 μl of 100 mM AMP, CMP, or UMP (instead of GMP), a reduced concentration of the corresponding NTP (only 1 μl of a 10-mM solution), and 4 microliters (not 1 μl) of 10 mM GTP. In addition, the efficient synthesis of such RNAs with T7 RNA polymerase requires the use of a class III ϕ2.5 T7 promoter (Coleman *et al.*, 2004).) After 4 h at 37 °C, the radiolabeled RNA is purified by electrophoresis on a 6% polyacrylamide/8 M urea gel. An excised gel fragment containing the radiolabeled transcript is crushed inside a microfuge tube, and the RNA is eluted by agitation for 18 h at 4 °C in 1 ml of 150 mM sodium chloride containing RNasin (20 U; Promega). The eluate is filtered by centrifugation at 13,000 rpm for 10 min in a Costar Spin-X column (0.45-μm pore size; Corning) and ethanol precipitated for 1 h at $-20\,^{\circ}$C in the presence of glycogen (25 μg). The RNA is redissolved in 50 μl of RNase-free water, RNasin (40 U; Promega) is added, and the RNA is stored at $-20\,^{\circ}$C.

The presence of a monophosphate at the 5'-end of the *in vitro* synthesized transcripts can be verified by demonstrating their susceptibility to destruction by Terminator (Epicentre), a 5'-exonuclease specific for monophosphorylated RNA. The reaction mixture (20 μl) should contain 2 μl of 10$\times$ reaction buffer (500 mM Tris Cl [pH 8.0], 20 mM MgCl$_2$, 1 M NaCl), 2 μl of 0.35 pmol/μl internally radiolabeled RNA, 0.1 μl of Terminator exonuclease (0.1 U), and 16 μl of water. After 2 h at 30 $^{\circ}$C, the reaction products are analyzed by electrophoresis on a 6% polyacrylamide/8 M urea gel beside a sample of the intact transcript.

8. ALTERNATIVE METHOD FOR EXAMINING THE PHOSPHORYLATION STATE OF RNA

Although PABLO is a powerful method for measuring the percentage of an RNA that is monophosphorylated, its successful application depends on efficient ligation, which can be jeopardized when the transcript of interest is highly structured near the 5'-end or when oligos X and Y are poorly designed. In addition, transcription initiation at multiple tandem sites can complicate the quantitative interpretation of PABLO ligation yields. These difficulties can be overcome by use of an alternative method of analysis that is based on the susceptibility of monophosphorylated RNA, but not triphosphorylated RNA, to digestion with a 5'-monophosphate–dependent exoribonuclease, such as Terminator (Fig. 5.5). Because this alternative procedure relies on measuring a reduction in band intensity, its usefulness is limited to RNAs that are at least 30 to 50% monophosphorylated. Therefore, it is much less sensitive and precise than PABLO, which can detect levels of monophosphorylation as low as $\sim$5%.

8.1. Probing the phosphorylation state of RNA with a 5'-monophosphate–dependent exonuclease

In a final volume of 20 μl, total cellular RNA (5 μg) is combined with 2 μl of 10$\times$ reaction buffer (500 mM Tris Cl [pH 8.0], 20 mM MgCl$_2$, 1 M NaCl), 1 μl of RNasin (40 U; Promega), and 1 μl of either Terminator Exonuclease (1 U; Epicentre) or water. After 3 h at 30 $^{\circ}$C, the samples are extracted with phenol/chloroform, ethanol precipitated in the presence of glycogen, and analyzed by Northern blotting.

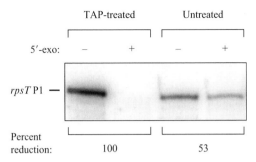

Figure 5.5 Examining the phosphorylation state of the *rpsT* P1 transcript by treatment with a 5′-monophosphate–dependent exonuclease. Total RNA isolated from *E. coli* was subjected to digestion with a 5′-monophosphate-dependent exonuclease and analyzed by gel electrophoresis and Northern blotting with an *rpsT*-specific radiolabeled probe. In addition, as a positive control, fully monophosphorylated RNA was generated by first treating total RNA with tobacco acid pyrophosphatase (TAP) to render it completely degradable. The percentage of the *rpsT* P1 transcript that was susceptible to exonucleolytic degradation was quantified to determine the fraction that was monophosphorylated in *E. coli* (53%, a percentage similar to that determined by PABLO) (Celesnik *et al.*, 2007).

For use as a positive control, fully monophosphorylated RNA can be generated by treating total cellular RNA with tobacco acid pyrophosphatase (TAP) before exonuclease digestion. RNA (5 μg) is combined with 1 μl of 10× TAP reaction buffer (0.5 M sodium acetate [pH 6.0], 10 mM EDTA, 1% β-mercaptoethanol, 0.1% Triton X-100), 0.5 μl of TAP (5 U; Epicentre), and 1 μl of RNasin (40 U; Promega). Water is added to a final volume of 10 μl. After 3 h at 37 °C, the mixture is extracted with phenol/chloroform, and the RNA is ethanol precipitated.

9. The Use of PABLO to Map the 5′-End of RNA

High-resolution mapping of RNA 5′-termini has traditionally relied on primer extension analysis. However, that method often gives inaccurate results because of the propensity of reverse transcriptases to terminate prematurely or to add untemplated nucleotides to primer extension products (Arnold *et al.*, 1998; Chen and Patton, 2001). Although premature termination caused by RNA secondary structure is a widely recognized property of those enzymes, a more insidious one that we have observed is their frequent tendency to terminate DNA synthesis 1 to 2 nucleotides before the 5′-end of an RNA template. A more dependable mapping procedure is RLM-RACE (RNA ligase-mediated rapid amplification of cDNA ends), in which transcripts are treated with TAP, ligated to an RNA oligonucleotide, amplified by

RT-PCR, and sequenced (Fromont-Racine *et al.*, 1993; Liu and Gorovsky, 1993; Sallie, 1993). This method works well for transcripts with unique 5'-ends, but those with multiple sites of transcription initiation produce mixed sequences that can be difficult to interpret unless a large number of amplification products are cloned and individually sequenced.

PABLO can be used as an alternate method for precisely mapping RNA 5'-ends. For example, monophosphorylated RNA I.613 synthesized *in vitro* gives a PABLO ligation yield of ~45% when its 5'-end and the 3'-end of oligo X are perfectly juxtaposed by base pairing to oligo Y_0, whereas no ligation occurs when those ends overlap or are separated by two or more nucleotides (oligo Y_{-1}, Y_{+2}, or Y_{+3}) (Fig. 5.6, left). Some ligation (~14%) also occurs when the bridging oligonucleotide leaves a one-nucleotide gap between the two ends (oligo Y_{+1}). This is not due to 5'-end heterogeneity, because the monophosphorylated RNA I.613 was synthesized by *in vitro* transcription with T7 RNA polymerase in the presence of excess AMP but in the absence of UMP. Consequently, only transcripts beginning at the expected position (A) would have been 5'-monophosphorylated and ligatable; any that might have begun at the preceding nucleotide (U) would have been triphosphorylated and, therefore, unreactive. A similar PABLO ligation pattern is observed when RNA I.613 or the *rpsT* P1 transcript encoding ribosomal protein S20 are instead produced *in vivo* by *E. coli* RNA polymerase and extracted from cells with total RNA. In each case,

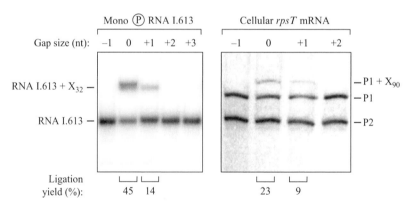

Figure 5.6 Mapping the 5'-end of RNA I.613 by PABLO analysis with a set of Y oligonucleotides. Left, PABLO analysis of internally radiolabeled 5'-monophosphorylated RNA I.613 synthesized *in vitro*. A set of bridging Y oligonucleotides that either perfectly juxtaposed the 3'-end of X_{32} with the unique 5'-end of RNA I.613 (gap size = 0), left them separated by 1 to 3 nucleotides (gap size = +1, +2, or +3), or caused them to overlap by one nucleotide (gap size = −1) was used. (Reprinted from Celesnik et al., 2007, with permission from Cell Press.) Right, PABLO mapping of the 5'-end of the *rpsT* P1 transcript extracted with total cellular RNA from *E. coli*. X_{90} and a P1-specific set of bridging oligonucleotides ($Y_{-1}, Y_0, Y_{+1}, Y_{+2}$) were used.

RLM-RACE reveals a unique $5'$-end, and ligation occurs not only for oligo Y_0 but also, to a lesser degree, for oligo Y_{+1} (Fig. 5.6, right).

To map the $5'$-end of a specific transcript by PABLO, the likely promoter and transcription initiation region must first be deduced by examining the DNA sequence of the corresponding gene. One then designs a set of bridging oligonucleotides that would juxtapose the $3'$-end of oligo X with either the predicted RNA $5'$-end (bridging oligo Y_0) or the nucleotides flanking it (oligos Y_{+1}, Y_{+2}, ..., and Y_{-1}, Y_{-2}, ...). Total cellular RNA is treated with TAP to convert $5'$-triphosphates to monophosphates and then examined in parallel PABLO reactions with each of the Y oligos. If the site of transcription initiation was correctly predicted, oligos Y_0 and Y_{+1} will both give ligation products (relative yield: $Y_0 > Y_{+1}$). On the other hand, if initiation instead occurs two nucleotides downstream, oligos Y_{-2} and Y_{-1} will give ligation products (relative yield: $Y_{-2} > Y_{-1}$). Transcription initiation at multiple adjacent sites renders $5'$-end mapping by PABLO analysis somewhat more complex because of the need to correct for the small contribution of Y_{N+1}-mediated ligation.

ACKNOWLEDGMENTS

The authors' research on pyrophosphate removal from bacterial transcripts is supported by a grant to J. G. B. from the National Institutes of Health (GM35769).

REFERENCES

Arnold, T. E., Yu, J., and Belasco, J. G. (1998). mRNA stabilization by the *ompA* $5'$-untranslated region: Two protective elements hinder distinct pathways for mRNA degradation. *RNA* **4,** 319–330.

Baker, K. E., and Mackie, G. A. (2003). Ectopic RNase E sites promote bypass of $5'$-end-dependent mRNA decay in *Escherichia coli. Mol. Microbiol.* **47,** 75–88.

Bouvet, P., and Belasco, J. G. (1992). Control of RNase E-mediated RNA degradation by $5'$-terminal base pairing in *E. coli. Nature* **360,** 488–491.

Bricker, A. L., and Belasco, J. G. (1999). Importance of a $5'$-stem-loop for longevity of *papA* mRNA in *Escherichia coli. J. Bacteriol.* **181,** 3587–3590.

Celesnik, H., Deana, A., and Belasco, J. G. (2007). Initiation of RNA decay in *Escherichia coli* by $5'$-pyrophosphate removal. *Mol. Cell* **27,** 79–90.

Chen, D., and Patton, J. T. (2001). Reverse transcriptase adds nontemplated nucleotides to cDNAs during $5'$-RACE and primer extension. *Biotechniques* **30,** 574–582.

Coleman, T. M., Wang, G., and Huang, F. (2004). Superior $5'$-homogeneity of RNA from ATP-initiated transcription under the T7 $\varphi 2.5$ promoter. *Nucleic Acids Res.* **32,** e14.

Deana, A., Celesnik, H., and Belasco, J. G. (2008). The bacterial enzyme RppH triggers messenger RNA degradation by $5'$-pyrophosphate removal. *Nature* **451,** 355–358.

Emory, S. A., Bouvet, P., and Belasco, J. G. (1992). A $5'$-terminal stem-loop structure can stabilize mRNA in *Escherichia coli. Genes Dev.* **6,** 135–148.

Fareed, G. C., Wilt, E. M., and Richardson, C. C. (1971). Enzymatic breakage and joining of deoxyribonucleic acid. *J. Biol. Chem.* **246,** 925–932.

Fromont-Racine, M., Bertrand, E., Pictet, R., and Grange, T. (1993). A highly sensitive method for mapping the 5'-termini of mRNAs. *Nucleic Acids Res.* **21,** 1683–1684.

Hambraeus, G., Karhumaa, K., and Rutberg, B. (2002). A 5'-stem-loop and ribosome binding but not translation are important for the stability of *Bacillus subtilis aprE* leader mRNA. *Microbiology* **148,** 1795–1803.

Jiang, X., Diwa, A., and Belasco, J. G. (2000). Regions of RNase E important for 5'-end-dependent RNA cleavage and autoregulated synthesis. *J. Bacteriol.* **182,** 2468–2475.

Kleppe, K., Van de Sande, J. H., and Khorana, H. G. (1970). Polynucleotide ligase-catalyzed joining of deoxyribo-oligonucleotides on ribopolynucleotide templates and of ribo-oligonucleotides on deoxyribopolynucleotide templates. *Proc. Natl. Acad. Sci. USA* **67,** 68–73.

Liu, X., and Gorovsky, M. A. (1993). Mapping the 5'- and 3'-ends of *Tetrahymena thermophila* mRNAs using RNA ligase mediated amplification of cDNA ends (RLM-RACE). *Nucleic Acids Res.* **21,** 4954–4960.

Mackie, G. A. (1998). Ribonuclease E is a 5'-end-dependent endonuclease. *Nature* **395,** 720–723.

Mathy, N., Bénard, L., Pellegrini, O., Daou, R., Wen, T., and Condon, C. (2007). 5' to 3'-exoribonuclease activity in bacteria: Role of RNase J1 in rRNA maturation and 5'-stability of mRNA. *Cell* **129,** 681–692.

Moore, M. J., and Sharp, P. A. (1992). Site-specific modification of pre-mRNA: The 2'-hydroxyl groups at the splice sites. *Science* **256,** 992–997.

Nath, K., and Hurwitz, J. (1974). Covalent attachment of polyribonucleotides to polydeoxyribonucleotides catalyzed by deoxyribonucleic acid ligase. *J. Biol. Chem.* **249,** 3680–3688.

Sallie, R. (1993). Characterization of the extreme 5'-ends of RNA molecules by RNA ligation-PCR. *PCR Methods Appl.* **3,** 54–56.

Santoro, S. W., and Joyce, G. F. (1997). A general purpose RNA-cleaving DNA enzyme. *Proc. Natl. Acad. Sci. USA* **94,** 4262–4266.

Sharp, J. S., and Bechhofer, D. H. (2005). Effect of 5'-proximal elements on decay of a model mRNA in *Bacillus subtilis*. *Mol. Microbiol.* **57,** 484–495.

Tock, M. R., Walsh, A. P., Carroll, G., and McDowall, K. J. (2000). The CafA protein required for the 5'-maturation of 16 S rRNA is a 5'-end-dependent ribonuclease that has context-dependent broad sequence specificity. *J. Biol. Chem.* **275,** 8726–8732.

CHAPTER SIX

A Proteomic Approach to the Analysis of RNA Degradosome Composition in *Escherichia coli*

Pierluigi Mauri* *and* Gianni Dehò[†]

Contents

Abstract

The RNA degradosome is a bacterial protein machine devoted to RNA degradation and processing. In *Escherichia coli,* it is typically composed of the endoribonuclease RNase E, which also serves as a scaffold for the other components: the exoribonuclease PNPase, the RNA helicase RhlB, and enolase. The variable presence of additional proteins, however, suggests that the degradosome is a flexible machine that may vary its composition in response to different conditions. Direct analysis of large protein complexes, together with simplified purification procedures, can facilitate qualitative and quantitative identification of RNA degradosome components under different physiologic and genetic conditions and can help to explain their role in the bacterial cell (see also

* Istituto di Tecnologie Biomediche, Consiglio Nazionale delle Ricerche, Segrate (Milan), Italy
† Dipartimento di Scienze biomolecolari e Biotecnologie, Università degli Studi di Milano, Milan, Italy

Methods in Enzymology, Volume 447
ISSN 0076-6879, DOI: 10.1016/S0076-6879(08)02206-4

Chapters 4, 11, 19, 20 and 22 regarding methods for the studying the degradosome and other multiprotein complexes in this volume.

Herewith we describe the application of multidimensional protein identification technology (MudPIT) in the rapid and quantitative identification of RNA degradosome components. RNA degradosome preparations obtained from specific conditions are enzymatically digested. The resulting peptides are fractionated using two-dimensional (ion-exchange and reversed-phase) chromatography and analyzed by tandem mass spectrometry. Bioinformatic analysis with the SEQUEST algorithm, which correlates experimentally obtained mass spectra with those predicted from peptide sequences in proteomic and translated genomic databases, allows identification of the corresponding proteins that compose the complex. The protein constituents of two or more degradosome samples are then compared to obtain a rapid evaluation of qualitative and quantitative differences in protein composition. Quantitative analysis is based on the observation that changes in relative protein abundance among different samples are reflected by statistical parameters (score values) assigned to each protein component of the RNA degradosome identified by the MudPIT approach. This correlation can be validated by independent methods such as Western blotting and determination of enzymatic activities. This fully automated procedure may be applied to the characterization of any complex protein mixture.

1. INTRODUCTION

1.1. The RNA degradosome

The *Escherichia coli* RNA degradosome is a multiprotein machine devoted to RNA degradation. It is typically composed by the endoribonuclease RNase E, which also serves as a scaffold for the assembly of the protein complex, the phosphorolytic 3′-to-5′-exoribonuclease polynucleotide phosphorylase (PNPase), the DEAD-box RNA helicase RhlB, and enolase (reviewed by Carpousis, 2007). RNase E, PNPase, and RhlB constitute the minimal core degradosome (Coburn *et al.*, 1999; Lin and Lin-Chao, 2005; Mackie *et al.*, 2001) that is thought to coordinate endonucleolytic and exonucleolytic activities to degrade highly structured RNA. However, other proteins such as polyphosphate kinase and the protein chaperons DnaK and GroEL have been found occasionally associated with the degradosome. At low temperatures, a different DEAD-box RNA helicase (DeaD alias CsdA) seems to replace (or add to) RhlB (Blum *et al.*, 1997; Miczak *et al.*, 1996; Prud'homme-Généreux *et al.*, 2004; Py *et al.*, 1996; Regonesi *et al.*, 2006). These (sometimes conflicting) findings suggest that the RNA degradosome is a highly dynamic protein machine and that its composition may vary according to the physiologic condition of the cell and/or the purification procedure. It remains unclear in some cases, however, whether such additional proteins are occasional contaminants or whether they are specific components with a particular function in the RNA degradosome. Moreover, in other bacterial species such as *Mycobacterium bovis* (Kovacs *et al.*, 2005) and *Rhodobacter capsulatus* (Jager *et al.*, 2001, 2004) or in eukaryotic organelles where an equivalent complex may be found (reviewed by Lin-Chao *et al.*, 2007), the RNA degradosome composition, and thus its structure, may differ from that of *E. coli*.

Notwithstanding recent advances in RNA degradosome purification by affinity coimmunoprecipitation, preparation of the RNA degradosome, identification of individual components, and assessment of their stoichiometric ratios is a demanding task. Preparation of the RNA degradosome is based on the copurification of proteins that remain associated with RNase E (Carpousis *et al.*, 2001; See chapter 4 by Carpousis *et al.* in this volume). The identification of the associated polypeptides has been traditionally performed by either N-terminal sequencing or mass spectrometry (MS) analysis of individual proteins separated by SDS-PAGE from purified RNA degradosome preparations (Blum *et al.*, 1997; Miczak *et al.*, 1996; Prud'homme-Généreux *et al.*, 2004; Py *et al.*, 1994, 1996). Once a new degradosome component has been identified, its presence and variation of stoichiometric amounts in different preparations can be monitored by measuring its enzymatic activity and/or by quantitative Western blotting (Rudolph *et al.*, 1999) with specific antibodies. However, given the dynamic properties of the complex, other unidentified proteins can be found under different environmental conditions and genetic backgrounds. A procedure that would directly and consistently determine the composition of the complex would more easily permit a comparison between different purification procedures and would be a desirable tool for the study of RNA degradosome biogenesis and function.

In this chapter, we describe simplified preparation procedures on the basis of the affinity purification of FLAG epitope-tagged RNase E (Miczak *et al.*, 1996), coupled to multidimensional protein identification technology (MudPIT) (Tyers and Mann, 2003; Washburn *et al.*, 2001) for the rapid and semiquantitative identification of different RNA degradosome components (Regonesi *et al.*, 2006).

1.2. Proteomic approaches to the analysis of complex protein mixtures

Proteomic analysis aims to identify individual components of a complex protein mixture. The classic approach for separating proteins present in a complex mixture (such as protein extracted from cells, tissues, or serum) is based on two-dimensional gel electrophoresis (2DE). In the first dimension, proteins are separated according to their isoelectric point with an immobilized pH gradient (IPG) (Bjellqvist *et al.*, 1982; Sanchez *et al.*, 1997). Separation in the second dimension by SDS-polyacrylamide gel electrophoresis (PAGE) is based on molecular weight. The separation ranges are typically approximately pI 4 to 10 and 10 to 250 kDa for first and second dimension, respectively.

In-gel detection of separated proteins by chemical (Coomassie or silver) staining or autoradiography of radiolabeled proteins provides two-dimensional (2D) maps in which the individual proteins are characterized simply by their pI and apparent MW. Comparison between 2D maps obtained from different experimental conditions would monitor variations in protein expression in response to a variety of developmental states and/or environmental conditions. A number of software applications have been developed that align the spots of multiple protein 2D maps and evaluate qualitative and quantitative changes in protein profiles (Marengo et al., 2005).

In-gel identification of individual proteins requires protein-specific reagents such as antibodies (for Western blotting) or in situ visualization of enzymatic activities. Amino acid sequencing of the N-terminal peptide of proteins cut out and eluted from the gel matrix was widely used until the 1990s. The application of mass spectrometry together with the development of proteomic and fully translated genomic databases have permitted the identification of individual proteins present in each 2DE spot by characterizing the peptides obtained using tryptic digestion of the 2DE-separated material recovered from the gel (Fig. 6.1). In particular, MALDI-TOF technology, which provides the molecular weights of tryptic peptides, allows peptide mass fingerprinting (PMF), whereas the peptide amino-acid sequences can be inferred with tandem mass spectrometry (MS/MS), a more recently adopted technology (Link et al., 1999).

An interesting innovation for 2DE proteomic analysis is differential in-gel electrophoresis (DIGE), which permits the direct evaluation of differences in protein abundance in up to three protein extracts (for example, samples obtained from different physiologic states) by labeling the proteins with different fluorescent dyes before gel electrophoresis. The labeled samples are then simultaneously separated in a single 2DE gel and imaged with mutually exclusive excitation/emission spectra. This permits quantitative comparisons of the intensities of comigrating fluorescence signals, thus avoiding distortions because of gel-to-gel variations (Zhou et al., 2002).

The 2DE approach has been important for developing proteomic studies, and for the past 10 years, it has been the method of choice for obtaining protein expression profiles. The gel-based approach presents advantages such as a relatively high resolution of proteins and low investments required. On the other hand, 2DE presents several limitations that may complicate quantification and high-throughput analysis, namely: (1) proteins with an either low ($<$10 kDa) or high ($>$200 kDa) molecular weight (MW), as well as those with an extreme isoelectric point (pI $<$4 or $>$10) cannot be detected; (2) hydrophobic proteins, typically membrane proteins, are not properly resolved and detected (Righetti and Boschetti, 2007); (3) the less

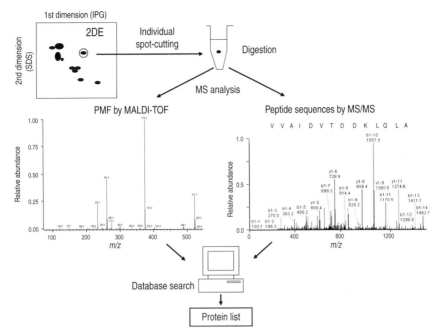

Figure 6.1 Steps involved in the classical two-dimensional electrophoresis (2DE) proteomic approach. Proteins are separated by two dimensional electrophoresis, which combines immobilized pH gradients (IPG, first dimension) and SDS-PAGE (second dimension). After staining and image data analysis, each spot of interest is cut out from the gel and digested by trypsin. The resulting peptide mixture is analyzed by mass spectrometry (MS) with either MALDI-TOF, which gives the peptide molecular weights (PMF), or tandem mass spectrometry (MS/MS), which gives the peptide sequences (shown over the MS/MS spectrum). Finally, computer data handling of spectra, on the basis of genomic or proteomic databases, permits the identification of protein(s) present in each electrophoretic spot.

represented proteins are difficult to analyze; (4) 2DE steps are time consuming and require extensive manual handling; and (5) reproducibility is low, and several experimental duplicates are, therefore, necessary.

The preceding problems may be overcome by proteomic methods on the basis of 2D microchromatography or nanochromatography coupled with tandem mass spectrometry (2DC-MS/MS), also known as multidimensional protein identification technology (MudPIT) (Tyers and Mann, 2003; Washburn et al., 2001).

2DC-MS/MS combines strong cation exchange chromatography (SCX) with reversed-phase (RP) separation of peptide mixtures obtained by enzymatic digestion of complex protein mixtures, followed by mass spectrometry analysis (Fig. 6.2). The entire protein mixture is digested, and the peptides thus obtained are usually loaded onto an SCX column and eluted by increasing (from 0 to 1 M) salt concentration steps. The eluate of each salt step is then directly loaded onto an RP column and fractionated with an acetonitrile gradient. Peptides eluted from the RP column are directly analyzed by a mass spectrometer. Typically, from 5 to 15 cycles (SCX salt steps and RP chromatograms) are applied, depending on sample complexity. 2DC-MS/MS can be performed with an offline setup that requires manual steps or the use of online configurations in which each step is automated (Swanson and Washburn, 2005).

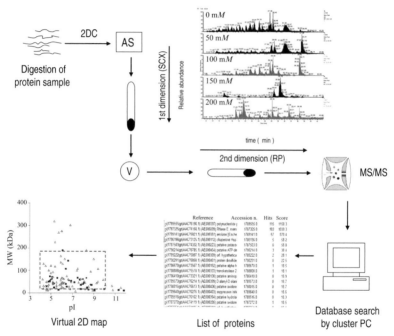

Figure 6.2   Pipeline for MudPIT proteomic approach. A trypsinized protein sample is directly loaded, by means of an auto sampler (AS), for two-dimensional chromatography (2DC) that involves peptide separation by strong cation exchange (SCX, first dimension) followed by reversed-phase (RP, second dimension) chromatography that is automatically coupled to SCX by a ten port valve (V). Separated peptides are directly analyzed with ion trap mass spectrometry (MS/MS). Data analysis of the resulting MS spectra, typically performed by a cluster PC, allows the identification of many proteins present in the original sample. The list of identified proteins can be plotted on a virtual 2D map. The dashed rectangle delimits the resolution range of conventional 2DE. (See Color Insert.)

Finally, peptide characterization and identification of the corresponding proteins is performed by matching the entire experimental mass spectra obtained from all RP chromatograms to theoretical peptide mass spectra deduced from public protein and/or fully translated genomic databases with specific software such as SEQUEST (Eng *et al.*, 1994). In addition, statistical values assigned by SEQUEST analysis may be used for a preliminary quantitative evaluation of the identified proteins (Liu *et al.*, 2004; Mauri *et al.*, 2005). All MudPIT steps after tryptic digestion are fully automated, and, thus, qualitative and quantitative analysis of the protein mixture may be obtained in a single run.

In this chapter, we discuss the application of the MudPIT approach to the analysis of the RNA degradosome. The general principles, however, may be applied to any complex protein mixture.

2. EXPERIMENTAL PROCEDURES

2.1. Experimental strategy

Proteomic approaches allow the rapid and unequivocal identification of the protein composition of the purified RNA degradosome. An interesting application is to compare the protein composition between different RNA degradosome preparations that may have been obtained under different physiologic conditions (e.g., cold or other environmental stresses), genetic backgrounds (e.g., mutants in specific components that may affect degradosome activity and/or assembly), and purification procedures.

Figure 6.3 illustrates the main steps used for the characterization of RNA degradosomes by MudPIT. Samples containing entire multiprotein complexes purified from different specific conditions are digested with trypsin and the peptide mixtures are fractionated by SCX and RP chromatographies and analyzed by MS. SEQUEST analysis of the MS/MS spectra allows the identification of the peptide sequences and the corresponding proteins in each mixture. The output protein lists of two or more degradosome samples are then compared to obtain a rapid evaluation of qualitative and semiquantitative differences in protein composition. Quantitative analysis is based on the observation that changes of relative protein abundance in different samples are reflected by the score values assigned by the SEQUEST analysis. This correlation may be validated by means of orthogonal independent methods.

2.2. RNA degradosome purification

A classical procedure for the purification of the RNA degradosome from bacterial cells by ammonium sulfate precipitation and SP-Sepharose chromatography followed by glycerol gradient sedimentation has been previously described in detail in this series (Carpousis *et al.*, 2001). More recently,

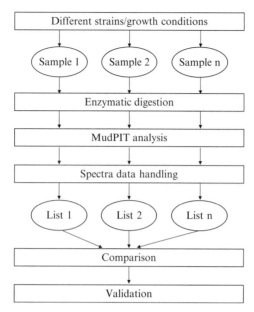

Figure 6.3 Outline of the MudPIT strategy used for investigating the protein composition of RNA degradosomes.

degradosome has been purified by affinity chromatography of tagged protein components such as FLAG-tagged RNase E or His-tagged PNPase (Miczak *et al.*, 1996; See chapter 4 by Carpousis *et al.* in this volume). These latter procedures are much less time and labor consuming, can be used to analyze small-scale preparations, and are thus suitable for preparing degradosome from several different genetic backgrounds (such as PNPase, RhlB, or enolase mutants), environmental conditions (e.g., temperature stress), or purification procedures (Lin and Lin-Chao, 2005; Prud'homme-Généreux *et al.*, 2004; Regonesi *et al.*, 2006). A caveat when recombinant proteins are used is that the abnormal and often nonregulated ectopic expression of the modified protein may alter the degradosome composition (Regonesi *et al.*, 2006). This could be overcome by replacing the wild-type gene on the chromosome with the allele encoding the modified protein by means of Red-promoted recombination (Datsenko and Wanner, 2000).

2.3. Enzymatic digestion

Digestion of RNA degradosome preparations is performed by means of modified trypsin (Promega or Sigma) that avoids protease autodigestion. Hydrolysis is made in 100 mM ammonium bicarbonate at pH 8.0. Usually, 10 to 20 μg of protein is digested with a substrate/enzyme ratio approximately 40 to 60:1 (w/w) in 30 to 50 μl final volume; after overnight incubation at 37 °C, the reaction is stopped by addition of either formic or trifluoreacetic acid so as to obtain a pH equal or less than 2. After acidification the samples can be stored at 4 °C.

The efficiency of peptide identification may be improved by additional specific operations. For example, the alkaline pH of the trypsin digestion facilitates the formation and shuffling of disulfide bridges between different tryptic peptides. This leads to the production of dipeptides, which are very difficult to identify. Cysteine-containing peptides are, therefore, poorly detected and characterized. This may be obviated by complete cysteine reduction before digestion so as to break the existing disulfide bonds in the native protein followed by alkylation to protect cysteines from new disulfide bridge formation during trypsin treatment (Hale *et al.*, 2004). Reduction of cysteines is performed by incubation of the degradosome samples at 37 °C in the dark with 15 mM DTT in tryptic buffer for 1 h. For the alkylation reaction, 50 mM iodoacetamide is added and incubation continued for 30 min at room temperature.

Sample desalting of tryptic digests permits to increase chromatography resolution and mass spectrometry sensitivity. This is performed by means of Pep-Clean (Pierce) C_{18} spin columns or ZipTip (Millipore) pipette tips that, in addition, allow concentration of the sample.

It has been reported that the use of organic solvents, typically, 5 to 20% acetonitrile, in the tryptic digestion mix increases the number of peptides identified for each protein (Russell *et al.*, 2001), thus improving the confidence of identification. This approach also reduces trypsin digestion time to a few minutes. This effect is related to a higher enzymatic activity or substrate solubility (Griebenow and Klibanov, 1995).

2.4. MudPIT analysis

Ten μl of trypsin-digested RNA degradosome are analyzed by 2DC-MS/MS with ProteomeX as a graphical interface (Thermo Electron Corporation, San José, CA) (Lin *et al.*, 2001). Peptides are first separated by ion exchange chromatography (Biobasic-SCX column, 0.32 i.d. × 100 mm, 5 μm, ThermoHypersil, Bellofonte, PA) by applying a seven-step ammonium chloride concentration gradient (0, 50, 100, 150, 200, 300, 600 mM) at a flow rate of 2 μl/min. Each salt step eluate is directly loaded, by means of a 10-port valve, into a C_{18} reversed-phase column (Biobasic-18, 0.180 i.d. × 100 mm, 5 μm; ThermoHypersil) and separated with an acetonitrile gradient (eluent A, 0.1% formic acid in water; eluent B, 0.1% formic acid in acetonitrile). The gradient profile is 5% eluent B for 3 min followed by 5 to 50% eluent B in 40 min at a 2 μl/ml flow rate. With smaller (e.g., 10 mM increase) salt steps, it is possible to increase the SCX resolution of peptides, but the analysis time increases too. Typically, MudPIT analysis of a digested RNA degradosome sample takes 6 to 7 h, corresponding to approximately four samples analyzed per day.

The peptides eluted from the C_{18} column are directly analyzed with an ion trap LCQXP mass spectrometer equipped with metal needle (10 μm i.d.) and with a limit of detection (LOD) approximately 150 fmol. The main MS parameters are as follow: the heated capillary is held at 160 °C, ion spray 3.2 kV, and capillary voltage 67 V. Full MS and MS/MS spectra are acquired in positive mode in the range 400 to 1600 m/z, with data-depending scan and dynamic exclusion for MS/MS analysis (35% collision energy). It is possible to use linear ion trap and high-resolution mass spectrometer to increase the sensitivity up to 30 and 0.1 fmol, respectively.

With the SEQUEST algorithm (University of Washington, licensed to ThermoElectron Corp.) (Eng et al., 1994), the experimental mass spectra produced (full MS and MS/MS) are then correlated to peptide sequences obtained by comparison with the theoretical mass spectra in the E. coli protein database downloaded from the NCBI ftp site (ftp://ftp.ncbi.nih.gov/repository/Eco/EcoProt/). Because the confidence of protein identification, particularly when data from a single peptide are used, depends on the stringency applied for the identification of the peptide sequence and peptide matching, a high stringency may be guaranteed as follows. First, the chosen minimum values of Xcorr should be greater than 1.5, 2.0, and 2.5 for single-, double-, and triple-charge ions, respectively. Second, the peptide mass search tolerance must be set to 1.0 Da, whereas this parameter is usually set to 2 or 3 Da. Whether the stringency applied is optimal may be confirmed by an additional confidence parameter, delCn (normalized correlation), which is considered optimal at values >0.07 (Durr et al., 2004). For protein assignment with data from a single peptide, stringent criteria must be applied according to recently published guidelines (Carr et al., 2004). In particular, only the first-best matching peptide is considered and only if it is found in multiple MS/MS spectra. In addition, protein identification with a single peptide is accepted as valid only if that peptide is found in at least three of four MudPIT analyses. The analysis of four independent replicas typically results in a coefficient of variation (CV) of approximately 3.2% and 10% for migration times and peak areas, respectively (Mauri et al., 2005).

As an example, Table 6.1 reports a typical list of proteins obtained from a MudPIT analysis of wild-type degradosome. In particular, 26 proteins were identified by at least two peptides. Although the proteins with the highest score correspond to those identified by the classical procedure, the other ones may represent low abundance proteins detected as background bands in silver-stained gels.

To visualize the protein list output data in a user-friendly format, we developed MAProMA (multidimensional algorithm protein map) software, which automatically plots MW vs pI for each identified protein (Mauri et al., 2005). A color code is automatically assigned according to a range of score values derived by SEQUEST data analysis. This provides an overview of the obtained protein profile in a virtual 2D map (Fig. 6.4) and allows a

Table 6.1 Output of MudPIT analysis of wild-type RNA degradosome of *E. coli*

Accession number[a]	Gene[b]	Protein[b]	Hits[c]	Score	pI	MW
3183553	*rne*	RNase E	136	1360.38	5.48	118,183
1172545	*pnp*	PNPase	120	1200.31	5.11	77,101
15832893	*eno*	Enolase	112	1120.34	5.32	45,656
16131636	*rhlB*	RhlB	20	200.24	7.29	47,127
16128449	*kefA*	Putative alpha helix protein	8	86.61	8.04	127,215
3916007	*yhjC*	Putative protein	5	64.27	8.48	33,330
16130390	*tktB*	Transketolase 2 isozyme	5	50.13	5.86	73,043
16129501	*ydfI*	Putative oxidoreductase	4	40.12	5.37	53,686
16130291	*b2359*	Putative protein	3	30.33	9.73	16,503
15803844	*rplB*	Ribosomal protein L2	3	30.23	10.93	29,861
16128555	*cusC*	Copper/silver efflux system	3	30.16	6.05	50,270
16131318	*yrhB*	Putative protein	3	30.16	4.73	10,614
13638202	*hrpA*	ATP-dependent RNA helicase	3	30.09	7.89	149,027
16129026	*yceB*	Putative protein	2	44.37	6.15	20,501
1729815	*eco47IIR*	Type II restriction enzyme	2	28.08	6.18	26,922
16131159	*yhdZ*	Putative protein	2	26.19	6.60	28,575
1169721	*focD*	Putative protein	2	20.55	6.72	96,211
15803580	*yqiB*	Putative protein	2	20.46	9.03	16,549
16128602	*dpiB*	Putative protein	2	20.46	5.77	61,685
15804336	*atpF*	F$_O$F$_1$ ATP synthase subunit B	2	20.44	5.99	17,265
16129400	*ydcT*	Putative protein	2	20.43	6.55	37,041
16131401	*yhjK*	Putative protein	2	20.41	5.81	73,081
16129356	*paaH*	3-Hydroxybutyryl-CoA dehydrogenase	2	20.39	5.65	51,733
15833433	*rpsC*	Ribosomal protein S3	2	20.38	10.27	25,984
15803822	*rpoA*	RNA polymerase, α-subunit	2	20.21	4.98	36,513
16128143	*fhuA*	Ferrichrome outer membrane transporter	2	20.12	5.47	82,183

[a] NCBI accession number.

[b] The accession number annotation has been manually replaced by the gene and protein name for clarity.

[c] Number of identified peptides. Only proteins with at least two hits are reported.

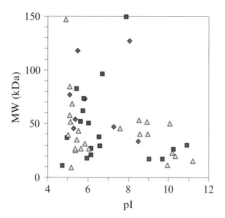

Figure 6.4 A virtual 2D map (theoretical pI vs MW) obtained by MAPROMA software for the set of proteins listed in Table 6.1 plus the others with at least one identified peptide. A color/shape code is assigned to each virtual protein spot according the confidence of identification (score value obtained by SEQUEST data handling): yellow/triangles, <20, blue/squares, from >20 to <40, and red/diamonds, >40. (See Color Insert.)

rapid evaluation of score values, a parameter related to identification confidence.

2.5. Differential analysis

One of the advantages of MudPIT analysis is that it not only identifies new RNA degradosome components under specific conditions but also provides a quantitative estimate of the relative amounts of RNA degradosome components in different preparations.

By analyzing different replicated samples, we observed that several proteins of the degradosome from wild-type compared with mutant cells presented a reproducible variation of score values assigned by SEQUEST software. This suggested that these score values might be related to quantitative changes in the amount of proteins detected in the two cell types. This has been verified as shown in Fig. 6.5, where the peak areas of the ion chromatograms corresponding to peptides identified by MudPIT for specific proteins (Fig. 6.5A) are plotted against the score value of the corresponding protein, normalized against the RNase E value for wild-type cells and *pnp-701* mutant (Regonesi *et al.*, 2004). As shown in Fig. 6.5B, a good linear correlation is obtained. Thus, the SEQUEST score value is predictive of the amount of peptides in each sample, and, with normalized score values, it is possible to obtain reproducible semiquantitative evaluations of the relative protein abundance in different samples. This is in agreement with recent reports that use spectral sampling (Liu *et al.*, 2004), peptide hits (Gao *et al.*, 2003), and other SEQUEST parameters to indicate protein relative abundance.

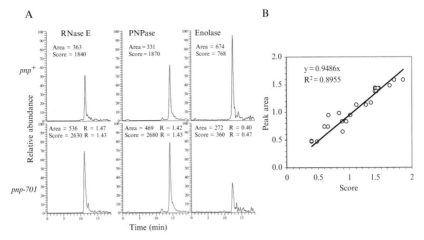

Figure 6.5 Correlation between peptide peak area and SEQUEST score value. (A) An example of extracted ion chromatograms of RNase E (IQISHISR), PNPase (GDISEFAPR), and enolase (DAGYTAVISHR) tryptic peptides identified in *pnp*⁺ (upper panels) and *pnp-701* (lower panels) degradosomes from cultures in which FLAG-RNase E was induced with IPTG. Peak areas, score values, and the ratio (R) between *pnp-701* and wild-type peak area or score are given (from Regonesi *et al.,* 2006, with permission). (B) Normalized score values (X) of identified proteins have been plotted versus the corresponding normalized peptide peak areas (Y). Experimental slope (approximately 0.95) is close to the theoretical one (1.00).

The relationship between score values and relative amount of protein in the samples has been confirmed by use of specific assays for PNPase and enolase, for which enzymatic activities can easily be assayed (Regonesi *et al.,* 2006). Obviously, for an absolute quantitative assay of proteins, stable isotope internal standards are necessary (Gygi *et al.,* 1999; Martin *et al.,* 2004). However, these techniques require higher costs and extensive sample handling.

On the basis of these findings, two algorithms that evaluate the relative protein abundance have been developed as tools of MAProMA software. In particular, the DAVE (differential average) value (Mauri *et al.,* 2005), which is based on the equation $(A - B)/0.5(A + B)$, estimates the relative changes between control (A, wild-type reference sample) and the sample under scrutiny (B, mutant). Namely, the difference $(A - B)$ between the score values of reference A and sample B is divided by the average of the same score values $0.5(A + B)$. According to this equation, when A equals B, the ratio is zero, which corresponds to no change in the abundance of a given protein between the two conditions. When the protein is not present in the reference $(A = 0)$, the ratio will be -2.0, whereas if the protein is not present in the sample $(B = 0)$, the ratio will be $+2.0$. Intermediate ratio values will, therefore, indicate different amounts of that protein under the two different conditions. To score only significant variations and avoid the possibility that variations might be an expression of the experimental error in the analysis, the threshold of DAVE value may be set to ±0.5, because this value corresponds to a score variation of 25% ($[0.5/2] \times 100$), which is higher than the calculated experimental variation of 15%. In particular, a DAVE value <-0.5 and $>+0.5$ indicates a potential down-representation and up-representation, respectively, of a given protein in the sample under scrutiny relative to the reference condition.

A second algorithm, DCI (differential coefficient index), is based on the equation $(A - B) \times 0.5(A + B)$ and is used for evaluating the absolute variation of score value of each protein in wild-type and mutant degradosome samples. In our experience, the significant variations are obtained when the DCI value is >400 and <-400 for potential up-representation and down-representation, respectively, in the reference sample. These values correspond to a variation of expression $>25\%$. To increase the confidence concerning the evaluation of potential variation of protein abundance, it is necessary that both indexes, DAVE and DCI, satisfy their respective thresholds.

On the basis of the relationship between score values and quantity of proteins, the normalized score values can be used for comparing the relative amount of each protein in the degradosomes obtained under specific conditions. Because RNase E serves as the scaffold for RNA degradosome assembly, it is taken as an internal standard, and the score of each protein identified in the same degradosome is normalized to the RNase E score. The relative stoichiometric variations of the other degradosome components can thus be appreciated. As an example, Fig. 6.6 reports the levels of "core degradosome" proteins, PNPase, enolase, and RhlB, and the associated DnaK, normalized to RNase E, in six different degradosome preparations from strains harboring different PNPase alleles (wild-type, a point, and a deletion mutant) and expressing different levels of tagged RNase E. It can be observed that the RNA degradosome composition is altered by FLAG-RNase E overexpression (RhlB is missing and DnaK is present) and/or PNPase availability (DnaK is present). These results are in good agreement with those obtained by Western or dot blotting and enzymatic analysis.

Alternately, it is possible to compare one to one the different lists of identified proteins for each specific degradosome preparation. In this case, it is first necessary to align each identified protein in the different samples. MAProMA software permits the comparison up to 10 different protein lists. DAVE and DCI parameters are then calculated for each protein in any pair of lists.

2.6. Validation

Validation of MudPIT analysis data may be performed by independent estimates of the abundance of two or more proteins present in the complex. This requires a biochemical assay and/or the availability of reagents, such as antibodies, that allow quantitative or semiquantitative determination of individual proteins in the sample.

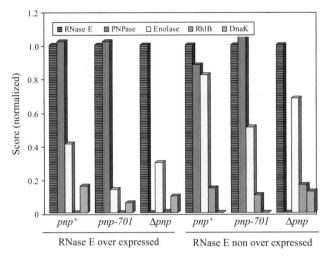

Figure 6.6   Normalized scores of "core RNA degradosome" and DnaK proteins in different degradosome preparations (modified from Regonesi *et al.*, 2006). SEQUEST scores of each degradosome sample obtained under two different conditions (basal or overexpressed levels of FLAG-RNase E) from wild-type, *pnp-701* (point mutant), and Δ*pnp* cultures have been normalized to the SEQUEST scores for RNase E in each sample. Histogram filling legend is reported on top of the figure Blue, RNase E: red, PNPase; yellow, enolase; green, RhlB; pink, Dnak. (See Color Insert.)

In the case of RNA degradosome, PNPase and enolase activities can be easily measured by simple photometric assays, as previously described in detail (Fontanella *et al.*, 1999; Regonesi *et al.*, 2004; Spring and Wold, 1975). Functional assays for the other main and accessory components are more difficult to set up. Moreover, such assays are obviously unsuitable to determine the presence of nonfunctional mutant protein, to determine the presence of nonfunctional mutant protein which would be desirable for studies of degradosome biogenesis. Quantitative Western blotting with specific antibodies (Rudolph *et al.*, 1999) may, in part, obviate such problems. Tagged derivatives (six histidines, FLAG, or other epitope tags) of one or more degradosome components not only simplify the degradosome purification by affinity chromatography but also allow the use of commercially available monoclonal antibodies to detect the tagged proteins.

3. PERSPECTIVE AND CONCLUSION

Many questions on the composition, structure, assembly, and function of the RNA degradosome are still open. Accessory proteins have been found to copurify in lower amounts together with or in substitution for the four core proteins RNaseE, PNPase, RhlB, and enolase. Among these, some have been detected only under specific conditions. For example, GroEL was reported only in a FLAG-purified degradosome preparation with an epitope-tagged Rne-3071–thermosensitive mutant protein (Miczak *et al.*, 1996), DeaD/CsdA in cold-adapted degradosomes (Prud'homme-Généreux *et al.*, 2004; Regonesi *et al.*, 2006), and DnaK seems to remain associated with abnormal degradosomes only (Regonesi *et al.*, 2006). Thus, the degradosome seems to be a flexible machine that may vary its composition in response to different stimuli, although it is not simple to discriminate between occasional contaminants and specific factors. Moreover, although the role of RNase E in degradosome assembly has been characterized in detail, little is known about the structural requirements of the other components and the biogenesis of this complex.

The relatively simple purification procedure based on affinity chromatography of FLAG epitope-tagged RNase E developed by Lin-Chao and collaborators (Miczak *et al.*, 1996) obviated the laboriousness of the classic purification (Carpousis *et al.*, 2001) and allows the analysis of the RNA degradosome composition under a variety of physiologic and genetic conditions. MudPIT analysis, a simple and reliable procedure for purification and characterization of complex protein mixtures, can be used not only for the rapid identification but also for quantification of the different components, thus providing a straightforward tool for the analysis of different RNA degradosome preparations.

The MudPIT proteomics approach provides a significant improvement over gel-based analysis, because it represents a fully automated technology that allows separation of digested peptides, the inference of their amino-acid sequence, and the identification of the corresponding proteins by a single analytical procedure. This approach permits a one-step identification of the proteins in RNA degradosomes from cells grown under different conditions and with different genetic backgrounds. In addition, with normalized score values from the SEQUEST algorithm, it is possible to easily obtain quantitative estimates of relative protein abundance. This seems to be particularly relevant for direct and high-throughput comparison of samples in a fully automated setting. Moreover, the normalized score values allow a quantitative comparison of a given protein in different preparations and correlates with the abundance of different proteins in the same sample. It should be stressed, however, that although the score value reflects the relative amounts of the different proteins within a sample, it does not provide absolute estimates and cannot be used directly to calculate absolute stoichiometric relationships between different proteins.

Proteomic analysis is a fast-moving field, and thus the procedure described herein should be taken as a basic outline that can be modified according to future developments.

ACKNOWLEDGMENTS

Support by joint grants from Ministero dell'Istruzione, dell'Università e della Ricerca and Università degli Studi di Milano (Programmi di Rilevante Interesse Nazionale 2005) and from Fondazione CARIPLO (Proteomic platform, NOBEL project) is gratefully acknowledged. The authors thank Marta G. Bitonti for MAProMA software.

REFERENCES

Bjellqvist, B., Ek, K., Righetti, P. G., Gianazza, E., Gorg, A., Westermeier, R., and Postel, W. (1982). Isoelectric focusing in immobilized pH gradients: Principle, methodology and some applications. *J. Biochem. Biophys. Methods* **6**, 317–339.

Blum, E., Py, B., Carpousis, A. J., and Higgins, C. F. (1997). Polyphosphate kinase is a component of the *Escherichia coli* RNA degradosome. *Mol. Microbiol.* **26**, 387–398.

Carpousis, A. J. (2007). The RNA degradosome of *Escherichia coli*: An mRNA-degrading machine assembled on RNase E. *Annu. Rev. Microbiol.* **61**, 71–87.

Carpousis, A. J., Leroy, A., Vanzo, N., and Khemici, V. (2001). *Escherichia coli* RNA degradosome. *Methods Enzymol.* **342**, 333–345.

Carr, S., Aebersold, R., Baldwin, M., Burlingame, A., Clauser, K., and Nesvizhskii, A. (2004). The need for guidelines in publication of peptide and protein identification data: Working Group on Publication Guidelines for Peptide and Protein Identification Data. *Mol. Cell Proteomics* **3**, 531–533.

Coburn, G. A., Miao, X., Briant, D. J., and Mackie, G. A. (1999). Reconstitution of a minimal RNA degradosome demonstrates functional coordination between a 3'-exonuclease and a DEAD-box RNA helicase. *Genes Dev.* **13**, 2594–2603.

Datsenko, K. A., and Wanner, B. L. (2000). One-step inactivation of chromosomal genes in *Escherichia coli* K-12 using PCR products. *Proc. Natl. Acad. Sci. USA* **97**, 6640–6645.

Durr, E., Yu, J., Krasinska, K. M., Carver, L. A., Yates, J. R., Testa, J. E., Oh, P., and Schnitzer, J. E. (2004). Direct proteomic mapping of the lung microvascular endothelial cell surface *in vivo* and in cell culture. *Nat. Biotechnol.* **22**, 985–992.

Eng, J. K., McCormack, A. L., and Yates, J. R. I. (1994). An approach to correlate tandem mass spectral data of peptides with amino acid sequences in a protein database. *J. Am. Soc. Mass Spectrom.* **5**, 976–989.

Fontanella, L., Pozzuolo, S., Costanzo, A., Favaro, R., Dehò, G., and Tortora, P. (1999). Photometric assay for polynucleotide phosphorylase. *Anal. Biochem.* **269**, 353–358.

Gao, J., Opiteck, G. J., Friedrichs, M. S., Dongre, A. R., and Hefta, S. A. (2003). Changes in the protein expression of yeast as a function of carbon source. *J. Proteome. Res.* **2**, 643–649.

Griebenow, K., and Klibanov, A. M. (1995). Lyophilization-induced reversible changes in the secondary structure of proteins. *Proc. Natl. Acad. Sci. USA* **92**, 10969–10976.

Gygi, S. P., Rist, B., Gerber, S. A., Turecek, F., Gelb, M. H., and Aebersold, R. (1999). Quantitative analysis of complex protein mixtures using isotope-coded affinity tags. *Nat. Biotechnol.* **17**, 994–999.

Hale, J. E., Butler, J. P., Gelfanova, V., You, J. S., and Knierman, M. D. (2004). A simplified procedure for the reduction and alkylation of cysteine residues in proteins prior to proteolytic digestion and mass spectral analysis. *Anal. Biochem.* **333**, 174–181.

Jager, S., Fuhrmann, O., Heck, C., Hebermehl, M., Schiltz, E., Rauhut, R., and Klug, G. (2001). An mRNA degrading complex in *Rhodobacter capsulatus*. *Nucleic Acids Res.* **29**, 4581–4588.

Jager, S., Hebermehl, M., Schiltz, E., and Klug, G. (2004). Composition and activity of the *Rhodobacter capsulatus* degradosome vary under different oxygen concentrations. *J. Mol. Microbiol. Biotechnol.* **7,** 148–154.

Kovacs, L., Csanadi, A., Megyeri, K., Kaberdin, V. R., and Miczak, A. (2005). Mycobacterial RNase E-associated proteins. *Microbiol. Immunol.* **49,** 1003–1007.

Lin, D., Alpert, A. J., and Yates, J. R. (2001). Multidimensional protein identification technology as an effective tool for proteomics. *American Genomic/Proteomic Technology* **1,** 38–46.

Lin, P. H., and Lin-Chao, S. (2005). RhlB helicase rather than enolase is the β-subunit of the *Escherichia coli* polynucleotide phosphorylase (PNPase)-exoribonucleolytic complex. *Proc. Natl. Acad. Sci. USA* **102,** 16590–16595.

Lin-Chao, S., Chiou, N. T., and Schuster, G. (2007). The PNPase, exosome and RNA helicases as the building components of evolutionarily-conserved RNA degradation machines. *J. Biomed. Sci.* **14,** 523–532.

Link, A. J., Eng, J., Schieltz, D. M., Carmack, E., Mize, G. J., Morris, D. R., Garvik, B. M., and Yates, J. R., III (1999). Direct analysis of protein complexes using mass spectrometry. *Nat. Biotechnol.* **17,** 676–682.

Liu, H., Sadygov, R. G., and Yates, J. R., III (2004). A model for random sampling and estimation of relative protein abundance in shotgun proteomics. *Anal. Chem.* **76,** 4193–4201.

Mackie, G. A., Coburn, G. A., Miao, X., Briant, D. J., and Prud'homme-Généreux, A. (2001). Preparation of *Escherichia coli* Rne protein and reconstitution of RNA degradosome. *Methods Enzymol.* **342,** 346–356.

Marengo, E., Robotti, E., Antonucci, F., Cecconi, D., Campostrini, N., and Righetti, P. G. (2005). Numerical approaches for quantitative analysis of two-dimensional maps: A review of commercial software and home-made systems. *Proteomics* **5,** 654–666.

Martin, D. B., Gifford, D. R., Wright, M. E., Keller, A., Yi, E., Goodlett, D. R., Aebersold, R., and Nelson, P. S. (2004). Quantitative proteomic analysis of proteins released by neoplastic prostate epithelium. *Cancer Res.* **64,** 347–355.

Mauri, P., Scarpa, A., Nascimbeni, A. C., Benazzi, L., Parmagnani, E., Mafficini, A., Della Peruta, M., Bassi, C., Miyazaki, K., and Sorio, C. (2005). Identification of proteins released by pancreatic cancer cells by multidimensional protein identification technology: A strategy for identification of novel cancer markers. *FASEB J.* **19,** 1125–1127.

Miczak, A., Kaberdin, V. R., Wei, C. L., and Lin-Chao, S. (1996). Proteins associated with RNase E in a multicomponent ribonucleolytic complex. *Proc. Natl. Acad. Sci. USA* **93,** 3865–3869.

Prud'homme-Généreux, A., Beran, R. K., Iost, I., Ramey, C. S., Mackie, G. A., and Simons, R. W. (2004). Physical and functional interactions among RNase E, polynucleotide phosphorylase and the cold-shock protein, CsdA: Evidence for a "cold shock degradosome." *Mol Microbiol.* **54,** 1409–1421.

Py, B., Causton, H., Mudd, E. A., and Higgins, C. F. (1994). A protein complex mediating mRNA degradation in *Escherichia coli*. *Mol. Microbiol.* **14,** 717–729.

Py, B., Higgins, C. F., Krisch, H. M., and Carpousis, A. J. (1996). A DEAD-box RNA helicase in the *Escherichia coli* RNA degradosome. *Nature* **381,** 169–172.

Regonesi, M. E., Briani, F., Ghetta, A., Zangrossi, S., Ghisotti, D., Tortora, P., and Dehò, G. (2004). A mutation in polynucleotide phosphorylase from *Escherichia coli* impairing RNA binding and degradosome stability. *Nucleic Acids. Res.* **32,** 1006–1017.

Regonesi, M. E., Del Favero, M., Basilico, F., Briani, F., Benazzi, L., Tortora, P., Mauri, P., and Dehò, G. (2006). Analysis of the *Escherichia coli* RNA degradosome composition by a proteomic approach. *Biochimie* **88,** 151–161.

Righetti, P. G., and Boschetti, E. (2007). Sherlock Holmes and the proteome—A detective story. *FEBS J.* **274,** 897–905.

Rudolph, C., Adam, G., and Simm, A. (1999). Determination of copy number of c-Myc protein per cell by quantitative Western blotting. *Anal. Biochem.* **269,** 66–71.

Russell, W. K., Park, Z. Y., and Russell, D. H. (2001). Proteolysis in mixed organic-aqueous solvent systems: Applications for peptide mass mapping using mass spectrometry. *Anal. Chem.* **73,** 2682–2685.

Sanchez, J. C., Rouge, V., Pisteur, M., Ravier, F., Tonella, L., Moosmayer, M., Wilkins, M. R., and Hochstrasser, D. F. (1997). Improved and simplified in-gel sample application using reswelling of dry immobilized pH gradients. *Electrophoresis* **18,** 324–327.

Spring, T. G., and Wold, F. (1975). Enolase from *Escherichia coli. Methods Enzymol.* **42,** 323–329.

Swanson, S. K., and Washburn, M. P. (2005). The continuing evolution of shotgun proteomics. *Drug Discov. Today* **10,** 719–725.

Tyers, M., and Mann, M. (2003). From genomics to proteomics. *Nature* **422,** 193–197.

Washburn, M. P., Wolters, D., and Yates, J. R., III (2001). Large-scale analysis of the yeast proteome by multidimensional protein identification technology. *Nat. Biotechnol.* **19,** 242–247.

Zhou, G., Li, H., DeCamp, D., Chen, S., Shu, H., Gong, Y., Flaig, M., Gillespie, J. W., Hu, N., Taylor, P. R., Emmert-Buck, M. R., Liotta, L. A., *et al.* (2002). 2D differential in-gel electrophoresis for the identification of esophageal scans cell cancer-specific protein markers. *Mol. Cell Proteomics* **1,** 117–124.

CHAPTER SEVEN

New Approaches to Understanding Double-Stranded RNA Processing by Ribonuclease III: Purification and Assays of Homodimeric and Heterodimeric Forms of RNase III from Bacterial Extremophiles and Mesophiles

Wenzhao Meng,* Rhonda H. Nicholson,[†] Lilian Nathania,* Alexandre V. Pertzev,* *and* Allen W. Nicholson*,[†]

Contents

Abstract

Ribonuclease III (RNase III) is a double-stranded (ds)-RNA–specific endonuclease that plays essential roles in the maturation and decay of coding and noncoding RNAs. Bacterial RNases III are structurally the simplest members of the RNase III family, which includes the eukaryotic orthologs Dicer and Drosha. High-resolution crystal structures of RNase III of the hyperthermophilic bacteria *Aquifex aeolicus* and *Thermotoga maritima* are available. *A. aeolicus* RNase III also has been cocrystallized with dsRNA or specific hairpin substrates.

* Department of Chemistry, Temple University, Philadelphia, Pennsylvania, USA
† Department of Biology and Chemistry, Temple University, Philadelphia, Pennsylvania, USA

Methods in Enzymology, Volume 447
ISSN 0076-6879, DOI: 10.1016/S0076-6879(08)02207-6

These structures have provided essential structural insight to the mechanism of dsRNA recognition and cleavage. However, comparatively little is known about the catalytic behaviors of *A. aeolicus* or *T. maritima* RNases III. This chapter provides protocols for the purification of *A. aeolicus* and *T. maritima* RNases III and also describes the preparation of artificial heterodimers of *Escherichia coli* RNase III, which are providing new insight on the subunit and domain interactions involved in dsRNA recognition and cleavage.

1. INTRODUCTION

Ribonuclease III was originally detected as an activity in *Escherichia coli* cell extracts, which selectively recognizes and cleaves double-stranded (ds) RNA *in vitro* (Robertson *et al.*, 1968). The role of RNase III in ribosomal RNA maturation was subsequently determined through the use of an *E. coli* strain that carried an inactivating mutation of the RNase III (*rnc*) gene (Dunn and Studier, 1973; Nikolaev *et al.*, 1973). Since then, much has been learned about the functions of RNase III in bacterial RNA processing, decay, and gene regulation, primarily through studies of the *E. coli* enzyme (reviewed by Court [1993] and Nicholson [2003]). Bacterial RNases III are phosphodiesterases and create short (~10 to 13 bp) products with two-nucleotide 3′-overhangs and 5′-phosphate, 3′-hydroxyl termini. The cleavage of cellular substrates occurs in a highly site-specific manner, with the target site identified by specific RNA sequence elements (Pertzev and Nicholson, 2006; Zhang and Nicholson, 1997). Moreover, cleavage of the target site (cleavage of both strands, or only one strand) is determined by local structural elements such as internal loops and bulges (Calin-Jageman *et al.*, 2003). RNase III family members can be classified according to polypeptide structure (Fig. 7.1). The Class 1 members include the bacterial RNases III, which contain a single dsRNA-binding domain (dsRBD) at the C-terminus and a nuclease domain that exhibits highly conserved acidic residues involved in metal binding and catalysis of phosphodiester cleavage. *E. coli* RNase III (and by inference all other Class 1 RNases III) functions as a homodimer and *in vitro* requires only a suitable divalent metal ion (Mg^{2+}, Mn^{2+}, Co^{2+}, Ni^{2+}) to catalyze accurate cleavage of cognate substrates (Court, 1993; Nicholson, 2003).

Current knowledge of the catalytic mechanism of RNase III is based largely on studies of the *E. coli* enzyme. However, this enzyme exhibits limited solubility, and a crystal structure is lacking. A landmark result was the publication in 2001 of an ~2-Å crystal structure of *A. aeolicus* (Aa) RNase III (Blaszczyk *et al.*, 2001). This study showed how the nuclease domain (NucD) self-associates to form the dimeric holoenzyme that contains two catalytic sites. Additional structural studies provided key data on

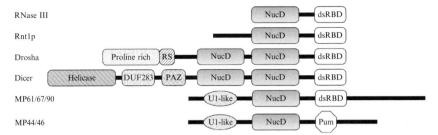

Figure 7.1 Ribonuclease III family members and their functional domains. Class 1 members include bacterial RNases III and *S. cerevisiae* Rnt1p (LaMontagne *et al.*, 2001). Class 2 includes Drosha, and Class 3 includes Dicer. MP61/67/90 and MP44/46 are functional components of the RNA editosomes of trypanosomatid kinetoplasts (Worthey *et al.*, 2003). Additional family members with different domain organizations are anticipated. Known domains are indicated by abbreviations (see below), with the black lines indicated linking polypeptide segments, or N- or C-terminal extensions with additional, yet-uncharacterized, domains. dsRBD, dsRNA-binding domain; DUF, domain of unidentified function; PAZ, Piwi-Ago-Zwille domain; NucD, nuclease domain; RS, Arginine/Serine-rich domain; U1, U1-like Zinc finger domain; Pum, Pumilio domain.

how RNase III recognizes dsRNA and the probable mechanism by which the scissile phosphodiester is cleaved (Blaszczyk *et al.*, 2004; Gan *et al.*, 2005, 2006, 2008). In addition, the Joint Center for Structural Genomics at the University of California, San Diego, solved the structure of *Thermotoga maritima* (Tm) RNase III (PDB entry code100W), and a crystal structure of *Mycobacterium tuberculosis* RNase III was described (Akey and Berger, 2005). The crystal structure of *Giardia intestinalis* Dicer provided further key insight as to how a eukaryotic RNase III family member generates short interfering RNAs and microRNAs (MacRae *et al.*, 2006). The availability of these structures—and those of Aa-RNase III and Tm-RNase III in particular—provides a new foundation for enzymologic and biophysical studies of the elementary steps in the a conserved mechanism of dsRNA processing. This chapter presents purification protocols and preliminary cleavage assays of Aa-RNase III and Tm-RNase III.

RNase III orthologs Drosha and Dicer are structurally more complex (MacRae and Doudna, 2007; MacRae *et al.*, 2006; Zhang *et al.*, 2004) and exhibit tandem nuclease domains (Fig. 7.1). The two domains associate in an intramolecular manner to form a pseudodimeric structure that contains a single dsRBD. This difference in the structures of Dicer and Drosha with that of bacterial RNases III prompted the preparation and analysis of artificial *E. coli* RNase III heterodimers that carry a single dsRBD (Meng and Nicholson, 2008). This chapter describes the production of bacterial RNase III heterodimers, including a heterodimer containing subunits from different species. The reader is referred to previous chapters that describe the affinity purification and assays for N-terminal (His)$_6$–tagged *E. coli*

RNase III (Amarasinghe *et al.*, 2001), Drosha (Lee and Kim, 2007), and Dicer (Kolb *et al.*, 2005).

2. Heterologous Expression, Affinity Purification, and Assays of *Aquifex aeolicus* and *Thermotoga maritima* RNases III

The procedure used to purify *E. coli* RNase III as an N-terminal (His)$_6$–tagged enzyme (Amarasinghe *et al.*, 2001) was also applied to Aa-RNase III and Tm-RNase III. *A. aeolicus* or *T. maritima* genomic DNA was used as a template for the amplification of the DNA sequences containing the respective single RNase III genes (Aa-*rnc* or Tm-*rnc*). The purified DNAs were cloned into the *Nde*1 and *Bam*H1 sites of pET-15b, and recombinants were identified as described previously (Amarasinghe *et al.*, 2001). The pET-15b (Aa-*rnc*) and pET-15b(Tm-*rnc*) plasmids were introduced into the expression host *E. coli* BL21(DE3)*rnc105,recA* (Amarasinghe *et al.*, 2001). This strain has an inactivating mutation in the chromosomal *rnc* gene to avoid contamination by the endogenous RNase III. A single colony of freshly transformed BL21 (DE3)*rnc105,recA* cells was inoculated into 5 ml LB broth (+100 μg/ml ampicillin). One milliliter of the 37 °C overnight culture was added to 250 ml LB broth (+100 μg/ml ampicillin). Cultures were grown at 37 °C with vigorous aeration to an OD (600 nm) of 0.3 to 0.4. Isopropyl-β-D-thiogalactopyranoside (IPTG) was added to a final concentration of 1 m*M*, and incubation with vigorous aeration continued for 4 h at 37 °C. Before the addition of the IPTG, a 1-ml aliquot was removed for SDS-PAGE analysis. The culture was cooled on ice, a 1-ml aliquot removed for SDS-PAGE analysis (see following), and the cells were collected by centrifugation at 4 °C. The cell pellet was stored at −20 °C until used for protein purification. Figure 7.2 shows the total protein profile of BL21(DE3)*rnc105,recA* containing pET-15b(Aa-*rnc*), either before (lane 1) or after IPTG (lane 2) induction. The Aa-RNase III polypeptide has a near-identical gel electrophoretic mobility as the Ec-RNase III polypeptide (Fig. 7.2, lane 3).

The affinity purification of Aa-RNase III and Tm-RNase III is based largely on the protocol provided in the HisBind resin system manual (Novagen). For convenience, Aa-RNase III purification will be described here, because the purification procedures for the two enzymes are essentially the same. It is not necessary to include a protease inhibitor in the purification buffers so long as the solutions are ice cold and affinity chromatography is performed as soon as possible after cell disruption. The cell pellet was thawed on ice and resuspended in 30 ml of prechilled buffer A (500 m*M* NaCl, 20 m*M* Tris-HCl [pH 8.0]) containing 5 m*M* imidazole (Im). The cell suspension was transferred to a 30-ml Corex centrifuge tube, and sonication was carried out at 4 °C with an ultrasonic homogenizer (Model 150

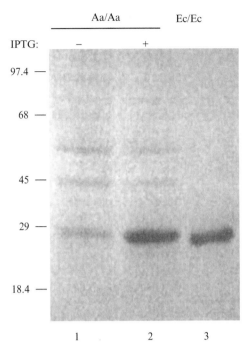

Figure 7.2 Overproduction of *Aquifex aeolicus* RNase III. BL21(DE3)*rnc105,recA* cells containing pET-15b(Aa-*rnc*) plasmid were grown and Aa-RNase III production induced by IPTG as described in the text. Aliquots of the cell culture were taken before and 4 h after IPTG addition, and total-cell protein analyzed by 12% SDS-PAGE. RNase III overproduction was examined by SDS-PAGE. The collected 1-ml cell culture aliquots were briefly centrifuged and the pellets resuspended in 100 μl of cell cracking buffer (50 mM Tris-HCl [pH 6.8], 20 mM EDTA, 10% [v/v] glycerol, 1% [v/v] β-mercaptoethanol, 1% [w/v] SDS, 0.04% [w/v] bromophenol blue). Samples were heated at 100 °C for 3 to 5 min, and aliquots electrophoresed in a 12% (w/v) polyacrylamide gel containing SDS. Prestained low-molecular-weight protein size markers were included in a side lane. The gel was stained with Coomassie brilliant blue R and then destained and dried by standard methods. The gel image was obtained by scanning. The His$_6$-RNase III polypeptide migrates in the gel as a species of ~28 kDa and is the major protein species in the cell by 4 h after induction. Lane 1, Cell protein pattern before IPTG addition; lane 2, cell protein profile 4 h after IPTG addition. Lane 3, purified *E. coli* (Ec) RNase III. The positions of protein molecular weight markers are indicated on the left side of the gel. The calculated molecular masses of (His)$_6$-tagged Aa-RNase III and Ec-RNase III are 27.6 kDa and 28.1 kDa, respectively.

XL2007; 4-W power setting; Misonix, Farmingdale, NY). Sonication was performed in 1-min bursts, repeated four times, with intermittent cooling on ice. Cell disruption was monitored visually and was judged complete when a significant decrease in viscosity to a final constant level was obtained. The sample was centrifuged at 7000 rpm for 20 min at 4 °C in a Sorvall SS34 rotor, and the supernatant was carefully removed to a separate centrifuge tube.

If necessary, the centrifugation step was repeated to obtain complete clarification. The Aa-RNase III or Tm-RNase III was largely present in the soluble fraction of the sonicated cell mixture.

Note: If an RNase III ortholog is purified for the first time, aliquots are taken from the soluble and insoluble fractions for SDS-PAGE analysis to determine which fraction contains the protein.

Aa-RNase III and Tm-RNase III were purified with a Ni^{+2} affinity column operated at 4 °C. The column was a 10-ml glass pipette and was prepared by applying a small plug of sterile glass wool as a column support; a short length of Tygon tubing to direct the eluent to the recipient tubes; and an adjustable clamp to control the flow rate. For Aa-RNase III, 1.5 ml of affinity resin was sufficient to purify protein from a 250-ml culture. The resin was first washed with 10 column volumes of buffer A supplemented with 5 mM Im, then charged with a 50 mM NiSO$_4$ solution. The clarified protein solution was slowly applied to the column. Approximately 1 h of loading time was required for a 30-ml volume. The resin was then washed with 20 column volumes of buffer A supplemented with 5 mM Im, followed by 10 column volumes of buffer A supplemented with 60 mM Im. The protein was eluted with five 1-ml aliquots of elution buffer (1 M NaCl, 20 mM Tris-HCl [pH 8.0], 400 mM Im). Most of the Aa-RNase III was present in the first three elute fractions, with the greatest amount in fraction 2. The fractions were combined, placed in dialysis tubing (Spectra-Por membrane, 8000 MW cutoff; Spectrum Industries, Laguna Hills, CA), and dialyzed against buffer I (1 M NaCl, 400 mM Im, 60 mM Tris-HCl [pH 8.0]) for ~2 h, and then dialyzed for an additional 2 h against buffer II (1 M NaCl, 60 mM Tris-HCl [pH 8.0]). Dialysis continued for 12 to 16 h against buffer III (1 M NaCl, 60 mM Tris-HCl [pH 8.0], 1 mM EDTA, 1 mM dithiothreitol [DTT]). Purified Aa-RNase III is stored at -20 °C in storage buffer (50% [v/v] glycerol, 0.5 M NaCl, 30 mM Tris-HCl [pH 8.0], 0.5 mM EDTA, 0.5 mM DTT). Aa-RNase III and Tm-RNase III are stable under these conditions. This protocol yielded ~1 mg of Aa-RNase III from a 250-ml bacterial culture. Aa-RNase III is free from contaminating nuclease activities.

2.1. Substrate cleavage assay

Figure 7.3 shows an assay that examined the ability of Aa-RNase III and Tm-RNase III to cleave internally ^{32}P-labeled R1.1[WC] RNA. The procedure for enzymatic synthesis and purification of R1.1[WC] RNA and related RNase III substrates is described elsewhere (Amarasinghe *et al.*, 2001). The assay shows that both Aa-RNase III and Tm-RNase III are active and cleave R1.1[WC] RNA with the same specificity as *E. coli* RNase III (Fig. 7.3A, compare lanes 10 to 12 with lanes 1 to 5, and Fig. 7.3B,

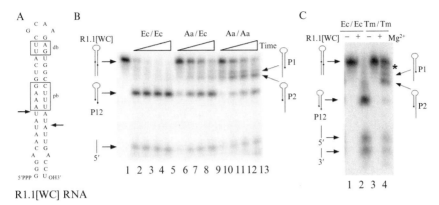

R1.1[WC] RNA

Figure 7.3 Cleavage of R1.1[WC] RNA by *Aquifex aeolicus* Aa-RNase III, *Thermotoga maritima* Tm-RNase III, and the putative *Aquifex aeolicus/Escherichia coli* Aa/Ec RNase III heterodimer. (A) Sequence and secondary structure of R1.1[WC] RNA. The arrows indicate the cleavage sites. "db" and "pb" indicate the distal box and proximal box, respectively, which contain recognition sequences for *E. coli* RNase III (Pertzev and Nicholson, 2006; Zhang and Nicholson, 1997). (B) Time-course cleavage assay with internally ^{32}P-labeled R1.1[WC] RNA. Purified enzyme (40 n*M* dimer concentration) was incubated with ^{32}P-labeled RNA (236 n*M*) in cleavage reaction buffer (Meng and Nicholson, 2007). Reactions were initiated by the addition of Mg^{2+} (10 m*M* final concentration), incubated at 37 °C for specific times (see below) then stopped by addition of excess EDTA (20 m*M*). Aliquots were electrophoresed in a 15% polyacrylamide gel containing 7 *M* urea in TBE buffer. Reactions were visualized by phosphor imaging. Lane 1, RNA incubated with enzyme in the absence of Mg^{2+}. Lanes 2 to 5, reaction with Ec-RNase III for 0.5, 1, 2, and 5 min, respectively. Lanes 6 to 9, reaction with putative Aa/Ec RNase III heterodimer for 1, 3, 5, and 10 min, respectively. Lanes 10 to 13, reaction with Aa-RNase III for 1, 3, 5, and 10 min, respectively. The positions of two products (P12, 5′) are indicated on the left side of the gel image, although the positions of the single-cleaved species (P1, P2) are indicated on the right. (C) Cleavage of R1.1[WC] RNA by Tm-RNase III. Internally, ^{32}P-labeled R1.1[WC] RNA (4000 dpm) was combined with enzyme (50 n*M* dimer concentration) in a 10-μl reaction containing 200 m*M* NaCl and 30 m*M* Tris-HCl (pH 8). MgCl$_2$ (10 m*M* final concentration) was added as indicated, and the reactions incubated at 45 °C for 15 min. Reactions were stopped with excess EDTA (20 m*M*), and then loaded and electrophoresed in a 15% polyacrylamide gel containing 7 *M* urea in 1× TBE buffer. Reactions were visualized by phosphor imaging. The positions of the products (P12, 5′, 3′) and single-cut intermediates (P1, P2) are shown. The asterisk indicates a cleavage product produced by Tm-RNase III but not Ec-RNase III. This cleavage site has not been mapped.

compare lanes 2 and 4). The putative Aa/Ec RNase III heterodimer exhibits a similar specificity (Figure 7.3B, lanes 6-9). The reactions involving Aa-RNase III and Tm-RNase III generate significant amounts of singly cleaved intermediates, suggesting a suboptimal affinity of R1.1[WC] RNA for each enzyme. This may reflect the fact that R1.1[WC] RNA is not a natural substrate for either enzyme. Efforts are under way to characterize cognate substrates for the two enzymes. Promising candidates are hairpins based on the 16S and 23S base-paired stem structures within the primary transcript of the 16S-23S-5S rRNA operon (R. H. N., L. N., and A. W. N., unpublished experiments). It also should be noted that assay temperatures of $\geq 45\,°C$ provide maximal activity for Aa- and Tm-RNase III, which remain active up to $\sim 90\,°C$. In contrast, temperatures $< 40\,°C$ significantly lower the activity of Aa- and Tm-RNases III, whereas *E. coli* RNase III is less affected (R. H. N., L. N., and A. W. N., unpublished).

3. PRODUCTION AND PURIFICATION OF *ESCHERICHIA COLI* RNASE III HETERODIMERS

The homodimeric structure of bacterial RNase III raises questions about the existence of specific intersubunit interactions important for substrate recognition and cleavage. In this regard, the study of mutant heterodimeric forms of DNA restriction endonucleases has provided mechanistic insight on the sequence-specific recognition and cleavage of DNA, as well as the involvement of intersubunit functional "cross-talk" (Wende *et al.*, 1996). A bacterial RNase III heterodimer was first reported by Klug and coworkers (Conrad *et al.*, 2002), who applied a dual plasmid *E. coli* expression system to coproduce RNase III subunits carrying different affinity tags, allowing formation of heterodimers *in vivo*. One tag was a hexa-histidine ([His]$_6$) tag, whereas the other tag was the maltose-binding protein. Serial affinity chromatography provided an RNase III heterodimer in which one subunit was catalytically inactive (Conrad *et al.*, 2002). Our approach uses a combination of the (His)$_6$ tag and the calmodulin-binding peptide (CBP) affinity tag (Zheng *et al.*, 1997), both of which are small in size (<2 kDa), to minimize effects on heterodimer behavior. We describe here the purification of an *E. coli* RNase III heterodimer (RNase III[NucD/wt]) in which one subunit lacks the C-terminal dsRBD.

E. coli BLR(DE3) cells (Novagen) were cotransformed by electroporation with plasmid pACYC184-*rnc*, which carries the wild-type *E. coli* *rnc* gene fused to the CBP sequence (Meng and Nicholson, 2008), and pET15b-*rnc*(ΔdsRBD), which contains the truncated *rnc* gene containing only the nuclease domain. A single colony was picked from an LB–agar plate containing ampicillin (Ap) and chloramphenicol (Cm) and grown overnight at 37 °C in 5 ml LB broth containing Ap (100 μg/ml) and Cm (40 μg/ml). The culture was diluted into LB broth (200 ml) containing Ap (100 μg/ml) and Cm (40 μg/ml), followed by further growth with vigorous shaking at 37 °C. At an optical density of ~ 0.4 (595 nm) IPTG (1 m*M*) was added, followed by continued incubation with shaking for 4 h. Cells were collected by low-speed centrifugation (6000 rpm for 20 min at 4 °C with a Sorvall

SLA-1500 rotor), and the cells pellets stored at $-20\,^{\circ}C$ until further use. As described earlier, an aliquot (1 ml) was removed from the culture immediately before IPTG addition, and a second aliquot was removed 4 h after IPTG addition for SDS-PAGE analysis. For better resolution of protein, the gel is run at 100 V until the bromophenol blue dye marker leaves the bottom of the 12% gel. *E. coli* cells obtained from 200 ml of culture were resuspended in 30 ml of Ni^{2+}-column binding buffer (20 mM Tris-HCl [pH 8.0], 500 mM NaCl, 5 mM Im) and subjected to repeated sonication at 0 to $4\,^{\circ}C$. An ultrasonic cell disrupter (Misonix, Inc.) was used at the 4 to 5 setting for 15 sec, with a 1-min pause, and repeated 20 times. The solution was clarified by low-speed centrifugation (6000 rpm, $4\,^{\circ}C$, 20 min with a Sorvall SLA-1500 rotor).

3.1. Affinity chromatography

Chromatography steps were performed at 4 to $5\,^{\circ}C$. The Ni^{2+}-NTA resin column (His · Bind resin, Novagen, 1.5 ml) was prepared in a 10-ml disposable plastic pipette, with a small amount of sterile glass wool as the column support, with a short piece of Tygon tubing to direct the eluent and a screw clamp to control the flow rate. Before sample addition, the resin was washed with 10 ml of Ni^{2+}-column binding buffer (see earlier) and then charged with 10 ml of 50 mM $NiSO_4$. The clarified extract (see earlier) was applied directly to the column at a rate of $\sim$1 ml/min. The column was washed with 30 ml of binding buffer, followed by 20 ml of wash buffer (20 mM Tris-HCl [pH 8.0], 500 mM NaCl, 60 mM Im) until protein could not be detected by standard assay (Bio-Rad Laboratories). Protein was eluted with successive 1-ml aliquots of elution buffer (20 mM Tris-HCl [pH 8.0], 1 M NaCl, 400 mM Im). The protein was present mainly in the first five elute fractions. The protein-containing fractions were combined, diluted twofold in Ca^{2+}-calmodulin(CaM)-column protein dilution buffer (65 mM Tris-HCl [pH 8.0], 3 mM $CaCl_2$), and loaded on a Ca^{2+}-CaM affinity column (Stratagene, Inc., 1 ml) at a rate of $\sim$1 ml/min. Before sample loading, the column was washed with 10 ml binding buffer (50 mM Tris-HCl [pH 8.0], 333 mM NaCl, 2 mM $CaCl_2$). After sample addition, the column was washed with 20 ml of Ca^{2+}-CaM-column binding buffer or until no protein could be visualized by protein assay. The protein was eluted with 5 × 1-ml aliquots of Ca^{2+}-CaM-column elution buffer (50 mM Tris-HCl [pH 8.0], 1 M NaCl, 2 mM EDTA). The protein-containing eluent fractions (usually present in the first five fractions) were combined, placed in dialysis tubing (Spectra-Por CE dialysis tubing, 10,000 MWCO), and dialyzed for 12 to 16 h at $4\,^{\circ}C$ against Ca^{2+}/CaM-column dialysis buffer (1 M NaCl, 60 mM Tris-HCl [pH 8.0], 1 mM EDTA, 1 mM DTT). The purified protein was stored at $-20\,^{\circ}C$ in 50% (v/v) glycerol, 0.5 M NaCl, 30 mM Tris-HCl (pH 8),

0.5 mM EDTA, and 0.5 mM DTT. Protein concentrations were determined by Bradford assay (Bio-Rad Laboratories) with bovine serum albumin as a standard.

3.2. Purification of RNase III heterodimers from inclusion bodies

RNase III heterodimers also can be purified from the inclusion body. The pellet collected by centrifugation after cell disruption (see earlier) is gently resuspended on ice in 4 ml of Ni^{2+}-NTA–column binding buffer supplemented with 6 M urea. Low-speed centrifugation (6000 rpm for 20 min at 4 °C in a Sorvall SLA-1500 rotor) provided a clarified supernatant that is combined with 25 ml of Ni^{2+}-NTA–column binding buffer, then loaded on a Ni^{2+}-NTA column. The column is washed and the protein eluted as described previously. The second Ca^{2+}-CaM column chromatography is performed as described previously. In this procedure, it is essential to carry out Ni^{2+}-NTA column chromatography first, because high concentrations of urea inactivate the calmodulin column (probably because of disruption of calmodulin structure). In contrast, 6 M urea does not affect the performance of the Ni^{2+}-NTA column.

REFERENCES

Akey, D. L., and Berger, J. M. (2005). Structure of the nuclease domain of ribonuclease III from *M. tuberculosis* at 2.1 Å resolution. *Protein Sci.* **14,** 2744–2750.

Amarasinghe, A. K., Calin-Jageman, I., Harmouch, A., Sun, W., and Nicholson, A. W. (2001). *Escherichia coli* ribonuclease III: Affinity purification of hexahistidine-tagged enzyme and assays for substrate binding and cleavage. *Methods Enzymol.* **342,** 143–158.

Blaszczyk, J., Tropea, J. E., Bubunenko, M., Routzahn, K. M., Waugh, D. S., Court, D. L., and Ji, X. (2001). Crystallographic and modeling studies of RNase III suggest a mechanism for double-stranded RNA cleavage. *Structure* **9,** 1225–1236.

Blaszczyk, J., Gan, J., Tropea, J. E., Court, D. L., Waugh, D. S., and Ji, X. (2004). Noncatalytic assembly of ribonuclease III with double-stranded RNA. *Structure* **12,** 457–466.

Calin-Jageman, I., and Nicholson, A. W. (2003). Mutational analysis of an RNA internal loop as a reactivity epitope for *Escherichia coli* ribonuclease III substrates. *Biochem.* **42,** 5025–5034.

Conrad, C., Schmitt, J. G., Evguenieva-Hackenberg, E., and Klug, G. (2002). One functional subunit is sufficient for catalytic activity and substrate specificity of *Escherichia coli* endoribonuclease III artificial heterodimers. *FEBS Lett.* **518,** 93–96.

Court, D. (1993). RNA processing and degradation by RNase III. *In* "Control of Messenger RNA Stability" (J. G. Belasco and G. Brawerman, eds.), pp. 71–116. Academic Press, New York.

Dunn, J. J., and Studier, F. W. (1973). T7 early RNAs and *Escherichia coli* ribosomal RNAs are cut from large precursor RNAs *in vitro* by ribonuclease III. *Proc. Natl. Acad. Sci. USA* **70,** 3296–3300.

Gan, J., Shaw, G., Tropea, J. E., Waugh, D. S., Court, D. L., and Ji, X. (2008). A stepwise model for double-stranded RNA processing by ribonuclease III. *Mol. Microbiol.* **67,** 143–154.

Gan, J., Tropea, J. E., Austin, B. P., Court, D. L., Waugh, D. S., and Xi, J. (2005). Intermediate states of ribonuclease III in complex with double-stranded RNA. *Structure* **13,** 1435–1442.

Gan, J., Tropea, J. E., Austin, B. P., Court, D. L., Waugh, D. S., and Ji, X. (2006). Structural insight into the mechanism of double-stranded RNA processing by ribonuclease III. *Cell* **124,** 355–366.

Kolb, F. A., Zhang, H., Jaronczyk, K., Tahbaz, N., Hobman, T. C., and Filipowicz, W. (2005). Human Dicer: Purification, properties, and interaction with PAZ PIWI domain. *Methods Enzymol.* **392,** 316–336.

LaMontagne, B., Larose, S., Boulanger, J., and Abou Elela, S. (2001). The RNase III family: A conserved structure and expanding functions in eukaryotic dsRNA metabolism. *Curr. Issues Mol. Biol.* **3,** 71–78.

Lee, Y., and Kim, V. N. (2007). *In vitro* and *in vivo* assays for the activity of the Drosha complex. *Methods Enzymol.* **427,** 87–106.

MacRae, I. J., and Doudna, J. A. (2007). Ribonuclease revisited: Structural insights into ribonuclease III family enzymes. *Curr. Opin. Struct. Biol.* **17,** 1–8.

MacRae, I. J., Zhou, K., Li, F., Repic, A., Brooks, A. N., Cande, W. Z., Adams, P. D., and Doudna, J. A. (2006). Structural basis for double-stranded RNA processing by Dicer. *Science* **311,** 195–198.

Meng, W., and Nicholson, A. W. (2008). Heterodimer-based analysis of subunit and domain contributions to double-stranded RNA processing by *Escherichia coli* RNase III *in vitro*. *Biochem. J.* **410,** 39–48.

Nicholson, A. W. (2003). The ribonuclease III superfamily: Forms and functions in RNA maturation, decay, and gene silencing. *In* "RNAi: A Guide to Gene Silencing" (G. Hannon, ed.), pp. 149–174. Cold Spring Harbor Laboratory Press, Cold Spring Harbor, New York.

Nikolaev, N., Silengo, L., and Schlessinger, D. (1973). Synthesis of a large precursor to ribosomal RNA in a mutant of *Escherichia coli*. *Proc. Natl. Acad. Sci. USA* **70,** 3361–3365.

Pertzev, A. V., and Nicholson, A. W. (2006). Characterization of RNA sequence determinants and antideterminants of processing reactivity for a minimal substrate of *Escherichia coli* ribonuclease III. *Nucl. Acids Res.* **34,** 3708–3721.

Robertson, H. D., Webster, R. E., and Zinder, N. D. (1968). Purification and properties of ribonuclease III from *Escherichia coli*. *J. Biol. Chem.* **243,** 82–91.

Wende, W., Stahl, F., and Pingoud, A. (1996). The production and characterization of artificial heterodimers of the restriction endonuclease EcoRV. *Biol. Chem.* **377,** 625–632.

Worthey, E. A., Schnaufer, A., Mian, I. S., Stuart, K., and Salavati, R. (2003). Comparative analysis of editosome proteins in trypanosomatids. *Nucl. Acids Res.* **31,** 6392–6408.

Zhang, H., Kolb, F. A., Jaskiewicz, L., Westhof, E., and Filipowicz, W. (2004). Single processing center models for human Dicer and bacterial RNase III. *Cell* **118,** 57–68.

Zhang, K., and Nicholson, A. W. (1997). Regulation of ribonuclease III processing by double-helical sequence antideterminants. *Proc. Natl. Acad. Sci. USA* **94,** 13437–13441.

Zheng, C.-F., Simcox, T., Xu, L, and Vaillancourt, P. (1997). A new expression vector for high level protein production, one step purification and direct isotopic labeling of calmodulin-binding peptide fusion proteins. *Gene* **186,** 55–60.

CHAPTER EIGHT

CHARACTERIZING RIBONUCLEASES *IN VITRO*: EXAMPLES OF SYNERGIES BETWEEN BIOCHEMICAL AND STRUCTURAL ANALYSIS

Cecília Maria Arraiano,[*,1,2] Ana Barbas,[*,2] *and* Mónica Amblar[†,2]

Contents

Abstract

The contribution of RNA degradation to the posttranscriptional control of gene expression confers on it a fundamental role in all biological processes. Ribonucleases (RNases) are essential enzymes that process and degrade RNA and constitute one of the main groups of factors that determine RNA levels in the cells. RNase II is a ubiquitous, highly processive hydrolytic exoribonuclease that plays an important role in RNA metabolism. This ribonuclease can act independently or as a component of the exosome, an essential RNA-degrading

[*] Instituto de Tecnologia Química e Biológica, Universidade Nova de Lisboa, Oeiras, Portugal
[†] Unidad de Investigación Biomédica, Instituto de Salud Carlos III (Campus de Majadahonda), Madrid, Spain
[1] Corresponding author
[2] Authors Contributed Equally

Methods in Enzymology, Volume 447
ISSN 0076-6879, DOI: 10.1016/S0076-6879(08)02208-8

multiprotein complex. In this chapter, we explain the general procedures normally used for the characterization of ribonucleases, using as an example a study performed with *Escherichia coli* RNase II. We present the overexpression and purification of RNase II recombinant enzyme and of a large set of RNase II truncations. We also describe several methods that can be used for biochemically characterizing the exoribonucleolytic activity and studying RNA binding *in vitro*. Dissociation constants were determined by electrophoretic mobility shift assay (EMSA), surface plasmon resonance (SPR), and filter binding assays using different single- or double-stranded RNA substrates. We discuss the synergies among the biochemical analyses and the structural studies. These methods will be very useful for the study of other ribonucleases.

1. INTRODUCTION

Escherichia coli RNase II is the prototype of the RNase II superfamily of exoribonucleases, whose homologues are present in all three domains of life (Grossman and van Hoof, 2006; Mian, 1997; Mitchell *et al.*, 1997; Zuo and Deutscher, 2001). In eukaryotes, RNase II (also called Dis3p/Rrp44p) is a component of both the nuclear and the cytoplasmic exosome, a complex of exoribonucleases crucial for RNA metabolism (Mitchell *et al.*, 1997). Recent reports have shown that Dis3/Rrp44 is the only catalytically active nuclease in the yeast core exosome (Dziembowski *et al.*, 2007) and plays a direct role in RNA surveillance contributing to the recognition and degradation of specific RNA targets (Schneider *et al.*, 2007). In addition, the human exosome possesses only hydrolytic activity, similar to the yeast exosome (Amblar *et al.*, 2006; LaCava *et al.*, 2005; Liu *et al.*, 2006) (Liu *et al.*, 2006). RNase II expression is differentially regulated at the transcriptional and posttranscriptional levels (Cairrão *et al.*, 2001; Zilhão *et al.*, 1993, 1996a, b) and can be regulated by environmental conditions (Cairrão *et al.*, 2001). Moreover, some family members have been shown to be involved in stress responses and virulence (Andrade *et al.*, 2006; Cairrão *et al.*, 2003). RNase II is a ubiquitous exoribonuclease that processively hydrolyzes RNA in the 3′ to 5′-direction, releasing 5′-monophosphates. RNA degradation in *E. coli* extracts is due mainly to the two major active exoribonucleases, RNase II and PNPase (Andrade *et al.*, 2008; Arraiano and Maquat, 2003; Arraiano *et al.*, 2007; Régnier and Arraiano, 2000). *E. coli* RNase II accounts for 90% of total hydrolytic activity in cell extracts (Deutscher and Reuven, 1991). RNase II binds RNA and DNA but is able to hydrolyze only RNA (Cannistraro and Kennell, 1994). The preferred substrates are poly(A) tails, and RNase II mutants have been instrumental in the characterization of poly(A)-dependent RNA-degradation mechanisms in prokaryotes (Arraiano *et al.*, 1988; Dreyfus and Régnier, 2002; Marujo *et al.*, 2000; Piedade *et al.*, 1995). This enzyme cleaves single-stranded RNA processively, one nucleotide at a time, up to 10 nucleotides. At fewer than 10 nucleotides, the enzyme becomes distributive. The end product of degradation of this enzyme is a 4-nucleotide RNA oligomer (Amblar *et al.*, 2006, 2007; Cannistraro and Kennell, 1994; Spickler and Mackie, 2000). The activity of RNase II is sequence independent but sensitive to RNA secondary structures (Amblar *et al.*, 2006; Arraiano and Maquat, 2003; Cannistraro and Kennell, 1999; Régnier and Arraiano, 2000). The three-dimensional structure of *E. coli* RNase II was recently determined (Frazão *et al.*, 2006; McVey *et al.*, 2006; Zuo *et al.*, 2006). The structure of RNase II RNA-bound complex (Frazão *et al.*, 2006; McVey *et al.*, 2006), together with biochemical data, gave new insight into the mechanisms of catalysis, translocation, and processivity of this important RNA-degrading enzyme.

To extensively characterize the biochemical properties of a ribonuclease, it is essential to perform *in vitro* enzymatic assays to analyze both ribonucleolytic activity and RNA-binding capability. We intend to provide detailed information about the different methodologies that have been and are currently being used to study RNase II. The methods described here will be very useful for the characterization of other ribonucleases. We also report the determination of dissociation constants using different methodologies and distinct single- or double-stranded RNA substrates.

The use of mutant derivatives provides useful information on the role of certain residues or different regions of the protein. A clear example is the single mutant D209N of *E. coli* RNase II, which allowed for identification of a residue essential for catalysis that is not involved in substrate binding (Amblar and Arraiano, 2005). The construction of a set of truncated RNase II proteins was also very useful for determining the functional domains of the enzyme and their contribution to the binding and degradation of different substrates. We present a detailed description of RNase II derivatives used in this study, how they were obtained or constructed, and the experiments performed with them. We also focus on the contribution of each domain in light of the structural properties of the enzyme. Finally, we discuss the synergies among the structural studies and biochemical analysis and how they contribute to explanations of the mechanism of RNA degradation by RNase II.

2. RNase II D209N Mutant

The *E. coli* strain SK4803 carries the mutant allele *rnb*296 that encodes an inactive RNase II enzyme. This mutant was isolated after nitrosoguanidine mutagenesis, and has less than 1% of the wild-type RNase activity against certain substrates, such as [³H] poly(U) (Nikolaev *et al.*, 1976). The phenotypic analysis of this mutant revealed that this strain had a slower growth, with a doubling time of about 2 h (Nikolaev *et al.*, 1976). This mutant was very useful as it allowed for the identification of other exoribo-nucleases in *E. coli* (Kasai *et al.*, 1977). The sequence of the *rnb*296 gene was determined, which showed that the mutant allele carried a point mutation in the coding sequence, which resulted in the single substitution of the aspartate 209 for an asparagine (D209N) (Amblar and Arraiano, 2005).

3. CONSTRUCTION OF RNASE II DERIVATIVES

On the basis of sequence analysis, three different domains were predicted for *E. coli* RNase II (Mian, 1997), including an N-terminal putative CSD, a central RNB catalytic domain, and a putative S1 domain at the C-terminus. To determine which regions of the enzyme actually correlated with the predicted functional domains of the protein, we constructed a set of 12 (His)$_6$-fusion truncated derivatives of the 644-amino-acid-long *E. coli* RNase II (Fig. 8.1). The RNase II–truncated proteins were constructed by removing different regions of *rnb* gene cloned in pFCT6.1 plasmid (Cairrão *et al.*, 2001). For this purpose, several restriction sites had to be introduced in *rnb* gene through site-directed mutagenesis using four primers and three PCR reactions (Higuchi, 1990). This recombinant PCR technique uses two PCR products that overlap in sequence, both containing the same mutation as part of the PCR primers. These overlapping primary PCR products can be denatured and subsequently reannealed and used as the template to produce the fragment bearing the mutation—the sum of the two overlapping products (Higuchi, 1990). The different mutagenic PCR reactions were performed and the resulting PCR fragments further cloned into pFCT6.1 to produce all the different RNase II mutant proteins (Amblar and Arraiano, 2005).

4. OVEREXPRESSION AND PURIFICATION OF RNASE II AND ITS DERIVATIVES

E. coli RNase II and the mutant derivatives (D209N single mutant and the truncated proteins) were overexpressed and purified as described subsequently. The corresponding plasmids were introduced into the BL21(DE3) *E. coli* strain to allow expression of the recombinant proteins upon IPTG induction. Cells were grown at 37 °C in LB medium supplemented with 150 μg/ml ampicillin to an A$_{600}$ of 0.45 and then induced by the addition of 1 mM of IPTG. The optimal induction conditions were standardized for each mutant by analyzing both protein production and protein solubility at different times after IPTG addition. For this purpose, samples were

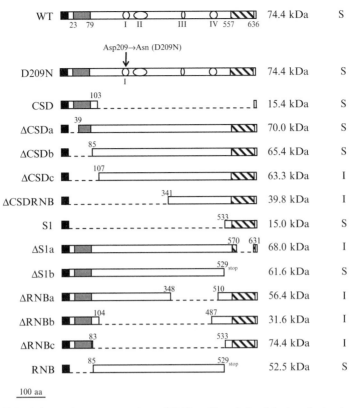

Figure 8.1 Schematic representation of RNase II and the deletion derivatives constructed. The CSD, RNB, and S1 putative domains predicted for *E. coli* RNase II are represented as gray, white, and diagonally striped boxes, respectively. The black box at the N-terminus of each enzyme represents the His-tag. Amino acids at the limits of each domain are indicated, where amino acid numbering corresponded to the sequence of untagged wild-type RNase II. The four highly conserved sequence motifs of the RNB domain (motif I to IV) are also depicted. The region deleted in each mutant is represented as dashed lines, and numbers correspond to the residues limiting the deleted region. The molecular weight of each mutant protein in kDa and whether the protein is more than 50% soluble (S) or less than 50% soluble (I) after IPTG induction is indicated to the right. The name used for each deletion protein is shown to the left. Adapted and reprinted from Amblar *et al.* (2006) with permission from Elsevier.

withdrawn at different induction times and the crude extracts were prepared as follows. Cells from a 1.5 ml culture were sedimented by centrifugation, washed by suspension in 1 ml of buffer A (20 m*M* Tris-HCl [pH 8], 100 m*M* KCl, and 1 mM MgCl$_2$), centrifuged again, and suspended in 150 μl of buffer A. Cells were lysed by adding 1 mg/ml BSA and, after 1 h of incubation at 0 °C, three cycles of freezing and thawing at −70 °C and 37 °C, respectively. After reducing lysate viscosity by passage through a

0.36-mm inner-diameter needle, samples were then centrifuged and the resulting supernatant stored at −70 °C.

The solubility of each protein during induction was tested by separating soluble and insoluble protein fraction using standard procedures as previously described (Amblar and López, 1998), followed by fractionation using SDS/PAGE. Only 7 of the 13 RNase II protein derivatives were shown to be more than 50% soluble after induction with IPTG. The CSD and S1 mutants, which correspond to the putative RNA-binding domains, were more than 90% soluble, and the same typified the ΔCSDb and ΔS1b mutants (i.e., mutant proteins that lack these domains). Similarly, the solubility of the RNB derivative, which lacks both putative RNA-binding domains, was also greater than 90%. However, the ΔCSDa mutant was only 60% soluble. This deletion mutant only lacks the RNP1 consensus sequence motif that has been found to be crucial for nucleic acid binding in all CSDs (Theobald *et al.*, 2003). The seven mutant proteins that were found soluble were purified for further analysis (Fig. 8.1).

Purification of all the recombinant proteins was performed by histidine affinity chromatography using HiTrap Chelating HP columns (GE Healthcare) and the AKTA HLPC system (GE Healthcare). Cells from 100-ml of IPTG-induced cultures were harvested by centrifugation, washed with 20 ml of buffer B (20 mM Tris-HCl, 0.5 M NaCl, pH8), and frozen at −80 °C. Just before purification, the frozen cells were thawed on ice and resuspended on 4 ml of buffer B in the presence of 1 mM of PMSF (phenylmethylsulphonyl fluoride, a protease inhibitor) and 20 mM of imidazol. Cell lysis was achieved by mechanical disruption using the French press at 9000 psi. The crude extract was further treated by incubation with 165 U of Benzonase endonuclease (Sigma) for 30 min at 0 °C, to remove the nucleic acids. The cell debris and insoluble material were pelleted by a 30 min centrifugation at 10,000×g, and the clarified extract was then added to a HiTrap Chelating Sepharose 1-ml column, previously equilibrated in buffer B plus 20 mM of imidazol and 2 mM of β-mercaptoethanol. Protein elution was achieved using a gradient of 20 mM to 500 mM imidazol in buffer B. The fractions containing the purified protein were pooled together and loaded into an ion-exchange 5 ml column (GE Healthcare) equilibrated with buffer C (20 mM Tris-HCl, 100 mM KCl, and 2 mM β-mercaptoethanol). Eluted protein was concentrated by centrifugation at 15 °C with Amicon Ultra Centrifugal Filter Devices (Millipore). Protein concentration was determined by spectrometry and 50% (v/v) glycerol was added to the final fractions prior to storage at −20 °C.

All the purified proteins were then characterized in detail with respect to exoribonucleolytic activity, RNA-binding ability, and substrate affinity. Different assays were performed with different substrates, all of them described subsequently, and the results provided valuable information regarding the RNase II activity.

5. ANALYSIS OF RIBONUCLEOLYTIC ACTIVITY

Proper characterization of a ribonuclease normally requires the performance of *in vitro* enzymatic assays. Like all kinds of enzymes, the enzymatic activity of a ribonuclease can be influenced in part by the substrate used in the assay, and their properties and degradation pattern will be different on different substrates. Moreover, the use of distinct substrates in *in vitro* assays can reveal different aspects of the enzyme that, finally, are a reflection of its behavior *in vivo*. In the case of RNase II exoribonuclease, several RNA substrates could be used. It is well known that RNase II has a strong preference for poly(A) stretches and that it is highly efficient in degrading poly(A) tails. The degradation of these tails can determine the half-life of the mRNA, especially if they are degraded by a poly(A) pathway of decay (Dreyfus and Régnier, 2002). For this reason, a poly(A) polymer is among the preferred substrates used to determine the activity of an RNase II sample. However, more specific substrates can also be employed. The use of synthetic mRNA substrates, either with or without a poly(A) tail, turned out to be a good choice for mimicking the actual *in vivo* enzyme substrates. Synthetic oligomers have also shown to be a powerful tool to study specific aspects of the RNase II degradation, such as single-stranded requirements at the 3′-end, base preferences, minimum length of substrates needed, among others. There is always an oligomeric substrate that can be designed for the study of every feature of RNase II degradation. Therefore, the use of these substrates is normally necessary for an extensive study of a ribonuclease.

All substrates mentioned previously were used in the study of the *E. coli* RNase II and different conclusions were extracted from each. The enzymatic assays performed with every type of substrate will be described in detail, as well as the advantages and special considerations that have to be taken into account when using each kind of approach.

5.1. RNase II exoribonucleolytic activity on poly(A) homopolymers

As mentioned previously, poly(A) tails are the preferred substrates for RNase II, and in *E. coli*, RNase II is responsible for 90% of the exonucleolytic degradation of synthetic RNA poly(A) homopolymers. This makes it possible to distinguish the activity of RNase II from the other ribonucleases present in a crude extract. Therefore, the main advantage of using poly(A) polymers as substrate is that it allows for selective detection of the activity of RNase II in crude extracts. For this reason, poly(A) homopolymers have been extensively used to study the activity of the wild-type RNase II and deficient strains in *E. coli* crude extracts (Donovan and Kushner, 1983,

1986). The substrate consists of a mix of poly(A) molecules of different sizes that are radioactively labeled (poly[8-^{3}H]adenylic acid). Upon degradation by RNase II, radioactive nucleotides monophosphate ([^{3}H]AMP) are released from the RNA polymer and the activity of the enzyme can be determined by measuring the soluble radioactivity. These assays were first used by Donovan and Kushner (1983), and, though several modifications have been introduced to make the assays quantitative (Amblar and Arraiano 2005), the procedures remained essentially the same. The reactions are carried out in 60 μl of volume of the optimal activity buffer for RNase II (100 mM KCl, 20 mM Tris, pH 8, 0.5 mg $\times$ ml^{-1} BSA) containing 1 mM of MgCl$_2$, 0.1 mM of substrate, and different protein concentrations (crude extracts or purified proteins). The mixtures are then incubated at 30 °C for 5 min and the reactions stopped by cooling at 4 °C. After that, the non-degraded substrate is precipitated by adding 120 μl of trichloroacetic acid (10%, v/v) followed by incubation at 4 °C for at least 30 min. The nondegraded substrate is then pelleted by centrifugation (15 min at 20,000g and 4 °C) and soluble [^{3}H]AMP is measured in a scintillation counter. One unit of enzymatic activity (UE) is the amount of enzyme required for the release of 10 nmol of [^{3}H]AMP in 15 min at 30 °C. Therefore, the units for enzymatic activity are defined by the following equation:

$$UE = ([\text{cpm} - BG] \times DF \times nmol \times 3 \times 1.8)/ (\text{total cpm} \times 10 \ nmol)$$

where cpm is the counts per minute of the sample, BG is the background cpm given by the buffer with no enzyme added to the reaction, DF is the dilution factor for the sample, and $nmol$ is the nmols of poly(A) used in the reaction. The factor by which the reaction time is multiplied to make it 15 min is 3. The multiplying factor of 1.8 results from the fact that only 100 μl from the total 180 μl were measured in the scintillation counter. *Total cpm* corresponds to the counts per minute of the total substrate used in the reaction. This value is obtained from the direct measuring of the control reaction (without enzyme added) before TCA precipitation. 10 *nmol* is the factor by which the UE has to be divided to be referred to the releasing of 10 nmol.

This kind of assay was useful, for example, to identify RNase II–defective strains such as SK4803, which carries the mutant allele *rnb296*. As detailed previously, crude extracts from this strain were totally inactive in the degradation of poly(A) (Donovan and Kushner, 1983). The mutation responsible for this phenotype was identified as the single amino acid substitution: Asp209 $\rightarrow$ Asn (D209N) (Amblar and Arraiano, 2005), and the corresponding protein was purified as a 6-histidine-tagged RNase IID209N protein, as described previously. The exoribonucleolytic activity of both the crude extracts from the IPTG-induced culture and the purified

Table 8.1 Specific exoribonucleolytic activity in crude extracts from BL21(DE3) overproducer strains on poly(A) substrate. The exoribonuclease activity was measured before and after 2 h of isopropyl thio-β-D-galactoside (IPTG) induction of BL21(DE3) containing the pMAA plasmid overproducing the D209N mutant or the pFCT6.9 plasmid overproducing the wild-type RNase II. BL21(DE3) cells without plasmid were used as a control. Each value is the mean of at least three independent experiments. Reprinted from Amblar and Arraiano (2005) with permission from Wiley-Blackwell Publishing.

	Specific Exoribonuclease Activity (UE/μg of protein)		
	BL21(DE3)	BL21(DE3) [pFCT6.9]	BL21(DE3) [pMAA]
Before IPTG induction	0.12	0.37	0.03
After 2 h IPTG induction	0.22	22.42	0.17

protein were assayed on poly[8-^{3}H]adenylic acid (Table 8.1). The experiments confirmed the total lack of activity of the RNase IID209N mutant, as the crude extracts from induced cultures presented a background activity (compared to 60-fold induction detected with wild-type protein), and the purified enzyme had no detectable exoribonucleolytic activity (Amblar and Arraiano, 2005).

Therefore, the ability to degrade a poly(A) homopolymer is an easy and rapid way to detect RNase II activity in crude extracts and to determine how active an RNase II sample is, either as a mix of proteins or as a purified enzyme.

5.2. Detection of ribonucleolytic activity on mRNA transcripts

The main role of RNase II consists of the exonucleolytic degradation of mRNAs either with or without a poly(A) tail at the 3'-end. Therefore, the best substrate to mimic the *in vivo* function of RNase II is a synthetic mRNA transcript. It is known that RNase II activity is blocked by the presence of double-stranded structures on the RNA molecule (Cannistraro and Kennell, 1999; Coburn and Mackie, 1996b; McLaren *et al.*, 1991; Spickler and Mackie, 2000), and so far various mRNA transcripts harboring stem-loop structures have been tested as RNase II substrates (Coburn and Mackie, 1996a,b, 1998; McLaren *et al.*, 1991; Miczak *et al.*, 1996; Spickler and Mackie, 2000). In all reported cases, the enzyme catalyzes the degradation of the single-stranded portion (ss) of the mRNA molecule from its 3'-end until it reaches the double-stranded (ds) region. Most recently, two different mRNAs have been used as substrates to study the activity of RNase

II and its D209N single mutant (Amblar and Arraiano, 2005): the malE–malF (McLaren et al., 1991) and the SL9A (Spickler and Mackie, 2000) transcripts. Both of them are small transcripts (malE–malF RNA is 375 nt, and SL9A RNA is 83 nt) with one or more stem-loop structures flanked by ss extensions at both ends. The malE–malF transcript contains two stem-loop structures: a large secondary structure formed by two inverted palindromic REP sequences, and a smaller and weaker secondary structure at the 3′-end of the molecule. In the SL9A transcript there is a stem-loop that contains a four-residue loop and a stem of 9 G–C base pairs. Moreover, the 3′-extension of this substrate mimics a typical bacterial poly(A) tail.

The mRNA substrates for these experiments are synthesized by in vitro transcription. For this purpose, the DNA sequence rendering the desired transcript has to be cloned under the control of a promoter specifically recognized by DNA-dependent RNA polymerases. The most common promoters are those derived from Salmonella typhimurium bacteriophage SP6 or from the E. coli bacteriophage T7 or T3, because of their specificity in initiating transcription. In these assays, preparation of the substrate is among the most important steps of the experiment and often takes longer than the activity assay itself. There are several considerations to be taken into account in the preparation of the transcripts. First, the DNA fragment to be transcribed needs to be linearized. The complete cleavage of the superhelical plasmid DNA by a restriction enzyme is essential. Small amounts of circular plasmid DNA will dramatically reduce the yield by producing multimeric transcripts. Second, transcription templates with 3′-protruding termini result in the synthesis of significant amounts of RNA molecules that are aberrantly initiated at the termini of the template. Therefore, restriction enzymes that generate protruding 3′-termini should be avoided. Once the template is prepared, the transcription reaction is performed as follows.

5.2.1. In vitro transcription reaction

In our case, SL9A and malE–malF RNA molecules were obtained by in vitro transcription using as templates the pSL9A plasmid linearized with XbaI (McLaren et al., 1991) or the pCH77 plasmid linearized with EcoRI (Spickler and Mackie, 2000). Transcription reactions were performed in a 20-μl volume containing 0.2 μg of DNA template; 10 mM of dithiothreitol; 500 μM of ATP, CTP, and GTP; 100 μM of UTP; 40 U of RNasin; 20 μCi of [α-^{32}P]-dUTP (GE Healthcare); and 10 U of T7 RNA polymerase (Promega). The reaction buffer used was recommended by the manufacturers (Riboprobe kit from Promega). The mixture was incubated at 37 °C for 30 min, after which an extra 10 U of T7 RNA polymerase were added and the mixture was incubated for an additional 30 min. The in vitro transcription reaction was terminated by the addition of 1 μl of 1 mg/ml RNase-free pancreatic DNase I and further incubation at 37 °C for 15 min.

DNase I was inactivated by adding 20 mM of EDTA and reactions were stored at −20 °C.

To analyze the RNase II degradation activity, it is important to have a homogeneous RNA substrate sample in which all mRNA molecules are full-length transcripts. Therefore, the radioactively labeled RNA transcripts need to be purified by polyacrylamide gel electrophoresis. Both SL9A and malE-malF transcripts were loaded in a 6% polyacrylamide/7 M urea gel and, after electrophoresis in 1× Tris-Borate-EDTA (89 mM Tris/borate, 8 mM EDTA, pH 8.5) buffer, the transcripts were audioradiographed, and the bands corresponding to the full-length transcript were cut from the gel. The RNAs were eluted from the gel slice by soaking in 0.5 M of NaAc, 1 mM of EDTA, pH 8, and 2.5% phenol, overnight (or ∼16 h) at room temperature (∼25 °C) with gentle agitation and further centrifugation and ethanol precipitation. Finally, the RNAs were stored at −70 °C until needed.

5.2.2. RNase II activity assay

Cleavage assays were performed at 37 °C in 15 μl of cleavage buffer containing 20 mM of Tris/HCl, pH 8, 2 mM of dithiothreitol, 100 mM of KCl, 0.5 mg/ml of BSA, and 1 mM of MgCl$_2$. The transcripts used as substrates have a strong tendency to form stem-loop structures, and the correct formation of such structures is required before the reaction to analyze the behavior of RNase II. Therefore, to allow formation of secondary structures before the reaction, the RNA substrates (10,000 cpm per reaction) were denatured for 10 min at 90 °C in the Tris component of the assay buffer and allowed to reanneal at 37 °C for 20 min. After reannealing, the other buffer components were added and the reaction was initiated by the addition of 2 nM of His(6)-RNase II or 540 nM of His(6)-RNase II D209N purified enzymes. Samples were withdrawn at different times and quenched in 3 volumes of formamide-containing dye. Reaction products were incubated at 90 °C for 5 min and analyzed on a 6% (w/w) or 8% polyacrylamide/7 M urea gel, for SL9A or malE-malF, respectively. Bands were detected by autoradiography and exoribonucleolytic activity was calculated by quantification of the relative intensities of the bands.

The results obtained with these mRNA transcripts revealed the degradation properties of both the wild-type RNase II and the mutant enzyme (Fig. 8.2). As expected, RNase II was able to rapidly degrade the 3′-ss portion of both transcripts and degradation stopped when encountering a stem-loop structure. The SL9A substrate was degraded in a two-step process (Fig. 8.2A) beginning with a rapid shortening of the RNA molecule from its 3′-end, generating a set of intermediates, followed by the further degradation of these intermediates at slower rates. These intermediates, partially resistant to degradation, presumably correspond to the stem-loop structure with a 3′-ss extension of ∼6–9 nucleotides. At longer reaction times (30 min) the enzyme was able to catalyze the limited digestion of such

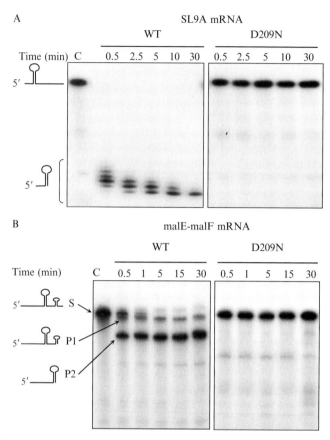

Figure 8.2 Exoribonuclease activity of RNase II and the D209N mutant enzyme on mRNA transcripts. The exoribonuclease activity of His(6)-RNase II (wild-type, WT) or His(6)-RNase IID209N (D209N) was assayed using SL9A mRNA (A) or malE-malF mRNA (B). Reactions were performed as described using 2 nM of WT enzyme or 540 nM of D209N mutant protein. Samples were taken at the time indicated, and the reaction products were analyzed in 8% (A) or 6% (B) polyacrylamide (PAA)/7 M urea gels. Schematic representations of the substrates and reaction products are depicted. Reprinted from Amblar and Arraiano (2005) with permission from Wiley-Blackwell Publishing.

intermediates and they were further converted into shorter molecules. With malE-malF, RNase II showed to be highly active and in only 30 sec 70% of the full-length transcript disappeared (Fig. 8.2B). Degradation of malE-malF generated two main intermediate products, P1 and P2, which were presumably produced by the stalling of the enzyme in the vicinity of the two secondary structures of the mRNA.

The RNase IID209N single mutant was unable to degrade either of these substrates and 100% of the full-length starting material remained intact, even after 30 min of incubation with as much as 540 nM protein (Fig. 8.2). Therefore, with these experiments we were able to confirm that the D209N mutation caused the complete inactivation of *E. coli* RNase II, suggesting an essential role of Asp209 for RNase II activity. These assays are highly sensitive and the substrates used mimic perfectly the actual substrates that RNase II encounters *in vivo*, being therefore, one of the best options to analyze the activity of purified exoribonucleases.

5.3. The usefulness of a set of RNA oligomers

Ribonucleases differ in their degradation properties, showing distinct double-stranded or single-stranded specificity, different behavior toward secondary structures, or diverse requirements regarding RNA compared to DNA. Knowing the mechanism of action of ribonucleases, it is essential to understand how they work *in vivo* and to unravel their cellular function. This is normally a difficult task and the use of oligoribonucleotides for *in vitro* studies has proven to be a powerful tool in this matter. In the case of *E. coli* RNase II, different types of oligonucleotides have been used to study distinct aspects of the enzymatic activity (Amblar *et al.*, 2006, 2007; Cannistraro and Kennell, 1994, 1999; Cheng and Deutscher, 2005). Recently, oligoribonucleotides were used as substrates to characterize the functional domains of RNase II (Amblar *et al.*, 2006).

As mentioned previously, three different domains were initially predicted for *E. coli* RNase II by sequence homology: a CSD domain at the N-terminal end of the protein, a central RNB catalytic domain, and an S1 RNA binding domain at the C-terminus. Both the CSD and the S1 domains are present not only in ribonucleases but also in many other proteins of unrelated function. All of them display a similar folding known as an oligosaccharide/oligonucleotide binding fold (OB-fold) (Murzin, 1993) but lack any primary sequence similarity, and proteins harboring these domains belong to the OB-clan. To characterize the role of each domain of *E. coli* RNase II, we constructed a set of deletion derivatives of RNase II by removing different portions of the protein. Six deletion derivatives were studied: RNB, S1, CSD, ΔS1b, ΔCSDa, and ΔCSDb (See Fig. 8.1) (Amblar *et al.*, 2006). Oligoribonucleotides were used to characterize the function of each domain in the cleavage activity of RNase II. Three synthetic oligonucleotides were designed: a 16-mer and a 30-mer oligoribonucleotide, and the complementary 16-mer oligodeoxyribonucleotide. The 30-mer oligoribonucleotide contains the same sequence as the 16-mer oligoribonucleotide plus an extra 14-nt poly(A) extension at its 3'-end. The combination of these oligonucleotides allows for the generation of three different RNA structures: a perfect double-stranded RNA substrate of 16-mer (16-16ds), a double-stranded RNA with an additional 14-nt single-stranded extension at the 3'-end (16-30ds), and a 30-mer single-stranded oligoribonucleotide

(30ss) (Fig. 8.3A). Both double-stranded substrates are in fact DNA:RNA ds molecules in which only the RNA strand will be degraded by the enzyme, thus facilitating the analysis of the results. In addition, the double-stranded structures formed by these oligonucleotides are quite strong and much more stable than the stem-loops of the transcripts used previously. Therefore, with use of this substrate we ensure that the double-stranded regions will not melt during the reaction, and we know the exact nucleotide in which the structures begin.

A

30ss RNA substrate

5'-CCCGACACCAACCACUAAAAAAAAAAAAAAA-3'

16-30ds RNA substrate (imperfect double strand)

5'-CCCGACACCAACCACUAAAAAAAAAAAAAAA-3'
3'-GGGCTGTCCTTGGTGA-5'

16-16ds RNA substrate (perfect double strand)

5'-CCCGACACCAACCACU-3'
3'-GGGCTGTCCTTGGTGA-5'

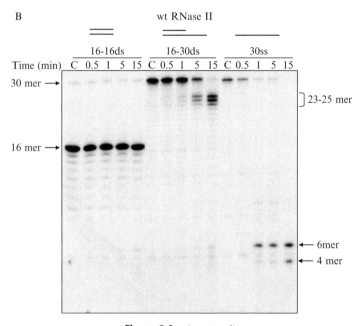

Figure 8.3   (*continued*)

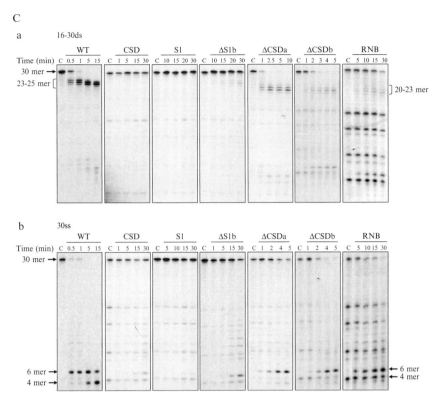

Figure 8.3 Exoribonuclease activity of RNase II using oligoribonucleotides. (A) Synthetic oligonucleotides used. Sequence of the three synthetic oligonucleotides (the 16- and 30-mer oligoribonucleotides, and the complementary 16-mer oligodeoxyribonucleotide) and the three RNA substrates resulting from their combination (16-16ds, 16-30ds, and 30ss) that were used for activity assays. Their schematic representation is depicted on the right. (B) RNase II activity on oligoribonucleotides. Activity assays were performed as described using 30 nM of 16-16ds, 16-30ds, or 30ss as substrates. Reactions were carried out with 1 nM of enzyme and samples were taken at the time points indicated in the figure. Reaction products were analyzed in a 20% PAA/7M urea gel. C–Control reactions (*C*) were incubated for 15 min with no enzyme added. Length of substrates and degradation products are indicated in the figure. (C) Comparison of the exoribonuclease activity of wild-type and deletion mutants using oligoribonucleotides. Activity assays were performed using 30 nM of 16-30ds (a) or 30ss (b) as substrate. Protein concentrations used in panel A were 5 nM of (His)$_6$-tag RNase II (WT), 100 nM of CSD, 1 μM of S1, 500 nM of ΔS1b, 4 nM of ΔCSDa, 0.2 nM of ΔCSDb, and 500 nM of RNB. Protein concentrations used in panel B were 1 nM of (His)$_6$-tag RNase II (WT), 100 nM of CSD, 1 μM of S1, 50 nM of ΔS1b, 4 nM of ΔCSDa, 0.06 nM of ΔCSDb, and 75 nM of RNB. Samples were taken during the reaction at the time points indicated in the figure and reaction products were analyzed in a 20% PAA/7 M urea gel. C–Control reactions with no enzyme added were incubated at the maximum reaction time for each protein. Length of substrates and degradation products are indicated in the figure. Reprinted from Amblar *et al.* (2006) with permission from Elsevier.

Exoribonucleolytic assays with these substrates were performed as follows. First, the 16-mer and the 30-mer oligoribonucleotides were labeled at their 5′-end with $[\gamma\text{-}^{32}P]$-ATP and T4 polynucleotide kinase. The Labeling reaction was carried out in a total final volume of 15 μl, at 37 °C for 1 h, containing 10 pmol of each oligomer. The substrates were then purified using MicroSpin G-25 columns (GE Healthcare) to remove the unincorporated nucleotides. The double-stranded substrates (16-30ds and 16-16ds) were generated by hybridization of the 16-mer and 30-mer to the complementary 16-mer oligodeoxynucleotide. The hybridization was performed in a 1:1 (mol/mol) ratio in 20 mM of Tris/HCl, pH 8 by for 5 min at 68 °C followed by 45 min at room temperature. Exoribonucleolytic reactions were carried out in a final volume of 12.5 μl containing 20 mM of Tris/HCl, pH 8, 100 mM of KCl, and 1 mM of MgCl$_2$. All cases used 30 nM of RNA substrate and the amount of enzyme was adjusted to obtain linear conditions. Reactions were started by the addition of the enzyme and incubated at 37 °C. Samples were withdrawn at different times and the reactions stopped by adding formamide-containing dye supplemented with 10 mM of EDTA. Finally, reaction products were resolved in a 20% (w/v) polyacrylamide/7 M urea gel and analyzed by autoradiography.

Using these RNA substrates, we demonstrated that, as expected, RNase II catalyzed the degradation of the single-stranded but not the double-stranded RNA (Fig. 8.3B). The 30ss substrates was totally degraded, generating an end product of 4–6 nt in length, while the 16-16ds substrate remained untouched. These results support the requirement of a 3′-single-stranded extension for RNase II degradation. With the 16-30ds substrate, RNase II catalyzed the rapid shortening of the 3′-single-stranded portion, generating 23–25 nt fragments as end products, which means that the enzyme stalls 7–9 nt before the double-stranded portion.

The 16-30ds and 30ss substrates were also used to test the activity of the deletion mutants of RNase II, and showed that, although with different efficiencies, almost all proteins were able to degrade both RNA substrates (Fig. 8.3C). Surprisingly, ΔCSDa and ΔCSDb were as efficient in degradation as the wild-type protein or even better, requiring 4 nM and 0.2 or 0.06 nM of protein, respectively, versus 5 nM or 1 nM of wild-type, to completely degrade both RNAs. This means that elimination of the whole CSD domain (ΔCSDb) or part of it (ΔCSDa) does not affect the exonucleolytic activity of RNase II, and even improves its activity significantly. Moreover, the products rendered by both ΔCSD proteins with 16-30ds substrate were slightly smaller than those of the wild-type protein (20–23 nt versus 23–25 nt, respectively), which suggests that the mutants can move closer to the double-stranded region, perhaps because their smaller size results in less steric hindrance. This phenomenon did not occur with the ΔS1b mutant, which showed low activity, detected only with a high concentration of protein (500 nM or 50 nM) and longer incubation times, but produced similar products than the wild-type. Similarly, the RNB mutant protein also required a high enzyme concentration to detect degradation activity of both RNAs (500 nM and 75 nM), but the intermediate products generated with the 16–30ds substrate were 20–23 nt in length, thus behaving as the ΔCSD deletion mutants.

From all these experiments we could extract several important considerations for RNase II enzyme. First, it seems that the S1 domain is highly important for RNase II activity and its contribution to productive RNA binding is much more important than that of the CSD domain. This domain somehow prevents the rapid degradation of RNA by RNase II. This could be highly important to overall mRNA decay in *E. coli*. RNase II is one of the major enzymes involved in mRNA processing and such an enzyme being out of control would probably be deleterious for the cell. If the CSD is, in fact, limiting the action of RNase II *in vivo*, it would be playing an important role working as a brake and thus preventing the massive degradation of RNA.

The use of synthetic oligonucleotides as substrate also enabled us to demonstrate how the double-stranded structure affects the activity of RNase II and its derivative mutants and determine the requirements at the 3'-end for degradation. Almost all enzymatic features could be analyzed using oligomers and the limiting step would probably be our ability to customize the appropriate substrate in each case.

6. Studying the RNA-Binding Abilities

Analysis of RNA-binding ability is an important aspect in the study of RNase II or any other ribonuclease. Investigating its capability to recognize different types of substrates, the nature of the RNA-protein complexes that is able to form, and/or the contribution of every region of the protein on RNA recognition are fundamental parts in the study of an enzyme.

6.1. Electrophoretic mobility shift assay

Among the most commonly used techniques to detect and study protein-nucleic acid interactions is the electrophoretic mobility shift assay (EMSA), also known as band-shift assay or gel-retardation assay. This procedure has shown to be very useful in the study of ribonucleases because it allows for direct visualization of the formation of stable RNA-protein complexes. The mobility shift assay involves the electrophoretic separation of an RNA-protein mixture in a nondenaturing polyacrylamide gel. This approach is very informative with respect not only to the types of RNA-protein complexes (one or more complexes) but also to their stability; however, it is normally quite difficult and must be standardized for each protein and substrate. Moreover, not all RNA-protein complexes can be analyzed with this method, as the complex needs to be stable enough to resist an electrophoresis step, and labile complexes normally result in a smear instead of a band because of dissociation during electrophoresis. Nevertheless, there are several parameters that can be optimized to improve the stability of the complex during gel electrophoresis, such as temperature, addition of divalent metal ions or other cofactors required for binding, and so on. It is also important to reach the equilibrium in the binding reactions prior to loading onto the gel; therefore, temperature and reaction times must be standardized for each particular protein.

In these assays, a control lane is always included in which the RNA probe without protein is added. As a result, the gel will present a single band corresponding to the unbound RNA fragment. In the remaining lanes the binding reactions with increasing amounts of protein are added. This will lead to the detection of retarded bands that represent the larger and less mobile complex as a result of the RNA probe being bound to the protein.

We have tested the RNA-binding ability of purified enzymes, wild-type RNase II, and mutants through band-shift experiments by using as substrates the *in vitro* transcribed mRNAs SL9A and malE-malF described herein. The reactions were carried out in 10 μl containing 2 fmol of the mRNA substrate (10,000 cpm), in a buffer of 100 mM of KCl, 2 mM of dithiothreitol, 20 mM of Tris/HCl, pH 8, and 10% (v/v) glycerol. To obtain the same final concentration of protein per assay, we added appropriate amounts of BSA to each reaction up to 0.1 μg of total protein. Because of the high exoribonucleolytic activity of *E. coli* RNase II enzyme, the binding reactions were supplemented with 10 mM of EDTA to prevent degradation of the substrate during running. Before carrying out the reactions, the mRNA substrates have to be denatured and renatured prior to being added to the mixtures, as described previously. The mixtures containing increasing concentrations of each enzyme were incubated for 10 min at 37 °C to allow formation of the RNA-protein complexes. The reactions were stopped by adding 2 μl of loading buffer containing 30% (v/v) glycerol, 0.25% (v/v) xylene cyanol, and 0.25% (v/v) bromophenol blue, and immediately analyzed in a 5% (w/v) nondenaturing polyacrylamide gel. Electrophoresis was performed with Tris-borate-EDTA at 20 mA and 4 °C. After 5 h of electrophoresis, the gel was fixed by incubation in 7% (v/v) acetic acid for 5 min and further dried. The RNA-protein complexes were subsequently detected with the PhosphorImager system (Molecular Dynamics).

6.1.1. *RNaseIID209N*

This band-shift experiment was used to investigate the effect of D209N mutation in the RNA-binding ability of RNase II. The results obtained with SL9A and malE-malF mRNA transcripts revealed that both wild-type and mutant proteins have a similar binding ability and are able to form stable RNA-protein complexes at 5 nM or 10 nM of protein (Fig. 8.4)

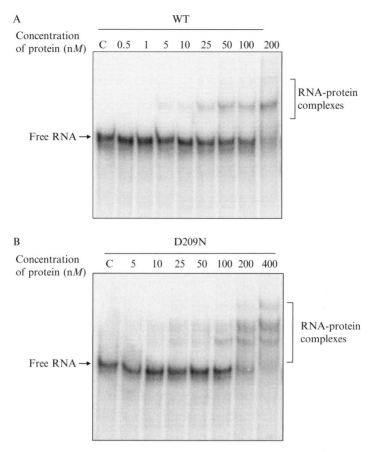

Figure 8.4 Effect of EDTA on RNA binding. malE-malF mRNA transcripts (2 fmol) were incubated at 37 °C with His(6)-RNase II (A) or His(6)-RNase IID209N (B) mutant enzyme in the presence of 10 mM EDTA. The enzyme concentration used is indicated in the figure. A control reaction without enzyme added was performed in all experiments (C). Binding reactions were applied to a 5% nondenaturing PAA gel. Free RNA and RNA-protein complexes were detected and quantified by using the Phosphor Imager system from Molecular Dynamics. Reprinted from Amblar and Arraiano (2005) with permission from Wiley-Blackwell Publishing.

(Amblar and Arraiano, 2005). In this case, EMSA was also used to determine the dissociation constants (K_D) of both proteins by quantification of the bands corresponding to the free and protein-bound RNA. Determination of K_D from a band-shift assay has several requirements that need to be taken into account. First, RNA-protein complexes have to be quite stable to give defined and quantifiable bands; in addition, it is necessary to perform binding reactions using an excess of protein versus RNA so that most of the protein molecules are free. Under these conditions, it may be assumed that the concentration of unbound protein is equivalent to the total protein concentration used in the reaction.

Therefore, the K_D of wild-type RNase II and D209N were estimated from the gel by quantification of the bands using ImageQuant software (Molecular Dynamics). The values obtained for the RNA-protein complex (C), free RNA (R), and free protein concentration (equivalent to the total protein added to the reaction in our conditions) [P], were plotted using the Hill representation ($\log([C]/([R] - [C])$) versus $\log[P]$). The Hill coefficient n for the protein tested ranged from 1.02 to 1.06. The apparent K_D (K) of wild-type and of D209N proteins was calculated from the equation $\log[K]$ ¼ $n \log[P] - \log([C] / ([R] - [C]))$, as described previously (Amblar *et al.*, 2001; Amblar and Arraiano, 2005). The K_D values obtained with malE-malF transcript were similar for wild-type RNase II and D209N mutant (382 nM and 344 nM, respectively) (Amblar and Arraiano, 2005), indicating that both enzymes have similar affinities for the substrate. Results showed that the lack of activity of the D209N mutant enzyme is due to a deficiency in the catalytic reaction but not in RNA binding, pointing to an essential role of the Asp209 in the reaction mechanism of RNase II.

6.1.1. RNase II deletion derivatives

EMSAs were also used to investigate the role of each putative domain of RNase II in RNA binding by comparing the abilities of the deletion derivatives and the wild-type protein to bind to malE-malF and SL9A transcripts (Fig. 8.5). We determined that the predicted CSD and S1 domains are able to bind RNA, thus confirming their role as RNA-binding domains. In addition, the binding experiments also revealed that elimination of S1 or CSD led to a reduced ability of the enzymes to form stable RNA-protein complexes. The reduction was much more pronounced in the absence of the S1, emphasizing that this domain seems to be more critical than CSD for the RNA binding of RNase II. Unexpectedly, our results showed that the RNB deletion derivative was still able to generate stable RNA-protein complexes, although it lacked the two putative RNA-binding domains.

6.2. Filter-binding assays

As mentioned previously, determination of K_D values from a gel-shift assay is not always possible. Quantification of the gel bands requires a defined band pattern that cannot always be achieved, especially when working with truncated proteins having altered binding abilities. Therefore, it is usually necessary to use other techniques to determine and study of RNA-protein complexes. Filter binding is a good choice because the binding mixture is rapidly filtered through the appropriate membrane, and only the substrate bound to the protein is retained. However, this technique is normally used in studies of DNA substrates rather than RNA substrates. We developed a procedure to determine the K_D values of RNase II and its deletion mutants

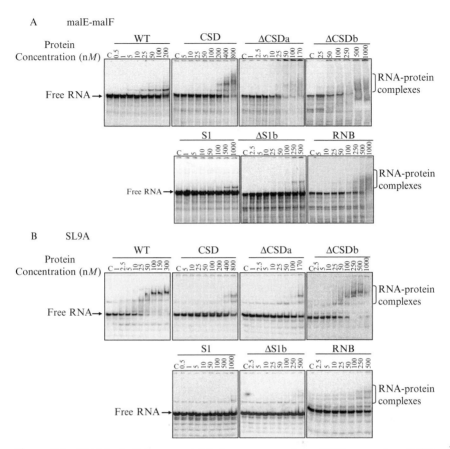

Figure 8.5 EMSA of wild-type and deletion mutants using mRNA transcripts. 10,000 cpm of malE-malF (A) or SL9A (B) were incubated with (His)$_6$-RNase II (WT) or the deletion mutants as indicated under the conditions described in the text. The enzyme concentration used is indicated. A control reaction without enzyme added was performed in all experiments (C). Binding reactions were applied to a 5% nondenaturing PAA gel. Free RNA and RNA-protein complexes was detected using the PhosphorImager (Molecular Dynamics). Reprinted from Amblar *et al.* (2006) with permission from Elsevier.

using ss or ds oligoribonucleotides. This procedure was based on a double-filter method previously described (Wong and Lohman, 1993) and adapted to our RNA substrates. This method uses a dot-blot apparatus with two membranes placed together, one beneath the other, to allow for the simultaneous trapping of both the bound and the free RNA substrate. In our method, the lower membrane was Hybond-N+ instead of the previously reported DEAE, because of the different nature of the substrates used (RNA instead of DNA). The treatment of the membranes was standardized to increase sensitivity and decrease background. The filtration of the binding reactions through the two-filter system allows the measuring of the RNA bound to the protein (as the complexes are retained in the upper nitrocellulose filter) and the free RNA (retained in the Hybond-N+ lower filter). This leads to a more accurate final result and minimizes the experimental errors (Fig. 8.6A).

In these assays, a nitrocellulose filter (Schleider & Schuell) had to be prepared by presoaking in 0.5 M of KOH for 20 min, followed by equilibration in binding buffer without EDTA for a minimum of 1 h before use. The Hybond-N$^+$ membrane (GE Healthcare) also had to be equilibrated in the same buffer prior to use. The binding reactions were performed in 30 μl of binding buffer (100 mM KCl, 20 mM Tris/HCl, pH 8, 2 mM DTT, 10 mM EDTA, and 10% [v/v] glycerol) containing 0.15 fmol of $5'$-[g-^{32}P]-ATP labeled poly(A)$_{35}$, 30-ribomer (hybridized or not hybridized to 16 nt RNA hybridized deoximer oligonucleotide) prepared as described previously. The protein concentrations were initially titrated using serial dilutions to span the desired concentration range for each specific protein. The reactions were incubated for 10 min at 37 °C and stopped by cooling at 4 °C, and the samples were then subjected to UV cross-linking (600 mJ, 10 min, 254 nm at 4 °C) and filtered using a dot-blot apparatus. Each well was flushed with 100 μl of binding buffer at 4 °C just before sample filtering and washed twice with the same buffer immediately after filtering. Nitrocellulose and Hybond-N$^+$ membranes were then dried for 10 min at 50 °C and exposed to a phosphorimaging screen for scanning densitometry.

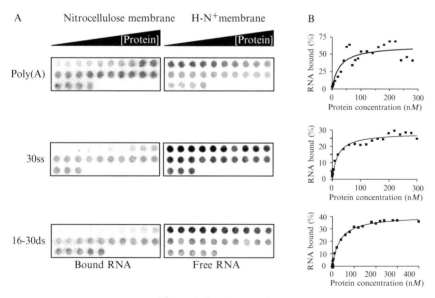

Figure 8.6 (*continued*)

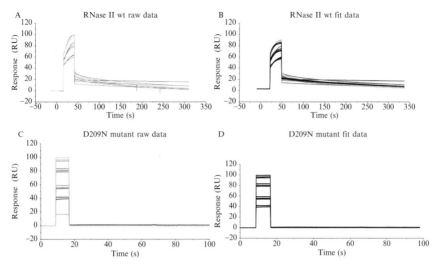

Figure 8.6 Different methods to determine the dissociation constants of wild-type RNase II and protein mutants. (A) Double-filter binding assay using oligoribonucleotides. A representative double-filter binding assay is shown. This experiment was carried out using different concentrations of wild-type enzyme, ranging from 0.01 nM to 1 μM, and the following substrates: poly(A), 30ss, and 16-30ds, as indicated. Dot blots obtained with each substrate. Protein concentration increases sequentially from the left to the right and from the top to the bottom. The RNA bound to the protein is retained in the nitrocellulose membrane (on the left) and the free RNA is trapped in the Hybond-N$^+$ (on the right). (B) Graphical representation of the filter-binding assays. From top to bottom: poly(A) substrate, 30ss substrate, and 16-30ds substrate. Reprinted from Amblar *et al.* (2006) with permission from Elsevier. (C) Determination of the dissociation constants (K_D) using surface plasmon resonance analysis: BIACORE 2000. The binding of wild-type RNase II and D209N mutant to 25-mer RNA substrate is shown. Panels on the left represent raw data and on the right fitted data. The concentrations assayed were 10 nM, 20 nM, 30 nM, 40 nM, and 50 nM, and all experiments included triple injections of each protein concentration to determine the reproducibility of the signal and control injections to assess the stability of the RNA surface during the experiment. Darker lines represent a global fit of each data set to a single site-interaction model. Rate constants and equilibrium constants were calculated using the BIA EVALUATION 3.0 software package, according to the fitting model 1:1 Languimir Binding.

Radiolabeled oligonucleotides retained within each dot were quantified with ImageQuant software (Molecular Dynamics). Equilibrium K_D values were obtained by nonlinear least-squares fitting of the data using graph-pad software (Fig. 8.6B).

Results obtained using the filter-binding assays revealed that wild-type RNase II has a similar K_D for the three substrates tested (Table 8.2). This indicates that the different efficiencies of degradation of single-stranded versus double-stranded RNAs is not related to the affinity of the enzyme for these substrates but to the catalytic properties of the enzyme.

Table 8.2 K_D values of wild-type and deletion derivatives for different RNA substrates. Dissociation constants were measured by using the double-filter binding assay as described in the text. Each value is the average of at least two independent experiments. ND, not detected. Reprinted from Amblar *et al.* (2006) with permission from Elsevier.

Enzyme	K_D for poly(A) (nM)	K_D for 30ss (nM)	K_D for 16–30ds (nM)
RNase II	47.75 ± 13.37	30.57 ± 14.44	37.70 ± 7.62
CSD	475.70 ± 112.70	333.80 ± 73.00	162.20 ± 50.74
ΔCSDa	2.73 ± 1.71	—	—
ΔCSDb	0.64 ± 0.08	40.49 ± 6.40	50.68 ± 21.43
S1	2493.00 ± 1029.00	ND	ND
ΔS1b	94.46 ± 58.03	423.20 ± 213.40	474.60 ± 66.80
RNB	1994.00 ± 344.40	1225.00 ± 630.70	240.50 ± 32.15

The affinity of the S1 deletion mutant relative to the wild-type protein was much lower, manifesting an ~10-fold increase in K_D when using the 30ss and 16-30ds substrates. However, the difference in affinity between the two proteins was not as drastic when using the poly(A) oligoribonucleotide as a substrate, as the K_D obtained using the S1 deletion mutant compared to the wild-type protein was only 2- to 3-fold greater. The S1 domain on its own presented a very low RNA-binding affinity, and it was only possible to estimate the K_D with poly(A) substrate. The most surprising result was obtained on elimination of the CSD, where a higher binding affinity for poly(A) substrate was obtained when compared to the wild-type. The K_D values obtained for the ΔCSDa and ΔCSDb proteins were 2.7 and 0.6 nM, respectively, versus 47.7 nM for the wild-type protein. However, the affinity of ΔCSDb was not affected when testing the other substrates. In agreement, the CSD domain on its own showed a lower binding affinity by poly(A) oligoribonucleotide (K_D = 475 nM) than by the other two substrates (K_D = 333.8 nM and 162.2 nM, respectively). This result led us to hypothesize that the low affinity for poly(A) tails presented by CSD may somehow prevent the tight binding of RNase II to this substrate. Moreover, the results obtained for the RNB mutant confirmed the previous EMSA analysis, revealing that this truncated protein was still able to bind different RNA substrates, though with a very low affinity.

Additional secondary-structure predictions of the RNB mutant revealed a region at its N-terminus with a folding similar to the CSD, thus indicating the presence of an extra RNA-binding domain (Amblar *et al.*, 2006). The presence of this new CSD domain (CSD2), which lacks the typical sequence motifs RNPI and RNPII (Theobald *et al.*, 2003), was later proven upon resolution of the RNase II structure (Frazão *et al.*, 2006).

6.3. BIACORE: Surface plasmon resonance analysis

Real-time binding data are essential to understand the dynamic interactions between proteins and other biomolecules (e.g., RNA) that drive and regulate biological processes. RNA-protein complexes are dynamic, and the kinetics of binding and release can influence such processes. By using BIACORE, the surface plasmon resonance analysis allows intermolecular interactions to be measured in real time and can provide both equilibrium and kinetic information about complex formation. This technology is a powerful tool that allows for the study of the dynamics of RNA-protein interactions (Park *et al.*, 2000).

There are, however, some differences that can be detected when comparing results obtained using EMSA or filter binding with those obtained using surface plasmon resonance analysis. Although association and dissociation are observed in real time when BIACORE is used, the equilibrium measurements obtained by the other two techniques rely on the maintenance of the intact complex, either during gel electrophoresis, where the complex might (partially) dissociate, or during filtering. In both cases, dissociation of the complex would lead to an underestimation of the substrate affinity.

For the BIACORE assays, the BIACORE SA chips were obtained from BIACORE Inc. (GE Healthcare). The flow cells of the SA streptavidin sensor chip were coated with a low concentration of the substrates. No substrate was added to flow cell 1, which served as a blank control. A 5'-biotinylated 25-nucleotide RNA (5'-CCC GAC ACC AAC CAC UAA AAA AAA A-3') was added ($\sim$2000 response units) to flow cell 2 for studying the protein-RNA oligonucleotide interaction. The target RNA substrate was captured on flow cell 2 by manually injecting 20 μl of a 500 nM solution of the RNA in 1 M NaCl at a 10 μl/min flow rate. The biosensor assay was run at 4 °C in the same buffer used for the EMSAs. The proteins were injected over flow cells 1 and 2 for 2 min at concentrations of 10, 20, 30, 40, and 50 nM using a flow rate of 20 μl/min.

All experiments included triple injections of each protein concentration to determine the reproducibility of the signal and control injections to assess the stability of the RNA surface during the experiment. Bound protein was removed using a 60 sec wash with 2 M of NaCl, which did not damage the substrate surface.

Data from flow cell 1 were used to correct for refractive index changes and nonspecific binding. Rate constants and equilibrium constants were calculated using the BIA EVALUATION 3.0 software package, according to the fitting model 1:1 Languimir Binding. This method was used to determine the K_D of the wild-type RNase II and several single mutants, such as the RNase II D209N enzyme (unpublished results). According to previous experiments, the results obtained by SPR indicated that wild-type RNase II and the D209N mutant have similar affinities for the RNA substrate analyzed, as the K_D values obtained were, respectively, 6.4 nM and 5.3 nM (Fig. 8.6C).

Therefore, three different techniques were used to analyze the RNA-binding ability of *E. coli* RNase II and its derivatives, and all provided valuable information, either qualitative or quantitative. Regarding the dissociation constants, the three approaches may be useful to estimate the K_D values. However, it is important to bear in mind that these K_D values are always apparent and never can be taken as absolute values measuring the actual affinity of the protein. Each method is affected by distinct parameters, and K_D values obtained through different techniques can never be compared. The key therefore resides in choosing the appropriate method for each particular case and using it to compare a protein of interest and its mutated derivatives.

7. STRUCTURAL STUDIES OF RNASE II

The combination of certain experimental approaches together with computational analysis made it possible to uncover some architectural features that may not be distinguished by sequence inspection (Amblar *et al.*, 2006). For instance, the analysis of RNase II RNB truncated mutant showed that it had characteristics of an RNA-binding domain. In fact, a deep characterization of an enzyme usually requires the determination of its three-dimensional structure. Knowledge of the structure of an enzyme, together with the experimental data, is extremely helpful when unraveling mechanisms of substrate binding and catalysis, as has been shown previously (Barbas *et al.*, 2008). This was the case for *E. coli* RNase II. The three-dimensional structure of the native RNase II and its RNA-bound complex were recently solved by X-ray crystallography (Frazão *et al.*, 2006; McVey *et al.*, 2006) (Fig. 8.7). The first important finding that arose from the structure was that, contrary to the sequence predictions, as noted above, RNase II harbors CSD2 in addition to CSD1. This result fits perfectly with our initial proposal based on sequence analysis and secondary-structure predictions that pointed to the presence of a putative OB-fold similar to the other CSD within the N-terminal region of the RNB domain (Amblar *et al.*, 2006).

In addition, the resolution of RNase II in complex with RNA allowed us to postulate a model for RNase II mediated-degradation based on the RNA-protein interactions. Thus, we were able to explain the molecular basis of

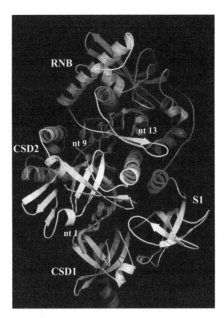

Figure 8.7 RNase II 3D structure in complex with the RNA substrate. Three-dimensional model of *E. coli* RNase II structure (Frazão *et al.*, 2006) (green to blue) bound to a 13-mer poly(A) chain (red). Labels indicate domains and nucleotides. Figure kindly provided by Dr. Nuno Micaelo from ITQB Portugal. (See Color Insert.)

several important features of RNase II, such as its specificity for binding RNA and not DNA, the processivity of RNA degradation, RNA translocation, and the release of 4 nts as a final end product. However, far from being an end point in the characterization of this enzyme, the resolution of a three-dimensional structure always opens many new and exciting questions.

The work summarized in this chapter shows how the synergy between structural and biochemical analysis may help understand the activity of an enzyme *in vitro*, which in turn, is necessary to explain its behaviour *in vivo*. Moreover, these techniques and results can be applied to the decay mechanisms of other ribonucleases, including all RNase II family members that have revealed a similar mode of action, namely those present in the exosome.

ACKNOWLEDGMENTS

We thank Rute Matos for the critical reading of this manuscript. We thank Gonçalo da Costa for advice on the surface plasmon resonance analysis (BIACORE). The authors would also like to thank Nuno Micaelo for the figure of the RNase II structure. Ana Barbas was a recipient of a Ph.D. fellowship and Mónica Amblar was a recipient of a postdoctoral fellowship, both funded by Fundação para a Ciência e a Tecnologia, Portugal. The work at the ITQB was supported by Fundação para a Ciência e a Tecnologia, Portugal.

REFERENCES

Amblar, M., and Arraiano, C. M. (2005). A single mutation in *Escherichia coli* ribonuclease II inactivates the enzyme without affecting RNA binding. *FEBS J.* **272,** 363–374.

Amblar, M., Barbas, A., Fialho, A. M., and Arraiano, C. M. (2006). Characterization of the functional domains of *Escherichia coli* RNase II. *J. Mol. Biol.* **360,** 921–933.

Amblar, M., Barbas, A., Gomez-Puertas, P., and Arraiano, C. M. (2007). The role of the S1 domain in exoribonucleolytic activity: Substrate specificity and multimerization. *RNA* **13,** 317–327.

Amblar, M., Lacoba, M. G., Corrales, M. A., and López, P. (2001). Biochemical analysis of point mutations in the 5′ to 3′-exonuclease of DNA polymerase I of *Streptococcus pneumoniae*. Functional and structural implications. *J. Biol. Chem.* **276,** 19172–19181.

Amblar, M., and López, P. (1998). Purification and properties of the 5′ to 3′-exonuclease D190 → A mutant of DNA polymerase I from *Streptococcus pneumoniae*. *Eur. J. Biochem.* **252,** 124–132.

Andrade, J. M., Cairrão, F., and Arraiano, C. M. (2006). RNase R affects gene expression in stationary phase: Regulation of *ompA*. *Mol. Microbiol.* **60,** 219–228.

Andrade J. M., Pobre, V., Silva, I. J., Domingues, S., and Arraiano C. M. (2008). The role of 3′ to 5′-exonucleases in RNA degradation. *Progress in Nucl. Acids Res. and Mole. Biol. Review*, in press.

Arraiano, C. M., Bamford, J., Brussow, H., Carpousis, A. J., Pelicic, V., Pfluger, K., Polard, P., and Vogel, J. (2007). Recent advances in the expression, evolution, and dynamics of prokaryotic genomes. *J. Bacteriol.* **189,** 6093–6100.

Arraiano, C. M., and Maquat, L. E. (2003). Post-transcriptional control of gene expression: Effectors of mRNA decay. *Mol. Microbiol.* **49,** 267–276.

Arraiano, C. M., Yancey, S. D., and Kushner, S. R. (1988). Stabilization of discrete mRNA breakdown products in *ams pnp rnb* multiple mutants of *Escherichia coli* K-12. *J. Bacteriol.* **170,** 4625–4633.

Barbas, A., Matos, R. G., Amblar, M., López-Viñas, E., Gomez-Puertas, P., and Arraiano, C. M. (2008). New insights into the mechanism of RNA degradation by ribonuclease II: Identification of the residue responsible for setting the RNase II end-product. *J. Biol. Chem.* **283,** 13070-13076.

Cairrão, F., Chora, A., Zilhão, R., Carpousis, J., and Arraiano, C. M. (2001). RNase II levels change according to the growth conditions: Characterization of *gmr*, a new *Escherichia coli* gene involved in the modulation of RNase II. *Mol. Microbiol.* **276,** 19172–19181.

Cairrão, F., Cruz, A., Mori, H., and Arraiano, C. M. (2003). Cold shock induction of RNase R and its role in the maturation of the quality control mediator SsrA/tmRNA. *Mol. Microbiol.* **50,** 1349–1360.

Cannistraro, V. J., and Kennell, D. (1994). The processive reaction mechanism of ribonuclease II. *J. Mol. Biol.* **243,** 930–943.

Cannistraro, V. J., and Kennell, V. J. (1999). The reaction mechanism of ribonuclease II and its interaction with nucleic acid secondary structures. *Biochem. et Biophys. Acta* **1433,** 170–187.

Cheng, Z. F., and Deutscher, M. P. (2005). An important role for RNase R in mRNA decay. *Mol. Cell* **17,** 313–318.

Coburn, G. A., and Mackie, G. A. (1996a). Differential sensitivities of portions of the mRNA for ribosomal protein S20 to 3′-exonucleases dependent on oligoadenylation and RNA secondary structure. *J. Biol. Chem.* **271,** 15776–15781.

Coburn, G. A., and Mackie, G. A. (1996b). Overexpression, purification and properties of *Escherichia coli* ribonuclease II. *J. Biol. Chem.* **271,** 1048–1053.

Coburn, G. A., and Mackie, G. A. (1998). Reconstitution of the degradation of mRNA for ribosomal protein S20 with purified enzymes. *J. Mol. Biol.* **279,** 1061–1074.

Deutscher, M. P., and Reuven, N. B. (1991). Enzymatic basis for hydrolytic versus phosphorolytic mRNA degradation in *Escherichia coli* and *Bacillus subtilis*. *Proc. Natl. Acad. Sci. USA* **88,** 3277–3280.

Donovan, W. P., and Kushner, S. R. (1983). Amplification of ribonuclease II (*rnb*) activity in *Escherichia coli* K-12. *Nucl. Acids Res.* **11,** 265–275.

Donovan, W. P., and Kushner, W. P. (1986). Polynucleotide phosphorylase and ribonuclease II are required for cell viability and mRNA turnover in *Escherichia coli* K-12. *Proc. Natl. Acad. Sci. USA* **83,** 120–124.

Dreyfus, M., and Régnier, P. (2002). The poly(A) tail of mRNAs: Bodyguard in eukaryotes, scavenger in bacteria. *Cell* **111,** 611–613.

Dziembowski, A., Lorentzen, E., Conti, E., and Seraphin, B. (2007). A single subunit, Dis3, is essentially responsible for yeast exosome core activity. *Nat. Struct. Mol. Biol.* **14,** 15–22.

Frazão, C., McVey, C. E., Amblar, M., Barbas, A., Vonrhein, C., Arraiano, C. M., and Carrondo, M. A. (2006). Unravelling the dynamics of RNA degradation by ribonuclease II and its RNA-bound complex. *Nature* **443,** 110–114.

Grossman, D., and van Hoof, A. (2006). RNase II structure completes group portrait of 3′-exoribonucleases. *Nat. Struct. Mol. Biol.* **13,** 760–761.

Higuchi, R. (1990). Recombinant PCR. *In* "PCR Protocols. A Guide to Methods and Applications" (M. A. Innis, D. H. Gelfand, J. J. Sninsky, and T. J. White, eds.) pp. 177–183. Academic Press, San Diego.

Kasai, T., Gupta, R. S., and Schlessinger, D. (1977). Exoribonucleases in wild-type *Escherichia coli* and RNase II-deficient mutants. *J. Biol. Chem.* **252,** 8950–8956.

LaCava, J., Houseley, J., Saveanu, C., Petfalski, E., Thompson, E., Jacquier, A., and Tollervey, D. (2005). RNA degradation by the exosome is promoted by a nuclear polyadenylation complex. *Cell* **121,** 713–724.

Liu, Q., Greimann, J. C., and Lima, C. D. (2006). Reconstitution, activities, and structure of the eukaryotic RNA exosome. *Cell* **127,** 1223–1237.

Marujo, P. E., Hajnsdorf, E., Le Derout, J., Andrade, R., Arraiano, C. M., and Regnier, P. (2000). RNase II removes the oligo(A) tails that destabilize the *rps*O mRNA of *Escherichia coli*. *RNA* **6,** 1185–1193.

McLaren, R. S., Newbury, S. F., Dance, G. S. C., Causton, H. C., and Higgins, C. F. (1991). mRNA degradation by processive 3′ to 5′-exoribonucleases *in vitro* and the implications for prokaryotic mRNA decay *in vivo*. *J. Mol. Biol.* **221,** 81–95.

McVey, C. E., Amblar, M., Barbas, A., Cairrão, F., Coelho, R., Romão, C., Arraiano, C. M., Carrondo, M. A., and Frazão, C. (2006). Expression, purification, crystallization and preliminary diffraction data characterization of *Escherichia coli* ribonuclease II (RNase II). *Acta Crystallogr. F* **62,** 684–687.

Mian, I. S. (1997). Comparative sequence analysis of ribonucleases HII, III, II, PH and D. *Nucl. Acid Res.* **25,** 3187–3195.

Miczak, A., Kaberdin, V. R., Wei, C. L., and Lin-Chao, S. (1996). Proteins associated with RNase E in a multicomponent ribonucleolytic complex. *Proc. Nat. Acad. Sci. USA* **93,** 3865–3869.

Mitchell, P., Petfalski, E., Shevchenko, A., Mann, M., and Tollervey, D. (1997). The exosome: A conserved eukaryotic RNA processing complex containing multiple 3′ to 5′-exoribonucleases. *Cell* **91,** 457–466.

Murzin, A. G. (1993). OB(oligonucleotide/oligosaccharide binding)-fold: Common structural and functional for non-homologous sequences. *EMBO J.* **12,** 861–867.

Nikolaev, N., Folsom, V., and Schlessinger, D. (1976). *Escherichia coli* mutants deficient in exoribonucleases. *Biochem. Biophys. Res. Commun.* **70,** 920–924.

Park, S., Myszka, D. G., Yu, M., Littler, S. J., and Laird-Offringa, I. A. (2000). HuD RNA recognition motifs play distinct roles in the formation of a stable complex with AU-rich RNA. *Mol. Cell Biol.* **20,** 4765–4772.

Piedade, J., Zilhão, R., and Arraiano, C. M. (1995). Construction and characterization of an absolute deletion mutant of *Escherichia coli* ribonuclease II. *FEMS Microbiol. Lett.* **127,** 187–194.

Régnier, P., and Arraiano, C. M. (2000). Degradation of mRNA in bacteria: Emergence of ubiquitous features. *Bioessays* **22,** 235–244.

Schneider, C., Anderson, J. T., and Tollervey, D. (2007). The exosome subunit Rrp44 plays a direct role in RNA substrate recognition. *Mol. Cell* **27,** 324–331.

Spickler, C., and Mackie, A. (2000). Action of RNases II and polynucleotide phosphorylase against RNAs containing stem-loops of defined structure. *J. Bacteriol.* **182,** 2422–2427.

Theobald, D. L., Mitton-Fry, R. M., and Wuttke, D. S. (2003). Nucleic acid recognition by OB-fold proteins. *Annu. Rev. Biophys. Biomol. Struct.* **32,** 115–133.

Wong, I., and Lohman, T. M. (1993). A double-filter method for nitrocellulose-filter binding: Application to protein-nucleic acid interactions. *Proc. Nat. Acad. Sci. USA* **90,** 5428–5432.

Zilhão, R., Camelo, L., and Arraiano, C. M. (1993). DNA sequencing and expression of the gene *rnb* encoding *Escherichia coli* ribonuclease II. *Mol. Microbiol.* **8,** 43–51.

Zilhão, R., Cairrão, F., Régnier, P., and Arraiano, C. M. (1996a). PNPase modulates RNase II expression in *Escherichia coli*: Implications for mRNA decay and cell metabolism. *Mol. Microbiol.* **20,** 1033–1042.

Zilhão, R., Plumbridge, J., Hajnsdorf, E., Régnier, P., and Arraiano, C. M. (1996b). *Escherichia coli* RNase II: Characterization of the promoters involved in the transcription of *rnb*. *Microbiology* **142,** 367–375.

Zuo, Y., and Deutscher, M. P. (2001). Survey and summary. Exoribonuclease superfamilies: Structural analysis and phylogenetic distribution. *Nucl. Acid Res.* **209,** 1017–1026.

Zuo, Y., Vincent, H. A., Zhang, J., Wang, Y., Deutscher, M. P., and Malhotra, A. (2006). Structural basis for processivity and single-strand specificity of RNase II. *Mol. Cell* **24,** 149–156.

CHAPTER NINE

The Role of RNA Chaperone Hfq in Poly(A) Metabolism: Methods to Determine Positions, Abundance, and Lengths of Short Oligo(A) Tails

Philippe Régnier *and* Eliane Hajnsdorf

Contents

Abstract

Polyadenylation is a posttranscriptional modification of RNA occurring in pro-karyotes, eukaryotes, and organelles. Long poly(A) tails help export eukaryotic mRNAs and promote mRNA stability and translation, whereas the short

CNRS UPR9073; Institut de Biologie Physico-Chimique; Université Paris Diderot, Paris7, Paris, France

Methods in Enzymology, Volume 447   © 2008 Elsevier Inc.
ISSN 0076-6879, DOI: 10.1016/S0076-6879(08)02209-X   All rights reserved.

bacterial tails facilitate RNA decay. The scarcity of polyadenylated RNAs is one of the obstacles for investigators studying bacterial polyadenylation.

The two methods described in this chapter were developed to determine how the poly(A) binding protein Hfq affects the polyadenylation of bacterial RNAs. The first is a 3′-RACE protocol specific to oligoadenylated RNA. This method was designed to rapidly collect a large amount of poly(A) containing 3′-terminal sequences to perform statistical analysis.

The second method is an RNA sizing protocol to analyze the polyadenylation status of primary transcripts that were not efficiently detected by 3′-RACE. The latter procedure is based on Northern blot analysis of 3′-RNA fragments generated by RNase H. In the presence of a gene-specific methylated chimeric RNA-DNA oligonucleotide, the enzyme is directed to a unique cleavage site. The 3′-RNA fragments, differing by just one nucleotide at their 3′-ends, are then separated in polyacrylamide gels.

1. INTRODUCTION

Polyadenylation is a posttranscriptional modification of RNA that occurs in prokaryotes, eukaryotes, and organelles. In eukaryotes, poly(A) tails contribute to the export to the cytoplasm and promote mRNA stability and translation. Nearly all functional eukaryotic mRNAs harbor long poly (A) tails (approximately 60 nucleotides in length in yeast or 250 nucleotides in mammals). The function and extent of bacterial polyadenylation are in marked contrast to those of eukaryotic mRNAs (Dreyfus and Régnier, 2002). In bacteria, it is usual to find tails containing only one to three A residues (Briani *et al.*, 2002; Hajnsdorf *et al.*, 1995; Le Derout *et al.*, 2003; Li *et al.*, 1998; van Meerten *et al.*, 1999; Xu *et al.*, 1993). In fact, a small fraction of bacterial RNAs has short oligo(A) tails that promote RNA degradation. A similar poly(A)-dependent RNA degradation mechanism has been found in organelles and nuclei. The destabilizing function of poly (A) is now considered to be a ubiquitous mechanism involved in quality control and the degradation of RNA (Anderson, 2005).

In *Escherichia coli*, poly(A) polymerase (PAP I) adds poly(A) extensions to the 3′-ends of many types of RNA, such as mRNA, precursor and mature forms of rRNAs and tRNAs, regulatory RNAs, and degradation intermediates. Models of poly(A)-dependent decay postulate that oligo(A) tails containing just a few As can very rapidly become the target of exoribonucleases, which may either shorten the tails, regenerate tail-less RNAs, or completely degrade the RNA (Folichon *et al.*, 2005). A likely explanation as to why poly(A) tails are scarce and short is provided by the opposing activities of PAP I and exoribonucleases. Polyadenylation facilitates the exonucleolytic degradation of RNA molecules that are protected from exonucleases by stable stem-loop structures. In this case, it is believed that poly(A) extensions offer an RNA toehold for polynucleotide phosphorylase (PNPase) and RNase R to bind and initiate their exonucleolytic degradation (Régnier and Arraiano, 2000; Vincent and Deutscher, 2006). Polyadenylation often affects the degradation of the fragments generated by endonucleases. It is probable that polyadenylation also facilitates the degradation of nonstructured RNA. RNase II fails to degrade secondary structures that protect the messenger but removes the poly(A) tails that facilitate the exoribonucleolytic activities of PNPase and RNase II. This paradox explains quite nicely why RNase II behaves like an RNA stabilizing factor (Marujo *et al.*, 2000). It is generally accepted that repeated cycles of poly(A) addition and exonuclease digestion are needed to overcome the resistance of structured RNA to exonucleolytic decay. Polyadenylation is also focused in other chapters in this volume (namely Chapters 1, 18, 21, 24 [Mohanty/ Kushner; Aiba *et al.* Holec/Gagliardi; Slomovic/Schuster] and see also chapters in RNA Turnover, Part B, ed. Maquat and Kiledjian).

In eukaryotic cells, the activity of poly(A) polymerase depends on many cofactors, including a poly(A) binding protein. In *E. coli* polyadenylation is modulated by the Hfq RNA-binding protein that preferentially interacts with A-rich regions of RNAs (Folichon *et al.*, 2003; Hajnsdorf and Régnier, 2000). Hfq was originally discovered as a host subunit of the RNA replicase of phage Qβ. Hfq has been demonstrated to be involved in other metabolic process, many of which involve small non coding RNAs. (ncRNAs) (Gottesman, 2004) (see also Chapter 18 by Aiba on the role of Hfq on small RNAs). It has been reported to facilitate base pairing between ncRNAs and their mRNA targets and to modify the conformation of RNA affecting its stability. Hfq belongs to the LSm family of RNA binding proteins that are also found in *eukarya* and *archaea*, (Arluison *et al.*, 2002; Moller *et al.*, 2002; Zhang *et al.*, 2002).

This chapter presents methods that can be used to study bacterial polyadenylation and describes our experimental approach to (1) determine the polyadenylation status of the small fraction of a bacterial mRNA harboring short oligo(A) tails, and (2) investigate how regulatory factors such as Hfq can affect polyadenylation.

2. GLOBAL ANALYSIS OF POLYADENYLATED TRANSCRIPTS

Many of the methods available to study eukaryotic poly(A) mRNAs cannot be used to characterize bacterial mRNAs, whose tails are much shorter and less abundant. For example, it is not possible to compare the electrophoretic mobility of a full-length mRNA to that of tail-less RNA, but selective binding of poly(A)-containing mRNAs on oligo(dT) cellulose has allowed the identification of bacterial polyadenylated RNA in bacteria (Sarkar, 1996). This method was a breakthrough—retains RNA harboring tails longer than 20 As—but it was not capable of capturing most polyadenylated bacterial RNA, whose tails are shorter than 20 As. Degradation of [^{32}P]pCp 3′-labeled RNA by RNase T1 and G, which respectively cleave after Gs and pyrimidines, has been used to detect the poly(A) track. Sizing the radioactive poly(A) tails is important to study how the modification of cellular metabolism (e.g., inactivation or overproduction of poly[A] polymerase and exoribonucleases) could affect the abundance and length of poly(A) tails (O'Hara *et al.*, 1995). Nevertheless, this method cannot precisely measure the abundance and length of the tails. Moreover, the methods outlined in the preceding provide only general information on poly(A)-containing RNAs. None of them allows us to evaluate the amounts of polyadenylated RNA present in comparison to tail-less transcripts or the polyadenylation status of a particular RNA species.

3. RT-PCR–BASED ANALYSIS OF 3′-RNA ENDS

RT-PCR–based methods with oligo(dT) or oligo(dT)-containing primers have been used to identify the polyadenylation sites in several genes. In these approaches, reverse transcription is initiated with primers harboring a 3′-terminal stretch of dT that can hybridize with the oligo(A) tails (Haugel-Nielsen *et al.*, 1996). Then full-length cDNAs initiated at the 3′-end of the poly(A) mRNA are amplified with gene-specific forward primers that hybridize in the messenger in combination with the oligo(dT)-containing RT primer. The mRNA-poly(A) junctions are identified in the sequence of the amplified cDNA fragments. Genome-wide transcriptome analysis with radiolabeled oligo(dT) primed cDNAs has been used to estimate the extent of polyadenylation in *E. coli* (Mohanty and Kushner, 2006).

However, these methods selectively detect poly(A) RNA, and they cannot provide information on the relative amounts of oligoadenylated RNA and tail-less molecules present or on the length and composition of the tails. Indeed, the number of As in amplified cDNA reflects the length of stretch of dT present in the RT primer. Another possible artefact arises from the fact that RT may also begin at an oligo(dT) primer that is base paired either with the nonencoded heterogeneous tails to be found in some bacterial RNAs or with encoded A–rich sequence. All the RT-PCR-based methods described above only allow to identify the 3′-part of polyadenylated transcripts located just upstream of the tail. It is worth mentioning that the cDNA-Smart method (Clonetech) developed to amplify full length polyadenylated RNA also begins by a reverse transcription step initiated with an oligo (dT) primer (Holec *et al.*, 2006).

In contrast, 3′-RACE methods (*rapid amplification of cDNA ends*) provide an overall view of the 3′-ends of RNAs originating from the same gene. In this procedure, all the adenylated and nonadenylated molecules of an RNA preparation are reverse transcribed from an RT primer complementary to an anchor oligonucleotide that has been ligated at the 3′-end. PCR is then carried out with an internal, gene-specific forward primer in combination with the RT primer, thus making it possible to determine the lengths of the tails, the locations of polyadenylation sites, and the proportions of oligo (A)-containing and tail-less RNAs (Le Derout *et al.*, 2003).

A similar method based on the circularization of RNA avoids addition of an alien anchor-oligo (Kuhn and Binder, 2002). In this case the 3′- and 5′-ends of RNA are ligated. Then the 3′ to 5′-junctions of circularized RNA are reverse transcribed from a primer annealed with the 5′-end of the RNA before to be amplified with a couple of gene specific oligonucleotides complementary to the 3′- and 5′-ends of the RNA. By this means, the 5′-end of the molecule can also be located. Real-time PCR has also been used to estimate the relative amounts of polyadenylated full-length *lpp* and *ompA* mRNAs (Mohanty and Kushner, 2006).

3.1. A poly(A) targeted 3′-RACE method to analyze 3′-ends of oligoadenylated RNAs

One of the obstacles faced by investigators studying bacterial polyadenylation is the scarcity of polyadenylated RNA. Indeed, in *E. coli* the proportion of polyadenylated RNA can be lower than 10% in *rpsO* messenger (see later) and as low as 0.011% (in bacteriophage f1 mRNA) (Goodrich and Steege, 1999). The polyadenylated RNA ratios reported for the *lpp* (0.4 to 0.7%) and *ompA* (0.74%) transcripts (Mohanty and Kushner, 2006) suggest that the *rpsO* transcript is more efficiently adenylated. One consequence of this scarcity is that a large number of RNA extremities from a particular transcript must be identified before one can obtain a sufficient sample of polyadenylated extremities for statistical analysis. We, therefore, developed a 3′-RACE method that can quickly collect a large sample of poly(A)-containing 3′-terminal sequences (Le Derout *et al.*, 2003). This is achieved by ligating an anchor oligonucleotide harboring a 5′-GCUU... terminal sequence to create a *Hind*III site (...AAGCTT...) in DNA fragments produced from the amplification of RNAs harboring at least two 3′-terminal As (Fig. 9.1).

Another 3′-RACE method with RT primers elongated of few 3′-terminal Ts can also be used to selectively amplify oligoadenylated mRNAs (Campos-Guillén *et al.*, 2005). Such RT primers should only initiate the reverse transcription of mRNAs harboring a few terminal As upstream of the anchor oligo ligated at their 3′-ends. However, to detect short stretches of oligo(A) in amplification experiments, we preferred to avoid the addition of oligo(dT) sequences associated with the RT-PCR reverse primer.

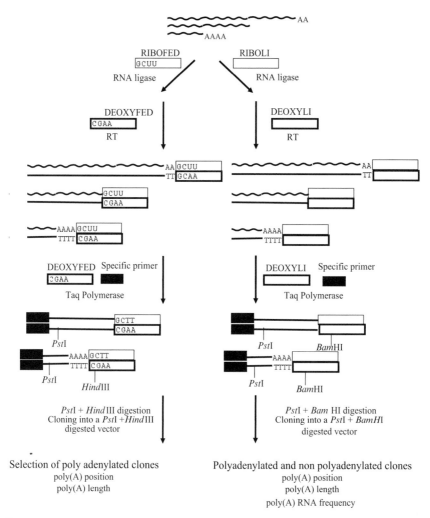

Figure 9.1 Experimental approach for detection and characterization of oligoadeny-lated transcripts. Total-cell RNA containing adenylated and tail-less molecules was tagged with synthetic oligoribonucleotides containing (RIBOFED) or missing (RIBOLI) the GCUU 3′-terminal sequence. The ligation mixture was then reverse transcribed with an oligonucleotide complementary to the tag (DEOXYFED or DEOXYLI) and amplified by PCR with an additional *rpsO*-specific primer. The PCR products were then cloned and sequenced. The ligation of the RIBOFED to RNA molecules ending with at least two As created an *Hind*III restriction site on the PCR products. This allowed us to select for ade-nylated 3′-ends by cloning the amplified DNA after digestion with *Hind*III and *Pst*I in a vector digested with the same enzymes (bottom left of the figure). In contrast digestion of RIBOLI-ligated amplified RNA with *Pst*I and *Bam*HI allowed to detect both adenylated and tail-less RNAs (Bottom right of the figure).

3.2. mRNA, anchors, and primers

The 3′-RACE method described in the following has been developed to analyze the polyadenylation status of the *rpsO* mRNA–coding ribosomal protein S15. We consider that *rpsO* mRNA is the *E. coli* transcript, whose mechanism of degradation is best characterized (Braun *et al.*, 1996; Hajnsdorf *et al.*, 1994, 1995, 1996; Marujo *et al.*, 2003). We have shown that this messenger, which is primarily inactivated by the endoribonuclease RNase E, can also be degraded by exonucleases in a polyadenylation-dependent exonucleolytic pathway that becomes active in the absence of RNase E. The involvement of polyadenylation in RNA stability prompted us to devise a tool to determine the polyadenylation status of the *rpsO* mRNA (Le Derout *et al.*, 2003) and to investigate how the poly(A) binding protein Hfq acts on the synthesis of the tails (Hajnsdorf and Régnier, 2000).

Two anchor oligoribonucleotides were used. The first—referred to as RIBOLI (5′-pUGGUGGU**GGAUCC**CGGGAUC)—contains an internal *Bam*H1 site (indicated in boldface letters) that was used to clone all adenylated and tail-less amplified mRNA 3′-ends (Fig. 9.1). The second one—RIBOFED (5′-p*GCUU*GGUGGU**GGAUCC**CG-3′)—contains the *GCUU* 5′-sequence responsible for creating the *Hind*III *site* (indicated in italics) that allowed us to selectively clone 3′-oligoadenylated RNA containing at least two terminal As. RIBOFED also contains an internal *Bam*H1 site (boldface) that can be used to clone all mRNA extremities.

Reverse transcription was initiated with two oligodeoxyribonucleotides, DEOXYLI (5′-GATCCCG**GGATCC**ACCACCA) and DEOXYFED (5′-CGGGATCCACCACC*AAGC*) that are complementary to RIBOLI and RIBOFED, respectively.

Selective PCR amplification of the *rpsO* cDNAs was performed with a forward primer that hybridizes *rpsO* between positions 202 and 221 upstream of the *Pst*I site that was subsequently used to clone the cDNAs. Amplified cDNAs were then cleaved by *Pst*I and *Bam*HI or *Pst*I and *Hind*III, depending on whether all or only the oligoadenylated *rpsO* mRNA extremities were selected. DNA fragments were then cloned and sequenced (see later). It should be noted that this technique allows the localization of polyadenylation sites in mRNAs that do not harbor *Hind*III sites in the distal part of the gene where polyadenylation is expected to take place. In the case of the *Hind*III-containing gene, the gene-specific restriction site used to clone the amplified DNA must be located downstream of the internal *Hind*III site. Alternately, the gene-specific cloning restriction site can be brought into the amplified DNA as a 5′-part of the forward PCR primer (Briani *et al.*, 2002). Finally, it must be pointed out that this 3′-RACE method, which is designed to detect adenylated mRNAs, cannot identify mRNAs that harbor only one single A added posttranscriptionally.

3.3. Bacterial strains and growth conditions

The *E. coli* strains used in all the experiments that follow contain a series of mutations that affect the genes involved in RNA decay. These include the *rne* gene encoding the endoribonuclease RNase E; the *pnp* and *rnb* genes coding for 3′ to 5′-exoribonucleases polynucleotide phosphorylase (PNPase) and RNase II; the *hfq* gene coding for the RNA binding protein Hfq; and the *pcnB* gene coding for the poly(A) polymerase (PAP I). Strains containing thermosensitive mutations in *rne* and *rnb* were grown at 30 °C and shifted to the nonpermissive temperature (44 °C) to inactivate the corresponding enzymes. To facilitate detection of the encoded message, the 3′-RACE experiments were performed with RNAs extracted from strains harboring the pFB1 high copy number plasmid carrying the *rpsO* locus.

Strains MG1693 *(thyA715 rph1)* (Arraiano *et al.*, 1988), SK5704 *(thyA715 rph1 rne1^{ts} pnp7 rnb500^{ts})* (Arraiano *et al.*, 1988) and IBPC922 *(thyA715 rph1 rne1^{ts} pnp7 rnb500^{ts} hfq1::Ω)* (Le Derout *et al.*, 2003) transformed with the pFB1 plasmid carrying the *rpsO* gene coding ribosomal protein S15 (Hajnsdorf *et al.*, 1995) were used for the 3′-RACE. These strains were designed WT, RNase II^{ts} PNPase-RNase E^{ts} and RNase II^{ts} PNPase-RNase E^{ts} Hfq$^-$. Northern blot analysis of oligoadenylated RNA described at the end of this chapter (Figure 9.6) were performed with a second set of strains harboring deletion alleles of *rnb* and *pcnB*. This include strains IBPC862 *(thyA715 rph1 Δrnb201 ΔpcnB)*, IBPC861 *(thyA715 rph1 rne1 Δrnb201 ΔpcnB)* (Folichon *et al.*, 2005), SK7988 *(thyA715 rph1 ΔpcnB)* (O'Hara *et al.*, 1990) and CMA201 *(thyA715 rph1 Δrnb201)* (Piedade *et al.*, 1995) designed RNase II$^-$ PAP$^-$, RNase II$^-$ PAP$^-$ RNE$^-$, PAP$^-$ and RNase II$^-$.

These strains were grown in LB medium supplemented with thymine (40 μg/ml) because of the *thy*A715 inactive allele of thymidylate synthase, and ampicillin (100 μg/ml) was added when required to maintain the plasmid. Aliquots were prepared either from cells growing exponentially at 37 °C (WT, RNase II$^-$, RNase II$^-$ PAP$^-$, PAP$^-$) or from cells grown at 30 °C and then shifted to 44 °C to inactivate thermosensitive RNase E and/ or RNase II (strains RNase II^{ts} PNPase$^-$ RNase E^{ts}, RNase II^{ts} PNPase$^-$ RNase E^{ts} Hfq$^-$, and RNase II$^-$ PAP$^-$ RNE^{ts}). Measurements of mRNA stability were performed after the inhibition of transcription initiation with rifampicin. The antibiotic was added to the exponentially growing cells to a final concentration of 500 μg/ml at the time of the temperature shift that inactivated the thermosensitive RNase E and RNase II.

3.4. RNA extraction

Ten-milliliter aliquots of the culture were rapidly mixed with an equal volume of ethanol preequilibrated at −70 °C, and cells were pelleted by centrifuging for 10 min at 6000 rpm in the JA20 rotor of a Beckman centrifuge. Bacteria were resuspended in 1.5 ml of cold 10 mM Tris-HCl (pH 7.3), 10 mM potassium chloride, and 5 mM magnesium chloride buffer before mixing with 1.5 ml of 20 mM Tris-HCl (pH 7.9), 200 mM NaCl, 40 mM EDTA, 1% SDS lysis buffer. The cells were lysed for 2 min at 95 °C, and the tubes were then chilled on ice. RNAs were extracted twice at 65 °C with 3 ml of phenol saturated with water and once with chloroform at room temperature. RNAs were then precipitated, washed with ethanol, and dissolved in water. The RNA concentrations were determined from the UV absorbance at 260 nm.

3.5. 3′-RACE amplification

Total-cell RNA prepared as described previously (2.5 μg) was ligated to 100 pmol oligo RIBOLI or RIBOFED dissolved in water, with 20 units of T4 RNA ligase (Promega) in a reaction buffer containing 12.5 mM ATP, 50 mM HEPES (pH 7.5), 20 mM MgCl$_2$, 3.3 mM DTT, 0.01 μg/μl BSA, and 10% DMSO (Donis-Keller, 1979).

After precipitation with ethanol, the pellet was resuspended in 20 μl water. Five μl of the ligation was annealed to 100 pmol DEOXY oligo complementary to the anchor in 10 μl 50 mM Tris-HCl (pH 8.5), 8 mM MgCl$_2$, 30 mM KCl, 100 mM DTT. The cDNA synthesis was performed by incubating the annealing mix with 10 units of AMV reverse transcriptase and 100 mM dNTP each at 42 °C for 1 h. The entire cDNA sample was directly subjected to PCR amplification with 100 pmol of the *rpsO* forward primer without any further addition of DEOXY oligo. PCR reactions were carried out under conditions specified by the manufacturer. The PCR products were digested with *Pst*1 and *Bam*HI or *Hin*dIII and cloned into the vector pT3T718U (Pharmacia) digested with the same enzymes. DH5α cells (Bethesda Research Laboratories, Inc., Gaithersburg, MD) were then transformed with the ligation mixture.

Single colonies were picked up, purified, resuspended in 20 μl H$_2$O and incubated at 95 °C for 5 min; 2 μl were directly subjected to PCR amplification with plasmid-specific primer U (GTAAAACGACGGCCAGT) and primer R (AACAGCTATGACCATG). After desalting on Sephadex G25 (Pharmacia) following the manufacturer's instructions, the PCR products were sequenced.

Polyadenylated clones were counted, considering those clones that possessed extra A residues between the encoded nucleotides and the terminal sequence of the anchor RIBO oligonucleotide. However, the method is not able to differentiate post-transcriptionally added As from encoded terminal A residues. Indeed, a …NNAAGCTT…mRNA-anchor junction (N = G, C, or T), where the NNAA sequence is encoded, could correspond either to a nonadenylated mRNA or to the posttranscriptional addition of one or two As. If we assume that these two As are encoded, this suggests that we are underestimating the number of polyadenylation sites and the size of the poly(A) tails.

4. Effect of Hfq on Polyadenylation of the *RPSO* Transcript

Approximately 10% of *rps*O mRNAs harbor an oligo(A) tail in wild-type cells. We used the 3′-RACE method to analyze the 3′-ends of this particular mRNA species in wild-type cells *in vivo* (Le Derout *et al.*, 2003) to determine the precise fraction of *rps*O transcripts that harbor poly(A) extensions and the length of these tails. We end-ligated the RIBOLI anchor nucleotide to RNA and initiated reverse transcription with the DEOXYLI primer complementary to RIBOLI as described above (Fig 9.6). PCR amplification was then performed with DEOXYLI and an *rps*O forward primer. The PCR fragments expected to reflect the polyadenylation status of the *rps*O transcripts were cloned and sequenced. Only four of them (10%) were found to harbor a very short oligo(A) extension ranging from one to three As (Fig. 9.2A, square symbols). Moreover, most of the 39 3′-ends analyzed by this method corresponded to transcripts lacking the transcription terminator hairpin (t1).

4.1. Locations of polyadenylation sites and lengths of the tails

We then used the oligo(A)-targeted 3′-RACE protocol to select for mRNA containing at least two consecutive As at their 3′-ends (Le Derout *et al.*, 2003). For this purpose, we ligated the RIBOFED oligoribonucleotide terminating with 5′-GCUU... to the total-cell RNA. The junctions between this oligoribonucleotide and the RNAs terminating with two As are shown in Fig. 9.1. Strikingly, 34 of the 49 3′-ends analyzed (69%) contained at least two As that could correspond either to nonadenylated transcripts ending in two As or to the posttranscriptional addition of one or two As (see earlier) (Fig. 9.2). The marked prevalence of these extremities in the population of mRNAs terminating in two As strongly suggests that they correspond primarily to the tail-less transcripts. An alternative, but less likely, explanation would be that these extremities correspond to preferential adenylation sites. On the other hand, the other 15 ends contained nonencoded As in stretches ranging from one to five As in length, confirming that oligo(A) tails are very short in the wild-type strain (Fig. 9.2B). Once again, only two of the RNA poly(A) junctions were located just downstream of the transcriptional terminator hairpin, indicating that this method generally detects polyadenylated degradation fragments lacking such stable terminal structures.

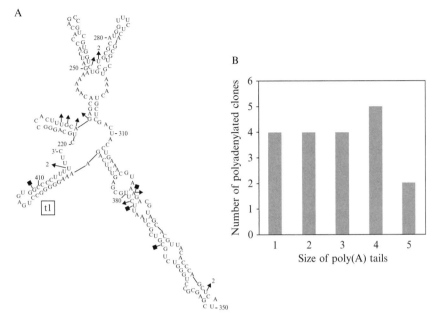

Figure 9.2   *rpsO* mRNA polyadenylation in wild-type cells. (A) Predicted structure of the *rpsO* mRNA 3′-region from position 320 (*PstI* cloning site) to position 420, which corresponds to the 3′-end of the transcription terminator t1. Folding was performed with *mfold* (http://www.bioinfo.rpi.edu/applications/mfold). The location of the *rpsO* 3′-extremities harboring poly(A) tails in RNAs isolated from WT are indicated on this secondary structure. Positions were deduced from RT-PCR analysis without (squares) (4 clones out of 39) or with selection of RNA extremities harboring at least two As at their 3′-ends (triangle) (13 extremities in the *rpsO* sequence of 49 identified). Number 2 (in boldface) indicates that two clones harbored poly(A) at the same position. Extremities terminated by two encoded As are not shown in this diagram. (B) Clones harboring posttranscriptionally added oligo(A) tails (shown in A) have been plotted as a function of the tail length. Two polyadenylated extremities located at the upstream RNase III site of the *rpsO-pnp* intercistronic region were included in the analysis. (Adapted from Le Derout *et al.* [2003] by permission of Oxford University Press.)

4.2. Hfq affects the frequency of oligoadenylated transcripts

The current model of polyadenylation in bacteria postulates that the length of poly(A) tails is the result of a dynamic equilibrium between the opposite activities of PAP I and 3′ to 5′-exonucleases—namely, PNPase and RNase II (Folichon *et al.*, 2005)—which can either shorten or completely remove single-stranded nucleotides from the 3′-end of RNAs. To investigate the role of Hfq in poly(A) synthesis, we compared the frequency of tails in isogenic *hfq⁺* and *hfq⁻* strains that are deficient for PNPase, RNase II, and RNase E. RNase E deficiency has the advantage of increasing the intracellular concentration of the *rpsO* transcript (Régnier and Hajnsdorf, 1991). Moreover, inactivation of RNase II and PNPase makes it possible to evaluate poly(A) synthesis in the absence of their exonucleolytic activities (Hajnsdorf *et al.*, 1994).

3'-RACE analysis of *rpsO* transcripts was performed by ligating the RIBOLI anchor ribonucleotide to the total-cell RNA, as described previously (Le Derout *et al.*, 2003). Interestingly, sequencing of the amplified *rpsO* cDNAs showed that 57% (41 of 72 clones) of 3'-mRNA extremities were adenylated in an *hfq*+ strain compared with only 20% (14 of 70 clones) of 3'-mRNA extremities in the strain lacking Hfq (Fig. 9.3). A chi-square test of the data indicates that the difference in the numbers of adenylated and nonadenylated clones in the *hfq*+ and *hfq*− strains was highly significant ($\chi^2 = 20.4$; $p < 0.001$). We concluded that Hfq enhances the polyadenylation status of RNA transcripts.

4.3. Hfq decreases the frequency of polyadenylation at the terminus of the primary transcript

We also used the oligo(A)-targeted 3'-RACE method to compare the locations and lengths of poly(A) tails in *hfq*+ and *hfq*− strains (Le Derout *et al.*, 2003). We isolated 71 and 70 oligoadenylated extremities from the *hfq*+ and *hfq*− strains, respectively, in the triple mutant strains deficient for PNPase, RNase II, and RNase E and found polyadenylation at different sites in both strains (Fig. 9.4). In the *hfq*− strain, we observed that 39 of 70 clones were located at position 420, just downstream of the transcription terminator hairpin, compared with 14 of 71 clones in the *hfq*+ strain. A chi-square test of data ($\chi^2 = 14.6$; $p < 0.001$) indicated that the difference in the number of adenylated clones located at the end of the primary transcript in *hfq*+ and *hfq*− strains was highly significant.

The large fraction of mRNA–oligo(A) junctions mapping downstream of t1 in the *hfq*− strain (56% of oligo[A] tails) could reflect a preferential polyadenylation of the primary transcripts, harboring a 3'-terminal stem-loop structure in the absence of Hfq. The fact that this fraction dropped to 20% in cells containing Hfq suggests that Hfq may facilitate adenylation of the 3'-end of molecules truncated within the coding sequence of *rpsO* as a result of both the endonucleolytic and exonucleolytic cleavage of the primary transcript. Alternately, Hfq may destabilize the polyadenylated transcripts in an *me-pnp-rnb*–independent manner.

4.4. Oligo(A) tails are longer in the presence of Hfq

Comparison of oligoadenylated 3'-extremities isolated from *hfq*+ and *hfq*− strains (71 and 70 clones, respectively) lacking exoribonucleases and RNase E demonstrates that Hfq also affects the length of poly(A) tails. Indeed, the number and average lengths calculated from the data reported in the histogram in Fig. 9.5 are 5.6 ± 0.9 and 9.1 ± 1.2 for the *hfq*− and the

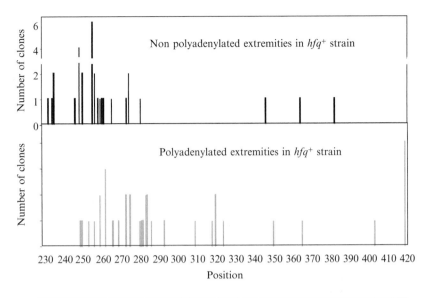

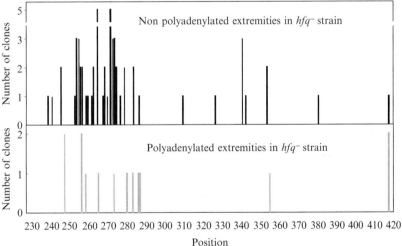

Figure 9.3 Location of the adenylated and nonadenylated 3′-extremities generated in strains deficient or proficient for Hfq. The distribution of nonadenylated and adenylated *rpsO* 3′-extremities in RNase II^ts PNPase⁻ RNE^ts and RNase II^ts PNPase⁻ RNE^ts Hfq⁻ strains (upper and lower panels, respectively) are shown. RNAs were isolated 15 min after the temperature shift that inactivates thermosensitive RNase E and RNase II. Positions were deduced from RT-PCR analysis during which all the RNA 3′-extremities with (grey) or without (black) nonencoded nucleotides were mapped. Extremities are indicated according to their positions in the *rpsO* sequence. (Adapted from Le Derout *et al.* [2003] by permission of Oxford University Press.)

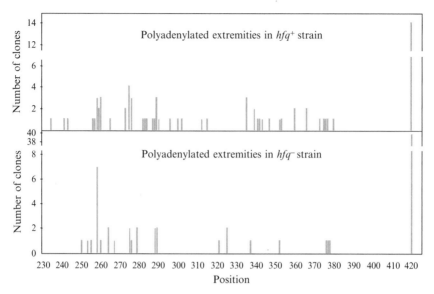

Figure 9.4 Comparison of the sites of *rpsO* poly(A) tails isolated from strains with or without Hfq. Amplified *rpsO* cDNA clones from strain RNase II^{ts} PNPase$^-$ RNE^{ts} and strain RNase II^{ts} PNPase$^-$ RNE^{ts} Hfq$^-$ (upper and lower panels, respectively) containing at least two As at the junction with the oligoribonucleotide were selectively cloned. The diagram shows 3′-extremities containing nonencoded riboadenylates on the sequence of the *rpsO* mRNAs isolated from the *hfq$^+$* and *hfq$^-$* cells (15 min after the temperature shift that inactivated thermosensitive RNase II and RNase E). Position 420 corresponds to *rpsO* mRNA terminating at the Rho-independent transcription terminator. (Adapted from Le Derout *et al.* [2003] by permission of Oxford University Press.)

hfq$^+$ strains, respectively. If tails longer than 14 (a rare occurrence) are excluded, the number and average lengths fall slightly to 5.1 ± 0.6 in the *hfq$^-$* and 7.3 ± 0.6 in the *hfq$^+$* strains. These results show that in the absence of RNase II, PNPase, and RNase E tails are longer in the *hfq$^+$* strain than in the mutant, thus confirming that Hfq stimulates poly(A) synthesis by PAP I.

5. A METHOD TO ANALYZE POLYADENYLATION OF PRIMARY TRANSCRIPTS

Investigations carried out on *rpsO* mRNA suggest that 3′-RACE preferentially amplifies nonbase-paired 3′-extremities resulting from endo-nucleolytic or exonucleolytic cleavages (Figs. 9.2 and 9.3). In contrast, primary transcripts terminating in the stable hairpins of the transcription terminator (probably the most abundant RNA species) are less efficiently detected. This bias in all likelihood reflects the lower efficiency of anchor ligation downstream of stable hairpins (E. Hajnsdorf, unpublished results)

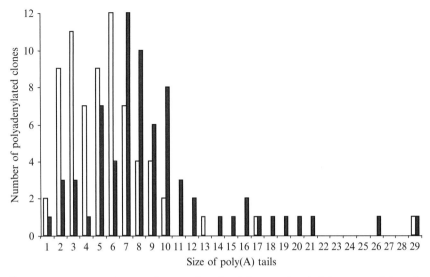

Figure 9.5 Length of *rpsO* poly(A) tails isolated from strains deficient or proficient for Hfq. 3′-RNA extremities containing posttranscriptionally added tails isolated from *hfq*⁺ (black bars) and *hfq*⁻ cells (white bars), as described in Fig. 9.4, are plotted as a function of poly(A) tail length. (Adapted from Le Derout *et al.* [2003] by permission of Oxford University Press.)

and the impediment of reverse transcription by the secondary structure of the terminator. Differences in ligation efficiency could explain why only 20% of the oligo(A) tails identified by the 3'-RACE method are located downstream of t1 compared with 52% when reverse transcription is initiated from oligo(dT$_{18}$) hybridized to poly(A) (Haugel-Nielsen *et al.*, 1996). This suggests that the relative number of clones obtained by 3'-RACE analysis and containing RNA extremities mapping downstream of t1 or in the coding sequence may not reflect the relative concentrations of full-length and truncated transcripts. Nevertheless, despite this bias, the 3'-RACE method allows us to compare the distribution of adenylated and tail-less mRNA extremities and the length of the tails in cells containing different sets of enzymes. The Northern blot analysis of 3'-RNA fragments generated by RNase H described in the following has been used to examine the polyadenylation status of *rpsO* primary transcripts.

5.1. Site-directed cleavage of *rpsO* mRNA by RNase H

rpsO mRNAs were cleaved by RNase H in the presence of a gene-specific methylated chimeric RNA-DNA oligonucleotide that directs the enzyme to a unique site of cleavage (Lapham *et al.*, 1997) localized approximately 50 nucleotides upstream of the transcription termination site. 3'-Fragments were separated on 10% polyacrylamide gels that were then transferred and probed for *rpsO* to identify the transcripts differing by only one nucleotide at their 3′-end (Marujo *et al.*, 2000).

The bacterial strains, growth conditions, and the RNA preparation method used were described previously. Site-directed cleavage of the *rpsO* transcript was carried out as previously described (Li *et al.*, 1999). Ten μg of total-cell RNA and 30 ng of chimeric oligonucleotide complementary to *rpsO* (see later) were mixed in 5 μl, incubated at 95 °C for 5 min and then at 50 °C for 15 min, and slowly cooled to 37 °C. One unit of RNase H (Pharmacia) and 2 units of RNase inhibitor (RNasin, Promega) were added to the reaction mixture to a final volume of 10 μl, and hydrolysis was carried out for 1 h at 37 °C. The reaction was stopped by the addition of 15 μl loading buffer. The 2′-O-methyl RNA-DNA chimeric oligonucleotide: 5′-GAdCdGdCdAGACCCAGGCGCUC-3′, is complementary to the 3′-portion of the *rpsO* mRNA between residues 352 and 370. A full-length *rpsO* mRNA was used as the control; this mRNA was transcribed *in vitro* from a DNA matrix amplified with the oligonucleotide 5′-<u>ATTAAGGT-GACACTATAG</u>$_1$CCGCTTAACGTCGCGTAAATTG-3′, containing the SP6 promotor (underlined), and the oligonucleotide 5′-G$_{420}$AAAAAAGGGGCCACTCAGG-3′. On incubation in the presence of the RNA-DNA chimeric oligonucleotide, this RNA is cleaved by RNase H into a 3′-fragment terminating in the UUUUUUC$_{420}$ sequence lying downstream of the transcription terminator. This 3′-fragment was used as a size marker in lanes 6 and M of Fig. 9.6A and C, respectively.

5.2. Northern blotting

Samples containing the RNA fragments generated by RNase H were loaded on 10% polyacrylamide, 50% urea sequencing gels, which were then electrotransferred to Hybond-N+ (Amersham) in 10 mM Tris-acetate (pH 7.8), 0.5 mM EDTA, and 5 mM sodium acetate buffer. RNA was fixed to the membrane in 50 mM NaOH for 5 min. The membrane was then neutralized for 5 min in 2× SSPE before being prehybridized in 50% formamide, 5× SSPE, 5× Denhart, and 0.5% SDS containing 200 μg/ml yeast RNA at 42 °C for 1 h, and hybridized in the same buffer to the ^{32}P-labeled RNA probe at 55 °C for at least 2 h. The membrane was then washed successively for 30 min at 68 °C in 2× SSPE and 0.1% SDS, 1× SSPE and 0.1% SDS, and finally in 0.1× SSPE and 0.1% SDS.

The probe was an anti-sense RNA fragment complementary to the 3′-extremity of the *rpsO* transcript between nucleotides 368 and 420, which was transcribed by T7 RNA polymerase from a PCR fragment amplified from plasmid pFB1 with the oligonucleotide 5′-<u>TAATACGACTC ACTATAGGG</u>$_{420}$AAAAAAGGGGCCACTCAGG-3′ containing the T7 promoter (underlined), and the oligonucleotide 5′-G$_{368}$TCGCTAA TTCTTGCG-3′. We used the 3′-RACE method to analyze the 3′-ends of this particular mRNA species in wild-type cells *in vivo* (Le Derout *et al.*, 2003).

6. Hfq STIMULATES POLY(A) SYNTHESIS

The Northern blot in Fig. 9.6A compares lengths of the 3'-fragments of *rpsO* mRNA isolated from the wild-type strain and from mutant bacteria lacking poly(A) polymerase, RNase II, and both RNase II and poly(A) polymerase. Lane 5 of Fig. 9.6A shows that a strain lacking PAP I and RNase II mostly contains two *rpsO* RNA species corresponding to primary transcripts, whose 3'-ends map at the C and the A located 7 and 8 nucleotides just downstream of the GC-rich hairpin of the transcriptional terminator (Folichon *et al.*, 2005). A time course analysis performed after the addition of rifampicin to block the initiation of transcription demonstrates that *rpsO* transcripts are not significantly modified posttranscriptionally when both RNase II and poly(A) polymerase are inactive (Fig. 9.6B). Cells containing PAP I and lacking RNase II (Fig. 9.6A, lane 4) accumulate *rpsO* mRNAs elongated by one or two As, which are almost completely absent in isogenic *rnb*$^+$ bacteria (Fig. 9.6A, lanes 2), presumably because they are shortened by RNase II. The short 3'-*rpsO* mRNA fragments (indicated by arrowheads on Fig. 9.6A), which are only detected if the cells have RNase II, are presumably the result of RNA nibbling by this enzyme.

The method described here, which permits the detection of primary transcripts elongated by PAP I, was used to investigate whether Hfq also affects the polyadenylation of *rpsO* primary transcripts (Le Derout *et al.*, 2003). For that purpose, we compared the elongation of the mRNA occurring in *hfq*$^+$ anf *hfq*$^-$ strains deficient for exonucleases after the blocking of transcription by rifampicin. Figure 9.6C shows that the *rpsO* transcript gains up to seven additional nucleotides at its 3'-extremity 4 minutes after the inactivation of RNase II and RNase E. In the strain lacking Hfq there are only three nucleotides added under these conditions. This suggests that Hfq stimulates the rate of polyadenylation at the end of the primary transcript in the absence of exonucleases and RNase E. These data demonstrate that this RNase H–targeted method is a powerful tool for the analysis of the polyadenylation status and the rate of poly(A) synthesis at the 3'-ends of primary transcripts terminating in the stable hairpins.

7. CONCLUSION

Our understanding of the role of polyadenylation in metabolism of bacterial mRNA is still very incomplete. Global investigations suggest that RNA transcribed from most open reading frames can be adenylated

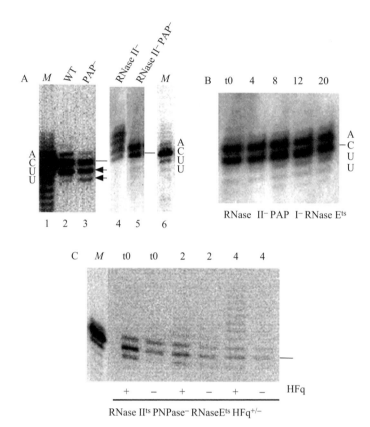

Figure 9.6   Length of *rpsO* poly(A) tails detected at the transcription terminator structure. Total bacterial RNA and *in vitro* synthesized runoff *rpsO* RNA ending in the UUUUUUC sequence of the terminator (see text) (lanes M) were treated with RNase H in the presence of the chimeric *rpsO* oligonucleotide and analyzed on a Northern blot probed for *rpsO* mRNA (see text). A and C mark the 3'-terminus of the *rpsO* primary transcript. (A) RNase II nibbles the *rpsO* transcript. Deficiency for PAP I and RNase II are indicated at the top of the lanes. WT, RNase II⁻, PAP⁻, and RNase II⁻ PAP⁻ strains were grown at 37 °C. (B) RNA polymerase generates two *rpsO* transcripts terminated at C420 and A421 located just downstream of the stretch of six Us of the terminator that was not posttranscriptionally modified in the absence of RNase II and PAP I. Strain RNase II⁻ PAP⁻ RNase E^{ts} was grown at 30 °C to an $OD_{650nm} = 0.4$, and shifted to 44 °C to inactivate RNase E (this was considered t_0). Rifampicin was added at the time of the shift and total-cell RNAs were prepared from bacteria withdrawn at different times after the shift. Times are indicated in minutes at the top of the autoradiograph. (C) Rate of poly(A) tail elongation by PAP I downstream of the *rpsO* transcription terminator in the presence and absence of Hfq. Strains RNase II^{ts} PNPase⁻ RNase E^{ts} and RNase II^{ts} PNPase⁻ RNase E^{ts} HFq⁻ grown at 30 °C were shifted to 44 °C at t_0 to inactivate RNase E and RNase II. Rifampicin was added at the time of the shift, and total-cell RNAs were prepared as in (B). (+) and (−) indicate whether cells contained the protein Hfq. (Adapted from Le Derout *et al.* [2003] with permission of Oxford University Press and from Folichon *et al.* [2005] by permission of Elsevier Masson SAS.)

(Mohanty and Kushner, 2006), and it is currently agreed that polyadenylation primarily affects the degradation of RNA fragments, whereas the stability of full-length transcripts is mainly controlled by RNase E (Dreyfus and Régnier, 2002). It has also been reported that polyadenylation negatively affects gene expression and that primary transcripts can be degraded by a poly(A)-dependent pathway (Hajnsdorf *et al.*, 1995; Joanny *et al.*, 2007). In contrast, the expression of some polypeptides is positively controlled by PAP I (E. Hajnsdorf, personal communication). In addition, polyadenylation is involved in the quality control of RNA, helping to remove nonfunctional bacterial tRNA (Li *et al.*, 2002) and many stable breakdown products. Another interesting possibility is that polyadenylation may contribute to the metabolism of the various small bacterial ncRNAs involved in the posttranscriptional control of gene expression (Viegas *et al.*, 2007). Our current notions regarding bacterial polyadenylation are based on research conducted on a limited number of model RNAs, and this work clearly needs to be broadened. We believe that the two methods described here to study the *rps*O mRNAs could be adapted to study polyadenylation of other transcripts.

ACKNOWLEDGMENT

We thank Lisa Chien for careful reading of the manuscript.

REFERENCES

Anderson, J. T. (2005). RNA turnover: Unexpected consequences of being tailed. *Curr. Biol.* **15,** R635–R638.

Arluison, V., Derreumaux, P., Allemand, F., Folichon, M., Hajnsdorf, E., and Regnier, P. (2002). Structural modelling of the Sm-like protein Hfq from *Escherichia coli*. *J. Mol. Biol.* **320,** 705–712.

Arraiano, M. C., Yancey, S. D., and Kuschner, S. R. (1988). Stabilization of discrete breakdown products in *ams pnp rnb* multiple mutant of *Escherichia coli J. Bact.* **170,** 4225–4263.

Braun, F., Hajnsdorf, E., and Régnier, P. (1996). Polynucleotide phosphorylase is required for the rapid degradation of the RNase E-processed *rps*OmRNA of *Escherichia coli* devoid of its 3′-hairpin. *Mol. Microbiol.* **19,** 997–1005.

Briani, F., Del Vecchio, E., Migliorini, D., Hajnsdorf, E., Regnier, P., Ghisotti, D., and Deho, G. (2002). RNase E and polyadenyl polymerase I are involved in maturation of CI RNA, the P4 phage immunity factor. *J. Mol. Biol.* **318,** 321–331.

Campos-Guillén, J., Bralley, P., Jones, G. H., Bechhofer, D. H., and Olmedo-Alvarez, G. (2005). Addition of poly(A) and heteropolymeric 3′-ends in *Bacillus subtilis* wild-type and polynucleotide phosphorylase-deficient strains. *J. Bacteriol.* **187,** 4698–4706.

Donis-Keller, H. (1979). Site specific enzymatic cleavage of RNA. *Nucleic Acids Res.* **7,** 179–192.

Dreyfus, M., and Régnier, P. (2002). The poly(A) tail of mRNAs: Bodyguard in eukaryotes, scavenger in bacteria. *Cell* **111,** 611–613.

Folichon, M., Arluison, V., Pellegrini, O., Huntzinger, E., Regnier, P., and Hajnsdorf, E. (2003). The poly(A) binding protein Hfq protects RNA from RNase E and exoribonucleolytic degradation. *Nucleic Acids Res.* **31,** 7302–7310.

Folichon, M., Marujo, P. E., Arluison, V., Le Derout, J., Pellegrini, O., Hajnsdorf, E., and Regnier, P. (2005). Fate of mRNA extremities generated by intrinsic termination: Detailed analysis of reactions catalyzed by ribonuclease II and poly(A) polymerase. *Biochimie* **87,** 819–826.

Goodrich, A. F., and Steege, D. A. (1999). Roles of polyadenylation and nucleolytic cleavage in the filamentous phage mRNA processing and decay pathways in *Escherichia coli.* *RNA* **5,** 972–985.

Gottesman, S. (2004). The small RNA regulators of *Escherichia coli*: Roles and mechanisms. *Annu. Rev. Microbiol.* **58,** 303–328.

Hajnsdorf, E., Braun, F., Haugel-Nielsen, J., Le Derout, J., and Regnier, P. (1996). Multiple degradation pathways of the *rpsO* mRNA of *E. coli.* RNase E interacts with the 5′- and 3′-extremities of the primary transcript. *Biochimie* **78,** 416–424.

Hajnsdorf, E., Braun, F., Haugel-Nielsen, J., and Regnier, P. (1995). Polyadenylation destabilizes the *rpsO* mRNA of *Escherichia coli.* *Proc. Natl. Acad. Sci. USA* **92,** 3973–3977.

Hajnsdorf, E., and Régnier, P. (2000). Host factor HFq of *Escherichia coli* stimulates elongation of poly(A) tails by poly(A)polymerase I. *Proc. Natl. Acad. Sci. USA* **97,** 1501–1505.

Hajnsdorf, E., Steier, O., Coscoy, L., Teysset, L., and Régnier, P. (1994). Roles of RNase E, RNase II and PNPase in the degradation of the *rpsO* transcripts of *Escherichia coli*: Stabilizing function of RNase II and evidence for efficient degradation in an *ams-rnb-pnp* mutant. *EMBO J.* **13,** 3368–3377.

Haugel-Nielsen, J., Hajnsdorf, E., and Regnier, P. (1996). The *rpsO* mRNA of *Escherichia coli* is polyadenylated at multiple sites resulting from endonucleolytic processing and exonucleolytic degradation. *EMBO J.* **15,** 3144–3152.

Holec, S., Lange, H., Kuhn, K., Alioua, M., Borner, T., and Gagliardi, D. (2006). Relaxed transcription in *Arabidopsis* mitochondria is counterbalanced by RNA stability control mediated by polyadenylation and polynucleotide phosphorylase. *Mol. Cell Biol.* **26,** 2869–2876.

Joanny, G., Le Derout, J., Bréchemier-Baey, D., Labas, V., Vinh, J., Régnier, P., and Hajnsdorf, E. (2007). Polyadenylation of a functional mRNA controls gene expression in *E. coli.* *Nucleic Acids Res.* **35,** 2494–2502.

Kuhn, J., and Binder, S. (2002). RT-PCR analysis of 5′ to 3′-end-ligated mRNAs identifies the extremities of cox2 transcripts in pea mitochondria. *Nucleic Acids Res.* **30,** 439–446.

Lapham, J., Yu, Y.-T., Shu, M.-D., Steitz, J. A., and Crothers, D. M. (1997). The position of site-directed cleavage of RNA using RNase H and 2′-O methyl oligonucleotides is dependent on the enzyme source. *RNA* **3,** 950–951.

Le Derout, J., Folichon, M., Briani, F., Dehò, G., Régnier, P., and Hajnsdorf, E. (2003). Hfq affects the length and the frequency of short oligo(A) tails at the 3′-end of *Escherichia coli rpsO* mRNAs. *Nucl. Acids Res.* **31,** 4017–4023.

Li, Z., Pandit, S., and Deutscher, M. P. (1998). Polyadenylation of stable RNAs *in vivo.* *Proc. Natl. Acad. Sci. USA* **95,** 12158–12162.

Li, Z., Pandit, S., and Deutscher, M. P. (1999). Maturation of 23S ribosomal RNA requires the exoribonuclease RNase T. *RNA* **5,** 139–146.

Li, Z., Reimers, S., Pandit, S., and Deutscher, M. P. (2002). RNA quality control: Degradation of defective transfer RNA. *EMBO J.* **21,** 1132–1138.

Marujo, P. E., Braun, F., Haugel-Nielsen, J., Le Derout, J., Arraiano, C. M., and Regnier, P. (2003). Inactivation of the decay pathway initiated at an internal site by RNase E promotes poly(A)-dependent degradation of the rpsO mRNA in *Escherichia coli.* *Mol. Microbiol.* **50,** 1283–1294.

Marujo, P. E., Hajnsdorf, E., Le Derout, J., Andrade, R., Arraiano, C. M., and Régnier, P. (2000). RNase II removes the oligo(A) tails that destabilize the *rpsO* mRNA of *Escherichia coli*. *RNA* **6,** 1185–1193.

Mohanty, B. K., and Kushner, S. R. (2006). The majority of *Escherichia coli* mRNAs undergo post-transcriptional modification in exponentially growing cells. *Nucleic Acids Res.* **34,** 5695–5704.

Moller, T., Franch, T., Hojrup, P., Keene, D., Bächinger, H. P., Brennan, R. G., and Valentin-Hansen, P. (2002). Hfq: A bacterial Sm-like protein that mediates RNA-RNA interaction. *Mol. Cell* **9,** 23–30.

O'Hara, E. B., Chekanova, J. A., Ingle, C. A., Kushner, Z. R., Peters, E., and Kushner, S. R. (1995). Polyadenylation helps regulate mRNA decay in *Escherichia coli*. *Proc. Natl. Acad. Sci. USA* **92,** 1807–1811.

Piedade, J., Zilhao, R., and Arraiano, M. C. (1995). Construction and characterization of an absolute deletion mutant of *Escherichia coli* ribonuclease II. *FEMS Microbiol. lett.* **127,** 187–193.

Régnier, P., and Arraiano, C. M. (2000). Degradation of mRNA in bacteria: Emergence of ubiquitous features. *BioEssays* **22,** 235–244.

Régnier, P., and Hajnsdorf, E. (1991). Decay of mRNA encoding ribosomal protein S15 of *Escherichia coli* is initiated by an RNaseE-dependent endonucleolytic cleavage that removes the 3′-stabilizing stem and loop structure. *J. Mol. Biol.* **217,** 283–292.

Sarkar, N. (1996). Polyadenylation of mRNA in bacteria. *Microbiology* **142,** 3125–3133.

van Meerten, D., Zelwer, M., Regnier, P., and Duin, J. (1999). *In vivo* oligo(A) insertions in phage MS2: Role of *Escherichia coli* poly(A) polymerase. *Nucleic Acids Res.* **27,** 3891–3898.

Viegas, S. C., Pfeiffer, .V., Sittka, A., Silva, I. J., Vogel, J., and Arraiano, C. M. (2007). Characterization of the role of ribonucleases in *Salmonella* small RNA decay. *Nucleic Acids Res.* **35,** 7651–7664.

Vincent, H. A., and Deutscher, M. P. (2006). Substrate recognition and catalysis by the exoribonuclease RNASE R. *J. Biol. Chem.* **281,** 29769–29775.

Xu, F., Lin-Chao, S., and Cohen, S. N. (1993). The *Escherichia coli pcnB* gene promotes adenylation of antisense RNAI of ColE1-type plasmids *in vivo* and degradation of RNAI decay intermediates. *Proc. Natl. Acad. Sci. USA* **90,** 6756–6760.

Zhang, A., Wassarman, K. M., Ortega, J., Steven, A. C., and Storz, G. (2002). The Sm-like Hfq protein increases OxyS RNA interaction with target mRNAs. *Mol. Cell* **9,** 11–22.

CHAPTER TEN

Assaying DEAD-box RNA Helicases and Their Role in mRNA Degradation in Escherichia coli

Agamemnon J. Carpousis, Vanessa Khemici, *and* Leonora Poljak

Contents

Abstract

The DEAD-box RNA helicases are a ubiquitous family of enzymes involved in processes that include RNA splicing, ribosome biogenesis, and mRNA degradation. In general, these enzymes help to unwind short stretches of double-stranded RNA in processes that involve the remodeling of RNA structure or of ribonucleoprotein complexes. Here we describe work from our laboratory on the characterization of the RhlB of *Escherichia coli*, a DEAD-box RNA helicase that is part of a multienzyme complex known as the RNA degradosome. RhlB interacts physically and functionally with RNase E and polynucleotide phosphorylase (PNPase), two other components of the RNA degradosome. We describe enzyme assays that demonstrated that the interaction between RhlB and RNase E is necessary for the ATPase and RNA unwinding activities of RhlB. We also describe an mRNA degradation assay that showed that RhlB facilitates the degradation of structured mRNA by PNPase. These assays are discussed in the context of how they have contributed to our understanding of the function of RhlB in mRNA degradation.

Laboratoire de Microbiologie et Génétique Moléculaire (LMGM), Centre National de la Recherche Scientifique (CNRS) and Université Paul Sabatier, Toulouse, France

Methods in Enzymology, Volume 447
ISSN 0076-6879, DOI: 10.1016/S0076-6879(08)02210-6

1. INTRODUCTION

The DEAD-box helicases are a ubiquitous family of ATPases involved in RNA metabolism (Linder, 2006; Tanner and Linder, 2001). They are characterized by nine conserved sequence motifs including the D-E-A-D motif involved in ATP binding and hydrolysis. These enzymes have a conserved catalytic core structurally related to RecA, a single-stranded DNA binding protein that has ATPase activity (Caruthers and McKay, 2002). The DEAD-box helicases are part of the larger DExD/H family that includes genuine RNA helicases capable of unwinding long stretches of double-stranded RNA. However, DEAD-box helicases are often only capable of opening a few RNA base pairs, and their activity seems to be directed at facilitating the remodeling of structured RNA or ribonucleoprotein (RNP) complexes during processes such as the biogenesis of the ribosome, the initiation of translation in eukaryotes, or the splicing of RNA. For this reason, some researchers prefer the term DEAD-box proteins to distinguish them from a classical helicases that can unwind long stretches of duplex RNA. The DEAD-box helicases generally have two characteristic biochemical activities *in vitro*: RNA-dependent ATP hydrolysis and ATPase-dependent RNA unwinding activity. Here we will describe protocols that we used to measure these activities during the characterization of the *Escherichia coli* DEAD-box helicase, RhlB. Nevertheless, because the *in vitro* activity of a DEAD-box RNA helicase rarely provides information regarding its function *in vivo*, we will also discuss work from our group that helped to establish a role for RhlB in the degradation of mRNA in *E. coli*.

E. coli encodes five DEAD-box helicases: RhlB, RhlE, SrmB, CsdA (also known as DeaD), and DbpA. We direct the reader to an excellent recent review by Iost and Dreyfus (2006) on these RNA helicases and their homologs in other gram-negative bacteria. Here we will describe what is known about their function in *E. coli*. SrmB, CsdA, and DbpA are involved in ribosome biogenesis. The function of RhlE is unknown. RhlB and CsdA are involved in mRNA degradation. Note that CsdA thus seems to have two distinct functions. In contrast to the eukaryotic microorganism *S. cerevisiae*, in which most, if not all, of the DEAD-box helicase are essential (Cordin *et al.*, 2006; Tanner and Linder, 2001), none of the *E. coli* DEAD-box RNA helicases are essential. Strains of *E. coli* in which two or more genes encoding DEAD-box helicases have been disrupted are also viable, suggesting that overlapping function is not the explanation for the viability of the singly disrupted strains. However, strains in which the *srmB* and *csdA* genes have been disrupted are cold sensitive because of severe problems with ribosome biogenesis (Charollais *et al.*, 2003, 2004). These strains also exhibit significant problems with ribosome biogenesis even at temperatures permissive for growth. Because many of the essential DEAD-box helicases in *S. cerevisiae* are also involved in ribosome biogenesis, these results suggest that either the ribosome biogenesis defects in *E. coli* are less severe or that the accumulation of incorrectly assembled ribosomes in *E. coli* is better tolerated.

RhlB was originally implicated in mRNA degradation because of its physical association by a protein–protein interaction with *E. coli* ribonuclease E (RNase E) (Mizak *et al.*, 1996; Py *et al.*, 1996; Coburn and Mackie, 1999). RNase E is an essential endoribonuclease important in the maturation of many different types of stable RNAs and the degradation of mRNA (Cohen and McDowall, 1997; Grunberg-Manago, 1999; Kushner 2002; Vanzo *et al.*, 1998). RNase E is also the keystone in a multienzyme complex known as the RNA degradosome (Carpousis *et al.*, 1994; Miczak *et al.*, 1996; Py *et al.*, 1996). Here we will introduce RNase E and the RNA degradosome. We direct the interested reader to a selection of reviews for further reading (Carpousis, 2007; Mercaida *et al.*, 2006; Symmons *et al.*, 2002) (see also chapters 4, 6, 11, 18, 19, 20, and 22 regarding methods for the studying the degradosome and other multiprotein complexes in this volume).

RNase E is a large and complex multidomain protein composed of 1061 residues. The N-terminal half of the protein folds into a compact structure that associates to form a tetrameric catalytic domain. The C-terminal half of the protein, which is mostly natively unstructured protein, contains a series of motifs that are involved in a variety of interactions including the protein–protein interactions necessary to form the RNA degradosome (Callaghan *et al.*, 2004; Chandran *et al.*, 2007; Vanzo *et al.*, 1998). The major proteins that associate with RNase E to form the RNA degradosome are RhlB, enolase, and polynucleotide phosphorylase (PNPase). Enolase is a glycolytic enzyme whose function in the RNA degradosome is not understood. The function of RhlB and PNPase in the RNA degradosome will be described in the following sections.

2. THE RNA-DEPENDENT ATPASE ACTIVITY OF RHLB

In the initial characterization of the RNA degradosome, we detected an ATPase activity that depended on the addition of exogenous RNA to the reaction (Py *et al.*, 1996). Evidence that the ATPase activity was due to RhlB came from an experiment in which affinity-purified antibody against RhlB inhibited the activity. RNA-dependent ATPase activity is typical of the DEAD-box RNA helicases. It is believed to be due to a cycle in which the ATP-bound form of the enzyme binds RNA followed by conversion to the ADP-bound form and release of the RNA. This reaction often works with a complex mixture such as total yeast RNA or synthetic RNAs such as poly(U) or poly(A). However, when we subsequently overexpressed and purified RhlB, we failed to detect any ATPase activity. After eliminating trivial possibilities such as the inadvertent isolation of a mutant during PCR amplification of the gene encoding RhlB, we considered the possibility that the interaction between RhlB and RNase E was necessary for the activity of RhlB. This turned out to be the case (Vanzo *et al.*, 1998): neither RhlB nor RNase E by themselves has significant RNA-dependent ATPase activity, whereas the two proteins mixed together have activity. Together with other work not described here, we mapped a site in the noncatalytic region of RNase E that binds RhlB and showed that polypeptides corresponding to this region can stimulate the RNA-dependent ATPase activity of RhlB. An example of the activation of RhlB by RNase E is shown in Fig. 10.1. The rate of ATP hydrolysis was measured in the presence of increasing amounts of RneHC2, a polypeptide that corresponds to residues 628 to 843 of RNase E. At saturating concentrations of RneHC2, the specific activity is comparable to the specific activity measured previously in the RNA degradosome. Note that in these reactions, the specific activity is normalized to the mass of RhlB, making direct comparisons possible. The stoichiometry of RneHC2 to RhlB at saturation suggests a 1:1 interaction on a molar basis. We also performed a negative control with catalytically defective RhlB in which the D-E-A-D motif was changed to D-K-A-D. The residual ATPase

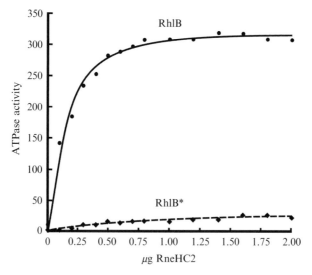

Figure 10.1 Determination of the RNA-dependent ATPase activity of wild-type RhlB (circles, solid line) and a catalytically defective variant (RhlB*), in which the DEAD motif was changed to DKAD (diamonds, dashed line). The reactions contained 1 μg of RhlB and 0 to 2 μg of RneHC2. The units of ATPase activity are expressed as mmol/min/mg of RhlB. Adapted from Vanzo *et al.* (1998).

activity detected with the DKAD variant at saturating concentrations of RneHC2 is at least 15-fold lower than wild-type RhlB. In subsequent work, a smaller polypeptide corresponding to residues 694 to 790 of RNase E was shown to be sufficient for the stimulation of the RNA-dependent ATPase activity of RhlB (Khemici *et al.*, 2004), and a minimal RhlB binding site of 64 residues has been defined by partial proteolysis experiments (Chandran *et al.*, 2007). In sequence alignments of RNase E homologs from *E. coli* and other gram-negative bacteria, the minimal RhlB binding site overlaps a conserved sequence motif (Chandran *et al.*, 2007). The conversion of a highly conserved arginine residue in this motif to alanine (R730A) has been shown to interfere with the protein–protein interaction between RhlB and RNase E, as well as the stimulation of the RNA-dependent ATPase activity of RhlB (Khemici *et al.*, 2004).

3. THE ATPASE-DEPENDENT RNA HELICASE ACTIVITY OF RHLB

A second characteristic of DEAD-box helicases is their capacity to unwind short RNA duplexes *in vitro* in a reaction that requires ATP hydrolysis. Figure 10.2 shows an example of such an assay. As is the case for the ATPase activity of RhlB, the addition of the RneHC2 polypeptide is necessary to detect RNA unwinding activity. In this assay, a short, self-annealing RNA oligonucleotide is used as substrate. The single-stranded and double-stranded forms of the oligonucleotide can be separated by electrophoresis in a native polyacrylamide gel. The following points are noteworthy: RNA unwinding is ATP-dependent (lane 10) and ATPase-dependent (lane 11), because the level of unwinding in both cases is significantly lower than in the presence of ATP (lane 7) . The RNA unwinding detected in the presence of ATPγS could be due either to slow hydrolysis of this ATP analog or to ATPγS dependent ssRNA binding driving the strand separation reaction. RhlB by itself has little, if any, RNA unwinding activity (lane 9). In the presence of saturating amounts of RneHC2, increasing the concentration of RhlB corresponds to increasing RNA unwinding activity (lanes 3 to 7). This experiment shows that the RneHC2–RhlB complex can unwind short RNA duplexes in an ATPase-dependent reaction. However, the experiment needs to be interpreted with caution, because, for technical reasons, it is performed with a large excess of enzyme to substrate (see "Discussion"). Additional experimental work has shown that some smaller polypeptides derived from RneHC2, which still bind to RhlB and stimulate the RNA-dependent ATPase activity, are less efficient in stimulating the RNA unwinding activity. These results suggest that an RNA binding site on RneHC2 might contribute to the RNA unwinding detected in this experiment. For further discussion see Candran *et al.* (2007).

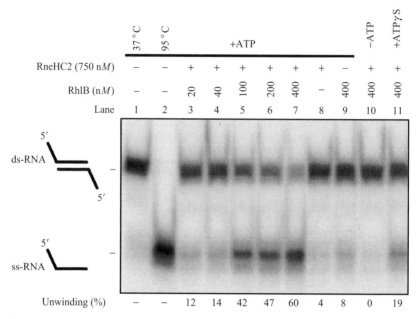

Figure 10.2 RNA unwinding activity of RhlB in the presence of RneHC2. Single-stranded and double-stranded RNA (ss-RNA and ds-RNA, respectively) were separated by native polyacrylamide gel electrophoresis. The substrate in this experiment is a chemically synthesized RNA oligonucleotide that was radiolabeled at its 5′-end with [^{32}P]-phosphate. The sequence of the oligonucleotide is 5′-GAAUGUACAUCAGA-GUGCGCACUC-3′. The underlined part of the sequence is self-complementary. Annealing forms a short RNA duplex with 5′-single-stranded extensions. Lane 1 is a control showing the input ds-RNA; lane 2, the input RNA after denaturation by heat. Lanes 3 to 11, RNA unwinding activity under various conditions in the presence or absence of RneHC2 and RhlB. The percent of RNA unwinding, indicated at the bottom of the gel, was quantified by phosphor imaging. Adapted from Chandran *et al.* (2007).

4. THE RHLB-PNPASE mRNA DEGRADATION PATHWAY

RhlB is now believed to facilitate mRNA degradation by interacting functionally with RNase E and PNPase as a component of the RNA degradosome. The discovery of the RhlB-RNase E mRNA degradation pathway is relatively recent (Khemici *et al.*, 2005; Leroy *et al.*, 2002), whereas the RhlB-PNPase mRNA degradation pathway has been well studied (Chandran *et al.*, 2007; Coburn *et al.*, 1999; Khemici and Carpousis, 2004; Py *et al.*, 1996; Vanzo *et al.*, 1998). For this reason, we will focus here on the RhlB-PNPase pathway. The degradation of mRNA in *E. coli* involves the combined action of endoribonucleases and exoribonucleases (Carpousis *et al.*, 1999; Coburn and Mackie, 1999; Grunberg-Manago, 1999; Kushner, 2002). Endoribonucleolytic cleavage, principally by RNase E, inactivates mRNA and breaks it into fragments that are then digested to nucleotides by the exoribonucleases. It is the initial attack by RNase E that determines mRNA stability, although recent work suggests that this event can be influenced by the phosphorylation state of the 5′-end of the message (Celesnik *et al.*, 2007). Regardless of this detail, for the work described here, it is important to note that mRNA decay intermediates are essentially undetectable except in genetic backgrounds where the exoribonucleases or accessory factors affecting their activity are knocked out. One of the original suggestions as to why RNase E and PNPase are part of the RNA degradosome was that their physical association might help to coordinate endoribonucleolytic and exoribonucleolytic activities in mRNA degradation (Carpousis *et al.*, 1994).

The exoribonuclease PNPase is a phosphorylase that degrades RNA to produce nucleotide diphosphates (NDPs). It is not a ribonuclease, because ribonucleases are hydrolytic by definition. Although PNPase can be considered to act like an exoribonuclease that degrades RNA in the 3′ to 5′-direction, there is an important difference: phosphorolysis is a near equilibrium reaction. At high NDP concentrations, the reverse synthetic reaction is possible, whereas hydrolysis by a ribonuclease is irreversible. Indeed, PNPase was originally discovered as an "RNA polymerase" (Grunberg-Manago *et al.*, 1955). It is now known that polynucleotide synthesis is template-independent, and, *in vivo*, the relatively high concentration of phosphate favors RNA degradation. The PNPase of *E. coli* is relatively resistant to thermal denaturation, and classical work on the characterization of PNPase showed that it can degrade structured RNAs such as tRNA in overnight digestions at 55 °C (Littauer and Soreq, 1982). Nevertheless, under physiologic conditions at lower temperatures, RNA secondary structure can slow or inhibit degradation by PNPase.

The classical biochemical work together with recent crystallographic studies determining the structure of PNPase suggests the following model of PNPase action (Symmons *et al.*, 2000, 2002). PNPase is a single-strand specific enzyme that degrades RNA in a processive 3′ to 5′-reaction. The catalytic site of the enzyme is buried in the interior of the enzyme, and access to the catalytic site requires the passage of the RNA through a channel that is only wide enough to accommodate single-stranded RNA. The channel between the catalytic core and the exterior is capped by a ring of RNA binding domains. RNA tethered to the external RNA binding domains can be "fed" into the catalytic core without having to release the substrate during the translocation step. Thus, PNPase can degrade long stretches of RNA without physically dissociating from the substrate. However, if the enzyme encounters a region of RNA secondary structure, it pauses because only single-stranded RNA can enter the channel to the catalytic site. There are two possible outcomes of pausing: either spontaneous thermal fluctuation opens RNA base pairs and the PNPase "nibbles" through the RNA structure or the PNPase dissociates from the substrate. After dissociation from the RNA, further degradation requires rebinding. Note, however, that PNPase requires 7 to 9 bases of single-stranded RNA at the 3'-end for binding and that molecules with fewer than 7 free bases resist further degradation by PNPase.

RNA degradation assays performed with the RNA degradosome suggested that RhlB could facilitate the degradation of structured RNA by PNPase in a reaction that requires the hydrolysis of ATP (Py *et al.*, 1996). The RNA used in these experiments was derived from the intergenic region between the *malE* and *malF* genes in *E. coli* (Fig. 10.3A,B). These genes are transcribed as part of a polycistronic mRNA covering the *malEFG* operon. The *malEFG* mRNA is then processed to the monocistronic *malE* mRNA. The intergenic region between *malE* and *malF* contains several copies of a DNA element (REP) that, when transcribed into RNA, folds into a stable RNA secondary structure that protects the *malE* mRNA from degradation by 3'-exonucleases such as PNPase. *In vitro*, PNPase pauses at the base of the REP stabilizer, leading to the accumulation of an RNA degradation intermediate (RSR) that is resistant to further digestion. However, if RhlB is present as a component of the RNA degradosome and ATP is added, PNPase can degrade the RSR. In Fig. 10.3C, the full-length substrate is converted to the RSR intermediate in the first 5 min of the reaction, and there is essentially no further degradation for the next 55 min. In the presence of ATP, the proportion of RSR after 5 min is significantly lower, and the remaining RSR is slowly degraded over the next 55 min. These results suggest that in the first phase of the reaction, ATP increases the processivity of PNPase, and that in the second phase, ATP facilitates degradation in a slow reaction involving the rebinding of RNA substrates that dissociated from PNPase in the first phase. Experiments performed over the past decade have shown that degradation of the REP stabilizer *in vitro* requires catalytically active RhlB and ATP hydrolysis (Khemici and Carpousis, 2004; Py *et al.*, 1996). Mixing and reconstitution experiments have shown that RhlB and PNPase have to be associated with RNase E as part of the RNA degradosome (Coburn *et al.*, 1999; Khemici *et al.*, 2004). Furthermore, RNase E is not involved in this reaction because, as in the experiment shown in Fig. 10.3C, it can be performed with variants of the RNA degradosome that do not contain the catalytic domain of RNase E. Finally, in *E. coli*, several hundred messages contain REP stabilizers, and we have shown that their degradation *in vivo* depends on two pathways: one requiring RhlB and PNPase as components of the RNA degradosome and the other requiring ribonuclease R (RNase R) and poly(A) polymerase (Cheng and Deutscher, 2005; Khemici and Carpousis, 2004).

A   The *malEFG* operon

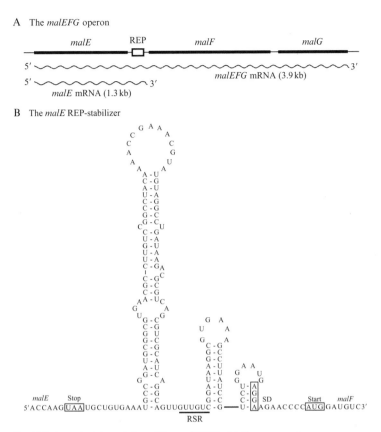

B   The *malE* REP-stabilizer

C   ATP-dependent degradation of the *malE* REP-stabilizer

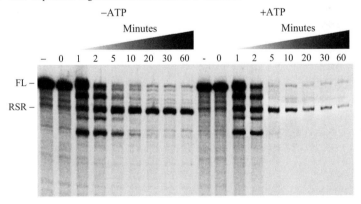

Figure 10.3 (A) Schematic diagram of the *malEFG* operon and its transcription. At the top, the thick lines represent the coding sequences, the thin lines the intergenic regions, and the box the repeated extragenic palindrome (REP) sequences. Below (wavy lines), the primary transcript (*malEFG*) and the *malE* transcript, which arises by processing of the primary transcript. (B) RNA structure of the *malE* REP stabilizer. The RNA

5. DISCUSSION AND PERSPECTIVE

The detailed molecular mechanism by which RhlB facilitates mRNA degradation by PNPase remains to be established. Nevertheless, it is worth making a few points on the basis of the work described previously. PNPase can degrade most of the substrates that it normally encounters without the assistance of RhlB. It is only exceptional RNA structures such as the REP stabilizer depicted in Fig. 10.3B that require RhlB. For this reason, we tend to think of RhlB as a factor that only acts when the processive degradation of an RNA substrate by PNPase is blocked. How could this work? Formally, the process is probably similar to the "nibbling" of RNA secondary structure by PNPase as described previously except that the free energy of ATP hydrolysis by RhlB helps to accelerate this process. Possible molecular mechanisms include an indirect acceleration by trapping free bases that arise by spontaneous thermal fluctuation or a direct acceleration by straining the RNA duplex in a way that promotes unwinding. In either case, there are potential experimental advantages in studying the mechanism of a DEAD-box RNA helicase involved in mRNA degradation. Biochemical and biophysical work with these helicases is often hampered, because the unwinding of a short RNA duplex is in competition with the reverse annealing reaction. This is the case in the RNA unwinding experiment described in Fig. 10.2, where a large excess of helicase is used relative to the RNA duplex. If the concentration of the RNA duplex is increased, the rate of the reverse annealing reactions becomes fast enough to mask the unwinding reaction. For this reason it is not feasible to work under conditions in which the substrate is in excess relative to the enzyme. Indeed, the concentration of the RNA duplex in this experiment (0.12 nM) is probably more than 100-fold lower than its K_d for binding to the RNA helicase (Chandran *et al.*, 2007). It is the high concentration of the RNA helicase in the experiment in Fig. 10.2 that drives the reaction. Note, however, that an RNA unwinding reaction coupled to the degradation of the substrate is

---

structure is based on experimental work (McLaren *et al.*, 1991). In this diagram, RSR (REP-stabilized RNA) indicates the position of the 3′-ends of the mature *malE* message extending 3 to 9 nt from the base of the REP stabilizer. The stop codon of *malE* and the start codon of *malF* are boxed. SD indicates the Shine–Dalgarno sequence for translation initiation of *malF*. (C) Degradation of REP-stabilized mRNA *in vitro*. A radiolabeled RNA substrate containing the *malE* REP stabilizer was digested with a mini-RNA degradosome containing the noncatalytic region of RNase E, RhlB, and PNPase. The reaction was performed in the absence or presence of ATP. To the left of the panel, FL indicates the position of the full-length RNA substrate and RSR the REP-stabilized mRNA decay intermediate. In each time course, (−) is a sample without enzyme that was incubated for 30 min; (0) is an aliquot of the reaction taken just before the addition of enzyme. Adapted from Khemici *et al.* (2004) and Khemici and Carpousis (2004).

effectively irreversible, permitting the study of the RhlB–PNPase complex under conditions that are closer to the physiologic situation. In the experiment described in Fig. 10.3, the RNA substrate and the mini-degradosome are present at the same concentration (20 nM), and we have performed assays in which the substrate is in 10-fold excess relative to the enzyme.

What is the physiologic relevance of the mRNA degradation reaction performed in Fig. 10.3? Indeed, the situation is more complicated than described here, and we do not yet have a complete understanding of how the degradation of the REP stabilizer is controlled. The *malEFG* mRNA is processed to the *malE* mRNA in a pathway that is likely to involve RNase E cleavage in the *malF* or *malG* coding sequence and exonucleolytic trimming to the REP stabilizer. The *malE* mRNA is long-lived, and the REP stabilizer protects the message from attack by exoribonucleases (Newbury *et al.*, 1987). We believe that the degradation of the monocistronic *malE* mRNA is initiated by an endoribonuclease, probably RNase E. It is only after the *malE* mRNA is fragmented by RNase E that the REP stabilizer is degraded. In the absence of the RhlB-PNPase pathway and a second pathway involving RNase R and poly(A) polymerase, mRNA fragments containing REP stabilizers seem to be completely stable *in vivo* (Cheng and Deutscher, 2005; Khemici and Carpousis, 2004). These considerations highlight the concept that the RhlB-PNPase pathway is involved in the degradation of mRNA fragments produced by RNase E, not intact messages.

An important finding in our work on RhlB was that its activity depends on the protein–protein interaction with RNase E. Understanding how the activity of RhlB is controlled by its interaction with RNase E is an important area of current research (Worrall *et al.*, 2007). The experimental evidence demonstrates that its interaction with RNase E as a component of the RNA degradosome is important for its biologic function. In general, how members of the DEAD-box family of RNA helicases are targeted to specific processes is still a largely open question. RhlB can be viewed as a generic DEAD-box helicase that was recruited into the mRNA degradation process by RNase E. This point of view might help to explain how members of a large family of highly conserved enzymes can be targeted to a variety of processes involving the remodeling of structured RNA or RNP complexes.

6. METHODS

6.1. ATPase assay

The ATPase assay is based on the colorimetric detection of inorganic phosphate (Pi) by use of the method of Chiflet *et al.* (1988). The detection of Pi will only briefly be discussed here. The assay is based on a reaction between Pi and molybdate producing a blue color. The assay can detect as little as 1 nmol of Pi. Because of this sensitivity, it is important to perform background controls, because water, glassware, and fine chemicals sometimes have significant traces of Pi. The assay is performed in high concentrations of SDS, which do not interfere with the development of the color. SDS stops the enzymatic reaction, and it solubilizes debris components, making it possible to use this assay with crude cell extracts. The fine chemicals used in the assay, which are relatively inexpensive, were obtained from Aldrich. Sodium meta-arsenate should be handled with care, because it is both toxic and carcinogenic.

This protocol is adapted from Py *et al.* (1996). In the experiment presented in Fig. 10.1, 50-μl reactions were assembled on ice by mixing RhlB and RneHC2 (10 μl, containing 10 mM Tris HCl, pH 7.5, 0.5 M NaCl, 0.5% Triton X-100, 5% glycerol, 1 mM EDTA, 1 mM DTT, and 0.5 mg/ml BSA) with RNA and ATP (40 μl, containing 10 mM Tris HCl, pH 7.5, 5 mM MgCl$_2$, 2.5 mM ATP, and 0.25 mg/ml yeast RNA). The reaction is incubated at 37 °C for 5 to 60 min (see following) and stopped by the addition of 50 μl of 12% SDS. The production of Pi was measured as follows. Just before use, 1 vol of 6% ascorbic acid and 1 vol of 1% ammonium molybdate are mixed, and 100 μl of this mix is added to the reaction, which is then incubated at 37 °C for 10 min. A solution (150 μl) containing 2% sodium citrate, 2% sodium meta-arsenate, and 2% acetic acid is added, and the reaction is incubated for another 10 min at 37 °C. At this point, the blue color, which is stable for several hours at room temperature, can be quantified by measuring the optical density at 850 nm. To calibrate the assay, a standard curve is constructed with 1 to 20 nmol of sodium phosphate, pH 7.0. For accurate determinations of ATPase activity, the reactions are performed in triplicate, and initial velocities are obtained by adjusting the time of incubation to limit the extent of hydrolysis to less than 15 nmol of ATP (15% of the total amount of ATP in the reaction).

6.2. RNA unwinding assay

This protocol is adapted from Chandran *et al.* (2007). An RNA oligonucleotide (see the legend of Fig. 10.2), chemically synthesized with an Applied Biosystems automated synthesizer, was phosphorylated with T4 polynucleotide kinase (Biolabs) and γ-[^{32}P]-ATP, and separated from the unincorporated radioactivity on a Sephadex G-25 column (Pharmacia). The specific activity of the radiolabeled RNA was $\sim$4 × 10^6 cpm/pmol. The labeled oligonucleotide was further purified by electrophoresis in a 12% DNA sequencing gel and eluted in 500 mM ammonium acetate, 10 mM magnesium acetate, 1 mM EDTA, 0.1 % SDS. After brief centrifugation to remove gel fragments, the oligonucleotide was precipitated with 2.5 volumes of ethanol and the pellet resuspended in 50 μl of hybridization buffer (20 mM

Tris-HCl, pH 7.5, 500 mM NaCl, 1 mM EDTA). The oligonucleotide was heated at 90 °C for 3 min and then hybridized overnight at room temperature (more than 90% of the RNA forms duplexes). From the sequence, the RNA duplex has a predicted ΔG of −22.2 kcal/mol (37 °C) and a T_m of 48 °C (0.1 nM) (Xia $et\ al.$, 1998).

RNA unwinding activity was assayed by measuring the conversion of duplex RNA to monomers. Purified proteins were incubated with 6 fmol (24,000 cpm) of duplex RNA in a 50-μl reaction (0.12 nM) containing 10 mM Tris-HCl, pH 7.5, 5 mM MgCl$_2$, 60 mM KCl, 40 μg of BSA, 100 ng of yeast tRNA, 50 U of RNasin (Promega), and, when indicated, ATP or [γ^{35}S]-ATP (2 mM). After incubation at 37 °C for 30 min, the reactions were stopped by the addition of 50 μl of 20% glycerol (0.2% SDS), 4 mM EDTA, and analyzed directly by electrophoresis in a native 0.5× TBE-12% polyacrylamide gel (29:1). The gel was dried and visualized with a PhosphorImager.

6.3. mRNA degradation assay

This protocol is adapted from Py $et\ al.$ (1996) and Khemici $et\ al.$ (2004). A mini-degradosome consisting of the noncatalytic region of RNase E, RhlB, and PNPase was reconstituted as described in Khemici $et\ al.$ (2004). The 375-nt uniformly radiolabeled RNA substrate covering the $malE$-$malF$ intergenic region was synthesized with T7 RNA polymerase, [^{33}P]-α-UTP and the plasmid pCH77 digested with EcoRI as the template (Py $et\ al.$, 1996). In the experiment in Fig. 10.3, the RNA substrate and the mini-degradosome were both present at a final concentration of 20 nM. The molar concentration of the mini-degradosome was calculated with the mass of PNPase in the reconstituted complex and 234 kDa as the molecular weight of the native trimeric PNPase. Just before performing the reaction, the RNA substrate was renatured in 20 mM Tris HCl, pH 7.5, by heating for 10 min at 90 °C and then incubating for 20 min at 37 °C. The final reaction mixture (40 μl) containing 20 mM Tris HCl, pH 7.5, 5 mM DTT, 1 mM MgCl$_2$, 10 mM K$_2$HPO$_4$, 40 units of RNasin (Promega), and 1 mM Mg-ATP (as indicated) was incubated at 30 °C. The reaction was initiated by the addition of 2 μl of enzyme and, at the indicated times, 4-μl aliquots were withdrawn, mixed with 4 μl of proteinase K solution (10 mM Tris HCl, pH 7.5, 10 mM EDTA, 0.2% SDS, 0.5 mg/ml proteinase K) and held at room temperature. At the end of the time course, the samples were incubated at 50 °C for 30 min, cooled to room temperature, and then 24 μl formamide dye loading mix was added. The samples were heat denatured, and 10 μl was separated by denaturing polyacrylamide gel electrophoresis (6%, 29:1, 0.5× TBE, 8 M urea). The gel was dried and visualized with a PhosphorImager.

ACKNOWLEDGMENTS

Our research in Toulouse is supported by the Centre National de la Recherche Scientifique (CNRS) with additional funding from the Agence Nationale de la Recherche (ANR, grant NT05_1-44659).

REFERENCES

Callaghan, A. J., Aurikko, J. P., Ilag, L. L., Gunter Grossmann, J., Chandran, V., Kuhnel, K., Poljak, L., Carpousis, A. J., Robinson, C. V., Symmons, M. F., and Luisi, B. F. (2004). Studies of the RNA degradosome-organizing domain of the *Escherichia coli* ribonuclease RNase E. *J. Mol. Biol.* **340,** 965–979.

Carpousis, A. J. (2007). The RNA degradosome of *Escherichia coli*: An mRNA-degrading machine assembled on RNase E. *Annu. Rev. Microbiol.* **61,** 71–87.

Carpousis, A. J., Van Houwe, G., Ehretsmann, C., and Krisch, H. M. (1994). Copurification of *E. coli* RNAase E and PNPase: Evidence for a specific association between two enzymes important in RNA processing and degradation. *Cell* **76,** 889–900.

Carpousis, A. J., Vanzo, N. F., and Raynal, L. C. (1999). mRNA degradation: A tale of poly (A) and multiprotein machines. *Trends Genet.* **15,** 24–28.

Caruthers, J. M., and McKay, D. B. (2002). Helicase structure and mechanism. *Curr. Opin. Struct. Biol.* **12,** 123–133.

Celesnik, H., Deana, A., and Belasco, J. G. (2007). Initiation of RNA decay in *Escherichia coli* by 5′-pyrophosphate removal. *Mol. Cell* **27,** 79–90.

Chandran, V., Poljak, L., Vanzo, N. F., Leroy, A., Miguel, R. N., Fernandez-Recio, J., Parkinson, J., Burns, C., Carpousis, A. J., and Luisi, B. F. (2007). Recognition and cooperation between the ATP-dependent RNA helicase RhlB and ribonuclease RNase E. *J. Mol. Biol.* **367,** 113–132.

Charollais, J., Dreyfus, M., and Iost, I. (2004). CsdA, a cold-shock RNA helicase from *Escherichia coli*, is involved in the biogenesis of 50S ribosomal subunit. *Nucleic Acids Res.* **32,** 2751–2759.

Charollais, J., Pflieger, D., Vinh, J., Dreyfus, M., and Iost, I. (2003). The DEAD-box RNA helicase SrmB is involved in the assembly of 50S ribosomal subunits in *Escherichia coli*. *Mol. Microbiol.* **48,** 1253–1265.

Cheng, Z. F., and Deutscher, M. P. (2005). An important role for RNase R in mRNA decay. *Mol. Cell* **17,** 313–318.

Chifflet, S., Torriglia, A., Chiesa, R., and Tolosa, S. (1988). A method for the determination of inorganic phosphate in the presence of labile organic phosphate and high concentrations of protein: Application to lens ATPases. *Anal. Biochem.* **168,** 1–4.

Coburn, G. A., and Mackie, G. A. (1999). Degradation of mRNA in *Escherichia coli*: An old problem with some new twists. *Prog. Nucleic Acid Res. Mol. Biol.* **62,** 55–108.

Coburn, G. A., Miao, X., Briant, D. J., and Mackie, G. A. (1999). Reconstitution of a minimal RNA degradosome demonstrates functional coordination between a 3′-exonuclease and a DEAD-box RNA helicase. *Genes. Dev.* **13,** 2594–2603.

Cohen, S. N., and McDowall, K. J. (1997). RNase E: Still a wonderfully mysterious enzyme. *Mol. Microbiol.* **23,** 1099–1106.

Cordin, O., Banroques, J., Tanner, N. K., and Linder, P. (2006). The DEAD-box protein family of RNA helicases. *Gene.* **367,** 17–37.

Grunberg-Manago, M. (1999). Messenger RNA stability and its role in control of gene expression in bacteria and phages. *Annu. Rev. Genet.* **33,** 193–227.

Grunberg-Manago, M., Oritz, P. J., and Ochoa, S. (1955). Enzymatic synthesis of nucleic acidlike polynucleotides. *Science* **122,** 907–910.

Iost, I., and Dreyfus, M. (2006). DEAD-box RNA helicases in *Escherichia coli*. *Nucleic Acids Res.* **34,** 4189–4197.

Khemici, V., and Carpousis, A. J. (2004). The RNA degradosome and poly(A) polymerase of *Escherichia coli* are required *in vivo* for the degradation of small mRNA decay intermediates containing REP-stabilizers. *Mol. Microbiol.* **51,** 777–790.

Khemici, V., Poljak, L., Toesca, I., and Carpousis, A. J. (2005). Evidence *in vivo* that the DEAD-box RNA helicase RhlB facilitates the degradation of ribosome-free mRNA by RNase E. *Proc. Natl. Acad. Sci. USA* **102,** 6913–6918.

Khemici, V., Toesca, I., Poljak, L., Vanzo, N. F., and Carpousis, A. J. (2004). The RNase E of *Escherichia coli* has at least two binding sites for DEAD-box RNA helicases: Functional replacement of RhlB by RhlE. *Mol. Microbiol.* **54,** 1422–1430.

Kushner, S. R. (2002). mRNA decay in *Escherichia coli* comes of age. *J. Bacteriol.* **184,** 4658–4665.

Leroy, A., Vanzo, N. F., Sousa, S., Dreyfus, M., and Carpousis, A. J. (2002). Function in *Escherichia coli* of the non-catalytic part of RNase E: Role in the degradation of ribosome-free mRNA. *Mol. Microbiol.* **45,** 1231–1243.

Linder, P. (2006). Dead-box proteins: A family affair—active and passive players in RNP-remodeling. *Nucleic Acids Res.* **34,** 4168–4180.

Littauer, U. Z., and Soreq, H. (1982). Polynucleotide phosphorylase. In "The Enzymes", third edit. (P. D. Boyer, ed.), Vol. 15. Academic Press, New York.

Marcaida, M. J., DePristo, M. A., Chandran, V., Carpousis, A. J., and Luisi, B. F. (2006). The RNA degradosome: Life in the fast lane of adaptive molecular evolution. *Trends Biochem. Sci.* **31,** 359–365.

McLaren, R. S., Newbury, S. F., Dance, G. S., Causton, H. C., and Higgins, C. F. (1991). mRNA degradation by processive 3′ to 5′-exoribonucleases *in vitro* and the implications for prokaryotic mRNA decay *in vivo. J. Mol. Biol.* **221,** 81–95.

Miczak, A., Kaberdin, V. R., Wei, C. L., and Lin-Chao, S. (1996). Proteins associated with RNase E in a multicomponent ribonucleolytic complex. *Proc. Natl. Acad. Sci. USA* **93,** 3865–3869.

Newbury, S. F., Smith, N. H., and Higgins, C. F. (1987). Differential mRNA stability controls relative gene expression within a polycistronic operon. *Cell* **51,** 1131–1143.

Newbury, S. F., Smith, N. H., Robinson, E. C., Hiles, I. D., and Higgins, C. F. (1987). Stabilization of translationally active mRNA by prokaryotic REP sequences. *Cell* **48,** 297–310.

Py, B., Higgins, C. F., Krisch, H. M., and Carpousis, A. J. (1996). A DEAD-box RNA helicase in the *Escherichia coli* RNA degradosome. *Nature* **381,** 169–172.

Symmons, M. F., Jones, G. H., and Luisi, B. F. (2000). A duplicated fold is the structural basis for polynucleotide phosphorylase catalytic activity, processivity, and regulation. *Structure Fold Des.* **8,** 1215–1226.

Symmons, M. F., Williams, M. G., Luisi, B. F., Jones, G. H., and Carpousis, A. J. (2002). Running rings around RNA: A superfamily of phosphate-dependent RNases. *Trends Biochem. Sci.* **27,** 11–18.

Tanner, N. K., and Linder, P. (2001). DExD/H box RNA helicases: From generic motors to specific dissociation functions. *Mol. Cell* **8,** 251–262.

Vanzo, N. F., Li, Y. S., Py, B., Blum, E., Higgins, C. F., Raynal, L. C., Krisch, H. M., and Carpousis, A. J. (1998). Ribonuclease E organizes the protein interactions in the *Escherichia coli* RNA degradosome. *Genes Dev.* **12,** 2770–2781.

Worrall, J. A., Howe, F. S., McKay, A. R., Robinson, C. V., and Luisi, B. F. (2008). Allosteric activation of the ATPase activity of the *Escherichia coli* RhlB RNA helicase. *J. Biol. Chem.* **283,** 5567–5576.

Xia, T., SantaLucia, J., Jr., Burkard, M. E., Kierzek, R., Schroeder, S. J., Jiao, X., Cox, C., and Turner, D. H. (1998). Thermodynamic parameters for an expanded nearest-neighbor model for formation of RNA duplexes with Watson-Crick base pairs. *Biochemistry* **37,** 14719–14735.

CHAPTER ELEVEN

PREPARATION OF THE *ESCHERICHIA COLI* RNASE E PROTEIN AND RECONSTITUTION OF THE RNA DEGRADOSOME

George A. Mackie, Glen A. Coburn, Xin Miao, Douglas J. Briant, Annie Prud'homme-Généreux, Leigh M. Stickney, *and* Janet S. Hankins

Contents

Abstract

The RNA degradosome is a multienzyme complex that plays a key role in the processing of stable RNAs, the degradation of mRNAs, and the action of small regulatory RNAs. Initially discovered in *Escherichia coli*, similar or related complexes are found in other bacteria. The core of the RNA degradosome is the essential endoribonuclease, RNase E. The C-terminus of this enzyme serves as a scaffold to which other components of the RNA degradosome bind.

Department of Biochemistry Molecular Biology, University of British Columbia, Vancouver, British Columbia, Canada

Methods in Enzymology, Volume 447
ISSN 0076-6879, DOI: 10.1016/S0076-6879(08)02211-8

These ligands include the phosphorolytic 3′-exonuclease, polynucleotide phosphorylase, the DEAD-box RNA helicase, RhlB, and the glycolytic enzyme, enolase. In addition, the DEAD-box RNA helicases CsdA and RhlE and the RNA binding protein, Hfq, may bind to RNase E in place of one or more of the prototypical components. This chapter describes purification of RNase E (the Rne protein), reconstitution of a minimal degradosome that recapitulates the activity of authentic degradosomes, and methods for the assay of the reconstituted complex.

1. Introduction

David Apirion and his colleagues (Ghora and Apirion, 1978; Misra and Apirion, 1979) discovered and characterized the endoribonuclease RNase E from *Escherichia coli*. It was initially described as an activity required for the penultimate step in the maturation of 5S rRNA, the processing of a 9S precursor to a 126-residue pre-5S RNA. It has since emerged as a major player in both stable RNA processing and RNA decay in *E. coli* (Belasco, 1993; Carpousis, 2007; Coburn and Mackie, 1999; Deutscher, 2006; Melefors *et al.*, 1993; Ow and Kushner, 2002). RNase E activity is readily demonstrated in crude extracts, but purification and characterization of the enzyme proved surprisingly difficult (Coburn and Mackie, 1999). Two approaches to this challenge ultimately succeeded. In the first, described in this series by Carpousis *et al.* (2001) and elsewhere (Carpousis *et al.*, 1994; Miczak *et al.*, 1996; Py *et al.*, 1996), RNase E activity was purified as part of a larger complex, the RNA degradosome. In the second, the complete *rne* gene was cloned and sequenced permitting overexpression of its product, the Rne protein (Cormack *et al.*, 1993). The Rne protein is surprisingly large (1061 aa residues) and acidic (pI $= 5.4$). In purified form, it manifests endonucleolytic activity identical to that of crude RNase E in the absence of other components of the RNA degradosome (Cormack *et al.*, 1993).

The discovery that a number of key components of the RNA processing and decay apparatus are organized into a multicomponent complex, the RNA degradosome (Carpousis *et al.*, 1994; Miczak *et al.*, 1996; Py *et al.*, 1996), begs the question of the function of the individual components. One avenue of investigation relies on reconstitution of the RNA degradosome from purified components (Coburn *et al.*, 1999; Mackie *et al.*, 2001). Ultimately, this should permit the dissection of the roles of the individual components and the assembly *in vitro* of complexes prepared from mutant proteins. Methods for the purification of Rne, the key scaffold in the assembly process, and for the reconstitution of active degradosome from individually purified enzymes were described in volume 342B in this series (Mackie *et al.*, 2001). This chapter updates the previous chapter.

2. Preparation of Crude RNase E

2.1. Buffers and solutions

M9ZB (per liter):

  10 g BDH peptone from casein (cat. # 1.07213) (Difco Bacto tryptone or Humco Sheffield NZ amine can be substituted).

  6 g Na_2PO_4.

  3 g KH_2PO_4.

  1 g NH_4Cl.

  Sterilize by autoclaving.

  Supplement with sterile $MgSO_4$ to 1 mM and $CaCl_2$ to 0.1 mM.

NZYCM (per liter):

  10 g BDH peptone (or equivalent).

  5 g Yeast extract.

  2.5 g Difco casamino acids.

  Sterilize by autoclaving.

  Add sterile $MgSO_4$ to 1 mM.

IPTG:

  0.1 M solution prepared by filter sterilization.

Buffer A:

  50 mM Tris–HCl.

  10 mM $MgCl_2$.

  60 mM NH_4Cl.

  0.5 mM EDTA.

  pH adjusted to 7.8.

  Sterilized by autoclaving.

  Supplemented with either 6 mM β-mercaptoethanol or 0.5 mM DTT and with protease inhibitors (see the text).

Buffer D:

  20 mM Tris–HCl.

  1 mM $MgCl_2$.

  0.1 mM EDTA.

  Adjusted to pH 8.0.

  Sterilized by autoclaving.

  Supplemented as described in the text.

Buffer G:

  50 mM Tris–HCl, pH 8.0.

  6 M Guanidine-HCl.

  150 mM NaCl.

  20% Glycerol.

  0.1 mM EDTA.

  Supplemented with DTT to 1 mM.

Buffer R:

   50 mM Tris-HCl, pH 8.0 (or 50 mM HEPES-KOH, pH 7.6)

   150 mM NaCl

   20% Glycerol

   0.1 mM EDTA

   Supplemented with DTT to 1 mM

Reconstitution buffer:

   20 mM Tris-HCl, pH 7.5

   1 mM MgCl$_2$

   20 mM KCl

   1.5 mM DTT

   10 mM Na-phosphate, pH 7.5 (optional)

   3 mM ATP (neutralized) (optional)

Formamide dye buffer:

   Deionized formamide

   0.5× TBE buffer (see below)

   0.01% Xylene cyanol FF

   0.01% Bromophenol blue

SDS running buffer:

   50 mM Tris base

   190 mM glycine

   0.5 mM neutralized EDTA

   0.1% SDS

TBE running buffer:

   90 mM Tris base

   90 mM Boric acid

   2 mM EDTA (disodium salt)

2.2. Growth of cultures

The following method applies to recombinant strains. To ensure the retention of plasmids, starter cultures must not be overgrown. The appropriate strain (e.g., strain GM402 [BL21(DE3) containing pGM102]; Cormack *et al.*, 1993) is used to inoculate into 10 ml of supplemented M9ZB medium containing: at least 50 mg/L ampicillin or 25 mg/L carbenicillin. Two serial 100-fold dilutions are made from this culture into identical 10-ml aliquots of the same medium. Cultures are grown overnight with shaking at 30 °C.

   A 1:100 dilution of a turbid, but not saturated, overnight culture is used to initiate growth of larger cultures in supplemented M9ZB medium containing carbenicillin. It is critical to maintain vigorous aeration during growth. For this reason we use the largest flasks available and add no more than one-tenth its nominal volume (e.g., 200 ml of culture in a 2-L flask). Cultures are grown at 29 to 30 °C with vigorous shaking. Growth is measured using turbidity (e.g., absorbance at 600 nm). When cultures reach an A_{600} of 0.4, IPTG is added to 1 mM. After 15 min, the cultures are diluted with an equal volume of warm NZYCM containing 50 mg/L carbenicillin and growth is continued at 29 to 30 °C with vigorous shaking for up to 5 h.

At harvest, each flask of culture is swirled for a minute in a slurry of ice and water to ensure rapid chilling. All subsequent manipulations are performed in a cold room or at 4 °C. Cultures are transferred to cold, tared centrifuge bottles and the cells harvested by centrifugation for 15 min at 5000 rpm in a Beckman JA-10 rotor or its equivalent. The supernatants are discarded, and the cell pellets are completely suspended with one-fourth the original culture volume of cold buffer A containing 6 mM β-mercaptoethanol. The suspension is pooled in a single bottle, and the cells harvested by centrifugation as before. The supernatant is discarded and any remaining liquid carefully wiped from the neck of the bottle. The pellet is weighed (yields are approximately 4 to 5 g/L) and is suspended with 3.5 ml/g wet weight of buffer A containing 0.1 mM DTT and 7.5% glycerol. Care should be taken to obtain a smooth suspension without excessive shearing. The suspended cells can be processed further or frozen quickly and stored at −70 °C. Frozen cells should be processed within a few days.

2.3. Preparation of S-30 extracts

If frozen, cell suspensions are thawed in cold water in a cold room then transferred to a chilled Aminco French pressure cell. The suspension is passed twice through the cell at a pressure of 8000 psi. Higher pressures seem to cause the loss of RNase E activity. The lysate is supplemented with freshly prepared DTT to 0.1 mM (experience suggests that 0.5 to 1.0 mM is preferable), with 1 to 2 U/mL DNase I (Sigma), and with 1 μg/ml of each of the following protease inhibitors: leupeptin, aprotinin, and pepstatin (all obtained from Sigma). Phenylmethylsulfonylfluoride (PMSF; Sigma-Aldrich) can also be added but is relatively ineffective in the buffers used. The lysate is left on ice for 10 min and then clarified by centrifugation for 45 min at 15,000 rpm in a Beckman JA-20 rotor or equivalent. The supernatant (S-30) is saved and the volume determined. The pellet is discarded. Inclusion bodies, a problem with some truncated forms of the Rne protein, can be found in this pellet.

2.4. Preparation of the AS-26 fraction

The clarified crude extract is diluted fourfold with buffer A containing 0.1 mM DTT and 7.5% glycerol and the cocktail of protease inhibitors enumerated previously. While the lysate is stirred on ice, enzyme grade $(NH_4)_2SO_4$ is added gradually over a period of 15 min to reach a concentration of 26% (weight/volume) or approximately 45% of saturation. It is not necessary to adjust the pH. After stirring for 15 min to ensure equilibration, the suspension is centrifuged for 20 min at 12,000 rpm in a Beckman JA-20 rotor. The supernatant is decanted, and the pellet is suspended with 1 ml of buffer A supplemented with DTT and protease inhibitors as earlier for each gram wet weight of the original cell pellet.

The suspended ammonium sulfate pellet is dialyzed at 4 °C against two changes of 100 volumes of buffer D containing 60 mM NH$_4$Cl, 0.1 mM DTT, 10% glycerol for 4 to 5 h in total. The dialyzed material, the AS-26 fraction, is divided into suitable portions (e.g., 1 ml) and quick frozen before storage at −70 °C. A small portion should be retained for the determination of protein concentration (typically 4 to 7 mg/ml), for assaying RNase E activity and for examining the quality of the preparation and the level of expression of Rne by SDS-PAGE. Characterization of this preparation has been described (Mackie, 1991).

3. Purification of RNase E

3.1. Preparative gel electrophoresis

The following is an expansion of a published method (Cormack *et al.*, 1993). We typically use a preparative slab gel (Gibco-BRL) with interior dimensions of 170 mm × 150 mm × 1.5 mm. The lower separating gel contains 5.5% acrylamide (49:1 acrylamide/bisacrylamide), 0.375 M Tris-HCl, pH 8.8, 5% glycerol, 0.1% SDS. The upper stacking gel contains 4.5% acrylamide (49:1 acrylamide/bisacrylamide), 0.06 M Tris-HCl, pH 6.8, 0.5 mM neutralized EDTA, and 0.1% SDS. The running buffer was described previously. A sample containing 3 mg of the AS-26 fraction is diluted with an equal volume of 120 mM Tris-HCl, pH 6.8, 3% SDS, 100 mM DTT, 0.5 mM EDTA, 10% glycerol, and traces of bromophenol blue. The sample is boiled for 5 min and applied to the gel. The initial voltage is set at 85 V until the tracking dye has entered the separating gel; the voltage is raised to 150 to 180 V, and the run is continued until the dye has exited the gel for 10 min. After the apparatus is disassembled, a vertical slice is removed from one or both edges of the gel and stained with 0.1% Coomassie blue in H$_2$O. Staining in an acidic fixing solution is also possible, but there is significant shrinking and swelling of the gel, making subsequent alignment difficult. The remainder of the gel is left adhering to one of the plates, wrapped with cellophane (e.g., Saran Wrap^{TM}) and is stored at 4 °C. After destaining, the guide strip is used to identify the position of the Rne band. We find it helpful to lay the plate with the gel on a sheet of finely ruled paper (e.g., graph paper) to align the guide strip and the main gel. The region of the main gel containing the Rne band is excised with a razor blade and cut into small cubes no more than 2 mm in any dimension.

Electroelution of the Rne protein can be accomplished in several ways. We have used a BioRad electroelution apparatus (Model No. 422) with success. Before use, the glass tubes and fritted discs are autoclaved in H_2O previously treated with 0.1% diethyl pyrocarbonate (Sigma–Aldrich) while the rubber bungs are soaked in 0.1% SDS at 60 °C for 60 min and rinsed with sterile H_2O. The dialysis caps are pretreated as recommended by the manufacturer. In this method, the cubed gel pieces are placed in a tube to a depth of approximately 1 cm above a fritted disc (see the manual for the BioRad Model 422). The tube is filled with running buffer (as previously) and placed in the BioRad apparatus. Elution is performed at 9 to 10 mA per tube for up to 7 h at ambient temperature. The lower reservoir is stirred, and buffer is continuously recirculated with a peristaltic pump, taking precautions not to short-circuit the current path. Bubbles tend to accumulate under the bottom of the dialysis caps. They can be removed with a stream of buffer delivered carefully from a curved disposable pipette.

After electroelution, the glass tube and frit are slipped carefully from the rubber bung and dialysis cap. This will cause foaming of the eluate, which can be suppressed with 2 to 5 μl of 2-butanol. The eluates are removed with a pipetter and transferred to a Corex centrifuge tube. The dialysis caps are rinsed sequentially with 0.3 ml of running buffer that is pooled with the eluates. DTT is added to 1 mM and the eluted protein precipitated by the addition of 4.5 volumes of acetone. SDS will precipitate immediately, but the suspension is left at -15 °C. The precipitated material is recovered by centrifugation at 12,000 rpm for 60 min in a Beckman JA-20 rotor. The supernatant is removed, and the pellet is washed three times with a total of 20 ml of 80% acetone containing 1 mM DTT. The final pellet is "dried" by allowing any acetone left to evaporate at ambient temperature for 10 min.

In an alternate method, the cubed pieces of gel containing Rne are placed in a sealed dialysis bag containing a minimal volume (<1 ml) of running buffer. The dialysis bag is immersed in running buffer and placed perpendicular to the direction of current flow in a small "submarine" gel apparatus that is used as an electrophoresis chamber. Sufficient current is applied to achieve a 100-V potential difference. Electroelution is continued for 6 to 7 h, followed by reversing polarity for 30 sec. The eluted protein is recovered from the dialysis bag, the bag is rinsed with 0.5 ml of running buffer, and the combined eluates are precipitated with acetone as described previously.

3.2. Renaturation of Rne

The pellet containing Rne (as an SDS-salt) is dissolved at room temperature in a minimal volume of buffer G supplemented with 1 mM DTT. Ideally, to avoid excessive dilution, no more than 0.5 ml of buffer should be used. This material is warmed to 37 °C for up to 60 min. In the original method, the dissolved material is then diluted 20- to 40-fold in buffer R supplemented with 1 m*M* DTT, 0.1 m*M* PMSF, 1 μg/ml of each of leupeptin, aprotinin, and pepstatin. The intent is to reduce the concentration of guanidine-HCl rapidly to less than 0.25 m*M* to promote refolding of the denatured protein (Hager and Burgess, 1980). As an alternative, we have dissolved the Rne-SDS pellet with 1.0 ml of buffer G, warmed it as earlier, and diluted this threefold with buffer R so that the guanidine-HCl concentration is reduced to 1.5 m*M*. Refolding is achieved during dialysis (see following). In either case, the diluted Rne is dialyzed against two changes of 100 volumes each of buffer D containing 60 m*M* NH_4Cl, 0.1 m*M* DTT, and 10% glycerol at 4 °C for at least 20 h in total. The dialyzed material is concentrated 10- to 20-fold in a centrifugal concentrator (e.g., a Millipore Centricon-30). The quality of the eluted protein and its approximate concentration can be determined by analysis on analytical SDS-PAGE with proteins of known concentration (e.g., BSA or β-galactosidase) as standards. Examples for two different preparations have been published (Coburn *et al.*, 1999; Cormack *et al.*, 1993). This method has also been applied successfully to soluble deletions of Rne (Cormack *et al.*, 1993). Methods for assaying RNase E activity are described in the following and elsewhere (Coburn *et al.*, 1999; Cormack *et al.*, 1993; Mackie *et al.*, 2001).

3.3. Nondenaturing purification of Rne

Virtually all recent strategies use overexpressing strains to facilitate purification, in some cases with purification "tags" (Miczak *et al.*, 1996). Taraseviciene and coworkers (1994) prepared extracts from an overexpressing strain by sonication and fractionated the crude extract by centrifugation at 200,000*g* in the presence of 0.4 m*M* $(NH_4)_2SO_4$ (S200). This step presumably released Rne from ribosomes or membrane fragments. Enrichment of the Rne protein in the S200 was achieved by precipitation with $(NH_4)_2SO_4$ to 40% of saturation. Proteins in the pellet were fractionated on a Toyopearl gel filtration column. Subsequent immunoaffinity purification on tandem sepharose columns charged with nonimmune and anti-Rne antisera, respectively, yielded a preparation of Rne that visually seemed to be >60% homogeneous. Some of the smaller protein species present in the purified material may represent proteolytic breakdown products. This procedure did result in a very significant loss of activity, mostly in the immunoaffinity step.

We have developed an independent alternative to the foregoing procedure. The principal difficulty is that only a portion, approximately 40%, of the overexpressed Rne protein behaves as a "soluble" protein, whereas most is either aggregated or in a complex, presumably a reflection of its membrane association (Liou *et al.*, 2001) or cytoskeletal-like structure (Taghbalout and Rothfield, 2007). Cultures of GM402 were grown and extracts prepared as described previously. The crude extract was made with

0.5 mM in NH$_4$Cl and centrifuged for 60 min at 200,000g in a Beckman Ti60 rotor at 4 °C to release Rne (Carpousis *et al.*, 1994) that was concentrated by precipitation by (NH$_4$)$_2$SO$_4$ to 40% of saturation. This material was "soluble" but eluted from a BioGel A 1.5-mM column in the flow through, ahead of β-galactosidase and RNA polymerase. Better results, however, were obtained by passing material from the ammonium sulfate pellet over a "Hi-trap Affi-Gel Blue" column (Pharmacia-Amersham Biotech, Ltd.). Rne was applied to this column in 10 mM HEPES, 1 mM EDTA, 2.5% glycerol, 0.25% Genapol X-080, 0.1 mM DTT, 0.1 mM PMSF, pH 7.6. Rne was retained on the column and eluted with a linear gradient of KCl (0 to 3 M in the loading buffer) at approximately 2 M KCl. Although not homogeneous, this fraction was almost eightfold more active than the ammonium sulfate pellet from which it is derived.

A significant portion of the Rne in the crude extract, up to 60%, was not soluble after extraction with NH$_4$Cl and remained in the 200,000g pellet unless a detergent such as 0.25% Genapol X-080 (Fluka) was included in the extraction buffer. The 200,000g supernatant in such an example was also fractionated with ammonium sulfate and passed over an Affi-Gel blue column as described previously. Roughly half the applied Rne flowed through this column and half was retained, eluting at approximately 2 mM KCl as earlier. The yield of Rne can be increased by applying the material in the flow-through to a second Affi-gel Blue column. The major contaminants in the fractions retained on Affi-gel Blue were of low molecular weight and presumably could be removed by gel filtration.

4. RECONSTITUTION OF THE RNA DEGRADOSOME

4.1. Purification of the components

The degradosome contains roughly stoichiometric quantities of Rne, a tetramer (Callaghan *et al.*, 2003), polynucleotide phosphorylase (PNPase), a trimer (Symmons *et al.*, 2000), but possibly in multiple copies (Callaghan et al., 2004), enolase, a dimer (Kühnel and Luisi, 2001), and RhlB, reported either as a monomer (Callaghan *et al.*, 2004; Chandran *et al.*, 2007) or a dimer in solution (Liou *et al.*, 2002). Some preparations contain lesser amounts of other proteins, including polyphosphate kinase (Blum *et al.*, 1997). The purification of Rne is described previously. Purifications of PNPase (Coburn and Mackie, 1998; Jones *et al.*, 2003) and RhlB (Coburn *et al.*, 1999) from recombinant strains overexpressing these enzymes have also been described. Purification tags were not used to facilitate purification, and the effect of their presence on reconstitution has not been determined. An N-terminal His-6 tag seems to inhibit PNPase significantly (unpublished data). A purification of enolase has been developed in the laboratory of Dr. Ben Luisi (Kühnel and Luisi, 2001). At least two other DEAD-box RNA helicases have also been reported to interact with RNase E and reconstitute ATP-dependent activity *in vitro*: CsdA (Prud'homme-Généreux *et al.*, 2004) and RhlE (Khemici *et al.*, 2004). The purification of His-6–tagged CsdA has been reported (Prud'homme-Généreux *et al.*, 2004); this preparation is active in reconstitution. Finally, alternative forms of the RNA degradosome have been reported: the "cold-shock degradosome" (Prud'homme-Généreux *et al.*, 2004) and a form in which the RNA binding protein, Hfq, substitutes for polynucleotide phosphorylase, RhlB, and enolase (Morita *et al.*, 2005; see Chapter 18 by Morita *et al.*, in this volume). The former, but not the latter, is described in the following.

4.2. Reconstitution

A method for reconstituting an active "minimal" degradosome has been described (Coburn *et al.*, 1999; Mackie *et al.*, 2001). In outline, incubations contain a suitable buffer, an RNA substrate, ATP and/or Na-phosphate, and recombinant proteins. This mixture is incubated, and samples are withdrawn to assay the disappearance of substrate. Alternately, the formation of protein–protein complexes can be assessed by coimmunoprecipitation.

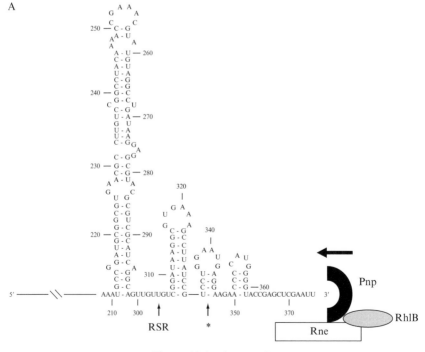

Figure 11.1   (*continued*)

The standard assay for activity after reconstitution is based on that described by Py *et al.* (1996). The buffer for reconstitution (see earlier) contains much lower ionic strengths than the standard buffer for assaying RNase E activity (Mackie, 1991). Na-phosphate and ATP are not necessary for the physical reconstitution of complete or partial degradosomes. The assay of coupled RNA helicase-polynucleotide phosphorylase activity requires that substrates possess a single-stranded 3′-extension (Blum *et al.*, 1999; Coburn *et al.*, 1999). Figure 11.1 illustrates two suitable substrates: (A) a 375-nt RNA containing the *malEF* intercistronic region (Coburn *et al.*, 1999; Py *et al.*, 1996) or (B) a 210-nt RNA containing 180 nt from the 3′-end of the *rpsT* mRNA extended by a 30-nt poly(A) tail (*rpsT*[268–447]-poly[A] (Coburn *et al.*, 1999). Methods for preparing these RNAs

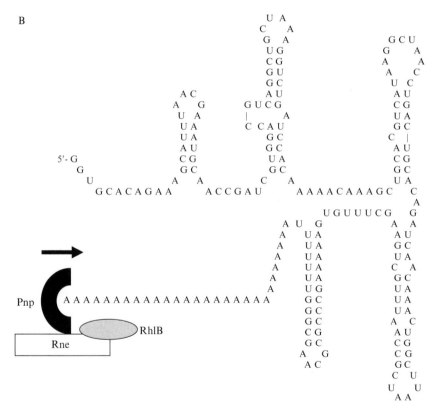

Figure 11.1 The secondary structures of substrates used to assay degradosome activity. (A) The 375-nt *malEF* RNA from the maltose operon (Coburn *et al.*, 1999; McLaren *et al.*, 1991; Mackie *et al.*, 2001; Py *et al.*, 1996). The 3′-attack of a minimal degradosome is shown schematically (Rne, open rectangle; PNPase, filled semi-circle; RhlB, shaded oval). RSR, REP-stabilized RNA (because of stalling of PNPase at the base of the stable hairpin); ⋆, an intermediate stalled product (Coburn *et al.*, 1999). (B) The 210-nt *rpsT*/ 180-poly(A)₃₀ RNA (Coburn *et al.*, 1999; Mackie *et al.*, 1997). The attack of a minimal degradosome on the 3′-poly(A) extension is shown schematically as in (A).

have been described (Coburn *et al.,* 1999; Mackie, 1991). A typical recon-stitution (40 µl) is assembled at 4 °C in a microcentrifuge tube by adding water, buffer, and cofactors first. Proteins are added to the incubation in the following amounts per 40 µl: Rne, 40 ng (~350 fmol of monomer); PNPase, 80 ng (~350 fmol of trimer); RhlB, 40 ng (~350 fmol of dimer); CsdA (substituted for RhlB), 40 ng (~600 fmol of monomer). Assays of activity are initiated by adding the RNA substrate to a final concentration of 20 n*M.* Incubations are performed at 30 °C. For the assessment of physical reassociation by coimmunoprecipitation, the quantities of protein and the volume of incubation should be increased by a factor of 25. Details of the procedures for immunoprecipitation are outlined elsewhere (Coburn *et al.,* 1999; Vanzo *et al.,* 1998).

Reconstituted activity is assessed by monitoring the ATP and phosphate-dependent disappearance of the full-length substrate in 6% poly-acrylamide gels containing 8 *M* urea (Coburn *et al.,* 1999; Py *et al.,* 1996) in TBE buffer. Samples, typically 4 µl, are withdrawn from the assay at the desired time, mixed with 12 µl of formamide dye buffer, and boiled for 60 sec before separation. The gels are run until the xylene cyanol dye has migrated approximately 75% of the length of the gel for *rps*T RNA sub-strates or until the dye has run off for *malEF* RNA substrates. Gels are fixed briefly in 5% acetic acid containing 5% ethanol, rinsed in H_2O, and dried before exposure. Typical examples of such assays are shown in Fig. 11.2. In the case of the *malEF* RNA, PNPase alone will shorten the substrate by approximately 35 residues to yield a 340-nt species termed the "RSR" (McLaren *et al.,* 1991; Py *et al.,* 1996). Further degradation of this interme-diate to mononucleotides and limit oligonucleotides depends on ATP and the presence of the RhlB helicase (Fig. 11.2A). Likewise, with the *rps*T RNA (Fig. 11.2B), PNPase alone will shorten the poly(A) tail on the substrate to yield an ~180-nt intermediate, but degradation past the termi-nal stem loop depends on ATP and RhlB (Coburn *et al.,* 1999). As an alternative to electrophoretic analysis, activity could be monitored by release of ribonucleoside 5'-diphosphates.

The method for reconstituting a minimal degradosome described here yields macromolecular complexes whose formation depends on Rne and whose activity depends on PNPase, RhlB or CsdA, phosphate, and ATP (Coburn *et al.,* 1999; Prud'homme-Généreux *et al.,* 2004). To this extent, the reconstituted complex recapitulates the properties of unfractionated degradosomes (Coburn *et al.,* 1999; Py *et al.,* 1996). However, the stoichi-ometry of the reconstituted complexes has not yet been determined accu-rately. Nonetheless, the development of a method for the reconstitution of the degradosome should open the path to a systematic investigation of the role of each of the components in the assembly process and to explaining the role of Rne in activating the helicase activity of RhlB, RhlE, or CsdA and coupling PNPase activity to that of RhlB, RhlE, or CsdA. The methods

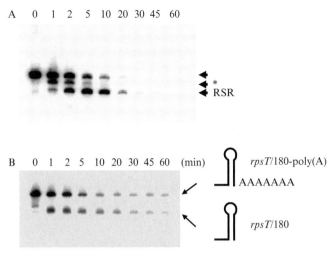

Figure 11.2   Activity of a minimal reconstituted degradosome against (A) the 375-nt *malEF* RNA or (B) the 210-nt *rpsT*/180-poly(A)$_{30}$ RNA. Reconstitutions were performed as described in the text and samples of 4 μl withdrawn at the time in minutes indicated above each panel. Samples were denatured in buffered formamide, separated in 6% polyacrylamide gels containing 8 M urea, and visualized by phosphorimaging. The triangles in the right margins of each panel denote the initial substrate and various intermediates; the asterisk refers to the "star" intermediate in Fig. 11.1A. These data are taken from Coburn *et al.* (1999) with permission.

described here have already shown that the C-terminal domain of Rne lacking endonuclease activity can fully support reconstitution of RhlB-dependent PNPase activity (Chandran *et al.*, 2007; Coburn *et al.*, 1999) and that CsdA or RhlE can substitute effectively for RhlB (Khemici *et al.*, 2004; Prud'homme-Généreux *et al.*, 2004).

A superior method for preparing quantities of reconstituted degradosomes suitable for biophysical work has been described recently. See Worrall, *et al.*, 2008.

ACKNOWLEDGMENTS

Previously unpublished work described in this chapter was supported by grants from the former Medical Research Council of Canada and its successor, the Canadian Institutes of Health Research.

REFERENCES

Belasco, J. G. (1993). *In* "Control of Messenger RNA Stability" (J. G. Belasco and G. Brawerman, eds.), pp. 3–12. Academic Press, San Diego.

Blum, E., Py, B., Carpousis, A. J., and Higgins, C. F. (1997). Polyphosphate kinase is a component of the *Escherichia coli* RNA degradosome. *Mol. Microbiol.* **26,** 387–398.

Blum, E., Carpousis, A. J., and Higgins, C. F. (1999). Polyadenylation promotes degradation of 3′-structured RNA by the *Escherichia coli* mRNA degradosome *in vitro*. *J. Biol. Chem.* **274,** 4009–4016.

Callaghan, A. J., Grossmann, J. G., Redko, Y. U., Ilag, L. L., Moncrieffe, M. C., Symmons, M. F., Robinson, C. V., McDowall, K. J., and Luisi, B. F. (2003). Quaternary structure and catalytic activity of the *Escherichia coli* ribonuclease E amino-terminal catalytic domain. *Biochem.* **42,** 13848–13855.

Callaghan, A. J., Aurikko, J. P., Ilag, L. L., Grossmann, J. G., Chandran, V., Kühnel, K., Poljak, L., Carpousis, A. J., Robinson, C. V., Symmons, M. F., and Luisi, B. F. (2004). Studies of the RNA degradosome-organizing domain of the *Escherichia coli* ribonuclease RNase E. *J. Mol. Biol.* **340,** 965–979.

Carpousis, A. J. (2007). The RNA degradosome of *Escherichia coli*: An mRNA-degrading machine assembled on RNase E. *Annu. Rev. Microbiol.* **61,** 71–87.

Carpousis, A. J., Van Houwe, G., Ehretsmann, C., and Krisch, H. M. (1994). Copurification of *E. coli* RNAase E and PNPase: Evidence for a specific association between two enzymes important in RNA processing and degradation. *Cell* **76,** 889–900.

Carpousis, A. J., Leroy, A., Vanzo, N., and Khemici, V. (2001). *Escherichia coli* RNA degradosome. *Methods Enzymol.* **342B,** 333–345.

Chandran, V., Poljak, L., Vanzo, N. F., Leroy, A., Miguel, R. N., Fernandez-Recio, J., Parkinson, J., Burns, C., Carpousis, A. J., and Luisi, B. F. (2007). Recognition and cooperation between the ATP-dependent RNA helicase RhlB and ribonuclease RNase E. *J. Mol. Biol.* **367,** 113–132.

Coburn, G. A., and Mackie, G. A. (1998). Reconstitution of the degradation of the mRNA for ribosomal protein S20 with purified enzymes. *J. Mol. Biol.* **279,** 1061–1074.

Coburn, G. A., and Mackie, G. A. (1999). Degradation of mRNA in *Escherichia coli*: An old problem with some new twists. *Prog. Nucl. Acids Res. Mol. Biol.* **62,** 55–108.

Coburn, G. A., Miao, X., Briant, D. J., and Mackie, G. A. (1999). Reconstitution of a minimal RNA degradosome demonstrates functional coordination between a 3′-exonuclease and a DEAD-box RNA helicase. *Genes Dev.* **13,** 2594–2603.

Cormack, R. S., Genereaux, J. L., and Mackie, G. A. (1993). RNase E activity is conferred by a single polypeptide: Overexpression, purification, and properties of the *ams/rne/hmp1* gene product. *Proc. Natl. Acad. Sci. USA* **90,** 9006–9010.

Deutscher, M. P. (2006). Degradation of RNA in bacteria: Comparison of mRNA and stable RNA. *Nucl. Acids Res.* **34,** 659–666.

Ghora, B. K., and Apirion, D. (1978). Structural analysis and *in vitro* processing to p5 rRNA of a 9S RNA molecule isolated from an *rne* mutant of *E. coli. Cell* **15,** 1055–1066.

Hager, D. A., and Burgess, R. R. (1980). Elution of proteins from sodium dodecyl sulfate-polyacrylamide gels, removal of sodium dodecyl sulfate, and renaturation of enzymatic activity: Results with sigma subunit of *Escherichia coli* RNA polymerase, wheat germ DNA topoisomerase, and other enzymes. *Anal. Biochem.* **109,** 76–86.

Jones, G. H., Symmons, M. F., Hankins, J. S., and Mackie, G. A. (2003). Overexpression and purification of untagged polynucleotide phosphorylases. *Protein Expr. Purif.* **32,** 202–209.

Khemici, V., Toesca, I., Poljak, L., Vanzo, N. F., and Carpousis, A. J. (2004). The RNase E of *Escherichia coli* has at least two binding sites for DEAD-box RNA helicases: Functional replacement of RhlB by RhlE. *Mol. Microbiol.* **54,** 1422–1430.

Kühnel, K., and Luisi, B. F. (2001). Crystal structure of the *Escherichia coli* RNA degradosome component enolase. *J. Mol. Biol.* **313,** 583–592.

Liou, G.-G., Jane, W.-N., Cohen, S. N., Lin, N.-S., and Lin-Chao, S. (2001). RNA degradosomes exist *in vivo* in *Escherichia coli* as multicomponent complexes associated with the cytoplasmic membrane via the N-terminal region of ribonuclease E. *Proc. Natl. Acad. Sci. USA* **98**, 63–68.

Liou, G. G., Chang, H. Y., Lin, C. S., and Lin-Chao, S. (2002). DEAD box RhlB RNA helicase physically associates with exoribonuclease PNPase to degrade double-stranded RNA independent of the degradosome-assembling region of RNase E. *J. Biol Chem.* **277**, 41157–41162.

Mackie, G. A. (1991). Specific endonucleolytic cleavage of the mRNA for ribosomal protein S20 of *Escherichia coli* requires the product of the *ams* gene *in vivo and in vitro*. *J. Bacteriol.* **173**, 2488–2497.

Mackie, G. A., Genereaux, J. L., and Masterman, S. K. (1997). Modulation of the activity of RNase E *in vitro* by RNA sequences and secondary structures 5′ to cleavage sites. *J. Biol. Chem.* **272**, 609–616.

Mackie, G. A., Coburn, G. A., Miao, X., Briant, D. J., and Prud'homme-Généreux, A. (2001). Preparation of *Escherichia coli* Rne protein and reconstitution of RNA degradosome. *Method Enzymol.* **342B**, 346–356.

McLaren, R. S., Newbury, S. F., Dance, G. S. C., Causton, H. C., and Higgins, C. F. (1991). mRNA degradation by processive 3′ to 5′-exoribonucleases *in vitro* and the implications for prokaryotic mRNA decay *in vivo*. *J. Mol. Biol.* **221**, 81–95.

Melefors, O., Lundberg, U., and von Gabain, A. (1993). *In* "Control of Messenger RNA Stability" (J. G. Belasco and G. Brawerman, eds.), pp. 53–70. San Diego, Academic Press.

Miczak, A., Kaberdin, V. R., Wei, C.-L., and Lin-Chao, S. (1996). Proteins associated with RNase E in a multicomponent ribonucleolytic complex. *Proc. Nat. Acad. Sci. USA* **93**, 3865–3869.

Misra, T. K., and Apirion, D. (1979). RNase E, an RNA processing enzyme from *Escherichia coli*. *J. Biol. Chem.* **254**, 11154–11159.

Morita, T., Maki, K., and Aiba, H. (2005). RNase E-based ribonucleoprotein complexes: Mechanical basis of mRNA destabilization mediated by bacterial noncoding RNAs. *Genes Dev.* **19**, 2176–2186. Erratum in: *Genes Dev.* (2006) **20**, 3487.

Ow, M. C., and Kushner, S. R. (2002). Initiation of tRNA maturation by RNase E is essential for cell viability in *E. coli*. *Genes Dev.* **16**, 1102–1115.

Prud'homme-Généreux, A., Beran, R. K., Iost, I., Ramey, C. S., Mackie, G. A., and Simons, R. S. (2004). Physical and functional interactions among RNase E, polynucleotide phosphorylase and the cold-shock protein, CsdA: Evidence for a "cold shock degradosome." *Mol. Microbiol.* **54**, 1409–1421.

Py, B., Higgins, C. F., Krisch, H. M., and Carpousis, A. J. (1996). A DEAD-box RNA helicase in the *Escherichia coli* RNA degradosome. *Nature* **381**, 169–172.

Symmons, M. F., Jones, G. H., and Luisi, B. F. (2000). A duplicated fold is the structural basis for polynucleotide phosphorylase catalytic activity, processivity, and regulation. *Structure* **8**, 1215–1226.

Taghbalout, A., and Rothfield, L. (2007). RNase E and the other constituents of the RNA degradosome are components of the bacterial cytoskeleton. *Proc. Nat. Acad. Sci. USA* **104**, 1667–1672.

Taraseviciene, L., Naureckiene, S., and Uhlin, B. E. (1994). Immunoaffinity purification of the *Escherichia coli* rne gene product. Evidence that the rne gene encodes the processing endoribonuclease RNase E. *J. Biol. Chem.* **269**, 12167–12172.

Vanzo, N. F., Li, Y. S., Py, B., Blum, E., Higgins, C. F., Raynal, L. C., Krisch, H. M., and Carpousis, A. J. (1998). Ribonuclease E organizes the protein interactions in the *Escherichia coli* RNA degradosome. *Genes Devel.* **12**, 2770–2781.

Worrall, J. A. R., Gorna, M., Crump, N. T., Phillips, L. G., Tuck, A. C., Price, A. J., Bavro, V. N., and Luisi, B. F. (2008). Reconstituion and analysis of the multienzyme Escherichia coli RNA degradosome. *J. Mol. Biol.* **382**, 870–883.

CHAPTER TWELVE

Identifying and Characterizing Substrates of the RNase E/G Family of Enzymes

Louise Kime, Stefanie S. Jourdan, *and* Kenneth J. McDowall

Contents

Abstract

The study of RNA decay and processing in *Escherichia coli* has revealed a central role for RNase E, an endonuclease that is essential for cell viability. This enzyme is required for the normal rapid decay of many transcripts and is involved in the processing of precursors of 16S and 5S ribosomal RNA, transfer RNA, the transfer-messenger RNA, and the RNA component of RNase P. Although there is reasonable knowledge of the repertoire of transcripts cleaved by RNase E in *E. coli*, a detailed understanding of the molecular recognition events that control the cleavage of RNA by this key enzyme is only starting to emerge. Here we describe methods for identifying sites of endonucleolytic cleavage and

Astbury Centre for Structural Molecular Biology, University of Leeds, Leeds, LS2 9JT, United Kingdom

Methods in Enzymology, Volume 447
ISSN 0076-6879, DOI: 10.1016/S0076-6879(08)02212-X

determining whether they depend on functional RNase E. This is illustrated with the *pyrG eno* bicistronic transcript, which is cleaved in the intergenic region primarily by an RNase E–dependent activity and not as previously thought by RNase III. We also describe the use of oligoribonucleotide and *in vitro*–transcribed substrates to investigate *cis*-acting factors such as 5′-monophosphorylation, which can significantly enhance the rate of cleavage but is insufficient to ensure processivity. Most of the approaches that we describe can be applied to the study of homologs of *E. coli* RNase E, which have been found in approximately half of the eubacteria that have been sequenced.

1. INTRODUCTION TO *E. COLI* RNASE E AND ITS HOMOLOGS

Ribonuclease E (RNase E) was discovered 30 years ago in *Escherichia coli* as an activity that generates the penultimate intermediate in the generation of 5S ribosomal RNA from a 9S precursor (Gegenheimer *et al.*, 1977). A temperature-sensitive (ts) strain unable to produce 5S RNA at a nonpermissive temperature was identified soon after (Apirion and Lassar, 1978; Ghora and Apirion, 1978) and shown to contain a thermolabile enzyme that cleaved the 9S precursor *in vitro* at two major sites (Ghora and Apirion, 1978; Misra and Apirion, 1979). Early work on RNase E also revealed that this enzyme cleaves RNAI (Tomcsanyi and Apirion, 1985), the antisense RNA regulator of ColEl-type plasmid replication (for review, see Cesareni *et al.*, 1991). The major site of RNase E cleavage in RNAI is located in a single-stranded region at the 5′-end (Tomcsanyi and Apirion, 1985).

The ts mutation that was central to the preceding studies was designated *rne-3071* (from ribonuclease *E*). Later studies revealed that this mutation maps to the same chromosomal locus as *ams-1* (Babitzke and Kushner, 1991; Melefors and von Gabain, 1991; Mudd *et al.*, 1990; Taraseviciene *et al.*, 1991), a ts mutation that was described initially as increasing mRNA stability (Arraiano *et al.*, 1988; Kuwano *et al.*, 1977; Ono and Kuwano, 1979). At a nonpermissive temperature, the *rne-3071* mutation also retards mRNA decay, and the *ams-1* mutation (from *a*ltered *m*RNA *s*tability) affects the maturation of 5S rRNA (Babitzke and Kushner, 1991; Melefors and von Gabain, 1991; Mudd *et al.*, 1990; Taraseviciene *et al.*, 1991). Furthermore, it is now known that both these mutations are located within the coding region of the *rne* gene that encodes RNase E (Cormack *et al.*, 1993; McDowall *et al.*, 1993).

More recently, the study of individual precursors has revealed that *E. coli* RNase E also plays a role in the processing of 16S rRNA (Li *et al.*, 1999b), transfer RNA (Li and Deutscher, 2002; Ow and Kushner, 2002), transfermessenger RNA (Lin-Chao *et al.*, 1999), and the RNA component of RNase P (Lundberg and Altman, 1995). RNase E is assisted in the maturation of 16S rRNA by a paralog called RNase G (Li *et al.*, 1999a), which has been shown to be required for the normal decay of several transcripts, including functional forms of *adhE* and *eno* mRNA (Kaga *et al.*, 2002a, 2002b; Umitsuki *et al.*, 2001). The sequence similarity between RNase E and RNase G extends throughout virtually the entire length of the latter (McDowall *et al.*, 1993), but is confined to the N-terminal half (NTH) of RNase E (McDowall and Cohen, 1996). The latter domain of RNase E is sufficient for cell viability (Kido *et al.*, 1996) and contains the site of catalysis (Callaghan *et al.*, 2005a; McDowall and Cohen, 1996). The C-terminal half of RNase E contains ancillary RNA-binding sites and segments that facilitate interaction with other enzymes as part of a complex called the *RNA degradosome* (for reviews, see Carpousis, 2002; Carpousis *et al.*, 2001; Marcaida *et al.*, 2006).

(see also chapters 4 and 10 by Carpousis *et al.*, in this volume regarding methods for the studying the degradosome in this volume.

When cells are grown under standard laboratory conditions, the other major components of the degradosome are the 3′-exoribonuclease polynucleotide phosphorylase (PNPase), the ATP-dependent RNA helicase RhlB, and the glycolytic enzyme enolase (Carpousis *et al.*, 1994; Miczak *et al.*, 1996). Although it has been shown that RhlB can assist decay by PNPase *in vitro* (Py *et al.*, 1996) and RNase E *in vivo* (Khemici *et al.*, 2005), the role of enolase is unclear. There is, however, evidence that it can, in concert with RNase E, influence the response to stress resulting from overabundant phosphosugar (for review, see Vanderpool, 2007). Other components seem to associate with the degradosome in a more transient manner; this includes the Hfq protein, a chaperone of small RNAs that bind particular RNAs thereby regulating their function (for review, see Aiba, 2007).

The repertoire of mRNAs that are cleaved by RNase E remains to be fully defined, despite the advent of gene-array technology. Although it would be relatively straightforward to determine those transcripts that increase when an *rne-3071* or *ams-1* (also called *rne-1*) mutant is shifted to a nonpermissive temperature, not all of the transcripts that accumulate would necessarily do so because their cleavage by RNase E is blocked. Complications arise because disruption of RNase E activity may also affect, for example, the number and stoichiometry of degradosomal complexes as a result of the level of RNase E being inversely related to its activity (Jain and Belasco, 1995; Mudd and Higgins, 1993), as well as the decay of the mRNA of other ribonucleases (Cairrão and Arraiano, 2006; Zilhão *et al.*, 1995). Therefore, other criteria must be met before an RNA that accumulates in an *rne*^ts strain can be widely accepted as an RNase E substrate. These criteria are demonstration that the half-life of the RNA is longer in an *rne*^ts strain and evidence that the primary transcript is cleaved to produce 5′-ends that are *rne*-dependent (i.e., detected at significantly higher levels in a wild-type compared with an *rne*^ts strain at a nonpermissive temperature). It is also preferable that the 5′-ends shown to be *rne*-dependent *in vivo* can be generated *in vitro* with purified components.

The detailed study of individual transcripts using the criteria described above has suggested that RNase E cleavage is required for the efficient decay of many primary transcripts (Arraiano *et al.*, 1993; Lin-Chao and Cohen, 1991; Loayza *et al.*, 1991; Mackie, 1991). It has also been shown that RNase E can segment polycistronic mRNA, thereby allowing different protein-coding regions to be degraded with different half-lives, which in turn affects their relative levels of expression (Arraiano *et al.*, 1997; Klug and Cohen, 1990; Newbury *et al.*, 1987a; Nilsson *et al.*, 1996; Patel and Dunn, 1995; Schirmer and Hillen, 1998; Smolke *et al.*, 2000; Yajnik and Godson, 1993). Moreover, the 3′-stem-loop structures that serve as transcription termination signals seem to be effective barriers to 3′-exonucleases (Chen *et al.*, 1988; Newbury *et al.*, 1987b). Thus, cleavage by RNase E and other endonucleases is generally considered to be the main factor that limits the lifetime of transcripts in *E. coli* (for review, see Belasco and Higgins, 1988). Following on from this, the factors that influence endonucleolytic cleavage should be considered major regulators of gene expression.

Study of the *E. coli ompA* transcript, which in rapidly growing cells has a half-life that is approximately fivefold longer than a typical *E. coli* mRNA (15 to 20 min compared with 2 to 3 min), has revealed that it owes its longevity to the presence of a stem loop no more than 2 to 4 nucleotides from its extreme 5′-end (Emory *et al.*, 1992). Moreover, it was shown that the half-life of RNAI (see earlier) can be extended by adding such a hairpin structure at its 5′-terminus (Bouvet and Belasco, 1992). This implied that RNase E has exceptional substrate specificity for an endoribonuclease; that is to say, it preferentially cleaves RNAs that have several unpaired nucleotides at the 5′-end (Bouvet and Belasco, 1992). In a separate line of investigation, it was found *in vitro* that RNase E cleavage can be enhanced by the presence of a 5′-monophosphate on the RNA, provided this group is immediately followed by a short segment that is single stranded (Mackie, 1998). Substrates of the same sequence, but with a hydroxyl or triphosphate group at the 5′-end, were cleaved less efficiently (Feng *et al.*, 2002; Jiang and Belasco, 2004; Mackie, 1998; Walsh *et al.*, 2001). This led to the proposal, which was confirmed when X-ray crystal structures were solved of NTH-RNase E in complexes with oligonucleotide substrates (Callaghan *et al.*, 2005a), that RNase E has a pocket that can bind a 5′-monophosphate group (Mackie, 1998). Recently, evidence has been obtained that *E. coli* cells contain a pyrophosphatase activity that converts the 5′-end of primary transcripts from a triphosphate to a monophosphate (Celesnik *et al.*, 2007). This offers an explanation as to why RNAs with unpaired 5′-terminal nucleotides *in vivo* may be more susceptible than those with a hairpin structure at their 5′-terminus, despite both being synthesized with a 5′-triphosphate that blocks efficient cleavage by RNase E (Celesnik *et al.*, 2007). It has been shown that sensing of 5′-monophosphate by *E. coli* RNase G can significantly enhance association with RNA and stimulate the decay of *adhE* and *eno* mRNA *in vivo* (Jourdan and McDowall, 2008).

The phosphorylation status and single-strandedness at the 5′-end of RNA are not the only factors that determine the rate of cleavage by RNase E. Although there is not a strong consensus sequence for sites cleaved by RNase E (Cohen and McDowall, 1997), differences in sequence can affect the rate of cleavage and number of phosphodiester bonds cleaved within a single-stranded segment (Ehretsmann *et al.*, 1992; Kaberdin, 2003; McDowall *et al.*, 1994). Although 5′-monophosphorylated oligonucleotide substrates with sequences corresponding to the single-stranded region at the 5′-end of RNAI are cleaved efficiently by RNase E *in vitro* (McDowall *et al.*, 1995), this is true of only the "a" site, one of two major points of cleavage in 9S RNA (Ghora and Apirion, 1978). Efficient cleavage at the "b" site requires that it is within the context of precursors of 5S RNA (Kaberdin *et al.*, 2000). This indicates that a factor(s) other than 5′-monophosphorylation and cleavage site sequence can promote cleavage by RNase E. It has been suggested that interaction between an as yet unidentified secondary structure and the arginine-rich RNA-binding domain within the C-terminal half of RNase E enhances the rate of cleavage at the "b" site, at least *in vitro* (Kaberdin *et al.*, 2000). The notion that factor(s) other than 5′-monophosphorylation and cleavage site sequence can promote RNase E cleavage is also supported by the findings that this enzyme requires a phylogenetically conserved hairpin within the 5′ untranslated region of its mRNA to mediate efficient autoregulation *in vivo* (Diwa and Belasco, 2002) and can efficiently cleave a transcript *in vivo* despite it being modified to block 5′-end interaction (Baker and Mackie, 2003).

Given the need for systematic study of the factors that control cleavage by RNase E, this chapter includes strategies and protocols that can be used to investigate the recognition and cleavage of RNA by RNase E. Furthermore, as much can be learned from comparative studies at the level of individual steps and pathways of processing and decay, we also describe methods for identifying major sites of RNase E cleavage *in vivo*. Most of the approaches described here can be applied to the study of homologs of *E. coli* RNase E.

2. IDENTIFICATION OF POTENTIAL *IN VIVO* SUBSTRATES OF RNASE E

2.1. Mutant strains and the quenching of RNA metabolism

Substrates of RNase E can be identified *in vivo* with strains that contain either the *rne-3071* (Apirion and Lassar, 1978) or *ams-1* (*rne-1*) ts mutation (Ono and Kuwano, 1979). The most widely used approach involves growing an *rne*ᵗˢ strain and an isogenic wild–type partner under identical conditions at a permissive temperature (most often between 30 and 35 °C) until exponential growth is achieved and then rapidly shifting the incubation temperature to a nonpermissive temperature (most often 45 °C). The cultures are then incubated for a further 30 to 40 min for mutant RNase E to be inactivated before the cells are harvested. In our laboratory, we change the temperature of incubation rapidly by switching culture flasks between shaking water baths. To preserve the cellular RNA, we quench its metabolism by adding an eighth volume of chilled 5% (v/v) phenol in absolute ethanol (Lin-Chao and Cohen, 1991), before harvesting the cells by centrifugation (5000g for 10 min). A commercially produced reagent that rapidly permeates the cells of a range of prokaryotic species and eukaryotes can also be used (RNA*later*, Ambion). Substrates of *E. coli* RNase G, which is not required for growth in standard LB medium, have been identified by comparing RNA isolated from an *rng* deletion strain and a congenic wild-type partner (Lee *et al.*, 2002; Wachi *et al.*, 1999). The use of *rne* and *rng* mutants to investigate the ribonuclease that directs a particular cleavage *in vivo* is illustrated with the intergenic region of the *pyrG eno* transcript (Fig. 12.1). This approach can be adopted for any bacterial species for which suitable mutants can be isolated. In the absence of such mutants, some groups have assessed the likelihood that an RNase E–like activity cleaves a particular transcript by cloning into *E. coli* and assessing its decay in mutant backgrounds (Condon *et al.*, 1997; Klug *et al.*, 1992).

2.2. Isolation of RNA and measurement of steady-state levels and half-lives

RNA can be isolated from bacteria with a number of commercial kits. These include RNAeasy® (Qiagen), RiboPure™-Bacteria (Ambion), and UltraClean™-Microbial (Mo Bio Lab). For those wishing a less expensive option for *E. coli*, we recommend the following protocol, which has been adapted from McDowall *et al.* (1994).

Resuspend a cell pellet corresponding to 5 to 8 OD_{600} units of culture in 1 ml of TE buffer; 10 mM Tris-HC1 (pH 8.0), 1 mM EDTA.

Lyze the cells immediately by adding 1 ml of 20 mM Tris-HC1 (pH 8.0), 40 mM EDTA, 300 mM NaCl, and 0.5% (w/v) SDS, which has been preheated to 100 °C. To isolate RNA from larger cell numbers, scale the above volumes accordingly.

Incubate mixture in a boiling water bath for 30 sec and cool on ice before extracting the lysate once with an equal volume of acid phenol saturated with 100 mM citrate buffer (pH 4.3).

Retain the aqueous phase and precipitate the total-cell nucleic acid by adding 2.5× volumes of absolute ethanol and then harvest by centrifugation at 7000g for 15 min.

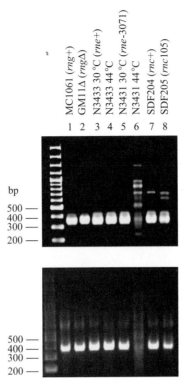

Figure 12.1 Investigating the origin of a 5′-end within the intergenic region of the *pyrG eno* bicistronic transcript. RLM-RT-PCR analysis was used to amplify the 5′-segment of the transcript (see Figs. 12.2 and 12.3). Total RNA was prepared from three different strains; GM11 with chromosomally deleted *rng* (RNase G) (lane 2), and its isogenic wild-type strain MC1061 (lane 1); a temperature-sensitive mutant of RNase E (N3431) (lanes 5 and 6), and its isogenic wild-type strain N3433 (lanes 3 and 4), grown at a permissive temperature of 30 °C (lanes 3 and 5) and grown at a nonpermissive temperature of 44 °C for 30 min (lanes 4 and 6); and a mutant lacking functional RNase III (SDF205) (lane 8), and its isogenic wild-type strain SDF204 (lane 7). The PCR step (which followed adapter ligation and reverse transcription) was performed with a primer specific for the RNA adapter (5′-CATGAGGATTACCCATGTCG-3′) and in separate reactions with two primers specific for the *eno* gene (5′-GTCAATGCCAGCCTGATCTT-3′-and 5′-TGCCGTCCAGGTCGATCATG-3′-corresponding to the upper and lower panels, respectively) that produce amplicons differing by 25 bp. As expected, the amplicons in the lower panel migrated more slowly. A single major species was detected in all strains except the *rne-3071* mutant at the nonpermissive temperature. Thus, the generation of the corresponding 5′-end is designated as being *rne*-or RNase E-dependent.

Resuspend the pellet in 200 µl of 10 m*M* Tris-HCl (pH 8.0), 1 m*M* EDTA, 150 m*M* NaCl, and reprecipitate by adding 500 µl of absolute ethanol. Remove the stringy chromosomal DNA with a sterile plastic tip and harvest the RNA in a benchtop centrifuge (14,000*g* for 15 min at 4 °C).

Resuspend the RNA pellet in 400 μl of TE, 150 mM NaCl, and extract further with an equal volume of acidic phenol (pH 4.3), an equal volume of acidic phenol: chloroform: isoamyl alcohol (25:24:1), and chloroform/ isoamyl alcohol (49:1).

Reprecipitate and harvest the RNA, washed twice with 70% (v/v) ethanol, dry briefly *in vacuo*, and resuspend in 100 μl water that has been made free of RNase activity by treating with diethylpyrocarbonate (DEPC) overnight (final concentration 0.1% [v/v]) and autoclaving.

Treat the RNA as necessary with RNase-free DNase I (Promega) (1 U per 100 μg of nucleic acid) in 400 μl of 40 mM Tris-HCl (pH 8.0), 10 mM MgSO$_4$, and 1 mM CaCl$_2$ for 1 h at 37 °C. Extract with an equal volume of acidic phenol:chloroform:isoamyl alcohol (25:24:1), add NaCl to 150 mM, and precipitate with 1 ml of absolute ethanol.

Quantitate the RNA by measuring the A$_{260}$ of a sample with a UV spectrophotometer and analyze by agarose gel electrophoresis to check that the ribosomal RNAs are intact and there is no obvious chromosomal DNA contamination.

The level of a particular transcript in RNA samples isolated from a wild-type strain and one defective in a gene encoding a member of the RNase E/G family can be compared using a variety of techniques that include Northern blotting, nuclease protection, primer extension, and quantitative-PCR assays. For transcripts that have not been studied previously, we recommend Northern blotting, because it is the only technique by which a transcript in its entirety can be readily analyzed. Northern blotting involves separating RNA molecules on the basis of their size using denaturing gel electrophoresis, blotting the RNA to a membrane support, and probing for a transcript(s) associated with a particular gene. Details of Northern blotting are not covered here, because this technique was first described 30 years ago (Alwine *et al.*, 1977), and protocols are widely available (Sambrook and Russell, 2001). One of the problems associated with Northern blotting is that the 23S and 16S rRNA, which have the highest abundance of all bacterial RNAs, can mask the presence of transcripts of similar size. In such a scenario, we would recommend the used of the MICROBExpress^{TM} kit (Ambion), which is reported to remove greater than 95% of the 16S and 23S rRNA from bacterial RNA samples. Transcripts that are found to be more abundant in a strain defective in an RNase E/G family member relative to a congenic wild-type partner are potentially substrates for the enzyme.

Bona fide substrates of a ribonuclease should not only accumulate in cells that have a mutation that disrupts the activity of the enzyme, they should also have a longer half-life. This value can be measured in *E. coli* and other eubacteria by adding rifampicin, a semisynthetic antibiotic, which binds and inhibits the endogenous RNA polymerase that synthesizes all categories of RNA. It has been reported that it is necessary to add rifampicin to a final concentration of 500 μg/ml to inhibit the transcription of every gene in

E. coli (Bernstein *et al.*, 2002). Once rifampicin is added, RNA is harvested at a number of different time points. Because the typical half-lives of bacterial mRNA tend to be on the order of a few minutes (Belasco, 1993; Bernstein *et al.*, 2002; Selinger *et al.*, 2003), suitable time points for RNA isolation would be 1, 2, 4, 8, 16, and 32 min. The RNA level in each of the samples can then be determined using the same techniques used to measure levels at steady state (i.e., in the absence of rifampicin when decay is balanced by transcription) as described previously. Because ribosomal RNA is the main constituent of RNA in bacteria (Neidhardt and Umbarger, 1996) and has a half-life in excess of 50 min (Yuan and Shen, 1975), the total amount of RNA in a cell does not diminish appreciably over the period that is typically used to estimate values of half-life. Therefore, the degradation of a particular RNA with time can be measured by comparing its relative level in samples obtained from successive time points each containing an equal mass of total RNA.

3. IDENTIFICATION OF SITES OF ENDONUCLEOLYTIC CLEAVAGE

3.1. Mapping the 5'-ends of specific transcripts

Primer extension and nuclease protection assays (Sambrook and Russell, 2001) can be used to detect 5'-ends that are diminished in a mutant strain and are, thus, potentially generated by the corresponding ribonuclease in wild-type cells. In nuclease protection assays, a labeled, single-stranded nucleic acid that is complementary to a specific transcript is mixed with a sample of RNA, and the two are allowed to hybridize. Excess probe and any segment of bound probe that is unprotected are removed by incubating with a nuclease such as S1 that can cut single-stranded, but not double-stranded nucleic acid. By analyzing the remaining probe using denaturing gel electrophoresis, it is possible to determine its length and thus the 5'-limit of the corresponding RNA. Primer extension assays involve incubating an RNA sample with a labeled oligonucleotide that is complementary to a particular transcript. When annealed to its target, the oligonucleotide serves as a primer that can be extended by an RNA-dependent DNA polymerase called reverse transcriptase to produce complementary DNA (cDNA). Extension will terminate at the 5'-end of the RNA; thus, the length of extension from the primer can be used to map the position of 5'-ends in an RNA sample. As for the nuclease protection assay, the products of primer extension can be analyzed by denaturing gel electrophoresis.

3.2. RNA ligase-mediated, reverse-transcription PCR assay

Primer extension and nuclease protection assays are widely used by laboratories that study bacterial RNA decay and processing, and detailed protocols are published (Sambrook and Russell, 2001); therefore, these methods will not be covered here. Instead, we describe a PCR-based protocol that we use in our laboratory to detect the 5′-ends of RNAs (Figure 12.3). The procedure, which was developed by others (Bensing *et al.*, 1996), uses RNA ligase to link a specific oligoribonucleotide to the 5′-ends of cellular RNAs that have a 5′-monophosphate group. Complementary DNA is produced with reverse transcriptase and random primers, and then the 5′-segment(s) of a particular transcript are amplified by PCR with a forward and reverse primer that are complementary to the ligated adapter and the transcript under investigation, respectively. The latter step can include nested primers (a second pair of primers that bind within the first PCR product) to increase the specificity, if required. We advocate the use of this method, because it is more sensitive than nuclease protection and primer extension assays, avoids the use of radioisotopes, and unlike nuclease protection and primer extension assays, can differentiate the 5′-triphosphorylated ends of newly synthesized transcripts from those that have been processed (Bensing *et al.*, 1996). Moreover, it does not require prior knowledge of the precise 5′-endpoint(s) of the RNA.

To identify those transcripts with 5′-monophosphorylated ends, undertake the following steps after growing cultures under the appropriate conditions and isolating RNA (see earlier).

Mix 20 pmol of an RNA adapter with 2 μg of total-cell RNA, 5 U of T4 RNA ligase (New England Biolabs) and 20 U of RNaseOUT^{TM} in 10 μl of RNA ligase buffer; 50 mM Tris-HCl (pH 7.8), 10 mM MgCl$_2$, 10 mM DTT, and 1 mM ATP. Incubate at 37 °C for 1 h. We use a custom-made, 54-mer adapter that was synthesized by Dharmacon and has the sequence 5′-ACAUGAGGAUUACCCAUGUCGAAGACAA-CAAAGAAGUUCAACUCUUUAUGUAUU-3′.

Anneal 50 ng of random hexamers (Amersham Biosciences) to 2 μl of ligated products in a total volume of 20 μl made up with DEPC-treated water. Heat at 65 °C for 5 min and chill quickly on ice.

The ligation products are then reverse transcribed into first strand cDNA. Add to the annealed mix (from the step above) 8.4 μl of a solution containing 250 mM Tris-HCl (pH 8.3), 375 mM KCl, 15 mM MgCl$_2$, and 50 mM DTT. Then add 4 μl of a 10 mM stock of dNTPs, 80 U of RNaseOUT^{TM} (Invitrogen) and DEPC-treated water to a final volume of 40 μl. From this mix 19 μl is removed and added to 1 μl (200 U) M-MLV reverse transcriptase RNase H minus (Promega). A further 19 μl from the same mix is added to 1 μl of DEPC-treated water to provide a negative control. Incubate at 25 °C for 15 min, 42 °C for 1 h, and finally 70 °C for 15 min to heat inactivate the reverse transcriptase. Add 80 μl of 10 ng/ml yeast tRNA (Ambion) to make a final volume of 100 μl.

Amplify the cDNA corresponding to the 5′-segments of the gene of interest with a primer that anneals to the RNA adapter (5′-CATGAGGAT-TACCCATGTCG-3′) and a primer designed to anneal to the gene of interest (optimal melting temperature of primer approximately 60 °C). The cycling parameters are primer specific, but we commonly use initial denaturation at 95 °C for 5 min and then 30 to 35 cycles of 95 °C for 30 sec, 50 °C for 30 sec, and 72 °C for 2 min. We have had most success with GoTaq® (Promega) or Immolase™ (Bioline) DNA polymerase and typically used as template 2 μl of the final 100 μl sample from the previous step.

Analyze the products of the PCR reaction by agarose gel electrophoresis.

Clone and sequence the amplicons to determine the precise 5′-ends of the transcript.

To identify transcripts with 5′-triphosphorylated ends, the RNA is treated with calf intestinal phosphatase (CIP) to generate 5′-hydroxylated ends and then with polynucleotide kinase (PNK) to produce 5′-monophosphorylated ends. Transcripts that are only detected after this treatment are likely to have had 5′-triphosphorylated ends. The possibility that this treatment reveals a transcript with a 5′-hydroxylated end can be excluded by omitting the CIP treatment before incubating with PNK; this will prevent the detection of 5′-triphosphorylated ends. The effect of these treatments is illustrated with, as an example, the intergenic region of the *pyrG eno* transcript; the results suggest that *rne*-dependent cleavage is the major process that generates a 5′-end within this region (Fig. 12.2).

Treat 5 μg total RNA with 20 U of CIP (New England Biolabs) for 1 h at 37 °C in 20 μl of 100 mM NaCl, 50 mM Tris-HCl (pH 7.9), 10 mM MgCl$_2$, and 1 mM DTT containing 40 U of RNaseOUT™ (Invitrogen).

Increase volume to 150 μl with DEPC-treated, RNase-free water and extract with an equal volume of a mix of acidic phenol (pH 4.3):chloroform:isoamyl alcohol (25:24:1) followed by extraction with chloroform:isoamyl alcohol (49:1).

Add to the aqueous phase 0.1 volume of 3 M sodium acetate (pH 5.5) and 1.0 volume of 100% (v/v) isopropanol, and incubate on ice for 10 min.

Harvest the RNA by centrifugation at 16,000g, 4 °C for 20 min, and wash the pellet with cold 70% (v/v) ethanol. Allow the pellet to air dry before resuspending in 5 μl of DEPC-treated RNase-free water.

Incubate the CIP-treated RNA with 20 U of PNK (New England Biolabs) at 37 °C for 45 min in 40 μl of 70 mM Tris-HCl (pH 7.6), 10 mM MgCl$_2$, and 5 mM DTT containing 1 mM ATP and 80 U of RNaseOUT™.

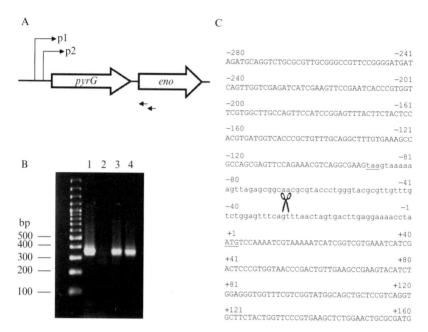

A

p1
p2

pyrG   *eno*

B   1 2 3 4

bp
500 —
400 —
300 —
200 —

100 —

C

```
-280                                      -241
AGATGCAGGTCTGCGCGTTGCGGGCCGTTCCGGGGATGAT

-240                                      -201
CAGTTGGTCGAGATCATCGAAGTTCCGAATCACCCGTGGT

-200                                      -161
TCGTGGCTTGCCAGTTCCATCCGGAGTTTACTTCTACTCC

-160                                      -121
ACGTGATGGTCACCCGCTGTTTGCAGGCTTTGTGAAAGCC

-120                                       -81
GCCAGCGAGTTCCAGAAACGTCAGGCGAAGtaagtaaaaa

-80                                        -41
agttagagcggcaacgcgtaccctgggtacgcgttgtttg

-40                                        -1
tctggagtttcagtttaactagtgacttgaggaaaaccta

+1                                        +40
ATGTCCAAAATCGTAAAAATCATCGGTCGTGAAATCATCG

+41                                       +80
ACTCCCGTGGTAACCCGACTGTTGAAGCCGAAGTACATCT

+81                                      +120
GGAGGGTGGTTTCGTCGGTATGGCAGCTGCTCCGTCAGGT

+121                                     +160
GCTTCTACTGGTTCCCGTGAAGCTCTGGAACTGCGCGATG
```

Figure 12.2 Screening of the intergenic region of the bicistronic *pyrG eno* transcript for 5′-ends that are hydroxylated or triphosphorylated with RLM-RT-PCR. (A) Schematic diagram illustrating the arrangement of the *pyrG* and *eno* genes in the chromosome of *Escherichia coli.* The promoters of the *pyrG* gene are indicated (p1 and p2), and the small arrows indicate the relative positions of primers near the 5′-end of the *eno* gene used for PCR amplification in Fig. 12.1 (not to scale). (B) Total RNA prepared from MC1061 cells (lane 1) was treated with CIP alone (lane 2), PNK alone (lane 3), and CIP followed by PNK (lane 4). After adapter ligation and reverse transcription, PCR was performed with a primer specific for the RNA adaptor (5′-CATGAGGATTACCCATGTCG-3′) and a primer specific for the *eno* gene (5′-GTCAATGCCAGCCTGATCTT-3′). The amplicons were then analyzed by agarose gel electrophoresis. No evidence was obtained for transcripts with 5′-ends that had a hydroxyl or triphosphate group. The sizes (bp) of certain DNA marker bands are given on the left of the panel. (C) Sequence of the intergenic region between *pyrG* and *eno.* The termination codon of *pyrG* (taa, underlined) and translation initiation codon of *eno* (ATG, underlined) are both indicated. The 5′-end of the *rne*-dependent, 5′-monophosphorylated transcript (see also Fig. 12.1) as determined by sequencing of the cloned PCR product is indicated by a pair of scissors (5′-UUUAA-CUA...). Nucleotide numbering was assigned by designating the first nucleotide of the translation start codon of *eno* as +1.

Add 2 μl of 0.5 *M* EDTA to quench the reaction and increase volume to 100 μl with RNase free water.

Extract with acidic phenol:chloroform:isoamyl alcohol and chloroform:isoamyl alcohol, and precipitate as described previously.

Resuspend the pellet in 10 μl of RNase-free water and analyze an aliquot by agarose gel electrophoresis to check the RNA had not degraded during treatment.

To enrich for transcripts that are 5′-triphosphorylated, the RNA can be treated with Terminator^{TM} (Epicentre Biotechnologies), a processive 5′-to 3′-exonuclease that will digest RNA with a 5′-monophosphate but not those with a 5′-triphosphate or 5′-hydroxyl group (Pak and Fire, 2007; Wu *et al.*, 2006). Subsequent treatment with CIP and PNK should reveal only those transcripts that had a 5′-triphosphate or hydroxyl group. A schematic diagram that summarizes the treatments required to differentiate the 5′-ends of processed transcripts from those of primary transcripts is shown in Fig. 12.3.

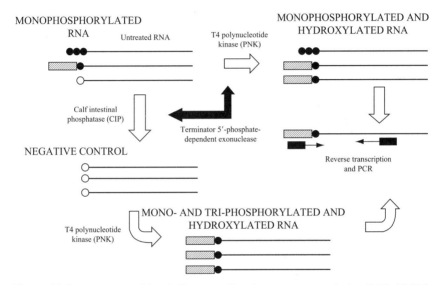

Figure 12.3 Summary of RNA ligase-mediated, reverse transcription PCR (RLM-RT-PCR). T4 RNA ligase is able to link efficiently the 3′-hydroxyl of an oligoribonucleotide (hatched box) to the 5′-end of a transcript provided it has a monophosphate (filled circle) and not a triphosphate (three filled circles) or hydroxyl (empty circle). After adapter ligation, the transcript of interest can be detected by reverse transcription and PCR with primers (filled boxes) specific for the adapter and gene. The status of the 5′-ends can be determined by combination of treatments with calf intestinal phosphatase (CIP) and T4 polynucleotide kinase (PNK). Text in Uppercase lettering to figure indicates the types of RNAs that will be detected before and after each treatment. Incubation with CIP removes mono- and triphosphate groups from 5′-ends to provide a negative control (i.e., no transcripts should be amplifiable with this treatment). Incubation of CIP-treated RNA with PNK should add a monophosphate group to the 5′-ends of all the RNAs in the sample, irrespective of their original 5′-status. Treatment with PNK alone identifies those transcripts with a hydroxyl or monophosphate group at the 5′-end. By comparing each of the RNA treatments, the original status of the 5′-end of a transcript can be determined. In addition, RNA can be treated with Terminator^{TM} 5′-phosphate–dependent exonuclease, which only degrades those transcripts that have a 5′-monophosphate group. This treatment is performed before either the CIP or the PNK step and reveals those transcripts that originally possessed a 5′-triphosphate or 5′-hydroxyl group without interference from those with a 5′-monophosphate.

4. Reconstitution of Cleavages *In Vitro*

Confirmation that a member of the RNase E/G family is responsible for generating a 5'-end detected *in vivo* can be obtained by demonstrating that it does so accurately *in vitro* with purified components. Although it may be preferable to study cleavage by *E. coli* RNase E *in vitro* with preparations of purified or reconstituted degradosome particles (see chapter by Mackie *et al.* in this volume) because this is more likely to reflect the situation *in vivo*, it has been shown for several transcripts that purified polypeptides of the RNase E/G family are able to cleave at sites identified *in vivo* (Cormack *et al.*, 1993; Kaberdin *et al.*, 2000; Lee *et al.*, 2002; Mackie, 1998; Mackie and Genereaux, 1993; McDowall and Cohen, 1996; Tock *et al.*, 2000).

4.1. Purification of RNase E and related enzymes

Recombinant polypeptides of RNase E from *E. coli* (Redko *et al.*, 2003), *Aquifex aeolicus* (Kaberdin and Bizebard, 2005), *Mycobacterium tuberculosis* (Kovacs *et al.*, 2005), and *Streptomyces coelicolor* (Lee and Cohen, 2003), as well as *E. coli* RNase G (Tock *et al.*, 2000), have been purified using immobilized metal affinity chromatography (Porath *et al.*, 1975). Moreover, active-site residues have been identified in *E. coli* RNase E that are highly conserved evolutionarily (Callaghan *et al.*, 2005a; Jourdan and McDowall, 2008). Thus, active site mutants can be constructed readily to confirm that an activity observed in a preparation originates from the RNase E/G family member that has been purified. Immobilized metal affinity chromatography is a routine laboratory procedure; thus, it will not be covered here. However, we would point out that in our experience exchanging the buffer immediately after the protein is eluted minimizes aggregation and batch-to-batch variation in specific activity. In particular, we exchange into buffer containing 20 mM Tris-HCl (pH 7.6), 0.5 M NaCl, 10 mM DTT, 10 mM MgCl$_2$, 0.5 mM EDTA, and 5% (v/v) glycerol. We suspect that the addition of DTT protects the "Zn link," a metal-sharing interface that helps organize the quaternary structure and catalytic site (Callaghan *et al.*, 2005b). Recombinant protein can then be purified further by gel filtration chromatography (e.g., Superdex 200 column, Amersham Biosciences) or concentrated by ultrafiltration (e.g., MacroSep device with a 10-kDa cutoff membrane, Flowgen). For those not wishing to use a tagged polypeptide, a published protocol for the purification of RNase G overproduced in *E. coli* (Briant *et al.*, 2003) may provide a starting point for purification by more traditional chromatography methods.

4.2. Synthesis of transcripts with 5′-monophosphorylated ends

RNA substrates can be synthesized by *in vitro* transcription with the RNA polymerase from bacteriophage T7. Moreover, a number of companies supply the necessary reagents in kit form (e.g., MEGAscript® [Ambion] and RiboMAX™ [Promega]). Given this situation, we focus here not on the transcription but on the generation of 5′-monophosphorylated substrate from *in vitro*–transcribed RNA. The approach is based on RNase H cleavage of a precursor RNA (Walsh *et al.*, 2001). We generate templates for *in vitro* transcription reactions by PCR. In the forward primer, we incorporate the sequence 5′-GGATCC<u>TAATACGACTCACTATAGG</u>-GATCCCGGG(N)$_{17-21}$-3′. The underlined residues form the T7 promoter, whereas "(N)$_{17-21}$" at the 3′-end represents the sequence specific to the target gene. The template is then transcribed with T7 RNA polymerase. Transcription starts within the promoter sequence on the 5′-side of the guanosine triplet, which is required for efficient initiation. A DNA/2′-O-methyl RNA chimera with sequence 5′-C$_m$C$_m$C$_m$dGdGdGd AU$_m$C$_m$C$_m$C$_m$-3′ is annealed to the 5′-extended RNA, and then the hybrid is incubated with RNase H. This enzyme cleaves the RNA at a single position, which is immediately 5′ to the first complementary deoxynucleotide (Lapham and Crothers, 1996; Walsh *et al.*, 2001). This generates a 5′-monophosphorylated RNA that starts with the three guanosines immediately before the start of the target sequence. To provide a 5′-triphosphorylated control, we transcribe RNA from a template that is synthesized with a forward primer with sequence 5′-GGATCC<u>TAATACGACTCACTATAGG</u>G(N)$_{17-21}$-3′. In this way, we are able to compare RNAs that are identical in sequence and only differ in the phosphorylation status at their 5′-ends (Walsh *et al.*, 2001). When a 5′-monophosphorylated substrate with no exogenous sequence is preferred, a DNA/2′-O-methyl RNA chimera can be designed to direct the cleavage of the phosphodiester bond immediately 5′ to the native sequence (Li *et al.*, 1999b). 5′-Triphosphorylated transcript without a guanosine triplet at the 5′-end can be generated *in vitro* with a preparation of *E. coli* RNA polymerase (associated with σ^{70}), which is supplied commercially (Epicenter Technologies, WI), and a template with an appropriate promoter sequence. However, in our experience, the yields are substantially lower than those obtained with T7 RNA polymerase.

Cleavage of 5′-extended RNA by RNase H produces a downstream product that is shorter and can thus be separated from the primary transcript by denaturing polyacrylamide gel electrophoresis. The 5′-monophosphorylated RNA is then gel purified, quantitated, and used as a substrate for *in vitro* cleavage assays. The advantage of generating substrates in this way is that one can be sure that most of the molecules are 5′-monophosphorylated. By comparison, the efficiency of treatment with CIP and PNK is variable and difficult to quantify. We also gel purify our 5'-triphosphorylated substrates, because shorter products can be produced as a result of transcription terminating prematurely. To our knowledge, all of the published assays for the cleavage of a transcript by a member of the RNase E/G family have used substrates that have been labeled with radioisotope. However, we have found that this is unnecessary, at least for the assay of *E. coli* RNase E and G. As shown in Fig. 12.4, cleavage of 9S RNA by the N-terminal catalytic half of RNase E can be monitored by ethidium bromide staining of gels in which reaction samples isolated at different time points have been separated. This result clearly shows that efficient cleavage of 9S RNA *in vitro* critically depends on the presence of a monophosphate group at the 5'-end. It also indicates that the production of p5S RNA occurs by a largely distributive mechanism under these conditions; most of the 9S RNA molecules, which are in excess over the enzyme, are cleaved at their "a" site to generate 7S RNA before cleavage occurs at the "b" site to produce p5S RNA. If greater sensitivity is required, RNA can be labeled internally by incorporating nucleotides that are labeled with a phosphorus radioisotope at the α position (Sambrook and Russell, 2001). For previously uncharacterized substrates, sites of cleavage can be mapped *in vitro* in the same way that they are determined *in vivo*. If it is desirable to avoid the use of radioisotopes, we recommend RLM-RT-PCR (see earlier). The amplicons so generated can be sequenced to determine the precise 5'-ends of the downstream cleavage products.

Our standard conditions for the cleavage of 5'-extended RNAs by RNase H are as follows.

Extract the products of a standard 20-μl transcription reaction with acidic phenol (pH 4.3):chloroform:isoamyl alcohol (25:24:1), precipitate and wash with ethanol, and resuspend in 40 μl RNase-free water (see above section on CIP treatment, for details).

Estimate the concentration of 5'-extended RNA by analyzing an aliquot (~ 1 μl) using gel electrophoresis (A_{260} measurements are only helpful if unincorporated nucleotides have been completely removed).

To 80 pmol of product, add the chimeric oligonucleotide in threefold excess. The amount of transcript in nanograms will vary depending on the length of the RNA.

Anneal RNA and chimeric oligonucleotide by heating for 2 min at 60 °C, cooling to 37 °C for 2 min, and placing on ice for 5 min.

Cleave RNA by incubating for 1 h at 37 °C in 60 μl of RNase H buffer; 20 mM Tris-HCl (pH 7.8), 40 mM KCl, 8 mM MgCl$_2$, 1 mM DTT containing 80 U RNaseOUT^{TM} and 10 U RNase H (Fermentas).

Extract the cleavage reaction with acidic phenol (pH 4.3):chloroform: isoamyl alcohol (25:24:1), precipitate, wash with ethanol, and resuspend in 40 μl RNase-free water.

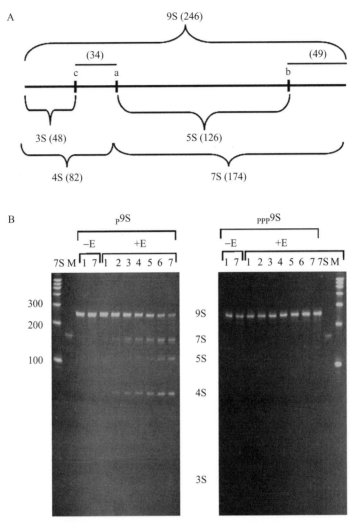

Figure 12.4 Discontinuous assay of RNase E activity comparing 5′-tri- and mono-phosphorylated 9S substrates. (A) Schematic representation of the 9S RNA and its cleavage sites. The cleavage sites are named "a," "b," and "c," according to the order they are thought to be cleaved. Brackets indicate RNA segments of previously defined size (9S, 7S, etc.). The approximate length in nucleotides of each possible fragment is given in parentheses. (B) Denaturing gel analysis of a discontinuous assay with 180 n*M* RNA and 5 n*M* RNase E (N-terminally His tagged, amino acids 1 to 529). The left panel shows 9S RNA with an RNase H–generated 5′-monophosphate. The right panel shows an assay containing triphosphorylated 9S. Both panels show assays containing RNase E (+E) and a negative control that did not contain enzyme (−E). Lanes numbered 1 to 7 contain samples from time points 0, 2, 5 10, 20, 40 and 80 min, respectively. The sizes of certain RNA marker bands are indicated on the left. As further control, triphosphorylated *in vitro* transcribed 7S RNA was loaded (7S). An enhancement of the ethidium bromide staining at later time points of the triphosphorylated 9S assay ensured that we did not miss any low-level cleavage.

Add 40 μl of 2× RNA loading dye; 95% (v/v) formamide, 0.025% (w/v) bromophenol blue, 0.025% (w/v) xylene cyanole, and 0.025% (w/v) SDS. Heat for 3 min at 95 °C and separate on a denaturing gel as described previously. Transcript that has not been cut with RNase H and triphosphorylated RNAs (not 5′-extended) can be loaded as controls.

After sufficient separation, isolate the shorter, 5′-monophosphorylated RNA from the gel by electroelution with D-tube dialyzers (Novagen) per the manufacturer's instructions. Cleavage efficiency of approximately 90% can be expected.

4.3. Cleavage conditions for *E. coli* RNase E and RNase G

Buffer conditions have been optimized for the N-terminal catalytic domain of *E. coli* RNase E; 25 mM *bis*-Tris-propane (pH 8.3), 15 mM MgCl$_2$, 100 mM NaCl, 0.1% (v/v) Triton X-100, and 1 mM DTT (Redko *et al.*, 2003). Moreover, K_M and k_{cat} values of 2.07 μM and 1.4 s^{-1}, respectively, were obtained for the cleavage of the model 5′-monophosphorylated substrate, BR13 (McDowall *et al.*, 1995). These values are similar to those obtained for *Eco*RV (Baldwin *et al.*, 1999; Waters and Connolly, 1994) and RNase HI (Kanaya *et al.*, 1990). Interestingly, with the same buffer conditions and substrate that were used to analyze NTH-RNase E, we have recently obtained K_M and k_{cat} values of 0.12 μM and 3.0 min^{-1}, respectively, for *E. coli* RNase G (Jourdan and McDowall, 2008). The lower K_M for RNase G may reflect the finding that this enzyme has to operate at a lower cellular concentration than RNase E in *E. coli* (Lee *et al.*, 2002). Our recent analysis of *E. coli* RNase G has also revealed that in the absence of a 5′-monophosphate, the K_M value for the cleavage of BR13 increases to 21 μM (Jourdan and McDowall, 2008). The change in this value contributes to a >100-fold reduction in the efficiency of cleavage (k_{cat}/K_M) (Jourdan and McDowall, 2008). Given this finding, the assay of the cleavage of 5′-hydroxylated substrates can be particularly vulnerable to RNase A-type ribonuclease, which are common laboratory contaminants. For this reason, we routinely add inhibitors of these potential contaminants.

Our standard conditions for the assay of members of the RNase E/G family use the optimized buffer listed earlier (25 mM *bis*-Tris-propane [pH 8.3], 15 mM MgCl$_2$, 100 mM NaCl, 0.1% [v/v] Triton X-100, and 1 mM DTT) and contains between 1 and 5 nM enzyme, and between 100 nM and 40 μM substrate. To inhibit contaminating RNase A-like ribonucleases, we consider ~30 U of RNaseOUT^{TM} (Invitrogen) to be sufficient for a 100-μl reaction. Before starting the reaction, which can either be initiated by addition of enzyme or substrate, all components should be prewarmed for 15 to 20 min at 37 °C. The assay is performed at 37 °C, and samples are removed immediately after the start of the reaction (zero time point) and at several time points thereafter, with increasing time between sampling. The reaction samples are quenched by immediately adding an equal volume of 2× RNA loading dye (see earlier), and are analyzed by denaturing gel electrophoresis.

4.4. Investigation of recognition and cleavage with oligoribonucleotide substrates

Oligoribonucleotides are proving to be a powerful tool in the investigation of the recognition and cleavage of RNA by members of the RNase E/G family (Feng et al., 2002; Jiang and Belasco, 2004; Jourdan and McDowall, 2008; Kaberdin and Bizebard, 2005; Tock et al., 2000). The cost of synthesizing oligoribonucleotides has decreased approximately tenfold over the last decade, making them affordable for most laboratories, and fluorophores can be added during the synthesis to provide an easily detectable and nonhazardous label (Muller et al., 2004). Moreover, with regard to the study of RNA cleavage by the RNase E/G family, 5′-monophosphates can be added chemically with high efficiency. Although, at this time, a 5′-triphosphate group cannot be added as part of the chemical synthesis of oligoribonucleotides, a 5′-hydroxyl can. Oligoribonucleotide substrates with the latter group provide a control for assessing whether a 5′-monophosphate is required for the efficient cleavage of a substrate. Because all of the members of the RNase E/G family that have been studied to date are 5′-end dependent, labels should be placed at the 3′-end. A variety of modified nucleotides can be inserted at internal positions to study the chemistry of cleavage (Redko et al., 2003), and 2′-O-methyl groups can be added to block cleavage by wild-type enzymes, thereby permitting the direct assay of RNA binding (Callaghan et al., 2003; Feng et al., 2002; Jourdan and McDowall, 2008). Although the insertion of certain modified nucleotides may require the expertise of a specialized laboratory (Venkatesan et al., 2003), the incorporation of 5′-monophosphate, 3′-labels, and 2′-O-methyl groups is routine for most companies (e.g., Dharamacon [CO, USA] and MWG [Germany]).

Oligoribonucleotides labeled with a 3′-fluorophore (e.g., fluorescein) are assayed discontinuously; that is to say, samples from a reaction are collected at different time points, quenched, and then, in a separate step, the amount of cleavage is determined. In the past, our laboratory has analyzed the amount of cleavage using automated ion-pair reverse-phase chromatography in line with an instrument that measures the fluorescence of labeled RNA species as they are eluted (Redko et al., 2003). However, unless one is familiar with the operation of such instruments, we recommend analyzing samples by denaturing polyacrylamide gel electrophoresis. When the oligoribonucleotide BR13-Fl (5′-GGGACAGUAUUUG-Fl-3′) is used, which will be cleaved into an 8- and 5-mer product of which only the 5-mer product carries the 3′-fluorescein, we advise separation of samples in a 15% (w/v) polyacrylamide gel (7 M urea, 19:1 acrylamide/bisacrylamide, 1× TBE). The gels can then be scanned in a flatbed imaging instrument, such as a Molecular Imager FX-Pro (BioRad). Others have described a continuous assay system for assaying *E. coli* RNase E activity using an extensively modified oligonucleotide substrate that contained a quencher of fluorescence and a fluorophore separated by the site of cleavage (Jiang and Belasco, 2004). On cleavage, the quencher is physically separated from the fluorophore; consequently, the level of fluorescence increases. This elegant approach, first described to our knowledge for the assay of pancreatic-type ribonucleases (Zelenko *et al.*, 1994), only offers savings in time, however, if instrumentation is available that can measure changes in fluorescence in multiple samples in parallel. Gels must still be run to identify the position(s) of cleavage.

To establish whether factors other than the sequence of the cleavage site influence the rate of cleavage of a newly identified substrate, we suggest synthesizing an oligoribonucleotide substrate that encompasses the single-stranded region in which cleavage occurs and comparing the rate of cleavage of this substrate with that of the full-length transcript. As described above, this revealed that the "b" site of 9S RNA is insufficient to direct efficient cleavage by RNase E and by implication suggests the presence of an uncharacterized effector of RNase E cleavage (Kaberdin *et al.*, 2000). More recently, we have identified an RNA structural arrangement that promotes *in vitro* the rapid cleavage by RNase E of substrates that lack a 5′-monophosphate (Jourdan, Stead, and McDowall, unpublished data). If it happens that an oligoribonucleotide substrate encompassing only a cleaved segment is insufficient to direct efficient cleavage, then additional segments can be added until efficient cleavage is restored. At present, oligoribonucleotides that approach 100 nt in length can be chemically synthesized. However, because the cost and efficiency of chemical synthesis is proportional to the length of RNA, we suggest synthesizing substrates longer than 50 nt by *in vitro* transcription (see earlier). After molecular dissection *in vitro*, the RNA elements thought to be important through biochemical studies should be disrupted within the context of the natural substrate *in vivo* to assess their biological relevance.

To assay the activity of a previously uncharacterized homolog of *E. coli* RNase E with the intention, for example, of determining the extent to which it is sensitive to 5′-phosphorylation, we suggest the use of oligoribonucleotides incorporating the sequence cleaved by *E. coli* RNase E at the 5′-end of RNAI (McDowall *et al.*, 1995). These have proved to be good substrates for analyzing *E. coli* RNase G (Feng *et al.*, 2002; Jiang and Belasco, 2004; Jourdan and McDowall, 2008), as well as homologs from *A. aeolicus* (Kaberdin and Bizebard, 2005), *M. tuberculosis* (Kovacs *et al.*, 2005), and *S. coelicolor* (Lee and Cohen, 2003). Oligoribonucleotides labeled with fluorescein at the 3′-end can be used to determine Michaelis–Menten parameters (Jourdan and McDowall, 2008; Redko *et al.*, 2003) and measure equilibrium binding in solution with fluorescence anisotropy (Jourdan and

McDowall, 2008). The latter will not be described in detail here, because it has been covered in previous volumes (Heyduk *et al.*, 1996; Jameson and Sawyer, 1995). Briefly, however, fluorescence anisotropy measures the rate of tumbling of a labeled macromolecule in solution; when another macromolecule binds, the rate of tumbling decreases and the anisotropy increases.

5. SUMMARY

In the preceding sections, methods have been described to identify and characterize substrates of members of the RNase E/G family of ribonucleases, which have a central role in the decay and processing of RNA in *Escherichia coli* and probably many other bacteria. This included methods that can be adopted to define elements that affect the binding and cleavage of RNA by these ribonucleases. These methods should be considered as a starting point and not necessarily inclusive of all approaches that can or should be taken. An important consideration when implementing these methods is that they are carefully evaluated for optimal conditions. For example, it should not be taken for granted that the conditions described for assaying the cleavage of RNA by *E. coli* RNase E will be optimal for homologs in other eubacteria, some of which are estimated to have last shared a common ancestor with *E. coli* hundreds of millions of years ago. Moreover, it should not be assumed that homologs would be as central to RNA processing and decay as RNase E is in *E. coli*. It is already known that the RNase E homolog in *S. coelicolor* is dispensable for growth (Lee and Cohen, 2003). One of the major challenges will be to determine the contribution of protein factors that bind members of the RNase E/G family to produce isolatable complexes or more transient associations. Although RNase E is primarily associated with PNPase, enolase, and the RhlB helicase in *E. coli* (Carpousis, 2002; Carpousis *et al.*, 2001; Marcaida *et al.*, 2006), it is associated, for example, with another 3'-exonuclease, RNase R, and another helicase, RhlE, in *Pseudomonas syringae* (Purusharth *et al.*, 2005) and two DEAD-box RNA helicases and the transcription termination factor Rho in *Rhodobacter capsulatus* (Jager *et al.*, 2001). In *E. coli,* the CsdA helicase is recruited to the degradosome during adjustment to cold-shock (Prud'homme-Genereux *et al.*, 2004), and two polypeptides named RraA and RraB (regulators of ribonuclease activity) are reported to bind and modulate cleavage by RNase E and alter the composition of the degradosome (Gao *et al.*, 2006; Lee *et al.*, 2003). Furthermore, it has been shown very recently that the RNA degradosome is a component of the *E. coli* cytoskeleton (Taghbalout and Rothfield, 2007).

Although an effort has been made to represent the numerous excellent contributions of investigators who furthered the study of RNA decay and processing by members of the RNase E/G family, we would direct readers with an interest in the RNase E/G family to the accompanying chapters on the degradosome and mRNA decay in *E. coli*.

ACKNOWLEDGMENTS

The corresponding author acknowledges funding from the Royal Society, the BBSRC, the Wellcome Trust, and Marie Curie Actions, the contribution of former postdocs and postgraduate students to the research, and the support of Simon Baumberg, recently deceased. We would also like to thank Vladimir Kaberdin and Masaaki Wachi for strains that were used here in the demonstration of methods.

REFERENCES

Aiba, H. (2007). Mechanism of RNA silencing by Hfq-binding small RNAs. *Curr. Opin. Microbiol.* **10,** 134–139.

Alwine, J. C., Kemp, D. J., and Stark, G. R. (1977). Method for detection of specific RNAs in agarose gels by transfer to diazobenzyloxymethyl paper and hybridization with DNA probes. *Proc. Natl. Acad. Sci. USA* **74,** 5350–5354.

Apirion, D., and Lassar, A. B. (1978). Conditional lethal mutant of *Escherichia coli* which affects processing of ribosomal RNA. *J. Biol. Chem.* **253,** 1738–1742.

Arraiano, C. M., Yancey, S. D., and Kushner, S. R. (1988). Stabilization of discrete mRNA breakdown products in *ams pnp rnb* multiple mutants of *Escherichia coli* K-12. *J Bacteriol.* **170,** 4625–4633.

Arraiano, C. M., Yancey, S. D., and Kushner, S. R. (1993). Identification of endonucleolytic cleavage sites involved in decay of *Escherichia coli* trxA messenger RNA. *J. Bacteriol.* **175,** 1043–1052.

Babitzke, P., and Kushner, S. R. (1991). The *ams* (altered messenger RNA stability) protein and ribonuclease E are encoded by the same structural gene of *Escherichia coli*. *Proc. Natl. Acad. Sci. USA* **88,** 1–5.

Baker, K. E., and Mackie, G. A. (2003). Ectopic RNase E sites promote bypass of 5′-end-dependent mRNA decay in *Escherichia coli*. *Mol. Microbiol.* **47,** 75–88.

Baldwin, G. S., Sessions, R. B., Erskine, S. G., and Halford, S. E. (1999). DNA cleavage by the EcoRV restriction endonuclease: Roles of divalent metal ions in specificity and catalysis. *J. Mol. Biol.* **288,** 87–103.

Belasco, J. G. (1993). mRNA degradation in prokaryotic cells: An overview. *In* "Control of Messenger RNA Stability" (J. G. Belasco and G. Brawerman, eds.), pp. 31–52. Academic Press, San Diego.

Belasco, J. G., and Higgins, C. F. (1988). Mechanisms of messenger RNA decay in bacteria: A perspective. *Gene* **72,** 15–23.

Bensing, B. A., Meyer, B. J., and Dunny, G. M. (1996). Sensitive detection of bacterial transcription initiation sites and differentiation from RNA processing sites in the pheromone-induced plasmid transfer system of *Enterococcus faecalis*. *Proc. Natl. Acad. Sci. USA* **93,** 7794–7799.

Bernstein, J. A., Khodursky, A. B., Lin, P. H., Lin-Chao, S., and Cohen, S. N. (2002). Global analysis of mRNA decay and abundance in *Escherichia coli* at single-gene resolution using two-color fluorescent DNA microarrays. *Proc. Natl. Acad. Sci. USA* **99,** 9697–9702.

Bouvet, P., and Belasco, J. G. (1992). Control of RNase E-mediated RNA degradation by 5′-terminal base-pairing in *Escherichia coli*. *Nature (Lond.)* **360,** 488–491.

Briant, D. J., Hankins, J. S., Cook, M. A., and Mackie, G. A. (2003). The quaternary structure of RNase G from *Escherichia coli*. *Mol. Microbiol.* **50,** 1381–1390.

Cairrão, N., and Arraiano, C. M. (2006). The role of endoribonucleases in the regulation of RNase R. *Biochem. Biophys. Res. Commun.* **343,** 731–737.

Callaghan, A. J., Grossmann, J. G., Redko, Y. U., Ilag, L. L., Moncrieffe, M. C., Symmons, M. F., Robinson, C. V., McDowall, K. J., and Luisi, B. F. (2003). Quaternary structure and catalytic activity of the *Escherichia coli* ribonuclease E amino-terminal catalytic domain. *Biochem.* **42,** 13848–13855.

Callaghan, A. J., Marcaida, M. J., Stead, J. A., McDowall, K. J., Scott, W. G., and Luisi, B. F. (2005a). Structure of *Escherichia coli* RNase E catalytic domain and implications for RNA turnover. *Nature (Lond.)* **437,** 1187–1191.

Callaghan, A. J., Redko, Y., Murphy, L. M., Grossmann, J. G., Yates, D., Garman, E., Ilag, L. L., Robinson, C. V., Symmons, M. F., McDowall, K. J., and Luisi, B. F. (2005b). "Zn-Link": A metal-sharing interface that organizes the quaternary structure and catalytic site of the endoribonuclease, RNase E. *Biochem.* **44,** 4667–4675.

Carpousis, A. J. (2002). The *Escherichia coli* RNA degradosome: Structure, function and relationship to other ribonucleolytic multienyzme complexes. *Biochem. Soc. Trans.* **30,** 150–155.

Carpousis, A. J., Leroy, A., Vanzo, N., and Khemici, V. (2001). *Escherichia coli* RNA degradosome. In "Ribonucleases, Pt B," (Allen W. Nicholson, ed.), Vol. 342, pp. 333–345. Academic Press Inc, San Diego.

Carpousis, A. J., Vanhouwe, G., Ehretsmann, C., and Krisch, H. M. (1994). Copurification of *E. coli* RNase E and PNPase: Evidence for a specific association between two enzymes important in RNA processing and degradation. *Cell* **76,** 889–900.

Celesnik, H., Deana, A., and Belasco, J. G. (2007). Initiation of RNA decay in *Escherichia coli* by 5'-pyrophosphate removal. *Mol. Cell* **27,** 79–90.

Cesareni, G., Helmercitterich, M., and Castagnoli, L. (1991). Control of ColE1 plasmid replication by antisense RNA. *Trends Genet.* **7,** 230–235.

Chen, C. Y. A., Beatty, J. T., Cohen, S. N., and Belasco, J. G. (1988). An intercistronic stem-loop structure functions as a messenger RNA decay terminator necessary but insufficient for *puf* messenger RNA stability. *Cell* **52,** 609–619.

Cohen, S. N., and McDowall, K. J. (1997). RNase E: Still a wonderfully mysterious enzyme. *Mol. Microbiol.* **23,** 1099–1106.

Condon, C., Putzer, H., Luo, D., and GrunbergManago, M. (1997). Processing of the *Bacillus subtilis* thrS leader mRNA is RNase E-dependent in *Escherichia coli*. *J. Mol. Biol.* **268,** 235–242.

Cormack, R. S., Genereaux, J. L., and Mackie, G. A. (1993). RNase E activity is conferred by a single polypeptide: Overexpression, purification, and properties of the *ams rne hmp1* gene product. *Proc. Natl. Acad. Sci. USA* **90,** 9006–9010.

Diwa, A. A., and Belasco, J. G. (2002). Critical features of a conserved RNA stem-loop important for feedback regulation of RNase E synthesis. *J. Biol. Chem.* **277,** 20415–20422.

Ehretsmann, C. P., Carpousis, A. J., and Krisch, H. M. (1992). Specificity of *Escherichia coli* endoribonuclease RNase E: *In vivo* and *in vitro* analysis of mutants in a bacteriophage T4 messenger RNA processing site. *Genes Dev.* **6,** 149–159.

Emory, S. A., Bouvet, P., and Belasco, J. G. (1992). A 5'-terminal stem-loop structure can stabilize messenger RNA in *Escherichia coli*. *Genes Dev.* **6,** 135–148.

Feng, Y. A., Vickers, T. A., and Cohen, S. N. (2002). The catalytic domain of RNase E shows inherent 3' to 5'-directionality in cleavage site selection. *Proc. Natl. Acad. Sci. USA* **99,** 14746–14751.

Gegenheimer, P., Watson, N., and Apirion, D. (1977). Multiple pathways for primary processing of ribosomal RNA in *Escherichia coli*. *J. Biol. Chem.* **252,** 3064–3073.

Ghora, B. K., and Apirion, D. (1978). Structural analysis and *in vitro* processing to p5 ribosomal RNA of a 9S RNA molecule isolated from an *rne* mutant of *Escherichia coli*. *Cell* **15,** 1055–1066.

Heyduk, T., Ma, Y. X., Tang, H., and Ebright, R. H. (1996). Fluorescence anisotropy: Rapid, quantitative assay for protein-DNA and protein-protein interaction. *In* "RNA Polymerase and Associated Factors, Pt B," (Sankar Adhya, ed.), Vol. 274, pp. 492–503. Academic Press Inc, San Diego.

Jager, S., Fuhrmann, O., Heck, C., Hebermehl, M., Schiltz, E., Rauhut, R., and Klug, G. (2001). An mRNA degrading complex in *Rhodobacter capsulatus*. *Nucleic Acids Res.* **29,** 4581–4588.

Jain, C., and Belasco, J. G. (1995). RNase E autoregulates its synthesis by controlling the degradation rate of its own messenger RNA in *Escherichia coli*: Unusual sensitivity of the *rne* transcript to RNase E activity. *Genes Dev.* **9,** 84–96.

Jameson, D. M., and Sawyer, W. H. (1995). Fluorescence anisotropy applied to biomolecular interactions. *In* "Biochemical Spectroscopy," (Kenneth Sauer, ed.), Vol. 246, pp. 283–300. Academic Press Inc, San Diego.

Jiang, X. Q., and Belasco, J. G. (2004). Catalytic activation of multimeric RNase E and RNase G by 5′-monophosphorylated RNA. *Proc. Natl. Acad. Sci. USA* **101,** 9211–9216.

Jourdan, S. S., and McDowall, K. J. (2008). Sensing of 5′-monophosphate by *Escherichia coli* RNase G can significantly enhance association with RNA and stimulate the decay of functional mRNA transcripts *in vivo*. *Mol. Microbiol.* **67,** 102–115.

Kaberdin, V. R. (2003). Probing the substrate specificity of *Escherichia coli* RNase E using a novel oligonucleotide-based assay. *Nucleic Acids Res.* **31,** 4710–4716.

Kaberdin, V. R., and Bizebard, T. (2005). Characterization of *Aquifex aeolicus* RNase E/G. *Biochem. Biophys. Res. Commun.* **327,** 382–392.

Kaberdin, V. R., Walsh, A. P., Jakobsen, T., McDowall, K. J., and von Gabain, A. (2000). Enhanced cleavage of RNA mediated by an interaction between substrates and the arginine-rich domain of *E. coli* ribonuclease E. *J. Mol. Biol.* **301,** 257–264.

Kaga, N., Umitsuki, G., Clark, D. P., Nagai, K., and Wachi, M. (2002a). Extensive overproduction of the AdhE protein by *rng* mutations depends on mutations in the *cra* gene or in the Cra-box of the *adhE* promoter. *Biochem. Biophys. Res. Commun.* **295,** 92–97.

Kaga, N., Umitsuki, G., Nagai, K., and Wachi, M. (2002b). RNase G-dependent degradation of the *eno* mRNA encoding a glycolysis enzyme enolase in *Escherichia coli*. *Biosci. Biotechnol. Biochem.* **66,** 2216–2220.

Kanaya, S., Kimura, S., Katsuda, C., and Ikehara, M. (1990). Role of cysteine residues in ribonuclease H from *Escherichia coli*: Site-directed mutagenesis and chemical modification. *Biochem. J.* **271,** 59–66.

Khemici, V., Poljak, L., Toesca, I., and Carpousis, A. J. (2005). Evidence *in vivo* that the DEAD-box RNA helicase RhlB facilitates the degradation of ribosome-free mRNA by RNase E. *Proc. Natl. Acad. Sci. USA* **102,** 6913–6918.

Kido, M., Yamanaka, K., Mitani, T., Niki, H., Ogura, T., and Hiraga, S. (1996). RNase E polypeptides lacking a carboxyl-terminal half suppress a *mukB* mutation in *Escherichia coli*. *J. Bacteriol.* **178,** 3917–3925.

Klug, G., and Cohen, S. N. (1990). Combined actions of multiple hairpin loop structures and sites of rate-limiting endonucleolytic cleavage determine differential degradation rates of individual segments within polycistronic *puf* operon messenger RNA. *J. Bacteriol.* **172,** 5140–5146.

Klug, G., Jock, S., and Rothfuchs, R. (1992). The rate of decay of *Rhodobacter capsulatus*-specific *puf* messenger RNA segments is differentially affected by RNase-E activity in *Escherichia coli*. *Gene* **121,** 95–102.

Kovacs, L., Csanadi, A., Megyeri, K., Kaberdin, V. R., and Miczak, A. (2005). Mycobacterial RNase E-associated proteins. *Microbiol. Immunol.* **49,** 1005–1009.

Kuwano, M., Ono, M., Endo, H., Hori, K., Nakamura, K., Hirota, Y., and Ohnishi, Y. (1977). Gene affecting longevity of messenger RNA: Mutant of *Escherichia coli* with altered messenger RNA stability. *Mol. Gen. Genet.* **154,** 279–285.

Lapham, J., and Crothers, D. M. (1996). RNase H cleavage for processing of *in vitro* transcribed RNA for NMR studies and RNA ligation. *RNA* **2,** 289–296.

Lee, K., Bernstein, J. A., and Cohen, S. N. (2002). RNase G complementation of *rne* null mutation identifies functional interrelationships with RNase E in *Escherichia coli*. *Mol. Microbiol.* **43,** 1445–1456.

Lee, K., and Cohen, S. N. (2003). A *Streptomyces coelicolor* functional orthologue of *Escherichia coli* RNase E shows shuffling of catalytic and PNPase-binding domains. *Mol. Microbiol.* **48,** 349–360.

Li, Z., Pandit, S., and Deutscher, M. P. (1999a). RNase G (CafA protein) and RNase E are both required for the 5′-maturation of 16S ribosomal RNA. *EMBO J.* **18,** 2878–2885.

Li, Z. W., and Deutscher, M. P. (2002). RNase E plays an essential role in the maturation of *Escherichia coli* tRNA precursors. *RNA* **8,** 97–109.

Li, Z. W., Pandit, S., and Deutscher, M. P. (1999b). RNase G (CafA protein) and RNase E are both required for the 5′-maturation of 16S ribosomal RNA. *EMBO J.* **18,** 2878–2885.

Lin-Chao, S., and Cohen, S. N. (1991). The rate of processing and degradation of antisense RNAI regulates the replication of ColE1-type plasmids *in vivo*. *Cell* **65,** 1233–1242.

Lin-Chao, S., Wei, C. L., and Lin, Y. T. (1999). RNase E is required for the maturation of *ssrA* RNA and normal *ssrA* RNA peptide-tagging activity. *Proc. Natl. Acad. Sci. USA* **96,** 12406–12411.

Loayza, D., Carpousis, A. J., and Krisch, H. M. (1991). Gene 32 transcription and messenger RNA processing in T4-related bacteriophages. *Mol. Microbiol.* **5,** 715–725.

Lundberg, U., and Altman, S. (1995). Processing of the precursor to the catalytic RNA subunit of RNase P from *Escherichia coli*. *RNA* **1,** 327–334.

Mackie, G. A. (1991). Specific endonucleolytic cleavage of the messenger RNA for ribosomal protein S20 of *Escherichia coli* requires the product of the *ams* gene *in vivo* and *in vitro*. *J. Bacteriol.* **173,** 2488–2497.

Mackie, G. A. (1998). Ribonuclease E is a 5′-end-dependent endonuclease. *Nature (Lond.)* **395,** 720–723.

Mackie, G. A., and Genereaux, J. L. (1993). The role of RNA structure in determining RNase E-dependent cleavage sites in the messenger RNA for ribosomal protein S20 *in vitro*. *J. Mol. Biol.* **234,** 998–1012.

Marcaida, M. J., DePristo, M. A., Chandran, V., Carpousis, A. J., and Luisi, B. F. (2006). The RNA degradosome: Life in the fast lane of adaptive molecular evolution. *Trends Biochem. Sci.* **31,** 359–365.

McDowall, K. J., and Cohen, S. N. (1996). The N-terminal domain of the *rne* gene product has RNase E activity and is non-overlapping with the arginine-rich RNA-binding site. *J. Mol. Biol.* **255,** 349–355.

McDowall, K. J., Hernandez, R. G., Chao, S. L., and Cohen, S. N. (1993). The *ams-1* and *rne-3071* temperature-sensitive mutations in the *ams* gene are in close proximity to each other and cause substitutions within a domain that resembles a product of the *Escherichia coli mre* locus. *J. Bacteriol.* **175,** 4245–4249.

McDowall, K. J., Kaberdin, V. R., Wu, S. W., Cohen, S. N., and Lin-Chao, S. (1995). Site-specific RNase E cleavage of oligonucleotides and inhibition by stem-loops. *Nature (Lond.)* **374,** 287–290.

McDowall, K. J., Linchao, S., and Cohen, S. N. (1994). A+U content rather than a particular nucleotide order determines the specificity of RNase E cleavage. *J. Biol. Chem.* **269,** 10790–10796.

Melefors, O., and von Gabain, A. (1991). Genetic studies of cleavage initiated messenger RNA decay and processing of ribosomal 9S RNA show that the *Escherichia coli ams* and *rne* loci are the same. *Mol. Microbiol.* **5,** 857–864.

Miczak, A., Kaberdin, V. R., Wei, C. L., and Lin-Chao, S. (1996). Proteins associated with RNase E in a multicomponent ribonucleolytic complex. *Proc. Natl. Acad. Sci. USA* **93**, 3865–3869.

Misra, T. K., and Apirion, D. (1979). RNase E, and RNA processing enzyme from *Escherichia coli*. *J. Biol. Chem.* **254**, 1154–1159.

Mudd, E. A., and Higgins, C. F. (1993). *Escherichia coli* endoribonuclease RNase E: Auto-regulation of expression and site-specific cleavage of messenger RNA. *Mol. Microbiol.* **9**, 557–568.

Mudd, E. A., Krisch, H. M., and Higgins, C. F. (1990). RNase E, an endoribonuclease, has a general role in the chemical decay of *Escherichia coli* messenger RNA: Evidence that *rne* and *ams* are the same genetic locus. *Mol. Microbiol.* **4**, 2127–2135.

Muller, S., Wolf, J., and Ivanov, S. A. (2004). Current strategies for the synthesis of RNA. *Curr. Org. Synth.* **1**, 293–307.

Neidhardt, F. C., and Umbarger, H. E. (1996). Chemical composition of *Esherichia coli*. In "*Escherichia coli* and *Salmonella*" (F. C. Neidhardt, ed.), Vol 1, pp. 13–16. ASM, Washington, D.C.

Newbury, S. F., Smith, N. H., and Higgins, C. F. (1987a). Differential messenger RNA stability controls relative gene expression within a polycistronic operon. *Cell* **51**, 1131–1143.

Newbury, S. F., Smith, N. H., Robinson, E. C., Hiles, I. D., and Higgins, C. F. (1987b). Stabilization of translationally active messenger RNA by prokaryotic REP sequences. *Cell* **48**, 297–310.

Nilsson, P., Naureckiene, S., and Uhlin, B. E. (1996). Mutations affecting mRNA processing and fimbrial biogenesis in the *Escherichia coli pap* operon. *J. Bacteriol.* **178**, 683–690.

Ono, M., and Kuwano, M. (1979). Conditional lethal mutation in an *Escherichia coli* strain with a longer chemical lifetime of messenger RNA. *J. Mol. Biol.* **129**, 343–357.

Ow, M. C., and Kushner, S. R. (2002). Initiation of tRNA maturation by RNase E is essential for cell viability in *E. coli. Genes Dev.* **16**, 1102–1115.

Pak, J., and Fire, A. (2007). Distinct populations of primary and secondary effectors during RNAi in *C. elegans. Science* **315**, 241–244.

Patel, A. M., and Dunn, S. D. (1995). Degradation of *Escherichia coli uncB* messenger RNA by multiple endonucleolytic cleavages. *J. Bacteriol.* **177**, 3917–3922.

Porath, J., Carlsson, J., Olsson, I., and Belfrage, G. (1975). Metal chelate affinity chroma-tography: A new approach to protein fractionation. *Nature (Lond.)* **258**, 598–599.

Prud'homme-Genereux, A., Beran, R. K., Iost, I., Ramey, C. S., Mackie, G. A., and Simons, R. W. (2004). Physical and functional interactions among RNase E, polynu-cleotide phosphorylase and the cold-shock protein, CsdA: evidence for a 'cold shock degradosome'. *Mol. Microbiol.* **54**, 1409–1421.

Purusharth, R. I., Klein, F., Sulthana, S., Jager, S., Jagannadham, M. V., Evguenieva-Hackenberg, E., Ray, M. K., and Klug, G. (2005). Exoribonuclease R interacts with endoribonuclease E and an RNA helicase in the psychrotrophic bacterium *Pseudomonas syringae* Lz4W. *J. Biol. Chem.* **280**, 14572–14578.

Py, B., Higgins, C. F., Krisch, H. M., and Carpousis, A. J. (1996). A DEAD-box RNA helicase in the *Escherichia coli* RNA degradosome. *Nature (Lond.)* **381**, 169–172.

Redko, Y., Tock, M. R., Adams, C. J., Kaberdin, V. R., Grasby, J. A., and McDowall, K. J. (2003). Determination of the catalytic parameters of the N-terminal half of *Escherichia coli* ribonuclease E and the identification of critical functional groups in RNA substrates. *J. Biol. Chem.* **278**, 44001–44008.

Sambrook, J., and Russell, D. W. (2001). "Molecular Cloning: A Laboratory Manual." Cold Spring Harbor Laboratories, New York.

Schirmer, F., and Hillen, W. (1998). The *Acinetobacter calcoaceticus* NCIB8250 *mop* operon mRNA is differentially degraded, resulting in a higher level of the 3'-CatA-encoding segment than of the 5′-phenolhydroxylase-encoding portion. *Mol. Gen. Genet.* **257,** 330–337.

Selinger, D. W., Saxena, R. M., Cheung, K. J., Church, G. M., and Rosenow, C. (2003). Global RNA half-life analysis in *Escherichia coli* reveals positional patterns of transcript degradation. *Genome Res.* **13,** 216–223.

Smolke, C. D., Carrier, T. A., and Keasling, J. D. (2000). Coordinated, differential expression of two genes through directed mRNA cleavage and stabilization by secondary structures. *Appl. Environ. Microbiol.* **66,** 5399–5405.

Taghbalout, A., and Rothfield, L. (2007). RNase E and the other constituents of the RNA degradosome are components of the bacterial cytoskeleton. *Proc. Natl. Acad. Sci. USA* **104,** 1667–1672.

Taraseviciene, L., Miczak, A., and Apirion, D. (1991). The gene specifying RNase E (*rne*) and a gene affecting messenger RNA stability (*ams*) are the same gene. *Mol. Microbiol.* **5,** 851–855.

Tock, M. R., Walsh, A. P., Carroll, G., and McDowall, K. J. (2000). The CafA protein required for the 5′-maturation of 16S rRNA is a 5′-end-dependent ribonuclease that has context-dependent broad sequence specificity. *J. Biol. Chem.* **275,** 8726–8732.

Tomcsanyi, T., and Apirion, D. (1985). Processing enzyme ribonuclease E specifically cleaves RNAI: An inhibitor of primer formation in plasmid DNA synthesis. *J. Mol. Biol.* **185,** 713–720.

Umitsuki, G., Wachi, M., Takada, A., Hikichi, T., and Nagai, K. (2001). Involvement of RNase G in *in vivo* mRNA metabolism in *Escherichia coli. Genes Cells* **6,** 403–410.

Vanderpool, C. K. (2007). Physiological consequences of small RNA-mediated regulation of glucose-phosphate stress. *Curr. Opin. Microbiol.* **10,** 146–151.

Venkatesan, N., Kim, S. J., and Kim, B. H. (2003). Novel phosphoramidite building blocks in synthesis and applications toward modified oligonucleotides. *Curr. Med. Chem.* **10,** 1973–1991.

Wachi, M., Umitsuki, G., Shimizu, M., Takada, A., and Nagai, K. (1999). *Escherichia coli cafA* gene encodes a novel RNase, designated as RNase G, involved in processing of the 5′-end of 16S rRNA. *Biochem. Biophys. Res. Commun.* **259,** 483–488.

Walsh, A. P., Tock, M. R., Mallen, M. H., Kaberdin, V. R., von Gabain, A., and McDowall, K. J. (2001). Cleavage of poly(A) tails on the 3′-end of RNA by ribonuclease E of *Escherichia coli. Nucleic Acids Res.* **29,** 1864–1871.

Waters, T. R., and Connolly, B. A. (1994). Interaction of the restriction endonuclease EcoRV with the deoxyguanosine and deoxycytidine bases in its recognition sequence. *Biochemistry* **33,** 1812–1819.

Wu, L. G., Fan, J. H., and Belasco, J. G. (2006). MicroRNAs direct rapid deadenylation of mRNA. *Proc. Natl. Acad. Sci. USA* **103,** 4034–4039.

Yajnik, V., and Godson, G. N. (1993). Selective decay of *Escherichia coli dnaG* messenger RNA is initiated by RNase E. *J. Biol. Chem.* **268,** 13253–13260.

Yuan, D., and Shen, V. (1975). Stability of ribosomal and transfer ribonucleic acid in *Escherichia coli* B/r after treatment with ethylenedinitrilotetraacetic acid and rifampicin. *J. Bacteriol.* **122,** 425–432.

Zelenko, O., Neumann, U., Brill, W., Pieles, U., Moser, H. E., and Hofsteenge, J. (1994). A novel fluorogenic substrate for ribonucleases: Synthesis and enzymatic characterization. *Nucleic Acids Res.* **22,** 2731–2739.

Zilhão, R., Regnier, P., and Arraiano, C. M. (1995). The role of endonucleases in the expression of ribonuclease II in *Escherichia coli. FEMS Microbiol. Lett.* **130,** 237–244.

CHAPTER THIRTEEN

Construction and Characterization of *E. coli* K12 Strains in Which the Transcription of Selected Genes Is Desynchronized from Translation

Florence Proux*,† *and* Marc Dreyfus*,†

Contents

Abstract

In *Escherichia coli*, synthesis and translation of individual mRNAs are usually synchronous, so that no long ribosome-free mRNA stretch exists between the RNA polymerase and the leading ribosome. By comparing situations in which the same mRNA (the *lacZ* mRNA) is synthesized either by the genuine *E. coli*

* École Normale Supérieure, Laboratoire de Génétique Moléculaire, Paris, France
† Centre National de la Recherche Scientifique (NRS), Paris, France

Methods in Enzymology, Volume 447
ISSN 0076-6879, DOI: 10.1016/S0076-6879(08)02213-1

RNA polymerase or the faster T7 RNA polymerase, we have previously shown that the outpacing of ribosomes by RNA polymerase destabilizes mRNAs, and more so as outpacing becomes larger. This destabilization requires the non-catalytic C-terminal region of RNase E; more generally, there is circumstantial evidence that this region is specifically involved in the fast decay of various untranslated mRNAs. The genetic system designed for desynchronizing transcription and translation with T7 RNA polymerase was originally designed in the *E. coli* B strain BL21(DE3). Here, we describe procedures for transferring this system to the more common *E. coli* K12 background. We also show that it can be used as a screen for identifying factors involved in the instability of untranslated mRNA. Protocols in use in this laboratory for RNA extraction, Northern blotting, and *β*-galactosidase assay are described and critically discussed.

1. INTRODUCTION

Prokaryotes lack a nuclear membrane. As a consequence, and because transcription and translation proceed in the same (5′ to 3′) direction, mRNAs can bind ribosomes as soon as ribosome binding sites (RBS) have been synthesized, and well before mRNA synthesis is complete. In *Escherichia coli*, this early binding of ribosomes has been documented for many years; moreover, in this organism, transcription and translation subsequently proceed with the same average elongation rate, as best illustrated by the direct visualization of nascent mRNAs covered with ribosomes by electron microscopy (Miller *et al.*, 1970). This synchronization presumably reflects both an active adjustment of the RNA polymerase (RNAP) rate to that of ribosomes over individual coding sequences (Landick *et al.*, 1985), and the existence of global regulatory mechanisms adjusting the rates of the two processes (Sorensen *et al.*, 1994; Vogel *et al.*, 1992).

What is the benefit, if any, of this situation for the *E. coli* cell? In the case of genes or operons that are regulated at the transcriptional level, the synchronization of transcription and translation minimizes the delay between the induction of transcription and the onset of protein synthesis, thus contributing to the efficiency of this mode of regulation. It has also been suggested that the presence of ribosomes on the nascent mRNA, by keeping it away from the DNA matrix, prevents the formation of RNA-DNA hybrids (R-loops) that would inhibit replication and/or transcription (Gowrishankar and Harinarayanan, 2004). However, the best-documented impact of transcription-translation synchronization is on mRNA stability. Naked mRNA is often prone to endonuclease attack (Baker and Mackie, 2003; Deana and Belasco, 2005; Dreyfus, 2009), and by ensuring that no gap exists between RNAP and the leading ribosome, the transcription–translation synchronization protects mRNAs from being degraded before having ever been translated. Thus, replacing the genuine *E. coli* RNAP by the T7 RNAP, which largely outpaces ribosomes (Iost *et al.*, 1992), causes a marked destabilization of transcripts *in vivo* (Iost and Dreyfus, 1995). The use of mutant T7 RNAPs that are affected in their catalytic turnover shows that, the greater is the difference in rate between the RNAP and the ribosomes, the greater is the destabilization (Makarova *et al.*, 1995). The destabilization is particularly marked with the *lacZ* gene and has been best characterized in this case, but it holds for other mRNAs as well (Lopez *et al.*, 1994). As for the mechanism of decay, the desynchronized *lacZ* mRNA is exceptionally sensitive to RNase E attack (Iost and Dreyfus, 1995), and indeed RNase E-mediated cleavages have been located in the 5′-proximal region of this RNA *in vivo* (Khemici *et al.*, 2005). More precisely, the nonessential C-terminal half (CTH) of RNase E, which serves as a scaffold for the assembly of the degradosome (Vanzo *et al.*, 1998), is required for the fast decay of this desynchronized mRNA, and CTH removal has been proposed as a simple mean for improving popular T7-based expression systems (Lopez *et al.*, 1999). Moreover, several other mRNAs that are genuinely untranslated also require the presence of the CTH for fast decay (Leroy *et al.*, 2002). However, the molecular mechanism underlying this requirement remains obscure.

So far, the instability of the desynchronized *lacZ* mRNA has been documented in *E. coli* B strains classically used for T7-based expression systems (Studier and Moffatt, 1986). Whether the same behavior also holds in *E. coli* K12, the first sequenced *E. coli* strain and the most commonly used for genetic experiments is not known. Of note, sequence comparisons of RNase E from *E. coli* K12 and *E. coli* B show 16 amino acid differences in the CTH (http://www.ncbi.nlm.nih.gov/BLAST/). Moreover, even in *E. coli* B, no systematic search for factors involved in the fast decay of the desynchronized *lacZ* mRNA has been undertaken.

Here we describe how the genetic system formerly used in *E. coli* B for desynchronizing *lacZ* transcription and translation can be transferred to *E. coli* K12. We observe that, in this background, the destabilization of the desynchronized *lacZ* mRNA is eventually so great that colonies appear white (Lac⁻) on X-gal plates, allowing for the construction of a simple screen for identifying factors that, beyond the CTH, are involved in the fast decay of this mRNA.

Many of the techniques used here are described in classical textbooks (for biochemical or molecular biology manipulations, see, e.g., Sambrook *et al.*, 1989; Sambrook and Russell, 2001; for buffers and growth media, see, e.g., Miller, 1972; for P1 transduction techniques, see, e.g., Silhavy *et al.*, 1984); other sources for more specialized techniques are cited in the text. Only protocols or experimental details that are more specific to our work are discussed here.

2. DESYNCHRONIZING THE TRANSCRIPTION AND TRANSLATION OF THE *LACZ* GENE IN *E. COLI* K12

2.1. Background

To desynchronize the transcription and translation of the *lacZ* gene, we initially started from the *E. coli* B strain BL21(DE3). This strain is a lysogen for phage λDE3 (*imm21 Δnin5 Sam7*), which carries the T7 gene *1* encoding the T7 RNAP under the IPTG-inducible P_{lacUV5} promoter (Studier and Moffatt, 1986; Fig. 13.1A). BL21(DE3) was made Lac⁻ by Tn*10* insertion within the genuine *lacZ* gene followed by imprecise Tn*10* excision (Lopez *et al.*, 1994), and an ectopic *lacZ* copy (followed by part of the *lacY* gene) was introduced on the chromosome between the divergently transcribed *malT* and *malP* genes. This ectopic *lacZ* gene was placed under the control of either the genuine *lac* promoter or a late T7 promoter for synchronized or desynchronized transcription, respectively (Fig. 13.2A). These constructions are subsequently referred to here as the P_{T7}-*lacZ* and P_{lac}-*lacZ* cassettes; they were designed so that the transcribed sequences are precisely the same in both cases. Moreover, the efficiency of the RBS that drives *lacZ* translation could be varied at will. The stability of the mRNA in both cases was inferred from mRNA steady-state levels, allowance being made for the difference in transcription frequencies between the two cassettes. Alternatively, the effect of inactivating RNase E (or of removing the CTH) upon the steady-state mRNA level and β-galactosidase expression was compared in both cases.

2.2. Procedure 1: Lysogenization of *E. coli* K12 with the λDE3 phage encoding T7 RNAP

To reconstruct the previous system in *E. coli* K12, we started from two Lac⁻ K12 isolates, MC1061 and HfrG6Δ12. MC1061, which has been extensively used by Dr. Carpousis and his colleagues in their studies on mRNA decay (see Carpousis *et al.*, 1994), carries a large deletion (Δ*lacX74*) encompassing the whole *lac* operon. HfrG6Δ12 (also referred to as HfrG6Δlac12 [Dreyfus, 1988] or ENS0 [Yarchuk *et al.*, 1992]) is a derivative of the male His⁻ strain HfrG6 (Suit *et al.*, 1964) that has been extensively used in our laboratory for studies on translation initiation. It carries a small *lac* deletion (from nt −13 to +58 with respect to the main *lacZ* transcription starting point) that removes the −10 promoter region and the *lacZ* translation start.

Lysogenization of MC1061 and HfrG6Δ12 with λDE3 is readily achieved with a kit available from Novagen (cat. 69734-3). Together with λDE3, which lacks the *int* gene and thus cannot lysogenize on its own, the kit contains titrated suspensions of (1) a helper phage (λB10) that also cannot

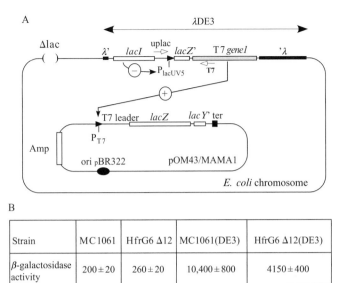

Figure 13.1 shown with labels:

A

λDE3

Δlac

λ'   *lacI*   uplac   *lacZ'*   T7 *gene1*   'λ

P$_{lacUV5}$ T7

T7 leader   *lacZ*   *lacY'* ter

P$_{T7}$

Amp

ori $_{pBR322}$ pOM43/MAMA1

E. coli chromosome

B

Strain	MC1061	HfrG6 Δ12	MC1061(DE3)	HfrG6 Δ12(DE3)
β-galactosidase activity	200 ± 20	260 ± 20	10,400 ± 800	4150 ± 400

Figure 13.1 Verification of λDE3 lysogens. (A) Schematic representation of the chromosome of MC1061(DE3) and HfrG6Δ12(DE3) containing the test plasmid pOM43/MAMA1. Chromosomal markers are shown in the conventional order (i.e., the clockwise orientation corresponds to increasing coordinates of the *E. coli* map). Brackets indicate a deletion in the genuine *lacZ* gene (7.8 min). The λDE3 phage is integrated at attλ (17 min). The λ *int* gene is interrupted by a 4.5 kb insertion consisting of (1) a 1724-nt fragment from the *lac* operon encompassing *lacI* and the first 146 codons of *lacZ* (open boxes) with the P$_{lacUV5}$ promoter in between; (2) the T7 *gene1* encoding the RNAP (light gray box), in the same orientation as *lacI* and *lacZ*. Positive (+) and negative (−) controls by the T7 RNAP and *lac* repressor, respectively, are indicated. Note that in HfrG6Δ12(DE3) the genuine *lacI* gene is still present, so that this strain carries two copies of this gene. Open arrows indicate the oligonucleotides used for PCR tests (see text and Table 13.1). Plasmid pOM43/MAMA1 is a pBR322 derivative carrying the promoter from the T7 late *gene10* (P$_{T7}$), which is recognized by T7 RNAP. P$_{T7}$ is followed by the gene *10* leader sequence and by the *lacZ* coding sequence (open box). Promoters are shown as closed arrows pointing to the relevant genes. (B) β-galactosidase activities in extracts from various strains carrying pOM43/MAMA1. Cells were grown without IPTG. Activities are averaged from two cultures (or two independent isolates of λDE3 lysogens).

lysogenize but provides the *int* function in *trans*; (2) a selection phage (λB482) that has the same immunity as λDE3 and thus kills cells that are not λDE3 lysogens, and (3) a derivative of bacteriophage T7 (T7 4107; tester phage) that fails to express the T7 RNAP and thus grows on IPTG-induced λDE3 lysogens but not on uninduced lysogens or on cells lacking λDE3 altogether. By plating the target cells with the three lambda phages according to the manufacturer's instructions, several hundred potential λDE3 lysogens were obtained with both MC1061 and HfrG6Δ12. Six MC1061(DE3) candidates were tested with the T7 4107 phage: all of

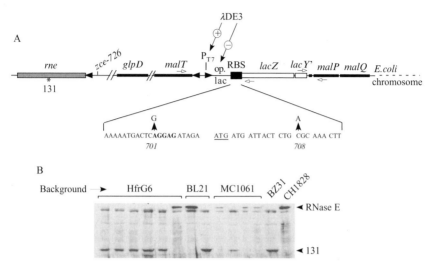

Figure 13.2 Schematic representation of the *rne* and *malT-malP* chromosomal regions from MC1061(DE3) and HfrG6Δ12(DE3) derivatives carrying the P$_{T7}$-*lacZ* cassette. (A) The *rne* gene (28 min on the chromosome) is shown as a dark gray box; also represented are the closely linked *zce-126* locus (see text) and the position of the *rne131* mutation (star). The P$_{T7}$-*lacZ* cassette is inserted within the *malP* promoter (76 min); it is bracketed on one side by the *malT* and *glpD* genes, and on the other side by the *malP* and *malQ* genes. The cassette itself consists of the T7 gene*10* promoter (P$_{T7}$), the *lac* operator (op. lac), an RBS (closed box), and the *lacZ* gene and part of the *lacY* gene (open boxes). In the P$_{lac}$-*lacZ* cassette, the P$_{T7}$ promoter is replaced by the genuine P$_{lac}$ promoter (not illustrated). The sequence of the *lamB* RBS that is used here, with the positions of the *701* and *708* mutations, is enlarged. Other symbols are as in Fig. 13.1. (B) Visualization of the full-length (RNase E) or truncated (131) RNase E on a Western blot. Derivatives of HfrG6Δ12(DE3), BL21(DE3), or MC1061(DE3) carrying the P$_{T7}$-*lacZ* cassette were transduced with a lysate grown on BZ31 (*zce-726::*Tn*10*), and transductants were tested with the anti-RNase E antibody. CH1828 and BZ31 (see text) are used as controls for full-length or truncated RNase E, respectively.

them gave very small plaques in the absence of IPTG and very large ones in its presence, indicating that they carry the IPTG-inducible T7 gene*1* from λDE3. The same test could not be used with HfrG6Δ12(DE3) candidates because bacteriophage T7 does not infect male strains. To test for the presence of the T7 gene*1* in this case, two HfrG6Δ12(DE3) candidates (and, as a control, two genuine MC1061(DE3) lysogens) were transformed with plasmid pOM43/MAMA1 (Chevrier-Miller *et al.*, 1990). This pBR322 derivative carries the *lacZ* gene under the control of the P$_{T7}$ promoter (Fig. 13.1A), so that β-galactosidase synthesis is dependent upon the presence of a functional T7 gene*1* in the host. With λDE3 lysogens, this synthesis can be observed in the absence of IPTG because the repression of P$_{lacUV5}$ by the *lac* repressor is leaky and allows some T7 RNAP synthesis; in the presence of IPTG the synthesis of β-galactosidase is so high that steady-state growth cannot be achieved (note that the *lacZ* gene from pOM43/ MAMA1 lacks the *lac* operator and thus does not titrate the repressor in the absence of IPTG). Measurement of the β-galactosidase expression (protocol 1) then unambiguously shows that the HfrG6Δ12(DE3) candidates are λDE3 lysogens (Fig. 13.1B). Two additional tests were done to establish this point. First, it was checked that the MC1061(DE3) and HfrG6Δ12(DE3) candidates are immune to the selection phage λB482, using strains BL21 and BL21 (DE3) as negative and positive controls, respectively. Second, the presence of the T7 gene *1* fused to the P_{lacUV5} promoter was checked directly by PCR on isolated colonies, using the oligonucleotides Uplac and T7 (Fig. 13.1A and Table 13.1). Amplification yields a fragment of 633 bp in λDE3 lysogens only.

2.3. Procedure 2: Introduction of the P_{T7}-*lacZ* and P_{lac}-*lacZ* cassettes in λDE3 lysogens

In our former work, the P_{T7}-*lacZ* (or P_{lac}-*lacZ*) cassettes were first assembled on a shuttle plasmid and then transferred to the *malP-malQ* intergenic region of the chromosome by a selection procedure (Raibaud *et al.*, 1984) that required that plasmidic genes downstream of the cassette were transcriptionally silent. For this reason, the transfer cannot be done directly in λDE3 lysogens; rather, it was first achieved in a strain (JM101) that does not produce T7 RNAP, and the cassettes were transferred to the BL21(DE3) background by P1 transduction (Iost and Dreyfus, 1995). The same procedure has been used here to introduce the cassettes into MC1061(DE3) and HfrG6Δ12(DE3). Many different P_{T7}-*lacZ* and P_{lac}-*lacZ* cassettes differing by the nature of the RBS in front of *lacZ* have been constructed (Iost and Dreyfus, 1995). Because our aim here is to uncover factors involved in the fast decay of the P_{T7}-*lacZ* mRNA, the RBS strength was tentatively adjusted so that stabilization of this mRNA results in a visible color change on indicator (e.g., X-gal) plates. As a first trial, we selected two weak alleles of the *lamB* RBS, *lamB701* and *lamB708* (Fig. 13.2A). Both carry point mutations that decrease RBS strength by creating inhibitory secondary structures, particularly with the *lamB701* RBS for which the structure is most stable (Iost and Dreyfus, 1995).

The P_{T7}-*lacZ* cassettes carrying the *lamB701* or *lamB708* RBS were transferred from JM101 to MC1061(DE3) or HfrG6Δ12(DE3) in two steps. Strain pop4068 (gift of Dr. Olivier Raibaud, Institut Pasteur) carries a mutation in the *glpD* gene, making it unable to grow on glycerol, and a Tn*10* inserted in the *malQ* gene, resulting in tetracycline resistance. Of note, the *glpD* and *malQ* genes, which bracket the *malT-malP* intergenic region (Fig. 13.2A) are located only 12 kb apart and can be cotransduced at high frequency (ca 75%) with bacteriophage P1. In a first step, a P1 lysate of pop4068 was used to transduce MC1061(DE3) or HfrG6Δ12(DE3) to tetracycline resistance. After isolation, Tet^R derivatives were plated on M63 minimal plates (Miller, 1972) supplemented with 0.2% glycerol and 5 μg/ml vitamin B1 and either 50 μg/ml histidine (for the His$^-$ strain HfrG6Δ12 [DE3]) or 0.2% casaminoacids (for the polyauxotrophic strain MC1061 [DE3]) to test their ability to grow on glycerol. One Tet^R, Gly$^-$ derivative was chosen in each case. In a second step, these derivatives were transduced with lysates grown on the JM101 derivatives carrying the P$_{T7}$-$lacZ$ cassette of interest, and the transductants were selected for growth on the same M63B1-glycerol plates as previously. The Gly$^+$ transductants were then tested for tetracycline resistance, and Tet^S candidates were retained: they must have received the P$_{T7}$-$lacZ$ cassette (Fig. 13.2A). These candidates were plated on McConkey maltose plates to check that they formed white (Mal$^-$) colonies, as expected from the disruption of the $malP$ promoter (Fig. 13.2A). Finally, the presence of the cassettes was checked by PCR, using the oligonucleotide pairs MalT/LacZ, and LacY/MalP (Fig. 13.2A and Table 13.1). Fragments of the expected size (268 and 394 bp, respectively) were obtained in each case. As controls, derivatives of BL21(DE3) (lac^-) carrying the same P$_{T7}$-$lacZ$ cassettes were independently reconstructed along the same lines, as were derivatives of MC1061(DE3) and HfrG6Δ12 (DE3) carrying P$_{lac}$-$lacZ$ cassettes; in this latter case, only the weaker $lamB701$ RBS was used.

We then compared the β-galactosidase synthesis from these different strains. Cells were grown in LB medium containing 0.5 mM IPTG, and β-galactosidase activity was assayed in the crude extracts. Within a given background, the $lamB701$ RBS yielded approximately 10-fold less β-galactosidase expression than the $lamB708$ RBS and the P$_{lac}$-$lacZ$ cassette yielded more expression than the P$_{T7}$-$lacZ$ cassette with the same RBS (Fig. 13.3A,B). These results are consistent with our previous reports (Iost and Dreyfus, 1995). More surprisingly, the expression from any given cassette differed significantly with the genetic background used. This was particularly true with the P$_{T7}$-$lacZ$ cassettes, where expression spanned a 3–4-fold range and ranked in the order MC1061(DE3) > BL21(DE3) > HfrG6Δ12(DE3) (Fig. 13.3A,B).

Table 13.1 Sequence of the oligonucleotides used in this work

MalT	GCT TTT TGA AAA TAC GCA ACG G
LacZ	GAT GTG CTG CAA GGC GAT TAA GTT GGG TAA
Uplac	GTT AGC TCA CTC ATT AGG CACC CC
T7	GCC AGT TCG ATG TCA GAG AAG TCG
LacY	CTT TGC TAC CGG TGA ACA GG
MalP	ACT GAC GTG AAA GCG CTT CC

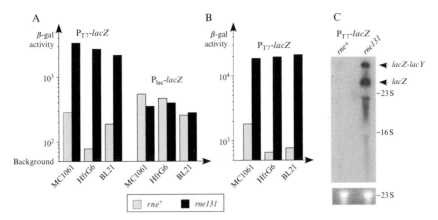

Figure 13.3   The effect of RNase E truncation. (A, B) Histograms showing β-galacto-sidase activities from the P_{T7}-*lacZ* or P_{lac}-*lacZ* cassettes in the presence of either the wild-type *rne* gene (light gray boxes) or the *rne131* allele (closed boxes). Note the log scale for the ordinates. The different genetic backgrounds used (i.e., the *E. coli* K12 strains MC1061 or HfrG6, or the *E. coli* B strain BL21) are noted below the corresponding diagrams. In A, the *lacZ* translation is driven by the *lamB701* RBS, whereas in B, it is driven by the stronger *lamB708* RBS (cf. Fig. 13.2A). (C) Northern blot showing the accumulation of mRNA from the P_{T7}-*lacZ* cassette, in the presence of either the wild-type or *rne 131* alleles. The strain used is the same as in the second leftmost histogram in panel A (HfrG6 derivative, RBS *lamB701*). The lower panel shows the ethidium bromide staining of 23S rRNA on the same gel, prior to the transfer.

2.4. Procedure 3: Removal of the RNase E CTH

We then tested whether the removal of the RNase E CTH stabilizes the P_{T7}-*lacZ* mRNA in the *E. coli* K12 background, as it does in the *E. coli* B background. The *rne131* allele (formerly *smbB131*; Kido *et al.*, 1996) carries a frameshift mutation removing residues 585–1061 of RNase E. Cells carrying this mutation are fully viable. Two transduction steps were used to transfer the *rne131* mutation from its initial isolate (strain BZ31; Kido *et al.*, 1996) into the MC1061(DE3), HfrG6Δ12(DE3), and BL21(DE3) derivatives. First, a P1 lysate was prepared on CH1828 (Mudd *et al.*, 1990). Together with an RNase E mutation (*rne1*) that prevents growth at 42 °C by making RNase E thermosensitive, this MC1061 derivative carries a Tn*10* inserted at locus *zce-726*, only 10 kb away from the *rne* gene (Nichols *et al.*, 1998); Fig. 13.2A). We have found that *rne* and *zce-26* can be cotransduced with a frequency of about 80%. This CH1828 lysate was used to transduce BZ31, and one Tet^R transductant able to grow at 42 °C was selected. This isolate, which retains the *rne131* allele, was then used to transduce the MC1061(DE3), HfrG6Δ12 (DE3), or BL21(DE3) derivatives carrying the different P_{T7}-*lacZ* and P_{lac}-*lacZ* cassettes. The Tet^R transductants were tested for the presence of the *rne131* allele by Western blotting, using an anti-RNase E rabbit antiserum donated by Dr. A. J. Carpousis (protocol 2). The presence of the full-length

(rne^+) or truncated ($rne131$) forms of RNase E is easily ascertained by this technique (Fig. 13.2B).

The levels of β-galactosidase expression from the different cassettes were then compared in the $rne131$ and rne^+ derivatives of MC1061(DE3), HfrG6Δ12(DE3), and BL21(DE3) (Fig. 13.3A,B). Whatever the RBS and genetic background used, the β-galactosidase expressions from the P_{T7}-$lacZ$ cassettes in LB medium was very much increased by the $rne131$ mutation (with the RBS $lamB701$, the increase ranged from 12-fold for MC1061 [DE3] to 35-fold for HfrG6Δ12[DE3], with BL21[DE3] lying in between). Interestingly, the increase was even larger when cells were grown in synthetic MOPS-glycerol medium supplemented with all amino acids, vitamins, and nucleic acid bases (Neidhardt $et~al.$, 1974) instead of LB (factors ranged from 30-fold for MC1061[DE3] to 80-fold for HfrG6Δ12 [DE3] in this case), suggesting that growth conditions can further modulate the extreme instability of the P_{T7}-$lacZ$ mRNA (not illustrated). In contrast, whatever the background used, the CTH removal had little effect on the β-galactosidase synthesis from the P_{lac}-$lacZ$ cassette (Fig. 13.3A), consistent with our former observations with the BL21(DE3) background (Lopez $et~al.$, 1999).

To further check that the increase in β-galactosidase expression from the P_{T7}-$lacZ$ cassette in the presence of the $rne131$ mutation reflects mRNA accumulation, total RNA was prepared from the rne^+ and $rne131$ derivatives of HfrG6Δ12(DE3) and the P_{T7}-$lacZ$ mRNA was visualized by Northern blot (protocols 3 and 4). As expected, the presence of the $rne131$ allele resulted in a very large increase of the mRNA signal (Fig. 13.3C).

We conclude that the RNase E CTH has a major and specific impact on the lifetime of the P_{T7}-$lacZ$ mRNA in the $E.~coli$ K12 background, as it has in the $E.~coli$ B background.

3. SPECIFIC PROTOCOLS

3.1. β-galactosidase assay

For β-galactosidase assay, many laboratories still use the classical protocol of Miller (1972), in which aliquots of cultures are first diluted in a suitable buffer (Z buffer), then treated with toluene or chloroform/SDS to permeabilize the cells, and finally incubated with o-nitrophenyl β-D-thiogalactopyranoside (ONPG) for a fixed time. After stopping the reaction, the β-galactosidase–mediated hydrolysis of ONPG to o-nitrophenate (ONP$^-$) is quantified by recording the OD at 420 nm. The β-galactosidase activity of the culture (in arbitrary Miller units) is finally obtained by dividing this OD_{420} by the reaction time and by the cell mass, estimated from the OD_{600}

of the culture and the volume of the aliquot. We routinely use an alternative protocol in which cleared lysates are first prepared from the cultures, and aliquots are then used for assaying separately protein concentration and ONPG hydrolysis. Activities can then be expressed in rational units (nmole ONPG hydrolyzed per min and per mg of total proteins). Although slightly more labor-intensive than the Miller protocol, this technique has two important advantages. First, it allows for comparisons between cultures independently of the shape or size of the cells, which may vary with the strain or with the growth conditions used. In contrast, the Miller assay will be sensitive to these parameters insofar as they affect the OD_{600} independently of the culture mass. Second, the technique yields a direct estimate of the percentage of β-galactosidase in total proteins, as pure β-galactosidase is known to have a specific activity of about 400.000 nmoles ONPG/min/mg (Fowler and Zabin, 1983).

Cells (1–5 ml) are grown at 37 °C with good aeration either in LB or in fully supplemented MOPS medium (Neidhardt *et al.*, 1974) containing glycerol (0.2%) as the carbon source. The media are eventually supplemented with IPTG (0.5 mM) and/or ampicillin (100 μg/ml). When the OD_{600} reaches 0.5, cells are centrifuged, resuspended in 100 μl of phosphate-buffered saline (PBS), and sonicated on ice with a cup-horn probe handling up to six Eppendorf tubes at a time. The clarified suspension is centrifuged (18.000g, 5 min, 4 °C). Total protein concentration in the supernatant is measured by adding an aliquot (volume v_1) to 1 ml of the protein assay reagent from Bio-Rad (cat. 500-0006) in a 1-cm pathway cuvette, and measuring the increase in OD_{595}. Known amounts of bovine serum albumin (BSA; BioLabs cat. B9001S) are used as a standard; the response of the assay is linear in the range of 0–15 μg protein, corresponding to approximately 0–0.75 OD_{595}. In parallel, another aliquot (volume v_2) is diluted in a 1-cm cuvette containing 1 ml of Z buffer (Miller, 1972) and 0.2 ml of ONPG (4 mg/ml in water). The hydrolysis of ONPG to ONP$^-$ is followed continuously by recording the increase of the OD_{420} versus time (mOD_{420}/min). The β-galactosidase activity (nmole ONPG/min/mg total proteins) is given by:

$$\text{Activity} = 22.6[(mOD_{420}/\text{min})/OD_{595}](v_1/v_2) \qquad (13.1)$$

The factor 22.6 takes into account the molar extinction coefficient of ONP$^-$ in Z buffer (2550 M^{-1}cm^{-1} at 420 nm), the correspondence between the OD_{595} and the absolute amounts of proteins (0.048 OD_{595}/μg BSA), and the difference in reaction volume between the two assays (1 ml vs. 1.2 ml).

This protocol was initially established for a conventional double-beam spectrophotometer equipped with a multicuvette holder thermostated at 28 °C (Uvikon 930, Kontron). With this device, variations as low as

0.1–0.2 mOD_{420}/min can be measured accurately during a 1-h kinetic experiment, allowing activities of 10–20 units to be measured without difficulty. Recently, the protocol has been successfully adapted to a 680XR microplaque reader from Bio-Rad, which allows the parallel processing of many samples in 96-well plates. The 595-and 415-nm built-in filters are used for protein and β-galactosidase assays, respectively, and the volumes are scaled down to 200 μl in both cases. In this case, the numerical coefficient to be used in the preceding formula is 18.8 instead of 22.6 because the volumes for the two assays are now identical.

3.2. Preparation of protein extracts for the detection of RNase E on Western blots

RNase E is metabolically stable *in vivo* (Lopez *et al.*, 1999), but it is very labile in crude cell extracts unless special precautions are taken to inhibit proteases (Carpousis *et al.*, 1994). For instance, very little RNase E can be detected in the sonicated extracts used for β-galactosidase assays (see section 3.1), using Western blotting or immunoprecipitation assays. In contrast, the protein is readily visualized when cells are lysed under denaturing conditions (Marchand *et al.*, 2001; Sousa *et al.*, 2001). For Western analysis, the following procedure gives satisfactory results (Fig. 13.2B). Aliquots of cultures growing exponentially in LB medium (200 μl at an OD_{600} of 0.5) are centrifuged, resuspended in 15 μl of Laemmli sample buffer, sonicated with a cup-horn probe until viscosity has decreased, heated at 95 °C for 5 min, and finally loaded on a standard 8% SDS gel (Laemmli, 1970). Transfer to nitrocellulose, reaction with rabbit anti-RNase E antibody and with anti-rabbit antibody coupled to HRP (Promega cat. W401B), and visualization of bound antibodies by chemiluminescence are done according to standard protocols (Sambrook *et al.*, 1989; Sambrook and Russell, 2001).

3.3. RNA isolation

For mRNA analysis and visualization, it is absolutely essential to choose an RNA extraction procedure that is fast compared to the synthesis/decay of the mRNA of interest so that its abundance in the extract reflects its abundance *in vivo*. This requirement is particularly stringent for the P_{T7}-*lacZ* mRNA, for which synthesis and decay presumably take only seconds at 37 °C (Iost *et al.*, 1992). It can be fulfilled either by instantly lysing the cells in a denaturating solution or by rapidly cooling the culture to block further mRNA metabolism. Although commercially available kits may be efficient in this respect, we currently use two home-adapted techniques, the boiling SDS method (Uzan *et al.*, 1988), and the hot phenol method (Aiba *et al.*, 1981). Of these, the latter technique is simpler, faster, and uses lower amounts of hazardous chemicals; moreover, it consistently gives high yields of undegraded RNA. Only this technique is described here.

Cultures (5 ml) are grown in LB supplemented with 0.5 mM IPTG in a 100-ml Erlenmeyer flask. When the OD_{600} reaches 0.6–0.8, the flasks are rapidly chilled near 0 °C (without freezing) by stirring for 5–7 sec in a dry ice–ethanol bath. Cultures are then transferred to prechilled 15-ml Falcon tubes and centrifuged at 0 °C. Pellets are rapidly resuspended in 0.5 ml of sodium acetate 20 mM, SDS 0.1%, EDTA 1 mM adjusted to pH 5.5 with acetic acid, and added to 0.5 ml of acidic phenol (Sigma P4682) preheated at 65 °C. The mixture is kept at 65 °C for 5 min with occasional vortexing, and the aqueous phase is collected. This extraction is repeated twice more, the last time at room temperature with phenol chloroform-isoamylalcohol (25:24:1). Finally, NaCl is added to 0.5 M, and the RNA is precipitated with ethanol, washed, resuspended in water, and quantified by spectrophotometry.

3.4. Northern analysis

RNA samples (5–15 μg) are denatured and electrophoresed through a 15 × 15 cm 1% agarose gel as described (Sambrook *et al.*, 1989), except that we now use TBE 0.5× as the running buffer instead of MOPS-formaldehyde buffer. During electrophoresis (100 V, 3 h, or 30 V overnight), the buffer is circulated with a peristaltic pump to prevent pH drifts. The gel is then vacuum- or capillary-blotted onto a Hybond N+ membrane (Amersham), using 20× SSC. After gentle washing with the same buffer, the RNA is cross-linked to the membrane with a UV cross-linker (Stratagene or Syngene), using default settings. Prehybridization (1 h) and hybridization (15 h) are done at 60 °C in Church buffer (Sambrook and Russell, 2001). The probe used is the 1828-bp *Hinc*II fragment internal to *lacZ*; 100 ng of this fragment are randomly labeled with deoxyadenosine 5′-[α-^{32}P] triphosphate (Amersham), using the hexanucleotide mix from Roche (cat. 1 277 081) together with cold dTTP, dGTP, and dCTP as recommended by the manufacturer. The probe is freed from unincorporated nucleotides by G50 filtration (Amersham ProbeQuant column; ref. 27-5335-01). After hybridization, the membrane is washed at 60 °C with 2×SSC, 0.5% SDS, then with 0.1×SSC, 0.5% SDS (20 min each). Finally, radioactivity is imaged with a Fuji FLA3000 imager.

4. CONCLUSION

It is clear from the results that the remarkable impact of CTH removal on the stability of desynchronized *lacZ* mRNA, first observed in *E. coli* B, also holds true for *E. coli* K12. Unexpectedly, we also observed that the

β-galactosidase synthesis from a given P_{T7}-$lacZ$ cassette varies significantly (up to 4-fold) from one strain to the next. Because these variations disappear in the presence of the $me131$ mutation, they must reflect differences in mRNA stability. That mRNA stability can vary among $E.$ $coli$ K12 isolates even if they are wild-type for enzymes known to participate in mRNA decay, has been observed by others (Bernstein et $al.$, 2004). However, these variations are of practical interest here. Indeed, in the MC1061 background, the P_{T7}-$lacZ$ cassette produces a Lac$^+$ (blue) phenotype on LB-Xgal-IPTG plates even with the weakest RBS used ($lamB701$). In contrast, the corresponding HfrG6Δ12(DE3) derivative, which produces 4-fold less β-galactosidase, scores Lac$^-$ (white). By mutagenizing this host and isolating mutants with a blue phenotype, it is possible to identify cellular factors that, beyond the CTH, participate in the instability of the P_{T7}-$lacZ$ mRNA. The results from this screen will be reported elsewhere.

ACKNOWLEDGMENTS

This work is funded by the Centre National de la Recherche Scientifique, by the École Normale Supérieure, and by the Agence Nationale pour la Recherche (grant NT05-1_44659).

REFERENCES

Aiba, H., Adhya, S., and de Crombrugghe, B. (1981). Evidence for two functional gal promoters in intact $Escherichia$ $coli$ cells. $J.$ $Biol.$ $Chem.$ **256,** 11905–11910.

Baker, K. E., and Mackie, G. A. (2003). Ectopic RNase E sites promote bypass of 5'-end-dependent mRNA decay in $Escherichia$ $coli.$ $Mol.$ $Microbiol.$ **47,** 75–88.

Bernstein, J. A., Lin, P. H., Cohen, S. N., and Lin-Chao, S. (2004). Global analysis of $Escherichia$ $coli$ RNA degradosome function using DNA microarrays. $Proc.$ $Natl.$ $Acad.$ $Sci.$ USA **101,** 2758–2763.

Carpousis, A. J., Van Houwe, G., Ehretsmann, C., and Krisch, H. M. (1994). Copurification of $E.$ $coli$ RNase E and PNPase: Evidence for a specific association between two enzymes important in RNA processing and degradation. $Cell$ **76,** 889–900.

Chevrier-Miller, M., Jacques, N., Raibaud, O., and Dreyfus, M. (1990). Transcription of single-copy hybrid $lacZ$ genes by T7 RNA polymerase in $Escherichia$ $coli$: mRNA synthesis and degradation can be uncoupled from translation. $Nucleic$ $Acids$ $Res.$ **18,** 5787–5792.

Deana, A., and Belasco, J. G. (2005). Lost in translation: The influence of ribosomes on bacterial mRNA decay. $Genes$ $Dev.$ **19,** 2526–2533.

Dreyfus, M. (1988). What constitutes the signal for the initiation of protein synthesis on $Escherichia$ $coli$ mRNAs? $J.$ $Mol.$ $Biol.$ **204,** 79–94.

Dreyfus, M. (2009). Killer and protective ribosomes. $Prog.$ $Nucleic$ $Acids$ $Res.$ $Mol.$ $Biol.$ In press.

Fowler, A. V., and Zabin, I. (1983). Purification, structure, and properties of hybrid ß-galactosidase proteins. $J.$ $Biol.$ $Chem.$ **258,** 14354–14358.

Gowrishankar, J., and Harinarayanan, R. (2004). Why is transcription coupled to translation in bacteria? *Mol. Microbiol.* **54**, 598–603.

Iost, I., and Dreyfus, M. (1995). The stability of *Escherichia coli lacZ* mRNA depends upon the simultaneity of its synthesis and translation. *EMBO J.* **14**, 3252–3261.

Iost, I., Guillerez, J., and Dreyfus, M. (1992). Bacteriophage T7 RNA polymerase travels far ahead of ribosomes *in vivo. J. Bacteriol.* **174**, 619–622.

Khemici, V., Poljak, L., Toesca, I., and Carpousis, A. J. (2005). Evidence *in vivo* that the DEAD-box RNA helicase RhlB facilitates the degradation of ribosome-free mRNA by RNase E. *Proc. Natl. Acad. Sci. USA* **102**, 6913–6918.

Kido, M., Yamanaka, K., Mitani, T., Niki, H., Ogura, T., and Hiraga, S. (1996). RNase E polypeptides lacking a carboxyl-terminal half suppress a *mukB* mutation in *Escherichia coli. J. Bacteriol.* **178**, 3917–3925.

Laemmli, U. K. (1970). Cleavage of structural proteins during assembly of the head of bacteriophage T4. *Nature* **227**, 680–685.

Landick, R., Carey, J., and Yanofsky, C. (1985). Translation activates the paused transcription complex and restores transcription of the trp operon leader region. *Proc. Natl. Acad. Sci. USA* **82**, 4663–4667.

Leroy, A., Vanzo, N. F., Sousa, S., Dreyfus, M., and Carpousis, A. J. (2002). Function in *Escherichia coli* of the non-catalytic part of RNaseE: Role in the degradation of ribosome-free mRNA. *Mol. Microbiol.* **45**, 1231–1243.

Lopez, P. J., Iost, I., and Dreyfus, M. (1994). The use of a tRNA as a transcriptional reporter: The T7 late promoter is extremely efficient in *Escherichia coli* but its transcripts are poorly expressed. *Nucleic Acids Res.* **22**, 1186–1193.

Lopez, P. J., Marchand, I., Joyce, S. A., and Dreyfus, M. (1999). The C-terminal half of RNase E, which organizes the *Escherichia coli* degradosome, participates in mRNA degradation but not rRNA processing *in vivo. Mol. Microbiol.* **33**, 188–199.

Makarova, O. V., Makarov, E. M., Sousa, R., and Dreyfus, M. (1995). Transcribing of *Escherichia coli* genes with mutant T7 RNA polymerases: Stability of *lacZ* mRNA inversely correlates with polymerase speed. *Proc. Natl. Acad. Sci. USA* **92**, 12250–12254.

Marchand, I., Nicholson, A. W., and Dreyfus, M. (2001). Bacteriophage T7 protein kinase phosphorylates RNase E and stabilizes mRNAs synthesized by T7 RNA polymerase. *Mol. Microbiol.* **42**, 767–776.

Miller, J. H. (1972). *Experiments in molecular genetics.* Cold Spring Harbor, NY: Cold Spring Harbor Laboratory.

Miller, O. L., Jr., Hamkalo, B. A., and Thomas, C. A., Jr. (1970). Visualization of bacterial genes in action. *Science* **169**, 392–395.

Mudd, E. A., Krisch, H. M., and Higgins, C. F. (1990). RNase E, an endoribonuclease, has a general role in the chemical decay of *Escherichia coli* mRNA: Evidence that *rne* and *ams* are the same genetic locus. *Mol. Microbiol.* **4**, 2127–2135.

Neidhardt, F. C., Bloch, P. L., and Smith, D. F. (1974). Culture medium for *Enterobacteria. J. Bacteriol.* **119**, 736–747.

Nichols, B. P., Shafiq, O., and Meiners, V. (1998). Sequence analysis of Tn*10* insertion sites in a collection of *Escherichia coli* strains used for genetic mapping and strain construction. *J. Bacteriol.* **180**, 6408–6411.

Raibaud, O., Mock, M., and Schartz, M. (1984). A technique for integrating any DNA fragment into the chromosome of *Escherichia coli. Gene* **29**, 231–241.

Sambrook, J., Fritsch, E. F., and Maniatis, T. (1989). *Molecular Cloning: A Laboratory Manual.* Cold Spring Harbor, NY: Cold Spring Harbor Laboratory Press.

Sambrook, J., and Russell, D. W. (2001). *Molecular Cloning: A Laboratory Manual,* 3rd ed. Cold Spring Harbor, NY: Cold Spring Harbor Laboratory Press.

Silhavy, T. J., Berman, M. L., and Enquist, L. W. (1984). *Experiments with Gene Fusions.* Cold Spring Harbor, NY: Cold Spring Harbor Laboratory Press.

Sorensen, M. A., Jensen, K. F., and Pedersen, S. (1994). High concentrations of ppGpp decrease the RNA chain growth rate: Implications for protein synthesis and translational fidelity during amino acid starvation in *Escherichia coli*. *J. Mol. Biol.* **236,** 441–454.

Sousa, S., Marchand, I., and Dreyfus, M. (2001). Autoregulation allows *Escherichia coli* RNase E to adjust continuously its synthesis to that of its substrates. *Mol. Microbiol.* **42,** 867–878.

Studier, F. W., and Moffatt, B. A. (1986). Use of bacteriophage T7 RNA polymerase to direct selective high-level expression of cloned genes. *J. Mol. Biol.* **189,** 113–130.

Suit, J. C., Matney, T. S., Doudney, C. O., and Billen, D. (1964). Transfer of the *Escherichia coli* K12 chromosome in the absence of DNA synthesis. *Biochem. Biophys. Res. Commun.* **17,** 237–241.

Uzan, M., Favre, R., and Brody, E. (1988). A nuclease that cuts specifically in the ribosome binding site of some T4 mRNAs. *Proc. Natl. Acad. Sci. USA* **85,** 8895–8899.

Vanzo, N. F., Li, Y. S., Py, B., Blum, E., Higgins, C. F., Raynal, L. C., Krisch, H. M., and Carpousis, A. J. (1998). Ribonuclease E organizes the protein interactions in the *Escherichia coli* RNA degradosome. *Genes Dev.* **12,** 2770–2781.

Vogel, U., Sorensen, M., Pedersen, S., Jensen, K. F., and Kilstrup, M. (1992). Decreasing transcription elongation rate in *Escherichia coli* exposed to amino acid starvation. *Mol. Microbiol.* **6,** 2191–2200.

Yarchuk, O., Jacques, N., Guillerez, J., and Dreyfus, M. (1992). Interdependence of translation, transcription and mRNA degradation in the *lacZ* gene. *J. Mol. Biol.* **226,** 581–596.

CHAPTER FOURTEEN

ANALYSIS OF mRNA DECAY IN
BACILLUS SUBTILIS

David H. Bechhofer,* Irina A. Oussenko,* Gintaras Deikus,*
Shiyi Yao,* Nathalie Mathy,[†] *and* Ciarán Condon[†]

Contents

Abstract

Studies of mRNA turnover in *B. subtilis* are less well known than in *E. coli*. Here we provide researchers who have an interest in gram-positive RNA processing with several protocols for RNA isolation, for 5′- and 3′-mapping of mRNAs and mRNA decay fragments, and we also include a comprehensive listing of *B. subtilis* mutants that are deficient in ribonucleases thought to be involved in mRNA decay.

* Mount Sinai School of Medicine of New York University, Department of Pharmacology and Systems Therapeutics, New York, USA
† CNRS UPR9073, Institut de Biologie Physico-Chimique, Paris, France

Methods in Enzymology, Volume 447
ISSN 0076-6879, DOI: 10.1016/S0076-6879(08)02214-3

259

1. INTRODUCTION

Much of the study of prokaryotic mRNA decay has been performed using *Escherichia coli*, the organism in which the instability of messenger (m)RNA was first described (Brenner *et al.*, 1961; Gros *et al.*, 1961). Fewer groups have studied mRNA turnover in gram-positive organisms, represented by *Bacillus subtilis*, perhaps because the choice of *B. subtilis* as a model organism for development focused initially on processes such as sporulation and competence development, which are driven primarily by regulation at the levels of transcription and protein turnover (Dubnau and Lovett, 2001; Piggot and Losick, 2001). Just as the processes of DNA replication, transcription, and translation were shown to involve similar enzymes and mechanisms in *E. coli* and *B. subtilis*, the same was assumed to be true for mRNA turnover. However, publication of the *B. subtilis* genome 10 years ago (Kunst *et al.*, 1997) revealed that the repertoire of ribonucleases and RNA-modifying enzymes that were known to exist and, in some cases, to be essential in *E. coli* were not identifiable by sequence homology in *B. subtilis*. Three examples of enzymes that are major players in mRNA decay in *E. coli* but have no sequence homologue in *B. subtilis* are RNase E (essential mRNA decay-initiating endonuclease), RNase II (major 3′ to 5′-exoribonuclease), and oligoribonuclease (essential RNA oligonucleotide turnover exonuclease). In addition, the gene encoding the poly(A) polymerase activity of *B. subtilis*, similar to the *pcnB* gene in *E. coli* that has significant effects on mRNA stability (Carpousis *et al.*, 1999), has so far eluded identification. In addition, biochemical and genetic experiments in *B. subtilis* over the past 15–20 years (Condon, 2003) have shown differences in the function of certain ribonucleases. For example, Deutscher and colleagues showed that PNPase is the major 3′-exonucleolytic activity in *B. subtilis* extracts, but the same enzyme plays a minor role in *E. coli* extracts (Deutscher and Reuven, 1991). Also, *E. coli* RNase III, a double-stranded specific endonuclease, can be deleted in *E. coli*, whereas the similar Bs-RNase III is essential in *B. subtilis* (Herskovitz and Bechhofer, 2000).

The 3′ to 5′-exoribonucleases PNPase (Deutscher and Reuven, 1991; Luttinger *et al.*, 1996; Mitra *et al.*, 1996), RNase R (Oussenko and Bechhofer, 2000), and RNase PH (Craven *et al.*, 1992) are present in both organisms and have functions in mRNA decay (Bechhofer and Wang, 1998; Oussenko *et al.*, 2005; Wang and Bechhofer, 1996) and stable RNA processing (Wen *et al.*, 2005; Yao *et al.*, 2007). (Interestingly, the *B. subtilis* PNPase gene was initially identified as a gene involved in competence development ((Luttinger *et al.*, 1996). The mechanistic basis of the requirement of PNPase activity for wild-type competence development remains to be elucidated.) More recently, ribonucleases that are unique to *B. subtilis* have been identified by biochemical means, their genes have been cloned, and strains have been constructed that are deleted for these genes. Examples include RNases J1 and J2 (Even *et al.*, 2005), RNase M5 (Condon *et al.*, 2001), YhaM (Oussenko *et al.*, 2002), and YhcR (Oussenko *et al.*, 2004).

The mechanism of mRNA decay, based on studies done in *E. coli*, is thought to proceed by a combination of a series of endonucleolytic cleavages, followed by exonucleolytic degradation of the resulting fragments from the 3′-end (Coburn and Mackie, 1999; Kushner, 2002). It is hypothesized that the native triphosphate 5′-end of an mRNA is resistant to decay, as is the 3′-terminal transcription terminator structure. Decay of mRNA is often initiated by an endonuclease cleavage performed by a 5′-end-dependent endonuclease (e.g., *E. coli* RNase E). It has recently been reported that conversion of the 5′-triphosphate end to a 5′-monophosphate end by a pyrophosphatase activity can occur in *E. coli*, and this may be required for efficient RNase E activity *in vivo* (Celesnik *et al.*, 2007). The initial RNase E cleavage generates an upstream fragment that has an unprotected 3′-end and a downstream fragment that has a monophosphate 5′-end. The rate of further endonucleolytic cleavages in the downstream fragment is enhanced by the presence of the 5′-monophosphate. On the basis of this model, analysis of the mechanism of mRNA decay in any organism would be greatly enhanced by the mapping of decay intermediates. Identification of upstream and downstream RNA decay fragments would constitute considerable evidence for the model of mRNA decay described previously. However, the lability of such fragments makes this difficult to achieve in a wild-type strain. Thus, the use of RNase mutant strains is vital for studying mRNA decay.

A *B. subtilis* strain that is lacking the phosphorolytic activity of PNPase, which appears to be the dominant mRNA turnover enzyme in this organism, has been used extensively to characterize specific mRNA decay intermediates (Bechhofer and Wang, 1998; Oussenko *et al.*, 2005). Although three other 3′-exoribonucleases (RNase R, RNase PH, and YhaM) can be shown to be involved in mRNA decay (Oussenko *et al.*, 2005), a strain that is missing only the *pnpA* gene, encoding PNPase, shows an accumulation of high levels of mRNA decay intermediates (see Fig. 14.1, lanes 1 and 2). Conversely, in a strain containing only PNPase and missing the other 3′-exoribonucleases, no decay intermediates are detected (Fig. 14.1, lane 3). Strains that contain only RNase R or only RNase PH or only YhaM show different levels of mRNA decay intermediates, suggesting the involvement of these ribonucleases in mRNA decay (Fig. 14.1, lanes 4–6).

Recent results with RNase J1 have implicated this enzyme also in mRNA decay. The fact that RNase J1 was initially described as an essential endonuclease whose activity was sensitive to the state of 5′-phosphorylation, similar to *E. coli* RNase E, suggested that RNase J1 might be a mediator of initiation of mRNA decay (Even *et al.*, 2005). Putzer and colleagues

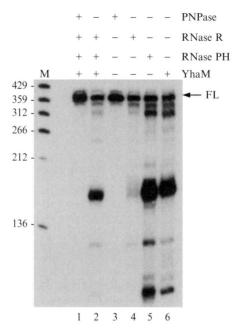

Figure 14.1 Analysis of *rpsO* mRNA decay intermediates in triple ribonuclease mutant strains. The presence or absence of the four known *B. subtilis* 3' to 5'-exoribonucleases is shown at the top. Migration of full-length *rpsO* mRNA is indicated by "FL." Marker lane (M) contains 5'-end-labeled *TaqI* DNA fragments of plasmid pSE420. (Reproduced from Oussenko *et al.*, 2005, with permission from American Society for Microbiology.)

reported that global mRNA half-life was increased slightly in an RNase J1 conditional mutant that was grown under conditions where RNase J1 was depleted (Even *et al.*, 2005). More recently, Condon and colleagues reported the surprising finding that, in addition to its endonuclease activity, RNase J1 has 5' to 3'-exoribonuclease activity (Mathy *et al.*, 2007). Previously, such a ribonuclease activity had not been known in the *Bacteria*. The existence of such an activity and its possible role in initiation of mRNA decay provides a plausible explanation for the importance of the 5'-end on mRNA decay (Condon, 2003). In separate studies, our groups have found an accumulation of mRNA decay intermediates under conditions of RNase J1 depletion (Deikus *et al.*, 2008; CC, unpublished). As such, we consider it likely that RNase J1 is generally involved in mRNA decay in *B. subtilis*.

In this chapter we present protocols for isolation of mRNA from *B. subtilis*, and protocols for fine-scale analysis of mRNA half-life and decay intermediates. We also provide a comprehensive listing of RNase mutants of *B. subtilis* that are useful for studying RNA processing. Details of these protocols should be helpful for investigators studying many aspects of prokaryotic RNA in *B. subtilis* and other organisms.

2. GROWTH MEDIA FOR ISOLATION OF RNA

For many applications, RNA is isolated from *B. subtilis* grown in rich media (e.g., Luria-Bertani medium or 2× YT medium). We find that we get higher-quality RNA (less background on Northern blots) in cells grown in a supplemented minimal medium, which we call RNA isolation medium (RIM), a modified Spizizen salts medium. For 100 ml of 10× Spizizen salts (SS) stock, dissolve 6 g KH_2PO_4, 14 g K_2HPO_4, 2 g $(NH_4)_2SO_4$, and 1 g Na cit·$2H_2O$ in 93.5 ml water. For 100 ml of 1× SS, add 0.1 ml of 1 M $MgSO_4$·$7H_2O$. For RIM, add the following components to 100 ml of 1× SS: 1 ml 50% glucose, 0.5 ml 20% casamino acids, 10 μl 10% yeast extract, and 0.5 ml of 10 mg/ml amino acid stock (for each required amino acid in auxotrophic strains).

3. INHIBITION OF *B. SUBTILIS* TRANSCRIPTION

For an mRNA decay experiment, rifampicin is added at time zero to prevent further transcription initiation. It is advisable to test the concentration of rifampicin required for rapid cessation of transcription, which is the first condition for doing mRNA decay studies. For this, we grow a 20-ml *B. subtilis* culture at 37 °C in RIM to mid-logarithmic phase. Ten ml of culture is added to each of two flasks containing 10 μl [5,6-^{3}H]-uridine (Perkin Elmer), and to one of the flasks rifampicin (Sigma) is added to the desired final concentration from a 10-mg/ml stock of rifampicin dissolved in methanol. Flasks are incubated with shaking at 37 °C and 1-ml aliquots are removed into 1 ml of cold 10% trichloroacetic acid (TCA) and let sit on ice for at least 10 min. The TCA precipitate is vacuum filtered slowly through a glass-fiber filter that has been presoaked in 10 mM of ATP to reduce background. The tube containing the TCA precipitate is washed out with 5% TCA, which is also passed through the glass-fiber filter. The filter is rinsed extensively with 5% TCA, dried, added to liquid scintillation fluid, and counted. In our hands, addition of rifampicin to a final concentration of 150 μg/ml results in 4-fold less [^{3}H]-uridine incorporation, relative to untreated cells, at the earliest time point taken. For 5 min after the rifampicin addition, we observe a slight rise in incorporation (likely reflecting transcription elongation), followed by a decline in tritium counts to a steady-state level that is about 10% of the level observed in untreated cells after 30 min. Other organisms may require higher levels of rifampicin for a complete block to transcription initiation.

4. RNA Isolation Protocol 1

The hot phenol method is a popular method for RNA isolation, which we now use extensively. The following protocol is for RNA isolation from 33 ml of cells, giving a yield of approximately 400 μg of total RNA. Acid phenol is prepared as follows: a 500-g bottle of phenol crystals (molecular biology grade) is melted in a 70 °C bath, and then 500 ml of 1× NAE buffer (50 mM of Na acetate pH 5.1, 10 mM of EDTA) is added and mixed at room temperature (RT). The mixture is placed at 4 °C overnight for phase separation, yielding phenol at pH 5.1. For RNA isolation, 5× NAES buffer is prepared just before starting the isolation by mixing 10× NAE (500 mM Na acetate, pH 5.1, 100 mM EDTA) with an equal volume of 10% SDS.

Grow a 40-ml culture of *B. subtilis* in RIM to an OD$_{600}$ of 0.4. Add 33 ml of cells to 1/2 volume of frozen 50 mM NaN$_3$ (prepared in 1× SS) in a 50-ml screw-cap tube. The NaN$_3$ is frozen in a slanted position to give maximum surface area for rapid cooling of the culture. Mix well with vortexing and harvest cells by centrifugation for 5 min at 3500 rpm at 4 °C in an SS34 rotor. Pour off the supernatant, resuspend cells in the remaining liquid, and transfer cells to a microcentrifuge tube. Centrifuge for 1 min at 14,000 rpm and remove the rest of the medium by aspiration. Resuspend cells in 400 μl of an RT 10-mg/ml lysozyme solution dissolved in 1 mM of EDTA, and keep at RT for 10 min. Add 100 μl of preheated (70 °C) 5× NAES and 500 μl of preheated (70 °C) phenol equilibrated with 1× NAE. Mix by vortexing and transfer to a 70 °C heat block. Incubate 10 min at 70 °C and mix by vortexing for 5–10 sec at 1-min intervals. We find that poor-quality microcentrifuge tubes do not hold up well (70 °C temperature and phenol) and that phenol leaks out under the cap during vortexing, a dangerous situation. It is best to use high-quality microcentrifuge tubes and keep firm pressure on the cap of the tube. Transfer to ice for 5 min; centrifuge at 14,000 rpm for 10 min at 4 °C. Transfer 450 μl of the aqueous phase to a microcentrifuge tube and add an equal volume of phenol:chloroform (1:1) solution. Mix by vortexing 10 sec and centrifuge at 14,000 rpm for 10 min at 4 °C. Transfer the aqueous phase (350 μl) to a new tube and add 1/10 volume 3 M Na acetate, pH 5.9, and 1 ml of ice-cold ethanol. Mix by inversion and precipitate for 20 min on ice. Centrifuge at 14,000 rpm for 20 min at 4 °C. Discard the supernatant and wash the pellet with 1 ml of 75% ethanol. Dry the pellet and dissolve in 200 μl of DNase I buffer (e.g., New England Biolabs restriction endonuclease buffer 3, which is 50 mM Tris pH 7.9, 10 mM MgCl$_2$, 100 mM NaCl, 1 mM DTT). Treat with 12 units of RNase-free DNase I (Roche) for 20 min at 37 °C. Add 1/2 volume phenol:chloroform (1:1). Vortex to mix well, spin down 3 min at RT, and transfer aqueous layer to a fresh tube. Add

1/10 volume of 3 M sodium acetate and mix. Add 2.5 volumes of ethanol, mix, and precipitate on ice for 20 min. Spin down 20 min at 4 °C. Remove the supernatant and wash the pellet with 1 ml of 75% ethanol. Dry the pellet and resuspend in 100 μl of 10 mM Tris, 0.1 mM EDTA ("TloE"). (As an extra precaution, buffers can be made with commercial RNase-free and DNase-free water and filtered through a 0.22-μM filter. However, we find that autoclaved deionized water is adequate when measures are taken to avoid introduction of RNases.) Add 2 μl of the RNA preparation to 1 ml of TloE and determine the concentration and purity of RNA spectropho-tometrically by measuring absorbance at 260 nm (A_{260}) and at 280 nm (A_{280}). The A_{260}/A_{280} ratio should be greater than 1.9. Store RNA in aliquots at −80 °C.

We routinely check the integrity of the RNA by electrophoresis on a 1% agarose–7% formaldehyde MOPS gel. The recipe to make one mini-horizontal MOPS gel is as follows: 0.25 g of agarose, 22.5 ml of water, and 3.1 ml 10× MOPS buffer (10× MOPS buffer = 200 mM MOPS, 50 mM Na acetate, 10 mM EDTA, adjust pH to 7.0 with NaOH). Micro-wave twice for 15 sec and once for 10 sec. Let agarose solution cool down so that the flask can be held, add 5.6 ml formaldehyde, and mix. Pour gel in a chemical hood or other well-ventilated area. We load 2–3 μg RNA per lane, with an equal volume of GenHunter dye (which contains ethidium bromide). RNA is run into the gel at 60 V and then electrophoresis con-tinues at 130 V until the bromophenol dye is three-quarters of the way down the gel. Destain gel in water and photograph. Approximately a 2:1 ratio of 23S to 16S RNA intensity should be observed.

5. RNA Isolation Protocol 2

A different RNA isolation protocol has been developed in our labo-ratory, derived from the protocol published by Ulmanen *et al.* (1985). This method does not use hot phenol, and it works well for multiple RNA isolations from 1 ml of cell culture, giving a yield of about 20 μg of RNA. The protocol can be scaled up for a 20-ml culture, in which 10-fold volumes of materials described subsequently are used.

Add 1 ml of mid-logarithmic phase cells to 0.5 ml frozen 50 mM NaN$_3$ in a microcentrifuge tube and vortex well to melt NaN$_3$. Spin down cells for 1 min at 14,000 rpm, shake off supernatant, resuspend cell pellet in 0.5 ml of ice-cold 1× SS by vortexing, spin down 1 min, and shake off supernatant. Resuspend cell pellet in 0.4 ml protoplasting buffer (15 mM Tris, pH 8.0, 6 mM EDTA, 0.45 M sucrose, 2 mg/ml lysozyme). Incubate on ice for 20 min. Spin down protoplasts at 14,000 rpm for 30 sec. Resuspend by vortexing very well in 0.2 ml of 80 mM Tris, pH 7.5,

10 mM MgCl$_2$, 10 mM β-mercaptoethanol. Add 0.2 ml of 1% SDS, 500 μg/ml proteinase K (Roche), 20 mM EDTA, 0.8 mg/ml 1–10 phenanthrolin, 0.4 mg/ml sodium-heparin. (Add buffer to proteinase K powder 10 min before use.) Incubate at 37 °C for 20 min. Pass cell extract through a 25G needle twice. Add 0.3 ml of phenol-chloroform (1:1), vortex well for 10 sec, and spin at 14,000 rpm at RT for 3 min. Remove aqueous layer with 1-ml pipette tip into fresh tube, add 0.3 ml of chloroform, vortex well for 10 sec, spin for 3 min, remove aqueous layer into fresh tube, add 0.8 ml of ethanol, and keep on ice for 30 min. Spin down for 30 min at 4 °C. Wash RNA pellet with 1 ml of 75% ethanol, spin down 2 min (RT), and dry pellet. Resuspend pellet in 100 μl DNase I buffer and continue with RNase-free DNase digestion and the rest of RNA isolation protocol 1.

6. RNA Isolation Protocol 3

A third protocol for RNA isolation uses glass beads to break open *B. subtilis* cells. The advantage of this protocol is a high RNA yield and the elimination of a lysozyme incubation step.

Dilute an overnight culture 1:100 in 25–30 ml fresh medium and grow cells to an OD$_{600}$ of 1.0. Spin down 20 ml of culture at 10,000 rpm for 10 min, if cells are grown in rich medium, whereas 6 min at 6000 rpm is sufficient if minimal medium is used. Decant the supernatant and freeze the pellet at −20 °C, even if proceeding immediately to the RNA isolation. Resuspend the pellet in 4 ml of ice-cold TE buffer (10 mM Tris, pH 7.5, 1 mM EDTA) and transfer to a 14-ml polypropylene tube on ice, containing 3 ml of water-saturated phenol, 0.5 ml of chloroform, 0.25 ml of 10% SDS, and 3.5 g of glass beads (0.1 mm; Touzart & Matignon). Vortex three times for 1 min, with at least 1-min intervals on ice in between vortexing. Centrifuge at 8,000 rpm for 10 min and transfer supernatant to 3 ml of water-saturated phenol on ice. Vortex as previously (three times for 1 min, with at least 1-min intervals on ice). Spin down again as previously and repeat vortexing procedure a third time. Add 1/10 volume of 10 M LiCl and one volume of isopropanol, and place at −20 °C for 10 min. Remove *all* isopropanol. Dry the pellet and resuspend in 200 μl of nuclease-free water, and transfer to a microcentrifuge tube. Extract three times with 400 μl of phenol/chloroform (1:1) and ethanol precipitate using 1/10 volume of 3 M sodium acetate, pH 5.0, and 3 volumes of 95% ethanol at −20 °C. Add 200 μl of 70% ethanol at −20 °C and spin down again. Remove ethanol and dry pellet for 3 min under vacuum. Resuspend in 100 μl of water. The OD$_{260/280}$ ratio should be about 2.0 and concentration should be 1 to 5 mg/ml.

7. GEL ELECTROPHORESIS AND BLOTTING

We describe here our protocols for Northern blotting to detect small RNAs (i.e., fewer than 500 nucleotides [nts]). Protocols for larger RNAs, involving agarose gels, are well known. We find that the use of denaturing polyacrylamide gels (PAGs) for Northern blotting is less well represented in the literature, although such gels are essential for analyzing differences in small RNAs that do not resolve well on agarose gels.

Run denaturing PAG (usually 5% or 6% polyacrylamide/7 M urea) in Tris-borate-EDTA buffer (TBE). Soak gel in 1× Tris-acetate-EDTA (TAE). (Recipe for 1 L of 50× TAE: 242 g of Tris base, 57.1 ml of glacial acetic acid, 100 ml of 0.5 M EDTA, pH 8.0.) Note that the gel size will increase slightly (2–3 mm) after soaking. Saturate two fiber pads with 1× TAE and avoid trapped air bubbles. Cut nylon membrane (we use GeneScreen; Perkin Elmer) to the dimensions of the gel with 2 mm left on both sides. Mark the side to which RNA will bind. Soak membrane in 1× TAE. Cut two pieces of Whatman 3 chromatography paper to the dimensions of the gel. Blotting sandwich (bottom to top) consists of a fiber pad, Whatman paper, gel, membrane, Whatman paper, fiber pad. The running buffer for the transfer is 1× TAE, and we use the Bio-Rad Mini Trans-Blot Cell. The transfer is done at 4 °C for 1 h at 20 V, followed by 2 h at 40 V (this requires the cooling tray). Alternatively, transfer is done overnight at 20 V, followed by 1 h at 30 V. The RNA is UV cross-linked on the membrane and can be stored at 4 °C.

8. LABELED SIZE MARKER FOR SMALL RNAs

5′-end-labeled DNA fragments are run on the gel in parallel with RNA samples and serve as size markers. For this, we typically use plasmid pSE420 (Brosius, 1992) that has been digested with $TaqI$ (at 65 °C), followed by calf-intestine phosphatase removal of 5′-phosphate and labeling with T4 polynucleotide kinase in the presence of $[\gamma^{32}P]$-ATP. The pSE420 DNA fragments that are used for size markers are shown in Fig. 14.2. Note that the 212-bp fragment is the result of partial digestion, so it appears in sub-stochiometric amounts. The size of this fragment has been determined from a DNA-sequencing gel. The larger $TaqI$ fragments (212–561) are markers for RNAs in the 200–500 nt size range when run for a long time on a 5% PAG; and the smaller $TaqI$ fragments (27–136) serve as markers for small RNAs run on 6% or higher-percentage PAG. Plasmid pSE420 also has two $TaqI$ fragments of 1444 and 828 base pairs, but these do not resolve well in the percentage PAG used for Northern blotting.

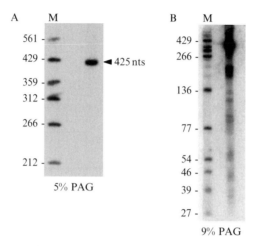

Figure 14.2 Range of labeled pSE420 *Taq*I DNA fragments used as size markers. (A) Northern blotting of a 5% PAG that was electrophoresed for an extended time, allowing separation of the markers (lane M) in the 200- to 500-nt range. The blot was probed with a gene-specific probe that detected an mRNA with a predicted size of 425 nt, as indicated. (B) Northern blotting of a 9% PAG that was electrophoresed for a short time, allowing the detection of small mRNA decay fragments. The blot was probed with a gene-specific probe that targets the 3′-terminus of the transcript. Numerous small fragments are detected, which suggests multiple endonuclease cleavages.

9. NORTHERN-BLOT ANALYSIS USING OLIGONUCLEOTIDE PROBES

5′-end-labeled oligonucleotide probes are used as highly specific probes for decay of different segments of an mRNA. Under the appropriate conditions, these probes, which are typically 24–36 nts, hybridize well to target RNA sequences, even those that are predicted to be base paired intramolecularly *in vivo*. Standard prehybridization, hybridization, and washing conditions are as follows: prehybridization/ hybridization buffer: 4× SSC (NaCl/Na-citrate buffer, where 20× SSC is 3 *M* NaCl, 0.33 *M* Na-citrate), 0.5% SDS, 1× Denhardt's solution (where 100 ml 100× Denhardt's contains 2 g of Ficoll 400, 2 g of polyvinylpyrrolidone, and 2 g of BSA), 0.1 mg/ml of salmon sperm DNA. Incubate membrane in prehybridization buffer at 42 °C for 1– 2 h in a hybridization oven. Add 5′-end-labeled oligonucleotide probe directly to buffer. Incubate at 42 °C as previously; for some probes, lower hybridization temperatures may be required. Wash twice at 42 °C for 20 min in 4× SSC, 1% SDS.

10. Northern-Blot Analysis Using Riboprobes

For hybridization with a uniformly labeled riboprobe, we use the following conditions: prehybridization/hybridization buffer is 50% deionized formamide, 0.75 M of NaCl, 50 mM of sodium phosphate (pH 7.4), 50 mM of EDTA, 0.2% Blotto (nonfat dry milk), 250 μg/ml of salmon sperm DNA, and 0.5% SDS. Degas buffer before use. Prehybridize for 1–2 h at 42 °C. Add probe and hybridize for several hours at 42 °C. Wash four times with 1× SSC, 0.2% SDS, for 20 min at 65 °C.

11. Quantitation of RNA by 5S RNA Probing

Although the MOPS gel described previously is sufficient for judging the intactness and relative concentrations of RNA preparations, the best method for quantitation of total RNA per lane is to probe for a stable RNA. We use 5S ribosomal RNA as a target for normalization of the quantity of RNA per lane. For this, the membrane that has been hybridized with the gene-specific probe is stripped of the probe. Stripping buffer consists of 0.1× SSC and 0.5% SDS. Heat stripping buffer to boiling in a small metal pan and place the membrane into the boiling buffer. Remove pan from heat and let stand 15 min. Remove membrane from buffer, reheat buffer to boiling, and repeat stripping. Rinse membrane in 2× SSC, and store wrapped in plastic at 4 °C. It is critical that the membrane be kept moist during all steps following the initial hybridization. Probing with the 5S rRNA probe follows the same protocol as previously for gene-specific oligonucleotide probes. The sequence of the *B. subtilis* 5S RNA probe is CCGACTAC CATCGGCGCTGA (complement to bases 61–80 of 5S RNA).

12. Mapping of 5′- and 3′-Ends

For 5′-end mapping, we routinely use a reverse transcriptase protocol. In our hands, the SuperScript III protocol (Invitrogen) works well to map the end of 5′-terminal decay fragments. Depending on the level of transcription of the target gene, 10–50 μg of total *B. subtilis* RNA can be used for the reaction.

An alternative, which gives both 5′- and 3′-ends, is to use reverse transcriptase in a RACE (rapid amplification of cDNA ends) reaction. Our protocol for abundant transcripts has the following three steps. First is RNA circularization: use 5 μg of total RNA in a 25-μl reaction containing 2.5 μl of 10× T4 RNA ligase buffer, 0.2 μl of RNasin (Promega), 1 μl of RQ DNase (Promega), and

2 μl of T4 RNA ligase (New England Biolabs). Incubate for 30 min at 37 °C, extract with phenol:chloroform (1:1), add 1/10 volume of 10 M LiCl to aqueous phase, and precipitate with ethanol. Resuspend in 10 μl of water. Second is reverse transcription across junction: use a gene-specific oligonucleotide, with an integrated restriction site for cloning, which is complementary to sequences downstream from the predicted 5′-end. Add 5 μl of the circularized RNA to 4 μl of dNTP mix (2.5 mM each), 0.2 μl of oligonucleotide, and 2.8 μl of water, for a total of 12 μl. Denature for 5 min at 65 °C and chill on ice for 5 min. Add 2 μl of 10× RT buffer (e.g., M-MuLV RNase H⁻ reverse transcriptase buffer from Finnzyme), 2 μl of 0.1 M DTT, and 3 μl of water. Incubate for 2 min at 42 °C. Add 1 μl of M-MuLV RNase H⁻ reverse transcriptase (200 U/μl) and incubate for 50 min at 42 °C. Heat kill enzyme by incubation for 15 min at 70 °C. Third is PCR across junction and cloning: use 3 μl of RT product in a standard PCR reaction using a second oligonucleotide that has a sequence near the predicted 3′-end of the target transcript. Purify PCR product, clone, and sequence several clones to determine the 5′- and 3′-ends.

For 5′-end mapping of transcripts, an S1 nuclease protocol may also be used. The probe is a PCR fragment that overlaps the predicted 5′-end of the transcript and is labeled on the complementary strand by an amplification reaction using a 5′-end-labeled downstream oligonucleotide. Then 10–30 μg of total RNA is mixed with 10 ng of PCR probe fragment (typically 10^5 Cerenkov cpm) in a 100-μl volume containing up to 50 μg of carrier yeast RNA. The mixture is precipitated with 1/10 volume 3 M sodium acetate, pH 5.0, 2.5 volumes of ethanol, and incubation at −80 °C for 30 min. After washing with 70% ethanol and drying, resuspend the pellet in 40 μl of deionized formamide (if necessary, heat to 75 °C and vortex gently). Add 10 μl of 5× hybridization buffer, which is 0.2 M PIPES, pH 6.4, 2 M NaCl, and 0.5 mM EDTA. (For 20 ml of 5× hybridization buffer, add 1.2 g PIPES, 2.34 g NaCl, and 1 ml of 0.1 M EDTA. The pH of the solution should be around 4.5. Adjust pH slowly to pH 6.4, using small volumes of 10 M NaOH. Sterilize by filtration through a 0.2-μm filter. Aliquot and store at −20 °C.) Incubate for 10 min at 75 °C (it is best to use a hybridization oven to avoid condensation). Transfer to a water bath set at 2–4 °C above the T_m ($T_m = 81.5 + 0.5[\%GC] + 16.6 \log[Na^+] − 0.6(\%$formamide). Hybridize overnight or for a minimum of 3 h. For S1 digestion, mix the following just before use and keep on ice: 400 μl of S1 buffer, 120 units S1 (Boehringer: 0.3 μl of 400 units/μl), 0.8 μl of denatured herring sperm DNA (10 mg/ml, boiled for 5 min). S1 buffer is 0.25 M of NaCl, 30 mM of sodium acetate, pH 4.6, 1 mM of $ZnSO_4$, and 5% glycerol. (For 10 ml of S1 buffer, add 0.5 ml of 5 M NaCl, 0.1 ml of 3 M sodium acetate, 0.1 ml of 0.1 M $ZnSO_4$, and 1 ml of 50% glycerol.) Add the S1 digestion mix to the hybridized sample and incubate at 37 °C for 30 min. Add 1 ml of cold ethanol to precipitate. Additional yeast tRNA (4 μg) can be added to aid in the precipitation. Wash pellet in 70% ethanol and resuspend in formamide-containing dye for gel electrophoresis next to a DNA sequencing ladder of the same gene, primed with the same oligonucleotide.

For 3′-end mapping by S1 nuclease, a 3′-end-labeled probe should be used that overlaps and is complementary to the predicted 3′-end of the transcript. The 3′-end is labeled using the proof-reading and fill-in activity of T4 DNA polymerase, in the presence of $[\alpha^{32}P]$-dCTP.

13. SEQUENCING-GEL NORTHERN BLOTTING

In our experience, the low GC content of many gram-positive genes may make it difficult to use S1 nuclease for 3′-end mapping. Instead, for small RNAs, we routinely do Northern blot analysis on RNA that has been separated on a high-resolution gel ("sequencing gel") to determine the precise size of an mRNA decay intermediate. For a 5′-terminal fragment, knowledge of the 5′-transcriptional start site is combined with fragment length to give the 3′-end location.

Standard DNA sequencing gels are run for the desired length of time. Transferring the gel to Whatman 3 chromatography paper is problematic for higher-percentage gels; therefore, we use gels that are 7.5% or less polyacrylamide. After separating the sequencing gel plates, carefully place Whatman 3 paper on the area of the gel that will be blotted. Cut away the gel not covered by the Whatman 3 paper, invert the glass plate, and press the gel against the Whatman 3 paper. Peel down the Whatman 3 paper while sliding the plate horizontally off the bench. The gel should remain attached to the paper. The procedure for electroblotting is the same as above, except that two pieces of Whatman 3 paper are used on either side of the gel. Electroblot at 10 V overnight, followed by an additional 30 min at 30 V.

One such experiment is shown in Fig. 14.3, in which *rps*O mRNA decay intermediates from various triple exoribonuclease mutant strains (lanes 2–4), as well as a quadruple mutant strain (lane 5), were analyzed by sequencing-gel Northern blotting. Because all of these intermediates are detectable by a 5′-terminal oligonucleotide probe, the size of the fragments gives a direct readout of the location of the 3′-ends. Although the gel in this figure is overexposed, to be able to visualize also faint bands, lighter exposures show a ladder of bands that are labeled the 180-,133- and 102-nt bands.

14. RIBONUCLEASE MUTANT STRAINS

Ribonuclease mutant strains are invaluable for studying RNA processing. Four 3′ to 5′-exoribonucleases (PNPase, RNase PH, RNase R, and YhaM) have been shown to be potentially involved in mRNA decay

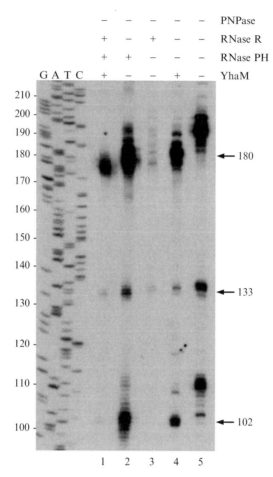

Figure 14.3 Northern blotting using a high-resolution sequencing gel. Mapping of *rpsO* mRNA decay intermediates in various PNPase mutant strains, by sizing using high-resolution Northern blotting. The 5′-end of all fragments maps to the transcriptional start site. Thus, precise sizing of decay fragments localizes the 3′-ends, which are at the downstream side of predicted secondary or tertiary RNA structures (Oussenko *et al.*, 2005). The presence or absence of the four known *B. subtilis* 3′ to 5′-exoribonucleases is shown at the top. The absence of all four known 3′ to 5′-exoribonucleases (lane 4) results in a shift in the size of decay intermediates, likely because the remaining 3′ to 5′-exonuclease activity in this strain cannot degrade the 3′-end of mRNA all the way up to a structural block. (Reproduced from Oussenko *et al.*, 2005, with permission from American Society for Microbiology.)

(Oussenko *et al.*, 2005). In Table 14.1, the available *B. subtilis* exoribonuclease mutant strains are listed. Single, double, triple, and quadruple mutants have been constructed. Included in Table 14.1 are strains that are conditional for RNase J1 expression, because, as mentioned in the introduction, the

Table 14.1 3'-exoribonuclease and RNase J1 mutant strains

Strain designation[a]	Genotype[b]	Missing ribonuclease(s)
Single mutants		
BG119[c]	Δ*pnpA*::Nm	PNPase
BG621[c]	Δ*pnpA*::Cm	PNPase
BG546[c]	Δ*pnpA*::Nm	PNPase
BG457[d]	*rph*ΩSp	RNase PH
BG388	Δ*rph*::Cm	RNase PH
BG295	Δ*rnr*::Sp	RNase R
BG664	Δ*rnr*::Tc	RNase R
BG389	Δ*rnr*::Cm	RNase R
BG376	Δ*yhaM*::Pm	YhaM
BG578[e]	*Pspac-ykqC* Em	RNase J1
Double mutants		
BG296	Δ*pnpA*::Nm Δ*rnr*::Sp	PNPase, RNase R
BG381	Δ*pnpA*::Cm Δ*rnr*::Sp	PNPase, RNase R
BG488	Δ*pnpA*::Cm *rph*ΩSp	PNPase, RNase PH
BG489	Δ*pnpA*::Nm *rph*ΩSp	PNPase, RNase PH
BG378	Δ*pnpA*::Nm Δ*yhaM*::Pm	PNPase, YhaM
BG393	Δ*rph*::Cm Δ*rnr*::Sp	RNase PH, RNase R
BG510	*rph*ΩSp Δ*rnr*::Tc	RNase PH, RNase R
BG511	*rph*ΩSp Δ*yhaM*::Pm	RNase PH, YhaM
BG395	Δ*rph*::Cm Δ*yhaM*::Pm	RNase PH, YhaM
BG377	Δ*rnr*::Sp Δ*yhaM*::Pm	RNase R, YhaM
BG421	Δ*rnr*::Tc Δ*yhaM*::Pm	RNase R, YhaM
BG645	Δ*pnpA*::Cm *Pspac-ykqC* Em	PNPase, RNase J1
Triple mutants		
BG508[f]	Δ*pnpA*::Cm *rph*ΩSp Δ*rnr*::Tc	PNPase, RNase PH, RNase R
BG517[f]	Δ*pnpA*::Nm *rph*ΩSp Δ*rnr*::Tc	PNPase, RNase PH, RNase R
BG512	Δ*pnpA*::Nm *rph*ΩSp Δ*yhaM*::Pm	PNPase, RNase PH, YhaM
BG379	Δ*pnpA*::Cm Δ*rnr*::Sp Δ*yhaM*::Pm	PNPase, RNase R, YhaM
BG422[f]	Δ*pnpA*::Cm Δ*rnr*::Tc Δ*yhaM*::Pm	PNPase, RNase R, YhaM
BG396	*rph*::Cm Δ*rnr*::Sp Δ*yhaM*::Pm	RNase PH, RNase R, YhaM
BG503	*rph*ΩSp Δ*rnr*::Tc Δ*yhaM*::Pm	RNase PH, RNase R, YhaM

(continued)

Table 14.1 (*continued*)

Strain designation[a]	Genotype[b]	Missing ribonuclease(s)
BG634	Δ*rph*::Cm Δ*rnr*::Tc Δ*yhaM*::Pm	RNase PH, RNase R, YhaM
Quadruple mutants		
BG505[f,g]	Δ*pnpA*::Cm *rph*ΩSp Δ*rnr*::Tc Δ*yhaM*::Pm	PNPase, RNase PH, RNase R YhaM
BG616[e]	P*spac-ykqC* Em *rph*ΩSp Δ*rnr*::Tc Δ*yhaM*::Pm	RNase J1, RNase PH, RNase R, YhaM
Quintuple mutant		
BG613[e,f,g]	Δ*pnpA*::Cm P*spac-ykqC* Em *rph*ΩSp Δ*rnr*::Tc Δ*yhaM*::Pm	PNPase, RNase J1, RNase PH RNase R, YhaM

[a] Strain designation are from the David H. Bechhofer (DHB) lab. Other labs have different designations for some of the strains. Except for BG578, all strains are derivatives of BG1, which is *trpC2 thr-5*. BG578 is a W168 derivative and is *trp*[+]. Order of strains is based on alphabetic sequence of missing RNase(s).

[b] Cm, chloramphenicol; Em, erythromycin; Nm, neomycin; Pm, phleomycin; Sp, spectinomycin; Tc, tetracycline.

[c] Grows slightly slower than the wild type; competence deficient (5- to 10-fold less than wild type); Tc-sensitive. BG119 and BG621 are glutamate auxotrophs. BG546 carries the same *pnpA*::Nm as BG119 but is a glutamate prototroph.

[d] Complete replacement of *rph* gene with Sp[R] gene.

[e] Contains pMAP65 (Petit *et al.*, 1998), a Nm[R] plasmid that carries the *lacI* gene and provides increased amount of *lac* repressor. Strain requires IPTG for growth.

[f] Not Tc-resistant due to PNPase mutation.

[g] Grows very slowly; colonies take two days to form.

5′-exonuclease/endonuclease activities of RNase J1 (encoded by the *ykqC* gene) are likely to play important roles in mRNA decay. Other *B. subtilis* endoribonucleases have been studied in our laboratories, including Bs-RNase III, EndoA, RNase J2, RNase M5, RNase P, RNase Z, and YhcR. We have constructed strains that are deleted for, or conditionally express, the genes encoding these activities. In many cases they have been combined with mutations in one or more of the four 3′ to 5′-exoribonuclease genes. These strains allow screening for the ribonuclease responsible for identified endonucleolytic cleavage events *in vivo*. For brevity's sake, they are not included in Table 14.1, but the strains are available on request.

REFERENCES

Bechhofer, D. H., and Wang, W. (1998). Decay of *ermC* mRNA in a polynucleotide phosphorylase mutant of *Bacillus subtilis*. *J. Bacteriol.* **180,** 5968–5977.

Brenner, S., Jacob, F., and Meselson, M. (1961). An unstable intermediate carrying information from genes to ribosomes for protein synthesis. *Nature.* **190,** 576–581.

Brosius, J. (1992). Compilation of superlinker vectors. *Methods Enzymol.* **216,** 469–483.

Carpousis, A. J., Vanzo, N. F., and Raynal, L. C. (1999). mRNA degradation: A tale of poly(A) and multiprotein machines. *Trends Genet.* **15,** 24–28.

Celesnik, H., Deana, A., and Belasco, J. G. (2007). Initiation of RNA decay in *Escherichia coli* by 5′-pyrophosphate removal. *Mol. Cell* **27,** 79–90.

Coburn, G. A., and Mackie, G. A. (1999). Degradation of mRNA in *Escherichia coli*: An old problem with some new twists. *Prog. Nucleic Acid Res. Mol. Biol.* **62,** 55–108.

Condon, C. (2003). RNA processing and degradation in *Bacillus subtilis*. *Microbiol. Mol. Biol. Rev.* **67,** 157–174.

Condon, C., Brechemier-Baey, D., Beltchev, B., Grunberg-Manago, M., and Putzer, H. (2001). Identification of the gene encoding the 5S ribosomal RNA maturase in *Bacillus subtilis*: Mature 5S rRNA is dispensable for ribosome function. *RNA* **7,** 242–253.

Craven, M. G., Henner, D. J., Alessi, D., Schauer, A. T., Ost, K. A., Deutscher, M. P., and Friedman, D. I. (1992). Identification of the rph (RNase PH) gene of *Bacillus subtilis*: Evidence for suppression of cold-sensitive mutations in *Escherichia coli*. *J. Bacteriol.* **174,** 4727–4735.

Deikus, G., Condon, C., and Bechhofer, D. H. (2008). Role of *Bacillus subtilis* RNase J1 endonuclease and 5′-exonuclease activities in *trp* leader RNA turnover. *J. Biol. Chem.* **238,** 17158–17167.

Deutscher, M. P., and Reuven, N. B. (1991). Enzymatic basis for hydrolytic versus phosphorolytic mRNA degradation in *Escherichia coli* and *Bacillus subtilis*. *Proc. Natl. Acad. Sci. USA* **88,** 3277–3280.

Dubnau, D., and Lovett, C. M., Jr. (2001). Transformation and recombination. *In* "*Bacillus subtilis* and its relatives: From genes to cells" (J. A. Hoch, R. Losick, and A. L. Sonenshein, eds.), pp. 453–471. American Society for Microbiology, Washington, DC.

Even, S., Pellegrini, O., Zig, L., Labas, V., Vinh, J., Brechemmier-Baey, D., and Putzer, H. (2005). Ribonucleases J1 and J2: Two novel endoribonucleases in *B. subtilis* with functional homology to *E. coli* RNase E. *Nucleic Acids Res.* **33,** 2141–2152.

Gros, F., Hiatt, H., Gilbert, W., Kurland, C. G., Risebrough, R. W., and Watson, J. D. (1961). Unstable ribonucleic acid revealed by pulse labelling of *Escherichia coli*. *Nature* **190,** 581–585.

Herskovitz, M. A., and Bechhofer, D. H. (2000). Endoribonuclease RNase III is essential in *Bacillus subtilis*. *Mol. Microbiol.* **38,** 1027–1033.

Kunst, F., Ogasawara, N., Moszer, I., Albertini, A. M., Alloni, G., Azevedo, V., Bertero, M. G., Bessieres, P., Bolotin, A., Borchert, S., Borriss, R., Boursier, L., *et al.* (1997). The complete genome sequence of the gram-positive bacterium *Bacillus subtilis*. *Nature* **390,** 249–256.

Kushner, S. R. (2002). mRNA decay in *Escherichia coli* comes of age. *J. Bacteriol.* **184,** 4658–4665.

Luttinger, A., Hahn, J., and Dubnau, D. (1996). Polynucleotide phosphorylase is necessary for competence development in *Bacillus subtilis*. *Mol. Microbiol.* **19,** 343–356.

Mathy, N., Benard, L., Pellegrini, O., Daou, R., Wen, T., and Condon, C. (2007). 5′ to 3′-exoribonuclease activity in bacteria: Role of RNase J1 in rRNA maturation and 5′-stability of mRNA. *Cell* **129,** 681–692.

Mitra, S., Hue, K., and Bechhofer, D. H. (1996). *In vitro* processing activity of *Bacillus subtilis* polynucleotide phosphorylase. *Mol. Microbiol.* **19**, 329–342.

Oussenko, I. A., Abe, T., Ujiie, H., Muto, A., and Bechhofer, D. H. (2005). Participation of 3′ to 5′-exoribonucleases in the turnover of *Bacillus subtilis* mRNA. *J. Bacteriol.* **187**, 2758–2767.

Oussenko, I. A., and Bechhofer, D. H. (2000). The *yvaJ* gene of *Bacillus subtilis* encodes a 3′ to 5′-exoribonuclease and is not essential in a strain lacking polynucleotide phosphorylase. *J. Bacteriol.* **182**, 2639–2642.

Oussenko, I. A., Sanchez, R., and Bechhofer, D. H. (2002). *Bacillus subtilis* YhaM, a member of a new family of 3′ to 5′-exonucleases in gram-positive bacteria. *J. Bacteriol.* **184**, 6250–6259.

Oussenko, I. A., Sanchez, R., and Bechhofer, D. H. (2004). *Bacillus subtilis* YhcR, a high-molecular-weight, nonspecific endonuclease with a unique domain structure. *J. Bacteriol.* **186**, 5376–5383.

Petit, M.-A., Dervyn, E., Rose, M., Entian, K.-D., McGovern, S., Ehrlich, D. S., and Bruand, C. (1998). PcrA is an essential DNA helicase of *Bacillus subtilis* fulfilling functions both in repair and rolling-circle replication. *Mol. Microbiol.* **29**, 261–273.

Piggot, P. J., and Losick, R. (2001). Sporulation genes and intercompartmental regulation. *In* "*Bacillus subtilis* and its relatives: from genes to cells" (J. A. Hoch, R. Losick, and A. L. Sonenshein, eds.), pp. 483–517. American Society for Microbiology, Washington, DC.

Ulmanen, I., Lundstrom, K., Lehtovaara, P., Sarvas, M., Ruohonen, M., and Palva, I. (1985). Transcription and translation of foreign genes in *Bacillus subtilis* by the aid of a secretion vector. *J. Bacteriol.* **162**, 176–182.

Wang, W., and Bechhofer, D. H. (1996). Properties of a *Bacillus subtilis* polynucleotide phosphorylase deletion strain. *J. Bacteriol.* **178**, 2375–2382.

Wen, T., Oussenko, I. A., Pellegrini, O., Bechhofer, D. H., and Condon, C. (2005). Ribonuclease PH plays a major role in the exonucleolytic maturation of CCA-containing tRNA precursors in *Bacillus subtilis*. *Nucleic Acids Res.* **33**, 3636–3643.

Yao, S., Blaustein, J. B., and Bechhofer, D. H. (2007). Processing of *Bacillus subtilis* small cytoplasmic RNA: Evidence for an additional endonuclease cleavage site. *Nucleic Acids Res.* **35**, 4464–4473.

CHAPTER FIFTEEN

Assay of *Bacillus subtilis* Ribonucleases *In Vitro*

Ciarán Condon,* Olivier Pellegrini,* Nathalie Mathy,*
Lionel Bénard,* Yulia Redko,* Irina A. Oussenko,[†]
Gintaras Deikus,[†] *and* David H. Bechhofer[†]

Contents

Abstract

Significant progress has been made recently regarding the identification of the ribonucleases involved in RNA maturation and degradation in *Bacillus subtilis*. More than half of these enzymes have no ortholog in *Echerichia coli*. To confirm that the *in vivo* effects of mutations in genes encoding RNases are direct, it is often necessary to purify the enzymes and assay their activity *in vitro*.

* CNRS UPR 9073 (affiliated with Université de Paris 7; Denis Diderot), Institut de Biologie Physico-Chimique, Paris, France
† Mount Sinai School of Medicine of New York University, New York

Methods in Enzymology, Volume 447
ISSN 0076-6879, DOI: 10.1016/S0076-6879(08)02215-5

Development of such assays is also necessary for detailed biochemical analysis of enzyme properties. In this chapter, we describe the purification and assay of 12 RNases of *B. subtilis* thought to be involved in stable RNA maturation or RNA degradation.

1. Introduction

The evolutionary separation of *Escherichia coli* and *Bacillus subtilis* is thought to have occurred between 1 and 3 billion years ago, before the separation of plants and animals. Thus, it should not come as a surprise that these two bacteria have found quite different solutions to the problem of RNA maturation and decay. Approximately three-quarters of all the currently identified ribonucleases (RNases) in these two organisms are unique to one or other of the species, and even the essential RNases are, for the most part, nonoverlapping (Condon, 2003, 2009). The chemistry of the RNA degradation reaction also differs between the two organisms, with exoribonucleolytic degradation in *E. coli* being primarily hydrolytic (consuming a molecule of water with each phosphodiester bond broken) and that of *B. subtilis* being primarily phosphorolytic (requiring inorganic phosphate to remove each successive nucleotide). Last, a 5′ to 3′-exoribonuclease activity has recently been discovered in *B. subtilis*, an activity previously thought to be confined to eukaryotes (Mathy *et al.*, 2007). Thus, although *E. coli* RNA decay relies primarily on endonucleolytic cleavage, followed by exonucleolytic degradation in the 3′ to 5′-orientation, *B. subtilis* has an additional strategy available for RNA decay. These observations have confirmed *B. subtilis* as a highly valuable alternative model for bacterial RNA degradation. In this chapter, we describe the purification and assay of twelve *B. subtilis* ribonucleases, both endoribonucleases (RNases M5, Z, P, J1, III, YhcR and EndoA) and exoribonucleases (RNases PH, R, PNPase, YhaM and Nano-RNase). These ribonucleases have been shown to be the key enzymes of RNA maturation and decay in this organism.

2. Purification and Assay of *B. subtilis* Endoribonucleases

2.1. RNase M5

RNase M5 catalyzes the maturation of 5S rRNA in low G + C gram-positive bacteria by cleaving on either side of a double-stranded stem to yield mature 5S rRNA in a single step (Fig. 15.1). The reaction requires ribosomal protein L18

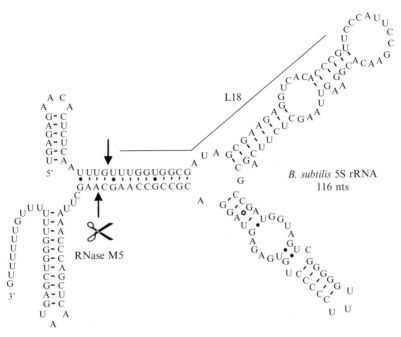

Figure 15.1 Maturation of 5S ribosomal RNA by RNase M5. The cleavage reaction requires prior binding of ribosomal protein L18 to the region indicated by the black line. Cleavage (indicated by the conventional scissors symbol) occurs on either side of a double-stranded region (arrows), leaving a 1-nt 3'-overhang.

as a cofactor, presumably to allow the 5S rRNA precursor to adapt the correct conformation for cleavage (Sogin and Pace, 1974).

2.1.1. Purification of RNase M5

RNase M5 (20.5 kDa) is present at approximately 100 copies per cell in *B. subtilis* (Condon *et al.*, 2001; Pace and Pace, 1990). Because of these low amounts, purification of RNase M5 from wild-type *B. subtilis* strains is quite intricate and almost inevitably results in a protein preparation that consists predominantly of ribosomal protein L6, because of its similar size, charge and much greater abundance. There are two ways to solve this problem. The first is to use *B. subtilis* strain SSB306 in which the wild-type L6 protein is replaced by a C-terminally His-tagged version, increasing its size sufficiently to allow separation of RNase M5 from L6 through many different chromatography steps and ultimately elution from 17.5% SDS–PAGE gels. This method was used to initially identify the gene encoding RNase M5 in *B. subtilis* and is described in some detail in (Condon *et al.*, 2001). Now that the gene has been identified, however, it is possible to purify either the native or His-tagged protein from overexpressing *E. coli* cells, which do not contain any contaminating RNase M5 activity.

The native protein is purified from overproducing *E. coli* strain BL21 (λDE3), containing plasmid pHMM1 (Condon *et al.*, 2001). Cultures (typically 2 L of 2× YT medium) are grown to OD_{600} of 0.6 in the presence of 200 μg/ml ampicillin, and expression of RNase M5 is induced by addition of 0.5 mM IPTG for 2.5 h. The final OD_{600} is typically approximately 1.7 and yields approximately 4 g of cell pellet that can be stored frozen. Cells are resuspended in 12 ml ice-cold SBII buffer (50 mM Tris-HCl, pH 7.4, 10 mM MgCl$_2$, 10 mM NH$_4$Cl, 1 mM DTT) containing an EDTA-free anti-protease tablet (Roche) and 1 μg/ml DNase I. Cells are lysed by two passages in a French press at 16,000 psi, and cell debris is removed by centrifugation at 9000 rpm for 20 min at 4 °C (Sorvall SS34 rotor). The supernatant is filtered at 0.45 μm and applied to an SP Sepharose HP 16/10 column (GE Healthcare) equilibrated in SBII buffer at 2 ml/min. The protein is eluted with a 200-ml linear gradient of 10 mM to 1 M NH$_4$Cl in the same buffer. The peak of RNase M5, identified on 17.5% SDS-PAGE gels, occurs at approximately 250 mM NH$_4$Cl. The peak fractions are pooled and dialyzed against buffer SBII and applied to a 5-ml HiTrap Heparin (GE Healthcare) column equilibrated in the same buffer at 1 ml/min. The protein is eluted with a 60-ml linear gradient of 10 mM to 1 M NH$_4$Cl in SBII, with peak fractions occurring at approximately 500 mM NH$_4$Cl. RNase M5 is concentrated by precipitation with 70% ammonium sulfate on ice for 3 h, followed by centrifugation at 9000 rpm for 20 min at 4 °C. The pellet is resuspended in 1 ml storage buffer (10 mM Tris-HCl, pH 7.5, 10 mM MgCl$_2$, 10 mM NH$_4$Cl, 1 mM DTT, 0.1 mM EDTA, 15% glycerol) and is stable for many years at −80 °C at a concentration of 1 mg/ml. At higher concentrations, RNase M5 tends to aggregate and precipitate.

C-terminal His-tagged RNase M5 is purified in a single step by applying overproducing extracts of strain CCE072 (Allemand *et al.*, 2005) to 5 ml His–Trap columns (GE Healthcare) equilibrated in 20 mM potassium phosphate buffer, pH 7.4, 250 mM NaCl, 10 mM MgCl$_2$, 5 mM imidazole, 1 mM β-mercaptoethanol and eluting with a linear gradient of 5 to 500 mM imidazole in the same buffer. His-tagged RNase M5 is dialyzed against the same storage buffer as the native protein. Protein yields are typically of the order of 10 to 15 mg from 400 ml of overproducing culture at OD_{600} of 1.7. There is no major difference in the specific activity of the native versus His-tagged RNase M5 proteins.

2.1.2. Assay of RNase M5 activity *in vitro*

Uniformly labeled (α-[^{32}P]-UTP) precursor 5S rRNAs are synthesized *in vitro* by T7 RNA polymerase, with plasmid pHMR1 cleaved by *Eco*RI as template, according to the protocol provided in "Standard Protocol 1." The construction of pHMR1, permitting expression of a 256-nt precursor fragment containing the *B. subtilis rrnB* 5S rRNA, is described in Condon *et al.* (2001).

RNase M5 activity is assayed in 5-μl reactions in 10 mM Tris-HCl, pH 7.4, 10 mM MgCl$_2$, 50 mM NH$_4$Cl, 2 mM DTT, containing 0.25 pmol of labeled 5S precursor, 2.5 μg of yeast competitor RNA (Sigma), and 1 ng *B. subtilis* ribosomal protein L18. The reaction buffer is usually prepared as a 5× mix (Table 15.1) that is filtered (0.22 μm) and stored frozen. L18 is purified according to Pace and Pace (1990), replacing the phosphocellulose columns with MonoS HR (GE Healthcare) columns. Reactions can also be performed in the absence of L18, however, by replacing it with 25 to 30% DMSO. It is thought that DMSO permits the folding of the substrate in a similar manner to that performed by L18; however, a direct effect of L18 (or DMSO) on RNase M5 conformation/activity has not been ruled out (Allemand *et al.*, 2005). Reaction mixes are incubated for 15 min at 37 °C, stopped by addition of 5 μl loading buffer (95% formamide, 20 mM EDTA, 0.05% bromophenol blue, 0.05% xylene cyanol) and applied directly to 5% polyacrylamide/7 M urea gels. The cleavage products are 116 (mature species), 85 (3′-extension) and 55 (5′-extension) nucleotides (nts) in length.

2.2. EndoA

EndoA is a member of the PemK/MazF family of toxins possessing ribonuclease activity (Pellegrini *et al.*, 2005). It cleaves in single-stranded regions containing a UAC triplet (Fig. 15.2). Subsequent experiments (N. M. and C. C., unpublished) showed a strong bias for a U-residue upstream and an A-residue downstream of the central triplet (i.e., UUACA). Cleavage by EndoA generates an upstream product with a 3′-phosphate and a downstream product with a 5′-hydroxyl group.

2.2.1. Purification of EndoA

EndoA (12.8 kDa), bearing a 6× His-Sumo N-terminal tag, is purified from *E. coli* BL21 (λDE3) cells containing plasmid pSMT3-YdcE, described in Gogos *et al.* (2003) and is purified to homogeneity by affinity chromatography and gel filtration. Cells are grown in LB at 37 °C to an A$_{600}$ = 0.6. The temperature is then reduced to 18 °C, induced with 0.1 mM IPTG, and grown for a further 16 h. Cells are spun down and resuspended in binding buffer (20 mM Tris-HCl, pH 8.0, 500 mM NaCl, 5 mM imidazole) at approximately 10 ml buffer per gram of pellet. Cells are lysed by sonication on high power for a total of 8 min (pulse 15 sec, let cool 45 sec) and centrifuged at 12,000 rpm in a Sorvall SS34 rotor for 1 h at 4 °C. The supernatant is loaded onto a nickel-chelating column (e.g., Ni-NTA agarose; Qiagen) equilibrated in binding buffer. The column is washed in binding buffer until absorbance stabilizes. A step gradient is performed with imidazole steps of 5 mM, 20 mM, 50 mM, and 250 mM in binding buffer with 10 column volumes of buffer for each step. EndoA usually

Table 15.1 Endoribonuclease reaction buffers (5×)

RNase M5	EndoA	RNase Z	RNase P	RNaseJ1	RNase III	YhcR
50 mM Tris-HCl, pH 7.4	50 mM Tris-HCl, pH 7.5	200 mM Tris-HCl, pH 8.4	100 mM HEPES-KOH, pH 7.4	100 mM HEPES-KOH, pH 8.0	50 mM Tris-HCl, pH 7.8	250 mM Tris-HCl, pH 8.0
50 mM $MgCl_2$	25 mM NH_4Cl	10 mM $MgCl_2$	10 mM $Mg(OAc)_2$	40 mM $MgCl_2$	15 mM $MgCl_2$	5 mM $CaCl_2$
250 mM NH_4Cl	0.5 mM EDTA	10 mM KCl	750 mM NH_4OAc	500 mM NaCl	500 mM NaCl	500 mM KCl
10 mM DTT		10 mM DTT	10 mM Spermidine			
			0.25 mM Spermine			
			20 mM DTT			

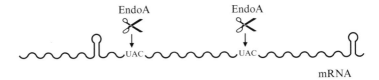

Figure 15.2 Endonucleolytic cleavage of mRNA at UAC single-stranded triplets by the EndoA toxin. The mRNA is indicated by a wavy line, and cleavage is specified by scissors symbols.

comes out in the 250 mM step. The fractions containing EndoA are pooled and dialyzed against binding buffer with no imidazole. After dialysis, the sample is loaded on a Talon Cobalt column (Clontech, Inc.) equilibrated in binding buffer without imidazole. The column is washed with at least 10 column volumes of the same buffer. EndoA is released from the 6His-SUMO tag (leaving an N-terminal serine) by addition of Ulp1 protease. One column volume of Ulp1 solution (1:100 w/w ratio of Ulp1 to EndoA, i.e. 0.2 mg of Ulp1 for approximately 20 mg EndoA) in no-imidazole binding buffer is run slowly into the column, which is then sealed so that it does not drip overnight. The gel bed is stirred to ensure uniform distribution of the protease solution. EndoA is eluted with 10 column volumes of no-imidazole buffer, and 10 fractions are collected, each corresponding to one column volume. The protein peak is identified by SDS-PAGE and dialyzed against 10 mM Tris-HCl, pH 8.0, 0.15 M NaCl, 5 mM DTT. EndoA is concentrated (Amicon Ultra, MWCO 5 kDa; Millipore) and loaded onto a Superdex 75 26/60 column (GE Healthcare). Peak fractions are pooled and concentrated to ~10 mg/ml. EndoA is quite unstable and rapidly loses activity with repeated freezing and thawing. It is, therefore, advisable to prepare single-use aliquots for storage at −80 °C.

2.2.2. Assay of EndoA activity *in vitro*

The first identified substrate for EndoA was an RNA fragment containing the *thrS* promoter and leader region. This artificial transcript was synthesized from a T7 RNA polymerase promoter placed upstream of the natural *thrS* promoter. EndoA cleaves in an AUACA sequence overlapping the transcribed −10 region of the *thrS* promoter. EndoA activity is assayed on 0.2 pmol uniformly labeled (α-[^{32}P]-UTP) *thrS* RNA fragment (380 nts) transcribed from plasmid pHMS17, cleaved by *Eco*NI (Gendron *et al.*, 1994). Transcription reactions are performed by T7 RNA polymerase as described in "Standard Protocol 1." EndoA assays are performed in 10 mM Tris-HCl, pH 7.5, 5 mM NH$_4$Cl, 0.1 mM EDTA in a 5-μl reaction volume. The reaction buffer is typically diluted from a 5× mix (Table 15.1) that is filtered (0.22 μm) and stored frozen. Assays are incubated for 20 min at 25 °C, stopped by addition of 5 μl loading buffer (95% formamide, 20 mM EDTA, 0.05% bromophenol blue, 0.05% xylene cyanol), and applied directly to 5% polyacrylamide/7 *M* urea gels. The primary reaction products are 270 and 110 nt in size.

2.3. RNase Z

RNase Z is a key enzyme in tRNA 3′-end processing in many organisms, and is essential in *B. subtilis* (Pellegrini *et al.*, 2003). It cleaves precursor tRNAs, for the most part directly after the discriminator base, the single unpaired nucleotide at the base of the acceptor stem, onto which is attached the universal CCA motif (Fig. 15.3). RNase Z preferentially cleaves tRNA precursors that lack a CCA motif, yielding a tRNA with a 3′-hydroxyl (ready for CCA addition) and a downstream fragment with a 5′-monophosphate group.

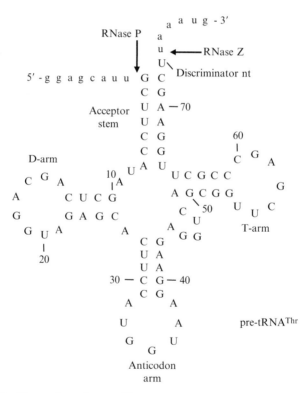

Figure 15.3 Maturation of tRNA^{Thr} precursor by RNase Z and RNase P. The cleavage sites are indicated by arrows. The mature tRNA species is shown in uppercase letters; sequences removed by RNase Z and RNase P cleavage are in lowercase. Some features of the tRNA (e.g., acceptor stem, discriminator nucleotide) are shown. Both RNase P and RNase Z recognize primarily the acceptor stem and T-arm of the tRNA.

2.3.1. Purification of RNase Z

RNase Z (33.9 kDa) is purified as a C-terminal His-tagged protein according to "Standard Protocol 2(a)." The fractions (1 ml) containing RNase Z are initially identified by measuring protein concentration (BioRad), pooled and immediately dialyzed in storage buffer containing 20 mM Tris-HCl, pH 7.8, 5 mM $MgCl_2$, and 15% glycerol. Even a few hours' exposure (e.g., the time to run a protein gel) to imidazole leads to precipitation of RNase Z. Protein purity, verified by SDS-PAGE analysis (12.5%), is typically >90%. The final yield is approximately 8 mg RNase Z from a 400-ml culture. The protein is stored at $-80\,°C$.

2.3.2. Assay of RNase Z activity *in vitro*

Uniformly labeled (α-[^{32}P]-UTP) precursor tRNAs are readily prepared by *in vitro* transcription of PCR templates, with the upper primer containing a T7 RNA polymerase promoter sequence. The template for the *B. subtilis* *trnI-Thr* precursor with a 6-nt 3′-trailer sequence is made by PCR amplification of plasmid pHMT1 (Luo *et al.*, 1998) with oligos HP560 and CC079 (Table 15.3). The T7 RNA polymerase *in vitro* transcription reactions are carried out according to "Standard Protocol 1."

RNase Z activity is assayed in 5-μl reaction volumes containing 40 mM Tris-HCl, pH 8.4, and 2 mM each $MgCl_2$, KCl, and dithiothreitol (DTT) and incubated for 15 to 20 min at 37 °C. The reaction buffer is usually prepared as a 5× mix (Table 15.1) that is filtered (0.22 μm) and stored frozen. Approximately 40,000 Cerenkov counts (approximately 1 pmol) of uniformly labeled precursor tRNA is used per assay, with concentrations of purified His-tagged RNase Z that can range from 0.15 to 150 ng/μl. The 5-μl reactions are stopped by addition of 5 μl loading buffer (95% formamide, 20 mM EDTA, 0.05% bromophenol blue, 0.05% xylene cyanol) and applied directly to 5% polyacrylamide/7 M urea gels. The RNase Z-cleaved tRNA^{Thr} is 73 nt in length.

2.4. RNase P

RNase P is a universal enzyme that catalyzes the endonucleolytic removal of extraneous sequences on the 5′-side of tRNA precursors (Fig. 15.3). It is a ribozyme, consisting of a catalytic RNA subunit (Guerrier-Takada *et al.*, 1983), encoded by the *rnpB* gene, and a protein subunit, encoded by *rnpA*. Like its *E. coli* counterpart, *B. subtilis* RNase P is essential for cell survival (Wegscheid *et al.*, 2006).

2.4.1. Purification of RNase P

RNase P can be purified either as holoenzyme, from specialized *B. subtilis* cells, or by reconstitution of protein subunit purified from *E. coli*, with the *in vitro*-synthesized *B. subtilis* RNA subunit.

Table 15.2 Exoribonuclease reaction buffers ($5\times$)

RNase R	RNase PH	PNPase[a]	YhaM	RNase J1	Nano-RNase
250 mM Na-Tricine, pH 8.0	250 mM Tris-HCl, pH 8.0	125 mM Tris-HCl, pH 8.0	250 mM Na-Tricine, pH 8.0	100 mM Tris-HCl, pH 8.0	250 mM HEPES-KOH, pH 7.5
5 mM MgCl$_2$	12.5 mM MgCl$_2$	250 mM KCl	5 mM MnCl$_2$	40 mM MgCl$_2$	25 mM MnCl$_2$
500 mM KCl	150 mM NaCl	1 mM EDTA	500 mM KCl	500 mM NH$_4$Cl	
	50 mM K$_2$HPO$_4$	12.5% glycerol		0.5 mM DTT	
	0.5 mM DTT				

[a] This is the $5\times$ prebinding buffer for PNPase reactions; degradation reactions are started by further addition of MgCl$_2$ to 1 mM and Na$_2$HPO$_4$ to 2 mM.

2.4.1.1. Holoenzyme purification N-terminally His-tagged RNase P holoenzyme is purified directly from *B. subtilis* strain d7 containing plasmid pB.s.[*mpA*-NH + *mpB*] that expresses both the His-tagged protein subunit and the RNA subunit under control of the IPTG-inducible *Pspac* promoter (Gossringer *et al.*, 2006). The d7 strain has the native *mpA* gene under control of a xylose-dependent promoter, allowing the expression of the native RNase P protein to be shut off. *B. subtilis* strain d7 bearing plasmid pB.s.[*mpA*-NH + *mpB*] is grown in 50 ml LB broth to mid-log phase, induced for 1.5 h with 1 mM IPTG, and harvested by centrifugation (5000g for 7 min at 4 °C). Approximately 0.3 g of cell pellet is washed once and resuspended in 4 to 5 ml of buffer R (10 mM HEPES, pH 6.2, 100 mM KCl, 8 mM MgCl$_2$, 9 mM imidazole). After preparation of cell lysates by sonication (10 min, duty cycle 50%, on ice), cell debris is removed by centrifugation at 18,000g for 12 min at 4 °C. A 1.3-ml volume of supernatant is then incubated with 0.6 ml of Ni-nitrilotriacetic acid (NTA) agarose slurry (preequilibrated in buffer R; binding capacity, 5–10 mg 6-His-tagged protein per ml of matrix; QIAGEN, Chatsworth, CA) for 45 min on an end-to-end shaker at 4 °C to allow the RNase P holoenzyme to adsorb to the matrix. The slurry is washed five times with buffer R and then four times with buffer R lacking imidazole. The RNase P holoenzyme is active while bound to Ni-NTA agarose. It can presumably be eluted with imidazole and dialyzed, but this has not been tested experimentally.

2.4.1.2. Holoenzyme reconstitution *B. subtilis* RNase P protein (13.7 kDa) with an N-terminal His-tag is purified from *E. coli* JM109 cells containing the *B. subtilis mpA* gene clone in plasmid pQE-30 (Rivera-Leon *et al.*, 1995). It is purified under denaturing conditions to fully remove the RNA subunit. Cultures are grown to an OD$_{578}$ of 0.6, and IPTG is added to a final concentration of 1 mM. Cells are harvested after 3 h (OD$_{578}$ ca. 2.5) of induction, by centrifugation for 10 min at 5000 rpm and 4 °C. The subsequent steps are performed at 4 °C or on ice, and all buffers are supplemented with 40 μg/ml of the protease inhibitor phenylmethylsulfonyl fluoride (PMSF). The cell pellet is resuspended in 10 ml of sonication buffer SB (50 mM Tris-HCl, pH 8.0, 0.3 M NaCl, 0.1% Triton X-100, 1 M NH$_4$Cl), and sonicated (Branson Sonifier 250, output 20, duty cycle 50%, 15 min on ice). After centrifugation for 30 min (4 °C, 14,500g), the supernatant is mixed with Ni-NTA agarose (400 μl for 2 L of cell culture) that has been prewashed twice with 10 ml SB buffer. Proteins are bound to the matrix under gentle mixing or rotation at 4 °C for 2 h, followed by three washes (centrifugation-resuspension cycles; centrifugation at 4 °C and 8500 rpm in a desktop centrifuge) with 2 ml ice-cold washing buffer (50 mM Tris-HCl, pH 8.0, 8 M urea, 0.1% Triton X-100, 30 mM imidazole). The RNase P protein starts to precipitate during this procedure;

therefore, the supernatant after each washing step is removed carefully to avoid release of protein aggregates from the matrix. The protein is then eluted with 500 μl elution buffer (50 mM Tris-HCl, pH 7.0, 10% glycerol, 7 M urea, 20 mM EDTA, 0.3 M imidazole) for 45 min at 4 °C under gentle shaking. The eluate is dialyzed twice for 1 h and subsequently overnight against 500 ml dialysis buffer (50 mM Tris-HCl, pH 7.0, 0.1 M NaCl, 10% glycerol) with 12- to 14-kDa molecular weight cutoff dialysis tubing. During dialysis, a white precipitate forms. The dialyzed material is transferred to a 2-ml microcentrifuge tube and centrifuged at 12,000 rpm for 20 min at 4 °C in a benchtop centrifuge. The supernatant contains RNase P protein devoid of any RNase P RNA contamination, whereas the pellet includes traces of RNase P RNA and is discarded. The pure protein supernatant is divided in aliquots, frozen in liquid nitrogen, and stored at -80 °C. All purification steps are checked by SDS-PAGE in 17% gels to confirm the purity and concentration of the RNase P protein.

The *B. subtilis* RNase P RNA subunit (401 nt) is prepared by *in vitro* transcription with T7 RNA polymerase (e.g., Ambion Mega ShortScript Kit), with plasmid pDW66 (Smith *et al.*, 1992) linearized with *Dra*I as template. The *in vitro* transcribed RNA is eluted from polyacrylamide gels, ethanol precipitated, resuspended in nuclease-free water, and stored frozen.

2.4.2. Assay of RNase P activity *in vitro*

RNase P is assayed on precursor tRNA substrates bearing 5′-extensions. Uniformly labeled (α-[^{32}P]-UTP) precursor tRNAs can be prepared by *in vitro* transcription of PCR templates, with the upper primer containing a T7 RNA polymerase promoter sequence. For example, the template for the *B. subtilis trn*I-*Thr* precursor with an 8-nt 5′-extension can be made by PCR amplification of plasmid pHMT1 (Luo *et al.*, 1998) with oligos CC274 and CC275 (Table 15.3). The T7 RNA polymerase *in vitro* transcription reactions are carried out according to "Standard Protocol 1." An alternative to uniformly labeled substrate tRNA is cold precursor tRNA mixed with trace amounts ($<$1 nM) of 5′-labeled precursor.

Although RNase P holoenzyme can be reconstituted, aliquoted for single use, and stored frozen at -80 °C, freshly reconstituted holoenzyme gives the most reproducible results. RNase P activity can be assayed in buffer KN2 (20 mM HEPES-KOH, pH 7.4, 150 mM NH$_4$OAc, 2 mM Mg (OAc)$_2$, 2 mM spermidine, 0.05 mM spermine, and 4 mM DTT. The reaction buffer can be prepared as a 5$\times$ mix (Table 15.1), filtered (0.22 μm), and stored frozen. *In vitro* synthesized RNase P RNA (10 nM) is incubated in KN2 buffer for 5 min at 55 °C, followed by at least 25 min at 37 °C. A fourfold excess (40 nM) of purified RNase P protein is added and incubation allowed to continue at 37 °C for a further 5 min. The volume of the holoenzyme mix is generally either 6 μl or 16 μl. The precursor tRNA substrate (100 nM; typically 10,000 Cerenkov cpm) in 4 μl KN2 buffer is

Table 15.3 Oligonucleotides

CC046: TATATCTAGACTATCGCCACCAAACAAATTGAGAG
CC079: CATTTAAGCTTCCAAGCGG
CC107: CATTGGAGCTTCCAAGCGG
CC169: *GCTCTAATACGACTCACTATA*GGGAAAAATCGCACA GCGATGTGCGTAGTC
CC170: TCCATCTGTAAGTGGTAGCCGAAGC
CC274: *ATTAATACGACTCACTATA*GGAGCATTGCTTCCATAG CTCAGCAGGTAG
CC275: AGCTTCCAAGCGGGCTCGAAC
CC455: *GCTCTAATACGACTCACTATA*GGGCCTTCGA AACGTGTTCTTTGAAAAC
CC460: GTCGTTTTCCTAACTTAACCGTTAAAAAGA ATCACTATG
CC461: GTTAGGAAAACGACTTAACCATATTTTTGAATGATG TCAC
DHB420: *ATCAAATTAATACGACTCACTATA*GGAGCCGCTGAGC
DHB422: AGCTTGCATGCCTGCAG
DHB734: *TTAATACGACTCACTATA*GGGAGCTTAGAAATA CACAAG
DHB882: AAATAAACCCAAAAGAAAGACT
HP027: CCTTGACTGCTCCATCAGGAAATG
HP128: *AGAATTCTAATACGACTCACTATA*GGGAGATTAAG AAAGACACACG
HP560: ATTAATACGACTCACTATAGCTTCCATAGCTCAG CAGGTAG
HP564: *ATTAATACGACTCACTATA*GCGGAGCATTGCTTC CATAGCTC

Nonhybridizing sequences, including integrated T7 promoters, are in italics.

prewarmed in parallel with the RNase P RNA in a separate tube. The reaction is started by combining the substrate and holoenzyme and incubated at 37 °C. Four-microliter aliquots are withdrawn at the desired times and mixed with an equal volume of loading buffer (2.7 M urea, 66% w/v formamide, 0.33× TBE (0.03 M Tris base, 0.03 M boric acid, 0.67 mM EDTA), 0.05% bromophenol blue, 0.05% xylene cyanol) and applied directly to 5% polyacrylamide/7 M urea gels. The RNase P–cleaved tRNA^{Thr} product is 73 nt in length.

2.5. RNase J1

RNase J1 has the rather unique property of being able to act as both an endoribonuclease and a 5′ to 3′-exoribonuclease (Fig. 15.4). It plays an important role in the endonucleolytic processing of the *thrS* leader mRNA

during tRNA-mediated antitermination (Even *et al.*, 2005), and its 5′ to
3′-exoribonuclease activity is exploited in the maturation of the 5′-end of
16S rRNA and some mRNAs (Britton *et al.*, 2007; Mathy *et al.*, 2007).
Although RNase J1 is an essential enzyme for *B. subtilis* viability, its paralog,
RNase J2, which has similar enzymatic properties and substrate specificity,
is nonessential. All of the purification and assay protocols for RNase J1
described in the following also apply to RNase J2.

2.5.1. Purification of RNase J1

Native *B. subtilis* RNase J1 (61.3 kDa) is overexpressed from BL21 Codon-
Plus cells (Stratagene) containing plasmid pET28-BsubYkqC (strain
CCE112 [Britton *et al.*, 2007]). A 400-ml culture (2× YT + 0.5%
glucose + 25 μg/ml kanamycin) is grown to an OD_{600} of 0.6 at 37 °C,
and RNase J1 expression induced by the addition of 0.5 mM IPTG. The
culture is harvested after 3 h, pelleted, and stored frozen at −80 °C until
further use. Cells are resuspended in 10 ml buffer B (40 mM bis–Tris, pH
6.5, 4 mM MgCl$_2$, 1 mM DTT, and 10% v/v glycerol) containing 50 mM
NaCl and 10 μg/ml DNase I, and disrupted by two passages through a
French press (15,000 psi). This and all subsequent steps are performed at
4 °C. Cell debris is removed by centrifugation (27,000g for 30 min at 4 °C),
and the supernatant is loaded on a 10-ml (5 × 1.6 cm) hydroxyapatite

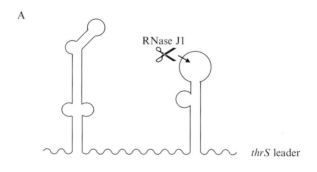

Figure 15.4 Endonucleolytic cleavage and 5′ to 3′-exonucleolytic degradation of
RNA by RNase J1. (A) Endonucleolytic cleavage (indicated by scissors symbol) of the
thrS leader RNA by RNase J1 in the loop of an antiterminator structure upstream of the
coding sequence. (B) 5′ to 3′-exonucleolytic maturation of the 5′-end of 16S rRNA by
RNase J1, indicated by a rightward-facing Pac-Man symbol.

(BioRad) column. The column is washed in buffer B containing 3 M NaCl and re-equilibrated in the same buffer containing 50 mM NaCl. A zero to 1 M K-phosphate, pH 6.8, gradient is applied in buffer containing 4 mM MgCl$_2$, 1 mM DTT, and 10% v/v glycerol. Fractions containing RNase J1, identified by 10% SDS-PAGE analysis, are pooled and dialyzed against buffer B containing 50 mM NaCl, and fractionated on a 5-ml HiTrap Heparin column (GE Healthcare) with a 50-mM to 2-M NaCl gradient in buffer B. RNase J1–containing fractions are pooled and loaded on a preparative Superdex 200 26/60 column (GE Healthcare) equilibrated in buffer B containing 150 mM NaCl. The enzyme is concentrated with 4 ml Amicon Ultra (30 kDa MWCO) filters (Millipore), glycerol added to 15%, and the enzyme is stored at $-80\,^{\circ}$C.

C-terminal His-tagged RNase J1 is overexpressed in BL21 CodonPlus cells containing plasmid pET28BsubYkqCHis (strain CCE093 [Mathy *et al.*, 2007]) and purified on 1 ml Cobalt Talon^{TM} resin (Clontech), according to "Standard Protocol 4." The peak fractions are pooled, dialyzed against 50 mM sodium phosphate buffer, pH 7.0, containing 300 mM NaCl and 10% glycerol, loaded on a 5-ml HiTrap Heparin column, and eluted with a gradient from 300 mM to 1.5 M NaCl in the same buffer. The peak fraction elutes at 600 mM NaCl and is dialyzed against 50 mM sodium phosphate buffer, pH 7.0, containing 300 mM NaCl and 25% glycerol. Typical concentrations are in the range of 1.5 to 2 mg/ml.

2.5.2. Assay of 5′ to 3′-exoribonuclease activity of RNase J1 *in vitro*

Essentially any RNA with an unprotected 5′-end (monophosphorylated and single stranded) can serve as a substrate for RNase J1 5′ to 3′-exonuclease activity *in vitro*. This activity was first described with a 5′-monophosphory-lated 280-nt fragment of *B. subtilis* 16S rRNA beginning 38 nt upstream of the mature 5′-end of the 16S sequence. The template for this so-called short +38 16S rRNA precursor is synthesized by PCR with the oligonucleotide pair CC169/CC170 (Table 15.3), where the upper primer has an integrated T7 RNA polymerase promoter sequence. The template for the PCR reaction is plasmid pUC19-*rrnW*16S, containing a cloned copy of the *B. subtilis rrnW* 16S gene (C. C., unpublished). Amplification from *B. subtilis* chromosomal DNA gives two products, differing by 64 nt, because of the heterogeneity in its *rrn* operons, but this mixed substrate also works well.

RNase J1 exonuclease activity is assayed in a 5-μl reaction volume containing 20 mM Tris-HCl, pH 8.0, 8 mM MgCl$_2$, 100 mM NH$_4$Cl, 0.1 mM DTT. The reaction buffer is prepared as a 5× mix (Table 15.1) that is filtered (0.22 μm) and stored frozen; 0.25 pmol of uniformly labeled 16S rRNA precursor is incubated with varying amounts of enzyme (0.06 to 6 μg, for example), for 30 min, or with a single enzyme concentration (e.g., 6 μg) for varying amounts of time (e.g., zero to 60 min). Reactions are incubated at 37 $^{\circ}$C and are stopped by addition of 5 μl loading buffer

(95% formamide, 20 mM EDTA, 0.05% bromophenol blue, 0.05% xylene cyanol). Degradation intermediates can be seen at lower temperatures, as low as 0 °C. Stopped reactions are loaded on 20% polyacrylamide/7 M urea gels. The mononucleotide reaction products migrate close to the bromophenol blue dye.

2.5.3. Assay of endoribonuclease activity of RNase J1 *in vitro*

The *thrS* leader region was the first identified substrate for the RNase J1 endonucleolytic cleavage reaction (Even *et al.*, 2005). This uniformly labeled substrate, bearing a 5′-triphosphate moiety, is transcribed *in vitro* with T7 RNA polymerase, as described in "Standard Protocol 1." The template for the transcription reaction is amplified by PCR from plasmid pHMS17 (Gendron *et al.*, 1994) with oligos HP128 and HP027 (Table 15.3), with the upper primer containing a T7 RNA polymerase promoter sequence. Transcription of this PCR fragment yields two products, 280 and 350 nt in length, corresponding to the prematurely terminated and runoff transcripts, respectively.

RNase J1 endonuclease assays are performed in a 10-μl reaction volume containing 20 mM HEPES-KOH, pH 8.0, 8 mM MgCl$_2$, 100 mM NaCl, 0.48 U/μl RNasin (Promega). The reaction buffer is prepared as a 5× mix (Table 15.1) that is filtered (0.22 μm) and stored frozen. Although the published buffers for the endo- and exo-activities of RNase J1 are slightly different, it is not yet known whether this significantly favors one activity over the other; 0.25 pmol labeled *thrS* leader transcript is incubated with 0.2 μg/μl RNase J1 for 20 min at 30 °C. Reactions are stopped by addition of 5 μl loading buffer and applied directly to 5% polyacrylamide/7 M urea gels. The cleavage products of the 350-nt runoff transcript are 245 and 105 nt in length.

2.6. RNase III

RNase III is best known for its role in ribosomal RNA maturation (Fig. 15.5). It is the founding member of the family of double-stranded specific ribonucleases that include Dicer and Drosha, known to play roles in the processing of miRNAs and siRNAs in eukaryotes (Drider and Condon, 2005). RNase III exists as a dimer. It can cleave on either the upstream or downstream side (or both) of double-stranded RNA helices, depending on features such as Watson–Crick base-pairing and bulges in certain positions (Calin-Jageman and Nicholson, 2003). Although RNase III is essential in *B. subtilis*, it is not essential in other firmicutes, or in *E. coli*.

2.6.1. Purification of RNase III

Native *B. subtilis* RNase III (28.4 kDa), even without a His-tag, can be purified from *E. coli* M15 (pREP4) cells (QIAGEN) carrying expression plasmid pQE60-*rncS* (Wang and Bechhofer, 1997), according to "Standard

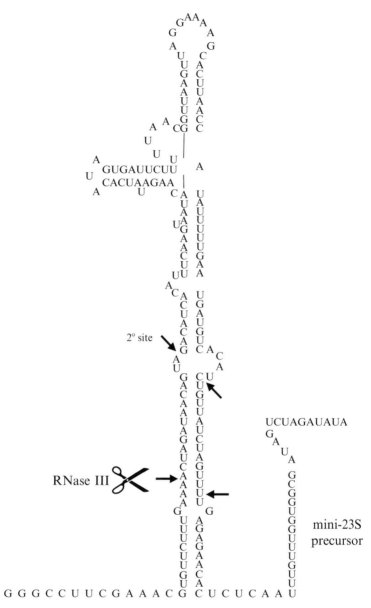

Figure 15.5 Cleavage of double-stranded RNA by RNase III. The structure shown is that of an artificial substrate mimicking the 23S rRNA processing stalk of the *B. subtilis* *rrnW* operon that we call the mini-23S precursor. RNase III cleaves at a primary site, indicated by the scissors symbol, and at a secondary site further up the stem, at higher enzyme concentrations.

Protocol 2" for His-tagged proteins. The protein apparently has a natural affinity for Ni-NTA agarose. Fractions containing eluted RNase III are determined by Bradford protein assay followed by SDS-PAGE analysis, pooled, and dialyzed against 200 ml storage buffer containing 20 mM Tris-HCl, pH 7.8, 500 mM NaCl, 50% glycerol, and 0.5 mM DTT for 4 h. Dialysis is continued in 500 ml storage buffer overnight. Aliquots are stored at $-80\,^{\circ}$C. The final yield is approximately 5 mg RNase III from 400 ml overexpressing culture.

2.6.2. Assay of RNase III activity *in vitro*

The assay involves the cleavage of a mini 23S precursor. A two-step PCR process synthesizes a template to produce an artificial substrate consisting of the 23S rRNA processing stalk with a 10-nt loop. PCR reactions with oligo pairs CC455/CC460 and CC461/CC046 (Table 15.3) are used to synthesize overlapping upstream and downstream halves of the *B. subtilis* *rrnW* processing stalk, respectively. The two overlapping PCR fragments are purified, and 5 ng is used to perform a third PCR reaction with the oligo pair CC455/CC046. The resulting PCR fragment is used as a template to synthesize uniformly labeled 23S minisubstrate (197 nt) with "Standard Protocol 1." RNase III assays are typically performed in a 5-μl reaction volume containing 10 mM Tris-HCl, pH 7.8, 3 mM MgCl$_2$, 100 mM NaCl, and 0.25 pmol labeled substrate RNA. The reaction buffer is prepared as a 5$\times$ mix (Table 15.1), filtered (0.22 μm), and stored frozen. Reactions containing from 4 to 400 ng RNase III are incubated at 37 $^{\circ}$C for 30 min, stopped with loading buffer, and applied to a 5% sequencing gel. The primary cleavage products are 126 nt (mature species), 26 nt (5′-extension), and 45 nt (3′-extension). At high concentrations of RNase III, cleavage of the mini-23S substrate occurs at a secondary site higher up on the processing stalk to produce a shorter "mature" species of 96 nt.

2.7. YhcR

YhcR is a Ca^{2+}-activated low-specificity endonuclease, with a similar mode of action to micrococcal nuclease (Oussenko *et al.*, 2004). It also has residual activity in the presence of Mn^{2+} ions. YhcR can act on both RNA and DNA with similar efficiency to generate low molecular-weight products, primarily nucleotide 3′-monophosphates (Fig. 15.6). It is one of the largest proteins in *B. subtilis* (1217 amino acids) and seems to be primarily attached to the cell wall through a C-terminal anchoring domain; it may, thus, function in the digestion and acquisition of extracellular nucleic acid.

2.7.1. Purification of YhcR

Native YhcR (132.4 kDa) was originally purified from *B. subtilis* strains lacking three exoribonucleases (PNPase, RNase R, and YhaM), through multiple chromatography steps, including ammonium sulfate precipitation, DEAE Sepharose, Affi-Gel Blue, gel filtration, and Mono-Q columns (Oussenko *et al.*, 2004). Now that the gene has been identified, however, it can be overexpressed and purified in *E. coli* as a C-terminal His-tagged derivative lacking its N-terminal signal sequence (full-length YhcR is apparently toxic to *E. coli*). A smaller derivative, from codons 36 to 529, also retains enzyme activity, albeit lower than full-length YhcR. His-tagged versions of YhcR are purified on Ni-NTA agarose columns under denaturing conditions, such as described in "Standard Protocol 3." Further purification of the His-tagged proteins is achieved by dialysis and precipitation with acetone, followed by preparative SDS-PAGE, blotting to a PVDF membrane, and elution. Purified YhcR is aliquoted and stored at $-80\,^{\circ}C$ in 25 mM Tris-HCl, pH 8.7, 5 mM β-mercaptoethanol, 0.5% Triton X-100, and 25% glycerol.

2.7.2. Assay of YhcR activity *in vitro*

Any uniformly labeled RNA can serve as a substrate for YhcR. YhcR activity was first described on 110- and 187-nt RNA fragments derived from SP82 phage (see Fig. 15.9) that have been described previously (Mitra *et al.*, 1996; Oussenko and Bechhofer, 2000). The standard YhcR reaction mixture contains 50 mM Tris or 50 mM Tricine, pH 8.0, 100 mM KCl, 1 mM MnCl$_2$, 0.2 to 0.5 pmol (2 to 4×10^5 cpm/pmol) of labeled substrate, and approximately 0.05 μg of YhcR protein. The reaction buffer is prepared as a 5$\times$ mix (Table 15.1), filtered (0.22 μm), and stored frozen. The reaction mixture is incubated for 20 min at 37 $^{\circ}C$ and then extracted with an equal volume of phenol-chloroform. Ten microliters of the aqueous phase is mixed with gel loading buffer and run in a 20% polyacrylamide/ 7 M urea gel to resolve the products, principally mononucleotides, with some small oligomers. Unlike YhaM and RNase R, whose activity will produce only one type of labeled mononucleotide (for example, [^{32}P]-UMP if substrate RNA was labeled with α-[^{32}P]-UTP), YhcR

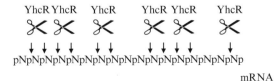

Figure 15.6 Endonucleolytic cleavage of RNA to mononucleotides and small oligonucleotide products by YhcR. Cleavage events are indicated by scissors symbols. The cleavage products bear 3′-monophosphate groups.

activity will result in a mixture of all four labeled mononucleotides, and this can be tested by thin-layer chromatography.

3. PURIFICATION AND ASSAY OF *B. SUBTILIS* EXORIBONUCLEASES

3.1. PNPase

Polynucleotide phosphorylase was the first bacterial ribonuclease to be discovered, although it was initially identified in 1955 for its capacity to synthesize RNA rather than degrade it (Grunberg-Manago *et al.*, 1955). PNPase and RNase PH are the two phosphorolytic 3′ to 5′-exoribonucleases commonly found in bacteria. PNPase is thought to be one of the principal enzymes of RNA turnover in *B. subtilis*, where up to 80% of RNA decay is phosphorolytic compared with just 30% in *E. coli* (Chaney and Boyer, 1972; Deutscher and Reuven, 1991; Duffy *et al.*, 1972).

3.1.1. Purification of PNPase

PNPase (77.3 kDa) is purified as a C-terminal His-tagged protein from 1 L *E. coli* BL21 (DE3) pLysS cells containing plasmid pNP21 (Deikus and Bechhofer, 2007; Wei *et al.*, 2006) grown in 2× YT medium to an OD_{600} of 0.6, followed by induction by addition of IPTG for 30 min. Cells are lysed in buffer containing 20 m*M* HEPES, pH 7.2, 0.1 *M* KCl, 5 m*M* $MgCl_2$, 10% glycerol, 0.1% Nonidet P-40, 100 U/ml DNase I (Grade II; Roche), 125 μl of protein inhibitor solution made by dissolving 1 tablet of Roche EDTA-free protease inhibitor in 500 μl H_2O, 1 m*M* PMSF, and 0.3 mg/ml lysozyme. His-tagged PNPase is applied to a HisTrap column (GE Healthcare) in binding buffer containing 20 m*M* HEPES, pH 8.4, 1 *M* KCl, 30 m*M* imidazole, and 1 m*M* PMSF and eluted in the same buffer containing 200 m*M* imidazole. HisTrap column elution buffer is 20 m*M* HEPES, pH 8.4, 1 *M* KCl, 200 m*M* imidazole, and 1 m*M* PMSF. The His-tag is removed from PNPase with biotinylated thrombin (Novagen) as follows: 2 U of biotinylated thrombin is added to 1 mg His-tagged PNPase in 5 ml thrombin cleavage buffer (20 m*M* Tris-HCl, pH 8.4, 0.15 *M* NaCl, and 2.5 m*M* $CaCl_2$) and incubated overnight at 4 °C. Thirty-two microliters of streptavidin agarose is added to the thrombin cleavage reaction and incubated for 30 min with gentle shaking. The reaction is transferred to the sample cup of the spin filter provided by Novagen and centrifuged in a microcentrifuge at 2000 rpm for 5 min. The filtrate contains the cleaved protein free of biotinylated thrombin. The protein is dialyzed against HisTrap binding buffer and reapplied to the HisTrap column. The flow-through material is collected, concentrated, and stored at 1 mg/ml in 50 m*M* Tris-HCl, pH 7.5, 100 m*M* KCl, 0.5 m*M* DTT, and 50% glycerol at −20 °C.

3.1.2. Assay of PNPase activity *in vitro*

PNPase can be assayed on any uniformly labeled and relatively unstructured RNA (Fig. 15.7). A 139-nt RNA fragment corresponding to the leader region of the *B. subtilis trp* operon has been used in recent studies (Deikus and Bechhofer, 2007). The template for synthesis of this fragment is amplified by PCR from *B. subtilis* chromosomal DNA with the oligonucleotide pair DHB734/DHB882 (Table 15.3). The T7 RNA polymerase promoter is contained in the upstream oligo. The first nucleotide in DHB882 is complementary to nt 139 of *trp* leader RNA. Uniformly labeled *trp* leader is synthesized *in vitro* by use of "Standard Protocol 1."

PNPase reactions are performed in two steps. In the first step, 2.5 n*M* PNPase is bound to 0.05 n*M* labeled substrate in 25 m*M* Tris-HCl, pH 8.0, 50 m*M* KCl, 1 m*M* DTT, 0.2 m*M* EDTA, and 2.5% glycerol in a 20-μl reaction volume. The prebinding buffer can be prepared as a 5× mix (Table 15.2), filtered (0.22 μm), and stored frozen. The phosphorolytic degradation reaction is initiated by addition of MgCl$_2$ to 1 m*M* and Na$_2$HPO$_4$ to 2 m*M*. Reactions are stopped by addition of an equal volume of gel loading buffer, and an aliquot is applied directly to polyacrylamide/ 7 *M* urea gels; 20% gels are used to see mononucleotide-diphosphate products; 6% gels are used to see degradation intermediates in the 40- to 150-nt range.

3.2. RNase PH

RNase PH plays a key role in tRNA (Fig. 15.8) and scRNA maturation in *B. subtilis* (Wen *et al.*, 2005; Yao *et al.*, 2007). It is the second of the two known phosphorolytic 3′ to 5′-exoribonucleases in *B. subtilis*. RNase PH is able to form homo-hexamers, and this is the foundation for assembly of the eukaryotic exosome, which contains a ring structure consisting of six different RNase PH homologs.

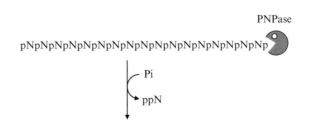

Figure 15.7 Phosphorolytic degradation of RNA by PNPase. 3′ to 5′-Exonucleolytic degradation by PNPase is indicated by a leftward-facing Pac-Man symbol. Each reaction cycle consumes one molecule of inorganic phosphate and generates a 5′-diphosphate nucleoside product.

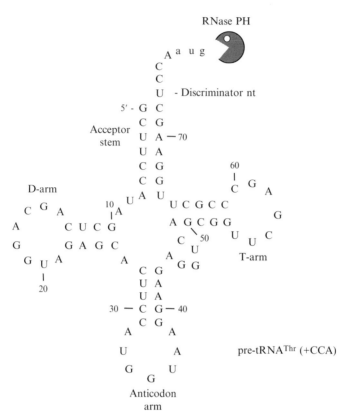

Figure 15.8 Maturation of tRNA^{Thr} (+CCA) precursor by RNase PH. RNase PH is shown as a leftward-facing Pac-Man symbol to indicate its 3′ to 5′-exonucleolytic progression. The mature tRNA species is shown in uppercase letters; sequences removed by RNase PH in lowercase. Removal of the nucleotide immediately 3′ to the CCA sequence is a slow step and may be catalyzed by another enzyme *in vivo*.

3.2.1. Purification of RNase PH

C-terminal His-tagged RNase PH is purified from a 400-ml culture (2× YT + 0.5% glucose) of BL21-CodonPlus λ(DE3)-RIL (Stratagene) carrying pET28-Rph according to "Standard Protocol 2(b)" (i.e., elution from Ni-NTA agarose by pH gradient). Fractions containing RNase PH (monomer mass, 26.5 kDa) are identified by 10% SDS-PAGE and dialyzed against 25 m*M* Tris-HCl, pH 7.5, 100 m*M* NaCl, 0.5 m*M* EDTA, 1 m*M* DTT, 10% glycerol. RNase PH rapidly loses activity with repeated freezing and thawing. It is, therefore, advisable to prepare single-use aliquots for storage at −80 °C. Alternatively, the protein can be stored at −20 °C in storage buffer containing 50% glycerol.

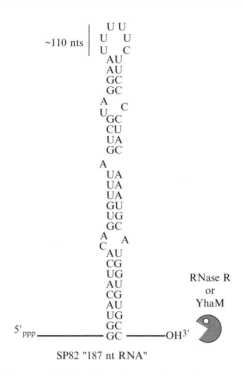

SP82 "187 nt RNA"

Figure 15.9 3′ to 5′-Exonucleolytic degradation of RNA by RNase R and YhaM. The RNA substrate shown is the "187-nt RNA" derived from SP82. The approximate 3′-end of the 110-nt premature termination product is indicated. RNase R and YhaM are shown as leftward-facing Pac-Man symbols.

3.2.2. Assay of RNase PH

Uniformly labeled (α-[^{32}P]-UTP) CCA-containing precursor tRNAs are prepared by *in vitro* transcription of PCR templates, with the upper primer containing a T7 RNA polymerase promoter sequence. The CCA moiety can be included in the lower primer if the tRNA being assayed does not naturally possess one. For example, the template for a CCA-containing *B. subtilis* *tmI- Thr* precursor with a 3-nt 3′-trailer sequence is made by PCR amplification of plasmid pHMT1 (Luo *et al.*, 1998) with oligos HP560 and CC107 (Table 15.3). The T7 RNA polymerase *in vitro* transcription reactions are carried out according to "Standard Protocol 1."

RNase PH assays are performed in a 5-μl reaction volume with 6 to 8 ng/μl (200 to 300 nM) labeled precursor RNA and 60 ng/μl protein (38 nM hexamer) in 50 mM Tris-HCl, pH 8.0, 2.5 mM MgCl$_2$, 30 mM NaCl, 10 mM K$_2$HPO$_4$, 0.1 mM DTT at 37 °C. The reaction buffer is prepared as a 5× mix (Table 15.2), filtered (0.22 μm), and stored frozen. Reactions are stopped by addition of 5 μl loading buffer (95% formamide, 20 mM EDTA, 0.05% bromophenol blue, 0.05% xylene cyanol) and run on

5% polyacrylamide/7 M urea gels. RNase PH-matured tRNA^{Thr} is 76 nt in length.

3.3. RNase R

RNase R is a hydrolytic $3'$ to $5'$-exoribonuclease that plays an important role in the degradation of structured RNA (Oussenko and Bechhofer, 2000). It is particularly active on 23S rRNA. It is the major Mg^{2+}-dependent exonuclease in *B. subtilis* and also functions, albeit less efficiently, in the presence of Mn^{2+} ions.

3.3.1. Purification of RNase R

B. subtilis RNase R (88.5 kDa) has not yet been successfully cloned and overexpressed in *E. coli*, presumably for reasons of toxicity. It can be partially purified from *B. subtilis* extracts lacking PNPase (strain BG119) as follows (Oussenko and Bechhofer, 2000). Cells are grown to an OD_{600} of approximately 0.8 to 1.0 in YT medium, centrifuged, washed in buffer A (20 mM Tris-HCl, pH 7.8, 100 mM KCl, 0.2 mM EDTA, 10% glycerol), and recentrifuged. The pellet can be stored frozen at $-80\,°C$. The cell pellet is thawed on ice and resuspended in buffer A (without glycerol) containing 1 mM PMSF and 1 mM DTT at 0.1 g of cells per ml. Lysozyme is added to 0.2 mg/ml and incubated at 37 °C for 20 to 25 min with occasional swirling until the cell suspension clarifies and becomes viscous. The chilled lysate is passed through a French press at 16,000 psi, and the cell debris is removed by centrifugation at 12,000g for 15 min at 4 °C. The supernatant is pelleted at 200,000g for 2 h at 4 °C. The supernatant is removed, and nucleic acids are precipitated by the addition of 130 mg/ml streptomycin sulfate with gentle stirring at 4 °C, followed by centrifugation at 27,000g. The nucleic acid pellet, which is poorly soluble, is dissolved in buffer A, dialyzed overnight against the same buffer, and diluted to 0.2 mg nucleic acid/ml. Nucleic acids are digested by addition of micrococcal nuclease (Worthington Biochemical) to 1.5 U/μg nucleic acid and incubation at 30 °C for 20 min in the presence of 1 mM CaCl$_2$. The sample is dialyzed against buffer A containing 1 mM CaCl$_2$, filtered (0.45 μm), and loaded on a DEAE-Sepharose CL-6B (Sigma) column, equilibrated in buffer A. RNase R is eluted with a linear gradient of 100 to 500 mM KCl in the same buffer. Fractions are assayed for RNase activity with [^{3}H]-labeled total RNA. [^{3}H]-labeled RNAs are synthesis by *in vitro* transcription using [^{3}H]-labeled ATP. Active fractions are adjusted to 30% glycerol and stored frozen at $-80\,°C$.

3.3.2. Assay of RNase R activity *in vitro*

Any uniformly labeled RNA, with a relatively unstructured region at its $3'$-end, can be used as a substrate for RNase R activity *in vitro*. The enzyme was initially characterized on the so-called "187-nt RNA" shown in

Fig. 15.9. This 187–nt RNA is generated by transcription of a PCR fragment generated on EG242 replicative form DNA with oligo pairs DHB420 and DHB422 (Table 15.3). Reactions (100 μl) are performed in 50 mM Na-Tricine, pH 8.0, 100 mM KCl, and 1 mM MgCl$_2$ containing 10 to 40 nmol substrate RNA and 0.2 μg protein. The reaction buffer can be prepared as a 5$\times$ mix (Table 15.2) that is filtered (0.22 μm) and stored frozen. Reactions are assembled on ice and incubated for 20 min at 37 °C. For [^{3}H]-labeled substrates, reactions are stopped by addition of 300 μl 0.5 mg/ml *E. coli* tRNA and 400 μl 20% TCA. Samples are incubated on ice for 10 min and then centrifuged in the cold for 10 min; 400 μl of supernatant is removed and added to 5 ml scintillation fluid for counting. Uniformly labeled (α-[^{32}P]-UTP) substrates can be assayed in 5 μl, the reaction stopped by addition of an equal volume of loading buffer (95% formamide, 20 mM EDTA, 0.05% bromophenol blue, 0.05% xylene cyanol), and the mixture loaded directly in a 20% polyacrylamide/7 M urea gel to resolve the mononucleotide products.

3.4. YhaM

YhaM is a Mn^{2+}-dependent hydrolytic 3' to 5'-exonuclease that functions on both single-stranded RNA and DNA (Oussenko *et al.*, 2002). It shows some activity in the presence of Co^{2+} (and Cd^{2+}; I. O. and D. H. B., unpublished) but is inactive in the presence of Mg^{2+}. YhaM can substitute for other exoribonucleases in the maturation of tRNA and in the degradation of mRNAs (Oussenko *et al.*, 2005; Wen *et al.*, 2005). Its principal function may be as a DNase, involved in DNA replication (Gennaro and Novick, 1986; Noirot-Gros *et al.*, 2002).

3.4.1. Purification of YhaM

YhaM (35.5 kDa) was originally purified from a *B. subtilis* strain lacking PNPase and RNase R, by selective ammonium sulfate precipitation, DEAE-Sepharose, Affi-Gel blue, and MonoQ column chromatography, followed by transfer and elution from PVDF membranes (Oussenko *et al.*, 2002). Its gene was subsequently cloned in pQE60 (Qiagen), permitting its overexpression and purification in *E. coli* as a C-terminal His-tagged protein. The protein is purified under denaturing conditions, by use of a procedure such as described in "Standard Protocol 3." The protein is stored at 4 °C in 8 M urea, 0.1 M Na$_2$HPO$_4$, 0.01 M Tris-HCl, pH 4.0, and diluted 1:10 in 25 mM Tris-HCl, pH 7.8, 25% glycerol before assaying.

3.4.2. Assay of YhaM activity *in vitro*

YhaM can be assayed on essentially any uniformly labeled and relatively unstructured RNA or single-stranded DNA. Its activity was originally assayed on the labeled 187-nt RNA (Fig. 15.9) and on a 3'-truncated

110-nt RNA, which is a premature termination product that is observed in high yield during the 187-nt RNA transcription reaction (Oussenko and Bechhofer, 2000). The standard (100 μl) YhaM reaction mixture contains 50 mM Na-Tricine, pH 8.0, 100 mM KCl, 1 mM MnCl$_2$, and 0.2 to 0.5 pmol labeled substrate (2 to 4 $\times$ 10^5 cpm/pmol). The reaction buffer can be prepared as a 5$\times$ mix (Table 15.2) that is filtered (0.22 μm), and stored frozen. Reactions are incubated for 20 min at 37 °C and extracted with an equal volume of phenol-chloroform. Ten microliters of the aqueous phase is mixed with 10 μl loading buffer (95% formamide, 20 mM EDTA, 0.05% bromophenol blue, 0.05% xylene cyanol) and applied to a 20% polyacrylamide/7 M urea gel to resolve mononucleotides.

3.5. Nano-RNase

Nano-RNase (34.9 kDa), encoded by the *nrnA* gene (formerly *ytqI*), was recently identified in *B. subtilis* as an enzyme with both oligoribonuclease and pAp-phosphatase activity (Mechold *et al.*, 2007). It can complement the essential function of the oligoribonuclease gene (*orn*) of *E. coli* and is capable of degrading small RNAs (<5 nts) *in vitro*, with a preference for 3-mers. It shows no activity on longer RNAs in the 20-nt range.

3.5.1. Purification of Nano-RNase

C-terminally His-tagged Nano-RNase can be purified from 200 ml of MG1655 cultures containing plasmid pUM412 (Mechold *et al.*, 2007) by use of "Standard Protocol 2a." The protein is dialyzed against 50 mM HEPES, pH 7.5, 150 mM NaCl, and stored at 4 °C. The yield is approximately 4 mg Nano-RNase from 200 ml overproducing culture.

3.5.2. Assay of Nano-RNase activity *in vitro*

Nano-RNase is assayed with commercially synthesized fluorescent RNA oligo 5-mers or 3-mers. This activity was first described on Cy5-CCCCC and Cy5-CCC oligos, for example. Reactions are performed in 50 mM HEPES, pH 7.5, and 5 mM MnCl$_2$. The reaction buffer can be prepared as a 5$\times$ mix (Table 15.2), filtered (0.22 μm), and stored frozen. Reactions are performed with 1.6 to 3.4 μM substrate and 0.1 to 0.3 μg/μl Nano-RNase at 37 °C. At the desired time intervals, 4.5-μl aliquots are withdrawn to microcentrifuge tubes containing an equal volume of 4$\times$ TBE (0.36 M Tris base, 0.36 M boric acid, 8 mM EDTA), 100 mM DTT, 16% glycerol, 20 mM EDTA, and frozen at −20 °C; 1.5- or 2.5-μl aliquots are run in a 22% polyacrylamide gel (no urea) containing 2$\times$ TBE with a 2$\times$ TBE running buffer. Fluorescent oligos are visualized with a Molecular Dynamics STORM apparatus in the 650-nm long-pass filter mode. *Note:* Because the Cy5 moiety has a lower net negative charge than RNA, removal of nucleotides will decrease the charge relative to the mass, causing the shortened oligo to migrate more slowly than the substrate!

4. CONCLUSION

We have presented here the protocols to purify and assay seven endo- and five exoribonucleases of *B. subtilis*. For the time being, these assays have primarily served to perform initial characterizations of properties of the different enzymes in question. Future exploitation of these assays will undoubtedly include detailed biochemical analyses of enzyme affinities and catalytic constants. For some of the essential RNases of *B. subtilis*, an evolution of these assays from radioactivity-based tests toward fluorescence-based alternatives would be an important first step toward the possibility of performing high-throughput screens for inhibitors and, ultimately, antimicrobial compounds.

ACKNOWLEDGMENTS

We thank R. Hartmann, U. Mechold, and A. Gogos for providing details of the protocols for the isolation and assay of RNase P, Nano-RNase, and EndoA, respectively. C. C. was supported by funds from the CNRS (UPR 9073), the Ministère de l'Education Nationale, and the Agence Nationale de la Recherche (ANR).

APPENDICES

Standard Protocol 1: T7 *In Vitro* Transcription Reactions

We synthesize uniformly [^{32}P]-labeled RNAs with either PCR (0.5 μg) or linearized plasmid templates (1 μg) bearing T7 RNA polymerase promoters. A typical 25-μl *in vitro* transcription reaction will contain 5 μl of 5× transcription buffer (Promega), 2.5 μl of 100 mM DTT, 1 μl of 40 U/μl RNasin (Promega), 5 μl of 5× label mix (2.5 mM ATP, CTP, GTP, and 60 μM UTP), 5 μl α-[^{32}P]-UTP (3000 Ci/mmol), template DNA, 2 μl of 20 U/μl T7 RNA polymerase (Promega), and nuclease-free water to 25 μl. The 5× reaction buffer should be diluted with the other components of the reaction before addition of the template DNA to avoid precipitation by spermidine in the reaction buffer. For RNAs longer than approximately 400 nt, we use 2.5 mM UTP in the label mix. Reactions are incubated at 37 °C for 1 to 3 h, depending on the length of the RNA to be synthesized. Two μl RQ DNase (Promega) is added and incubation at 37 °C continued for 15 min. RNAs are separated from unincorporated nucleotides by purification on G50 spin columns (GE Healthcare). This step is critical if mononucleotide production by exoribonucleases is being assayed.

T7 RNA polymerase-transcribed RNAs are generally initiated with between one and three G-residues, with the first nucleotide bearing a 5′-triphosphate group. RNAs bearing 5′-monophosphate or 5′-hydroxyl groups can be synthesized by adding an eightfold excess of GMP or guanosine, respectively, to the *in vitro* transcription reaction from 50 or 100 mM stock solutions (i.e., 4-mM final concentration). This procedure does not seem to significantly alter the RNA yield.

Standard Protocol 2: Native Purification of His-Tagged Proteins on Ni-NTA Agarose

(a) Elution by imidazole

400 ml ($2 \times$ YT + 0.5% glucose) of culture containing the overexpression plasmid are grown to an OD_{600} of 0.6 at 37 °C, and protein expression is induced by the addition of 0.5 mM IPTG. The culture is harvested after 3 h, pelleted, and frozen at −80 °C until further use. The frozen cells are resuspended in 10 ml of buffer containing 20 mM Tris-HCl, pH 7.8, 0.5 M NaCl, 10% glycerol, and 0.1% Triton X-100 at 4 °C, to which is added 10 µg/ml DNase I and an EDTA-free anti-protease tablet (Roche). The suspension is passed twice through a French press (20,000 psi) and the lysate centrifuged for 30 min at 15,000g. Imidazole-HCl (pH 7.8) is added to the supernatant to give a final concentration of 1 mM. The resulting mix is applied to 1 ml Ni^{2+}-NTA resin (Qiagen) in a small column (or 1-ml syringe with a Whatman paper filter), attached to a peristaltic pump, and previously equilibrated with approximately 15 ml cell resuspension buffer at 10 ml/h. The resin is then washed with 10 ml 20 mM Tris-HCl, pH 7.8, 0.3 M NaCl, 20 mM imidazole, followed by 10 ml 50 mM sodium phosphate, pH 6.0, 0.3 M NaCl at 4 °C. The His-tagged protein is eluted in 10 ml buffer containing 20 mM Tris-HCl, pH 7.8, 0.3 M NaCl, and 250 mM imidazole, and 1-ml fractions are collected. Fractions are analyzed by SDS-PAGE for protein purity, and peak fractions are pooled and dialyzed against the relevant storage buffer.

(b) Elution by pH gradient

The protocol for elution of His-tagged proteins by pH gradient is identical to that described in the previous section, except that elution is by a 15-ml gradient of 50 mM sodium phosphate pH 6.0, 0.3 M NaCl, to 50 mM sodium acetate, pH 4.0, 0.3 M NaCl, with a standard gradient maker.

Standard Protocol 3: Purification of His-Tagged Proteins under Denaturing Conditions on Ni-NTA Agarose

This protocol is used for proteins that form inclusion bodies or for soluble proteins whose His-tag is not accessible for interaction with Ni–NTA agarose under native conditions. In the first case, most of the overproduced protein is found in the pellet of cellular debris after the first centrifugation step after cell lysis. In the second case, the overproduced protein remains in the supernatant after the first centrifugation step but is found primarily in the flowthrough and wash fractions after application to the Ni–NTA agarose column.

The culture and cell lysis conditions for His-tagged proteins to be purified under denaturing conditions are identical to those for native proteins, described in Standard Protocol 2. Proteins that form inclusion bodies are solubilized from the pellet of cell debris in 5 ml 6 M guanidine-HCl, 0.1 M NaH$_2$PO$_4$, 10 mM Tris-HCl adjusted to pH 8.0 with NaOH at room temperature. Cell debris is recentrifuged at 10,000 rpm for 20 min at room temperature. The supernatant is absorbed to 1 ml Ni–NTA agarose resin in batch conditions for at least 1 h and then applied to a small column attached to a peristaltic pump. The column is washed successively in 10 ml 6 M guanidine-HCl, 0.1 M NaH$_2$PO$_4$, 10 mM Tris-HCl, pH 8.0; 10 ml 8 M urea, 0.1 M NaH$_2$PO$_4$, 10 mM Tris-HCl adjusted to pH 8.0 with NaOH; 10 ml 8 M urea, 0.1 M NaH$_2$PO$_4$, 10 mM Tris-HCl adjusted to pH 6.3 with HCl; and finally eluted in 8 M urea, 0.1 M NaH$_2$PO$_4$, 10 mM Tris-HCl, 250 mM imidazole adjusted to pH 8.0 with NaOH. The protein-containing fractions are renatured by dialyzing against relevant storage buffer overnight at 4 °C.

Soluble proteins whose His-tag is not available for interaction with Ni–NTA agarose are denatured by dialyzing the first-spin supernatant against 8 M urea, 0.1 M NaH$_2$PO$_4$, 10 mM Tris-HCl adjusted to pH 8.0 with NaOH. The dialyzed sample is absorbed to 1 ml Ni–NTA agarose resin in batch conditions for at least 1 h and then applied to a small column attached to a peristaltic pump. The column is washed successively in 10 ml 8 M urea, 0.1 M NaH$_2$PO$_4$, 10 mM Tris-HCl adjusted to pH 8.0 with NaOH; 10 ml 8 M urea, 0.1 M NaH$_2$PO$_4$, 10 mM Tris-HCl adjusted to pH 6.3 with HCl; and finally eluted in 8 M urea, 0.1 M NaH$_2$PO$_4$, 10 mM Tris-HCl, 250 mM imidazole adjusted to pH 8.0 with NaOH. The protein-containing fractions are renatured by dialyzing against relevant storage buffer overnight at 4 °C.

Standard Protocol 4: Native Purification of His-Tagged Proteins on BD-Talon

Cobalt-Talon resin (Clontech) has been used to purify His-tagged proteins to a higher degree of purity than Ni-NTA agarose in some cases. The culture conditions for His-tagged proteins to be purified with BD-Talon resin are identical to those for Ni-NTA agarose, described in Standard Protocol 2. The cell pellet is lysed in 10 ml buffer containing 50 mM NaH$_2$PO$_4$, 300 mM NaCl, 10% glycerol, 0.1% Triton-X100 at 4 °C, to which is added 10 μg/ml DNase I and an EDTA-free anti-protease tablet (Roche). The suspension is passed twice through a French press (20,000 psi) and the lysate centrifuged for 30 min at 15,000g. Imidazole-HCl, pH 7.8, is added to the supernatant to give a final concentration of 1 mM. The resulting mix is applied to 1 ml BD-Talon resin in a small column (or 1-ml syringe), attached to a peristaltic pump, and previously equilibrated with approximately 15 ml cell resuspension buffer at 10 ml/h. The resin is then washed with 10 ml 50 mM NaH$_2$PO$_4$, 300 mM NaCl, 20 mM imidazole, and eluted in 10 ml buffer containing 50 mM NaH$_2$PO$_4$, 300 mM NaCl, 150 mM imidazole. One-milliliter fractions are collected. Fractions are analyzed by SDS-PAGE for protein purity, and peak fractions are pooled and dialyzed against the relevant storage buffer.

REFERENCES

Allemand, F., Mathy, N., Brechemier-Baey, D., and Condon, C. (2005). The 5S rRNA maturase, ribonuclease M5, is a Toprim domain family member. *Nucleic Acids Res.* **33,** 4368–4376.

Britton, R. A., Wen, T., Schaefer, L., Pellegrini, O., Uicker, W. C., Mathy, N., Tobin, C., Daou, R., Szyk, J., and Condon, C. (2007). Maturation of the 5′-end of *Bacillus subtilis* 16S rRNA by the essential ribonuclease YkqC/RNase J1. *Mol. Microbiol.* **63,** 127–138.

Calin-Jageman, I., and Nicholson, A. W. (2003). Mutational analysis of an RNA internal loop as a reactivity epitope for *Escherichia coli* ribonuclease III substrates. *Biochemistry* **42,** 5025–5034.

Chaney, S. G., and Boyer, P. D. (1972). Incorporation of water oxygens into intracellular nucleotides and RNA. II. Predominantly hydrolytic RNA turnover in *Escherichia coli.* *J. Mol. Biol.* **64,** 581–591.

Condon, C. (2003). RNA processing and degradation in *Bacillus subtilis*. *Microbiol. Mol. Biol. Rev.* **67,** 157–174.

Condon, C. (2009). RNA processing in bacteria. *In* "The Encyclopedia of Microbiology," (M. Schaecter, ed.), Elsevier. In Press.

Condon, C., Brechemier-Baey, D., Beltchev, B., Grunberg-Manago, M., and Putzer, H. (2001). Identification of the gene encoding the 5S ribosomal RNA maturase in *Bacillus subtilis*: Mature 5S rRNA is dispensable for ribosome function. *RNA* **7,** 242–253.

Deikus, G., and Bechhofer, D. H. (2007). Initiation of decay of *Bacillus subtilis trp* leader RNA. *J. Biol. Chem.* **282,** 20238–20244.

Deutscher, M. P., and Reuven, N. B. (1991). Enzymatic basis for hydrolytic versus phosphorolytic mRNA degradation in *Escherichia coli* and *Bacillus subtilis*. *Proc. Natl. Acad. Sci. USA* **88,** 3277–3280.

Drider, D., and Condon, C. (2005). The continuing story of endoribonuclease III. *J. Mol. Microbiol. Biotechnol.* **8,** 195–200.

Duffy, J. J., Chaney, S. G., and Boyer, P. D. (1972). Incorporation of water oxygens into intracellular nucleotides and RNA. I. Predominantly non-hydrolytic RNA turnover in *Bacillus subtilis*. *J. Mol. Biol.* **64,** 565–579.

Even, S., Pellegrini, O., Zig, L., Labas, V., Vinh, J., Brechemmier-Baey, D., and Putzer, H. (2005). Ribonucleases J1 and J2: Two novel endoribonucleases in *B. subtilis* with functional homology to *E. coli* RNase E. *Nucleic Acids Res.* **33**, 2141–2152.

Gendron, N., Putzer, H., and Grunberg-Manago, M. (1994). Expression of both *Bacillus subtilis* threonyl-tRNA synthetase genes is autogenously regulated. *J. Bacteriol.* **176**, 486–494.

Gennaro, M. L., and Novick, R. P. (1986). cmp, a *cis*-acting plasmid locus that increases interaction between replication origin and initiator protein. *J. Bacteriol.* **168**, 160–166.

Gogos, A., Mu, H., Bahna, F., Gomez, C. A., and Shapiro, L. (2003). Crystal structure of YdcE protein from *Bacillus subtilis*. *Proteins* **53**, 320–322.

Gossringer, M., Kretschmer-Kazemi Far, R., and Hartmann, R. K. (2006). Analysis of RNase P protein (*rnpA*) expression in *Bacillus subtilis* utilizing strains with suppressible *rnpA* expression. *J. Bacteriol.* **188**, 6816–6823.

Grunberg-Manago, M., Oritz, P. J., and Ochoa, S. (1955). Enzymatic synthesis of nucleic acidlike polynucleotides. *Science* **122**, 907–910.

Guerrier-Takada, C., Gardiner, K., Marsh, T., Pace, N., and Altman, S. (1983). The RNA moiety of ribonuclease P is the catalytic subunit of the enzyme. *Cell* **35**, 849–857.

Luo, D., Condon, C., Grunberg-Manago, M., and Putzer, H. (1998). *In vitro* and *in vivo* secondary structure probing of the *thrS* leader in *Bacillus subtilis*. *Nucleic Acids Res.* **26**, 5379–5387.

Mathy, N., Benard, L., Pellegrini, O., Daou, R., Wen, T., and Condon, C. (2007). 5′ to 3′-exoribonuclease activity in bacteria: Role of RNase J1 in rRNA maturation and 5′-stability of mRNA. *Cell* **129**, 681–692.

Mechold, U., Fang, G., Ngo, S., Ogryzko, V., and Danchin, A. (2007). YtqI from *Bacillus subtilis* has both oligoribonuclease and pAp-phosphatase activity. *Nucleic Acids Res.* **35**, 4552–4561.

Mitra, S., Hue, K., and Bechhofer, D. H. (1996). *In vitro* processing activity of *Bacillus subtilis* polynucleotide phosphorylase. *Mol. Microbiol.* **19**, 329–342.

Noirot-Gros, M. F., Dervyn, E., Wu, L. J., Mervelet, P., Errington, J., Ehrlich, S. D., and Noirot, P. (2002). An expanded view of bacterial DNA replication. *Proc. Natl. Acad. Sci. USA* **99**, 8342–8347.

Oussenko, I. A., Abe, T., Ujiie, H., Muto, A., and Bechhofer, D. H. (2005). Participation of 3′ to 5′-exoribonucleases in the turnover of *Bacillus subtilis* mRNA. *J. Bacteriol.* **187**, 2758–2767.

Oussenko, I. A., and Bechhofer, D. H. (2000). The *yvaJ* gene of *Bacillus subtilis* encodes a 3′ to 5′-exoribonuclease and is not essential in a strain lacking polynucleotide phosphory-lase. *J. Bacteriol.* **182**, 2639–2642.

Oussenko, I. A., Sanchez, R., and Bechhofer, D. H. (2002). *Bacillus subtilis* YhaM, a member of a new family of 3′ to 5′-exonucleases in gram-positive bacteria. *J. Bacteriol.* **184**, 6250–6259.

Oussenko, I. A., Sanchez, R., and Bechhofer, D. H. (2004). *Bacillus subtilis* YhcR, a high-molecular-weight, nonspecific endonuclease with a unique domain structure. *J. Bacteriol.* **186**, 5376–5383.

Pace, N. R., and Pace, B. (1990). Ribosomal RNA terminal maturase: Ribonuclease M5 from *Bacillus subtilis*. *Methods Enzymol.* **181**, 366–374.

Pellegrini, O., Mathy, N., Gogos, A., Shapiro, L., and Condon, C. (2005). The *Bacillus subtilis ydcDE* operon encodes an endoribonuclease of the MazF/PemK family and its inhibitor. *Mol. Microbiol.* **56**, 1139–1148.

Pellegrini, O., Nezzar, J., Marchfelder, A., Putzer, H., and Condon, C. (2003). Endonu-cleolytic processing of CCA-less tRNA precursors by RNase Z in *Bacillus subtilis*. *EMBO J.* **22**, 4534–4543.

Rivera-Leon, R., Green, C. J., and Vold, B. S. (1995). High-level expression of soluble recombinant RNase P protein from *Escherichia coli*. *J. Bacteriol.* **177,** 2564–2566.

Smith, D., Burgin, A. B., Haas, E. S., and Pace, N. R. (1992). Influence of metal ions on the ribonuclease P reaction. Distinguishing substrate binding from catalysis. *J. Biol. Chem.* **267,** 2429–2436.

Sogin, M. L., and Pace, N. R. (1974). *In vitro* maturation of precursors of 5S ribosomal RNA from *Bacillus subtilis. Nature* **252,** 598–600.

Wang, W., and Bechhofer, D. H. (1997). *Bacillus subtilis* RNase III gene: Cloning, function of the gene in *Escherichia coli*, and construction of *Bacillus subtilis* strains with altered *rnc* loci. *J. Bacteriol.* **179,** 7379–7385.

Wegscheid, B., Condon, C., and Hartmann, R. K. (2006). Type A and B RNase P RNAs are interchangeable *in vivo* despite substantial biophysical differences. *EMBO Rep.* **7,** 411–417.

Wei, Y., Deikus, G., Powers, B., Shelden, V., Krulwich, T. A., and Bechhofer, D. H. (2006). Adaptive gene expression in *Bacillus subtilis* strains deleted for *tetL. J. Bacteriol.* **188,** 7090–7100.

Wen, T., Oussenko, I. A., Pellegrini, O., Bechhofer, D. H., and Condon, C. (2005). Ribonuclease PH plays a major role in the exonucleolytic maturation of CCA-containing tRNA precursors in *Bacillus subtilis. Nucleic Acids Res.* **33,** 3636–3643.

Yao, S., Blaustein, J. B., and Bechhofer, D. H. (2007). Processing of *Bacillus subtilis* small cytoplasmic RNA: Evidence for an additional endonuclease cleavage site. *Nucleic Acids Res.* **35,** 4464–4473.

CHAPTER SIXTEEN

STAPHYLOCOCCUS AUREUS ENDORIBONUCLEASE III: PURIFICATION AND PROPERTIES

Clément Chevalier,* Eric Huntzinger,*,† Pierre Fechter,*
Sandrine Boisset,‡ François Vandenesch,‡ Pascale Romby,*
and Thomas Geissmann*

Contents

Abstract

Staphylococcus aureus ribonuclease III (Sa-RNase III) belongs to the enzyme family known to process double-stranded RNAs consisting of two turns of the RNA helix. Although the enzyme is thought to play a role in ribosomal RNA processing and gene regulation, the deletion of the *rnc* gene in *S. aureus* does not affect cell growth in rich medium. *S. aureus* RNase III acts in concert with regulatory RNAIII to repress the expression of several mRNAs encoding virulence factors. The action of the RNase is most likely to initiate the degradation of repressed mRNAs leading to an irreversible repression. In this chapter, we

* Architecture et Réactivité de l'ARN, Université de Strasbourg, CNRS, IBMC, Strasbourg, France
† Max Planck Institute for Developmental Biology, Tübingen, Germany
‡ INSERM U851, Centre National de Référence des Staphylocoques, Université de Lyon I, Lyon, France

Methods in Enzymology, Volume 447   © 2008 Elsevier Inc.
ISSN 0076-6879, DOI: 10.1016/S0076-6879(08)02216-7

describe the overexpression and purification of recombinant RNase III from *S. aureus*, and we show that its biochemical properties are similar to the orthologous enzyme from *Escherichia coli*. Both enzymes similarly recognize and cleave different RNA substrates and RNA-mRNA duplexes.

1. INTRODUCTION

Among ribonucleases, the double-stranded RNA (dsRNA) endoribonucleases came into the spotlight with the discovery of their involvement in the RNA interference machinery and miRNA processing, which are now considered major regulators of gene expression in higher eukaryotes (Bernstein *et al.*, 2001). These endoribonucleases have been classified across phylogeny in three main groups on the basis of their domain organization (for reviews, see Conrad and Rauhut, 2002; Drider and Condon, 2004; Nicholson, 1999). Bacterial RNase III belongs to the group I family, which contains only one characteristic ribonuclease domain and a dsRNA-binding domain (dsRBD). *Escherichia coli* (Ec) RNase III, which was first described and purified by Robertson *et al.* (1968), is one of the most extensively studied members of the family. The enzyme is active as a homodimer, with only one subunit sufficient for catalytic activity and substrate recognition (Conrad *et al.*, 2002). Two specific domains at the N- and C-terminal parts of the protein carry out the two functions of the enzyme: catalysis and dsRNA binding, respectively. The catalytic domain, which carries a 9-amino acid conserved sequence ERLEFLGDS/A, is sufficient for homodimerization and retains full activity (Sun *et al.*, 2001). Furthermore, point mutations in this region inhibit the catalytic properties but do not alter binding to dsRNA (Sun *et al.*, 2004). Ec-RNase III requires Mg^{2+} as a cofactor for catalysis (Mn^{2+} supports activity as well) through the activation of water molecules to perform a nucleophilic attack on a phosphodiester bond, and it is inhibited by other divalent cations such as Ca^{2+} (Li *et al.*, 1993; Gan *et al.*, 2007). Hydrolysis produces a characteristic terminal dsRNA structure consisting of a $5'$-phosphate group and a two-base overhang at the $3'$-end (see Fig. 16.1). Despite the large number of reports, it is difficult to clearly define substrate specificity. Although the dsRNA substrate can be reduced to 11 base pairs (Court, 1993; Nicholson, 1999), footprinting and interference assays revealed that the homodimeric enzyme makes multiple essential contacts with the ribose-phosphate backbone that span two turns of a regular A-form helix (Li and Nicholson, 1996). Although there is no defined consensus sequence for the RNase III binding site and catalysis, it was shown that particular base pairs located at defined positions act as antideterminants. This behavior is of particular interest as it

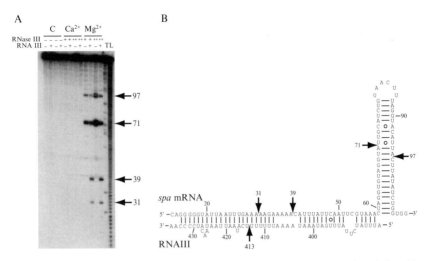

Figure 16.1 Recombinant Sa-RNase III requires $MgCl_2$ for activity and is inhibited by Ca^{2+}. (A) RNase III hydrolysis of 5′-end-labeled *spa* mRNA that is free of (−) or bound by RNAIII (+). Hydrolysis reactions were carried out in the presence of $CaCl_2$ (Ca^{2+}) or $MgCl_2$ (Mg^{2+}). Incubation controls were carried out in parallel using mRNA free (−) or bound to RNAIII (+) in the two different buffers and in the absence of RNase III (C). (*Lane T*): RNase T1 ladder under denaturing conditions. (*Lane L*): Alkaline hydrolysis ladder. Arrows indicate the RNase III cleavages. (B) Secondary structure model of the RNAIII-*spa* mRNA complex. Cleavage at positions +31/+413 and +71/+97 produces the typical two-base overhang at the 3′-end. Cleavage at position +39 occurs in an internal loop and is not accompanied by a concomitant cleavage in the RNAIII strand (not shown).

confers to these RNAs protection from RNase III–mediated hydrolysis (Pertzev and Nicholson, 2006; Zhang and Nicholson, 1997). The enzyme is nevertheless quite adaptable as coaxially stacked short helices can be cleaved (Franch *et al.*, 1999; Gan *et al.*, 2006; Pertzev and Nicholson, 2006), as can dsRNA that is interrupted by internal loops. In this latter case, cleavage occurs only at one of the two strands (Calin-Jageman and Nicholson, 2003; Fig. 16.1).

In several bacteria, such as *E. coli* and *Bacillus subtilis*, RNase III has been shown to be involved in the processing of both 16S and 23S rRNAs from a 30S precursor (for a review of rRNA processing, see Srivastava and Schlessinger, 1990). Perhaps a main function of RNase III is its involvement in the regulation of gene expression in phages, plasmids, and cellular genes (reviewed in Nicholson, 1999). In *E. coli*, the enzyme represses its own synthesis and cleaves a stem-loop structure in the 5′ untranslated region (5′ UTR) of *rnc* mRNA resulting in its rapid degradation (Bardwell *et al.*, 1989) and is also involved in the first step of the degradation of *pnp* mRNA encoding polynucleotide phosphorylase (Jarrige *et al.*, 2001).

RNase III–dependent mRNA degradation can also be mediated through *trans*-acting antisense RNA (Darfeuille *et al.*, 2007; Wagner *et al.*, 2002), as the binding of the antisense RNA to the target mRNA creates a dsRNA that is a substrate for RNase III. This is particularly the case for antisense RNA regulation of plasmids, phages, and transposons (Wagner *et al.*, 2002). However, in many cases, RNase III–mediated degradation of the repressed mRNA is a consequence of translational inhibition rather than of the primary control event (Deana and Belasco, 2005). Although much work has been carried out using two system models, it remains to be determined whether the substrate specificity and cellular function of this ubiquitous enzyme are common to all prokaryotes. In this chapter, we describe the overexpression of recombinant *S. aureus* RNase III, its purification, and its comparison with Ec-RNase III using *in vitro* probing experiments. The role of *S. aureus* RNase III in virulence gene expression and further functional implications will also be discussed.

2. *S. AUREUS* RNASE III IS NOT ESSENTIAL FOR CELL GROWTH BUT REGULATES VIRULENCE GENE EXPRESSION

S. aureus RNase III (Sa-RNase III, EC 3.1.26.3) is a homodimer of 243-amino acid (27.9 kDa) polypeptides. Sa-RNase III shares 83% amino acid identity with *Staphylococcus epidermidis*, 49% with *B. subtilis*, and 33% with *E. coli* RNase III. A *S. aureus rnc* mutant (obtained by homologous recombination) is viable, and growth is not affected in rich medium (Huntzinger *et al.*, 2005). In contrast, the growth of an *E. coli rnc* mutant is abnormally slow (Babitzke *et al.*, 1993; Studier, 1975), whereas *B. subtilis* RNase III is essential (Herskovitz and Bechhofer, 2000). Although the specific essential role of the RNase in *B. subtilis* has yet to be addressed, this phenotype is nevertheless not related to rRNA maturation (Herskovitz and Bechhofer, 2000). It was shown that in an *E. coli rnc* null strain, the 30S rRNA precursor is easily detected if total-cell RNA is fractionated in an agarose gel stained with ethidium bromide (Wang and Bechhofer, 1997). This experiment was performed using an *S. aureus rnc* mutant and isogenic RN6390 strains. We were unable to detect a significant amount of 30S rRNA precursor using total-cell RNA and ethidium bromide–stained gels (data not shown). Although Northern blotting experiments must be performed with a specific labeled probe complementary to the spliced region of the rRNAs, the lack of detection of the 30S rRNA precursor strongly suggests that *S. aureus* RNase III plays a minor role in rRNA processing, as has been shown to be the case for *B. subtilis* (Herskovitz and Bechhofer, 2000).

It was previously proposed that an alternative pathway for rRNA processing is activated in *rnc* mutant Gram-positive bacteria (Herskovitz and Bechhofer, 2000). A novel enzyme (mini-III) which contains an RNase III-like catalytic domain but lacks the dsRNA binding domain was recently discovered in gram positive bacteria. The *B. subtilis* mini-III enzyme is involved in the final step of the 23S rRNA maturation (Redko *et al.*, 2008).

We have recently shown that Sa-RNAIII plays a role in virulence regulatory pathways and acts in coordination with the regulator RNAIII, which is the intracellular effector of the quorum-sensing system. *S. aureus* RNAIII has multiple functions: it behaves as an mRNA encoding δ-hemolysin and acts as a translational activator (Morfeldt *et al.*, 1995; Novick *et al.*, 1993) and repressor (Boisset *et al.*, 2007; Geisinger *et al.*, 2006; Huntzinger *et al.*, 2005) of gene expression. The structure of RNAIII is characterized by 14 stem-loop motifs and 2 distant interactions that delimitate structural domains (Benito *et al.*, 2000). We recently showed that *spa* mRNA encoding protein A, a major virulence factor, is regulated by RNAIII at the transcriptional and posttranscriptional levels (Huntzinger *et al.*, 2005). In the latter case, RNAIII mediates the translational inhibition and rapid degradation of *spa* mRNA. The 3′-end domain of RNAIII rapidly binds the ribosome binding site (RBS) of *spa* mRNA, to which it is partially complementary, and forms an extended duplex of more than 30 base pairs interrupted by internal loops. *In vitro*, this annealing is sufficient to inhibit the formation of the translation initiation ribosomal complex. However, *in vivo* translational protein fusion experiments demonstrated that regulation by RNAIII in a *rnc* mutant background was not efficient and was accompanied by the reappearance of *spa* mRNA. *In vitro* RNase III cleavage assays show that *spa* mRNA and RNAIII are both cleaved within the duplex (Fig. 16.1). Moreover *spa* mRNA carries a long stem-loop structure downstream of the RBS that is efficiently recognized and cleaved by RNase III (Fig. 16.1). In the absence of RNAIII, *spa* mRNA is highly stable, which is enhanced in the *rnc* mutant (Huntzinger *et al.*, 2005). Thus, the coordinated action of RNase III and the regulatory RNA might be essential *in vivo* to degrade the stable mRNA and irreversibly arrest translation. This model is not restricted to the regulation of protein A expression but was further generalized to other mRNAs encoding virulence factors (SA1000, a fibrinogen binding protein, secretory antigen precursor, and the transcriptional regulatory protein Rot, (Boisset *et al.*, 2007). Of interest, RNase III was found specifically enriched on affinity purification beads coated with biotinylated RNAIII, but no significant hydrolysis of RNAIII was detected (Boisset *et al.*, 2007). Thus, the enzyme may be guided by RNAIII to induce rapid degradation of the target mRNAs, coupling the two regulatory events (i.e., inhibition of initiation of translation and degradation).

3. BIOCHEMICAL PROPERTIES AND SUBSTRATE SPECIFICITY OF *S. AUREUS* RNASE III

S. aureus RNase III has been purified using a protocol similar to that of *E. coli* RNase III (Franch *et al.*, 1999). Both enzymes tend to aggregate at low salt. Therefore, purification and storage should be carried out in the presence of high salt (see below). *S. aureus* RNase III also requires $MgCl_2$ for full activity and is inhibited if Mg^{2+} is replaced by Ca^{2+} (Fig. 16.1). The effect of *S. aureus* RNase III was also examined *in vitro* using several *S. aureus* RNA substrates, and on a 33 base-pair RNA helix (Figs. 16.1–16.3). The sequence of this helix was derived from the CopA-CopT complex that inhibits the translation of *repA* mRNA, which encodes the initiator protein and regulates the copy number of plasmid R1 in *E. coli* (Blomberg *et al.*, 1990). Both enzymes induce strong cleavage at both sides of the helix leaving a two-base overhang at the 3′-ends (Fig. 16.1). Interestingly, the cleavages were identical to those found in the whole stable CopA-CopT complex both *in vitro* (Malmgren *et al.*, 1997) and *in vivo* (Wagner *et al.*, 2002). Taken together, the experiments strongly suggest that the two enzymes have similar properties.

We further show that *S. aureus* RNase III is able to cleave *in vitro* several mRNAs bound to the regulatory RNAIII (Boisset *et al.*, 2007; Huntzinger *et al.*, 2005). As for *spa* mRNA, RNAIII forms a long duplex comprising the RBS of the mRNA. Interestingly, in the absence of RNAIII, *spa* mRNA was efficiently cleaved by RNase III at two main positions in a long regular helix (Fig. 16.1). As for the regular helix (Fig. 16.2), a two-base overhang at the 3′-end is generated. In the RNAIII-*spa* mRNA duplex, two other cleavages in *spa* mRNA are detected even if the formed duplex is inter- rupted by internal loops. Unfortunately, we were not able to detect any RNase III–dependent cleavages in *spa* mRNA *in vivo* under repressed conditions, as the mRNA rapidly depleted from the cell. This rapid deple- tion also is due to *spa* mRNA being regulated by RNAIII simultaneously at the transcriptional and the posttranscriptional levels (Huntzinger *et al.*, 2005). Nevertheless, in the *rnc* mutant, the mRNA was detected and the half-life was enhanced, strongly arguing for the participation of the RNase III in the degradation pathway of *spa* mRNA. It was previously shown that *E. coli* RNase III is able to cleave a large variety of RNA structure motifs, such as a three-way junction or short helices that are coaxially stacked (Franch *et al.*, 1999). This may have some functional importance for antisense regulation: in many cases, only short duplexes are formed (Brantl, 2007; Wagner *et al.*, 2002). This is apparently the case for *S. aureus* RNase III. Indeed, we have previously shown that two distant domains of RNAIII bind to two hairpin loops of *rot* mRNA, forming loop-loop

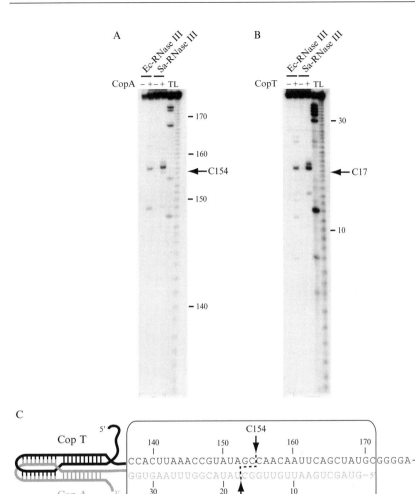

Figure 16.2 Comparative analysis of the cleavage patterns in a dsRNA induced by RNase III from *S. aureus* and from *E. coli*. (A, B) The experiment was performed using 5'-end-labeled CopT (which contains only the last 33 nucleotides of CopT) (A) and 5'-end-labeled CopA (which contains only the first 33 nucleotides of CopA) (B). Cleavage reactions were carried out using the CopA-CopT dsRNA incubated without RNase III (−) or with (+) either *E. coli* (Ec-RNase III) or *S. aureus* RNase III (Sa-RNase III). (*Lane T*): RNase T1 under denaturing conditions. (*Lane L*): Alkaline hydrolysis ladder. (C) Schematic view of the secondary structure model of CopA/CopT with the RNase III cleavage sites. The sequences of the interacting RNAs (dsRNA) that were used in the cleavage assays are squared.

interactions (Boisset *et al.*, 2007; see Fig. 16.3). The intermolecular base pairings are very short (7 and 8 nucleotides, respectively), and are *a priori* not considered RNase III substrate. However, a specific RNase III cleavage was found in the mRNA bound to RNAIII located at the same position of each loop-loop interaction. The strongest cleavage was found located in the SD sequence of *rot* mRNA (Fig. 16.3). Conversely, only one cleavage was detected in RNAIII bound to *rot* mRNA (Boisset *et al.*, 2007). These RNase III cleavages might result from coaxial stacking of the two helices induced by the loop-loop interaction forming a long helical structure recognized by the homodimeric enzyme. The absence of cleavage on the other side of the newly formed helix in RNAIII (H14) might be prevented by the long crossing loop that connected the loop-loop interaction (Fig. 16.3). Northern blot analysis of total RNAs extracted in late exponential phase revealed several species of *rot* mRNA, one of the shorter fragments was absent in the *rnc* mutant and in the strain deleted of the regulatory RNAIII. However, we were not able to map precisely the RNase III–dependent cleavage site in the 5′ UTR region of *rot* mRNA, which presented a rather complex pattern of degradation (Boisset *et al.*, 2007; Geisinger *et al.*, 2006; Hsieh *et al.*, 2008). *Rot* mRNA presented three redundant UGGGA sequences that could potentially base pair with three redundant sequences UCCCA found in hairpin motifs 7, 13, and 14 of

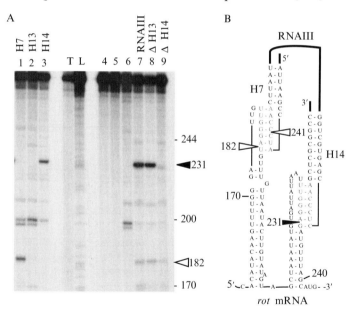

Figure 16.3 *S. aureus* RNase III specifically cleaves *rot* mRNA bound to RNAIII. (A) RNase III hydrolysis of 5′-end-labeled *rot* mRNA in the absence (no RNAIII) or presence of the isolated hairpins (H7, H13, H14) of RNAIII, of wild-type RNAIII (RNAIII), or of RNAIII mutants lacking the hairpin H13 (ΔH13) or H14 (ΔH14). (*Lane C*): Incubation controls of *rot* mRNA in the absence of RNase III. (*Lane T*): RNase T1 under denaturing conditions. (*Lane L*): Alkaline hydrolysis ladder. Arrows indicate RNase III cleavages. (B) Secondary structure model of the RNAIII–*rot* mRNA complex with the RNase III cleavages denoted by arrows (▷ moderate, ▶ strong cleavage). The structure model is derived from Boisset *et al.* (2007).

RNAIII (Boisset *et al.*, 2007). This raises the question of how to find the partners for each loop-loop interaction. Using isolated hairpin motifs of RNAIII, we could show that the RNase III-dependent cleavage at the SD sequence of *rot* mRNA results from the binding of hairpin 14 of RNAIII, whereas the other cleavage in *rot* mRNA is induced by the binding of hairpin 7 of RNAIII (Fig. 16.3). Thus, cleavage assays by RNase III can be considered a useful tool for probing *in vitro* RNA-RNA complexes.

4. RNase III Overexpression and Purification

Chemicals and reagents are of the highest commercially available quality. Distilled water is treated using a Millipore Milli-Q Synthesis system and can be further sterilized. The RNase III coding sequence was amplified from *S. aureus* RN6390 genomic DNA using primers rnase980/rnase1670 (5′-TAA AGG ATC CTC TAA ACA AAA GAA AAG TGA GAT AGT TAA TC-3′- and 5′-TAC CCC CGG GTT TAA TTT GTT TTA ATT GCT TAT AGG CAC TTT TAG-3′; underlined are the *Bam*HI and *Sma*I restriction sites, respectively). The PCR product was digested with *Bam*HI and *Sma*I, ligated to the linearized pQE30 (Qiagen) vector and transformed into *E. coli* M15[pRep4] (Qiagen) resulting in M15[pRep4/pLUG515] overexpressing *S. aureus* His$_6$-RNase III (Huntzinger *et al.*, 2005). pRep4 carries a resistance to kanamycin and is required to donate sufficient *LacI*. *E. coli* His$_6$-RNase III was purified from strain W3110, transformed freshly with the plasmids pRNC and pMS421 (Franch *et al.*, 1999). pRNC is an ampicillin-resistant pBR322 derivative that carries the His$_6$-RNase III coding sequence under control of the *LacI*-repressed P$_{A1/04/03}$ promoter (Lanzer and Bujard, 1988). pMS421 carries a resistance to spectinomycin and is required to donate sufficient *LacI* (Graña *et al.*, 1988).

Caution must be taken during the purification procedure. Overexpression of His$_6$-RNase III results in the accumulation of the recombinant protein in inclusion bodies. To prevent the formation of inclusion bodies, the culture was incubated at 30 °C in the presence of glucose. For *S. aureus* RNase III, a significant amount of the protein remained in a soluble form. Because the enzyme has a tendency to aggregate at low ionic strength, the RNase has been stored in a buffer containing a high concentration of monovalent ions.

M15[pRep4/pLUG515] cells are grown in 500 ml LB medium containing 0.2% glucose and in the presence of 100 μg/ml ampicillin and 25 μg/ml kanamycin at 30 °C. For strain W3110, 100 μg/ml spectinomycin is added in the place of kanamycin. At an OD (600 nm) of 0.2, 1 mM IPTG is added and incubation continued for 4 h at 30 °C with vigorous aeration. Cells are cooled on ice, collected by centrifugation at 4 °C, and washed with 10 ml lysis buffer (30 mM Tris-HCl, pH 8.0, 25% sucrose). The cells are treated for 15 min on ice in the presence of 0.5 mg/ml lysozyme (dissolved in the lysis buffer). The lysis of the cell is followed by a DNase treatment (10 min on ice) in 10 ml lysis buffer containing 5 mM MgCl$_2$. The sample is centrifuged at 10,000 rpm (15 min, 4 °C) in a Beckman JA-14 rotor, and the supernatant is transferred to a sterile tube.

At this step of the purification process, evaluation is advised of the yield of the expression of the recombinant RNase III and of the soluble fraction by analyzing the protein (present in the pellet and the supernatant) in a SDS-12% polyacrylamide gel electrophoresis. If His$_6$-RNase III is predominantly present in the inclusion bodies, the protein (in the pellet) can be solubilized by incubation for 30 min at 4 °C in 10 ml buffer A (25 mM Tris-HCl, pH 8.0 (4 °C), 8% ammonium sulfate, and 0.1 mM EDTA). In parallel, the soluble fraction the protein (supernatant) is precipitated by gentle addition of 4.58 g ammonium sulfate (2 M final concentration) for 1 h on ice. After centrifugation (30 min, 10,000 rpm, 4 °C), the pellet is resuspended in 10 ml 25 mM Tris-HCl (pH 8.0), 0.1 mM EDTA.

Meanwhile 600 μl Ni-NTA agarose (Qiagen) is first equilibrated twice with 1 ml buffer A. After a brief centrifugation, the supernatant is removed and the beads are mixed with 1 ml of the same buffer and kept at 4 °C. The beads are then transferred to 10 ml of the extract containing His$_6$-RNase III, and incubated for 1 h at 4 °C with gentle agitation. The RNase III is then purified using a chromatographic support (poly-prep chromatography columns, Bio-Rad). The sample is first transferred to an empty poly-prep support and is washed successively with 1 ml buffer A and 1 ml buffer A containing 5 mM imidazole. His$_6$-RNase III is eluted with 1 ml buffer B (25 mM Tris-HCl (pH 8), 1 M NH$_4$Cl, 1 mM DTT) in the presence of 250 mM imidazole (prepared in buffer B). This step is followed by three elutions with 1 ml buffer B in the presence of 500 mM imidazole. Aliquots (5 μl) of each fraction are then controlled by electrophoresis on SDS-12% polyacrylamide gel. The fractions containing His$_6$-RNase III are pooled. If the absorbance spectra revealed the presence of nucleic acids, a further purification step is carried out. The fractions containing the RNase III are directly loaded on an affinity chromatography linked to polyI-polyC (Pharmacia) pre-equilibrated with buffer B. After three washes in Buffer B, the enzyme is eluted with buffer B containing 2 M NH$_4$Cl.

The purified RNase III is then dialyzed at 4 °C against a large volume of the storage buffer (30 mM Tris-HCl [pH 8.0], 500 mM KCl, 0.1 mM DTT, 0.1 mM EDTA, and 50% bidistillated glycerol). The buffer is changed after 3 h and the dialysis process is continued overnight. Concentration of proteins is estimated by measuring its absorbance at 280 nm ($\varepsilon_M =$ 13,300 Mcm^{-1}). Mass spectrometry analysis and N-terminal sequencing have been carried out on the purified enzyme to ensure that there are no heterodimers formed by Sa-RNase III and endogenous Ec-RNase III.

5. RNA Substrate Preparation

The different RNAs used as potential substrates for RNase III (dsRNA, mRNA, RNAIII) were all produced by *in vitro* transcription of DNA templates using T7 phage RNA polymerase. The DNA templates are usually pUC18-derived plasmid in which a PCR fragment containing the T7 promoter fused to the gene of interest has been inserted into the polylinker (Huntzinger *et al.*, 2005). Short RNAs were transcribed directly from the PCR fragment. All RNA fragments start with two guanines (Milligan *et al.*, 1987). The RNAs were purified from shorter RNA fragments (abortive transcription, hydrolysis), the DNA template, and unincorporated NTPs by electrophoresis on a denaturing 10% polyacrylamide-8 M urea slab gel (Milligan *et al.*, 1987). If the RNAs are homogeneous, they are purified using chromatography, such as gel filtration (GSW2000, Tosohaas) (Romaniuk *et al.*, 1987), anion exchange monoQ column (Jahn *et al.*, 1991), or ion-pairing reversed-phase HPLC (Gelhaus *et al.*, 2003). A typical RNA synthesis and purification protocol is given next.

For a transcription scale of 100 μl, a yield of about 100 μg RNA can be expected. The reaction is performed in a total volume of 100 μl, in which the different products are added in the following order: 10 μl 10 × T7 transcription buffer (0.4 M Tris-HCl, pH 8; 0.5 M NaCl), 1 μl 10 mg/ml BSA, 1 μl RNase inhibitor (RNasin, 40 U, Promega), 1 μl 0.1 M DTE (1,4-dithioerythritol), 1 μl 0.5 M spermidine, 4 μl each 100 mM of NTP, DNA template (5 μg), and 2 μl homemade T7 RNA polymerase (5 U). The reaction mix is then incubated for at least 2 h at 37 °C. A white precipitate of inorganic pyrophosphate can be observed if the yield of transcription is high. For low-yield transcription, incubation might be extended to overnight in an incubator to prevent condensation. The precipitated pyrophosphate can be removed by a short centrifugation or by the addition of 50 μl 0.1 M EDTA. If necessary, the DNA template can be removed at the end by the addition of 10 U RNase-free DNase 1 (Amersham). However, this is usually not necessary as transcript and DNA template differ in size.

The RNAs are then submitted to phenol-chloroform extraction and precipitated with 2.5 vol cold ethanol in the presence of 0.4 M ammonium acetate. The RNAs can be stored in the precipitate form at −20 °C for several months. After centrifugation, the pellet is washed with 70% ethanol, air-dried for 2 min, and dissolved in 100 μl gel-loading buffer (8 M urea, 0.025% xylene cyanol, and 0.025% bromophenol blue). The RNAs are purified by electrophoresis on a preparative denaturing polyacrylamide gel. Depending of the size of the RNA, 5% to 15% polyacrylamide: bis-(19:1)-8 M urea-1× TBE are used. After electrophoresis, the transcript is visualized by UV shadowing.

The gel slices are covered with the elution buffer (0.5 M ammonium acetate, 1 mM EDTA), and the RNA is eluted in the presence of 10% (vol) phenol for overnight at 4 °C. After ethanol precipitation, the pellet is dissolved in 90 μl RNase-free H_2O. Concentration of the RNA is estimated by measuring its absorbance at 260 nm.

Before use, the RNA is denatured for 1 min at 90 °C and quickly cooled on ice. Then, 10 μl 10× TMK buffer (100 mM Tris-HCl, pH 7, 100 mM $MgCl_2$, 1 M KCl) are added, followed by an incubation step at 20 °C for 15 min.

6. END-LABELING OF RNA

For 5′-end labeling, the RNA (5 μg) is previously dephosphorylated at its 5′-end by the calf intestinal phosphatase (CIP). The reaction is done at 37 °C for 1 h in 50 μl of reactional mixture containing 5 μl CIP (1 U/ml, Roche) and 5 μl commercial reaction 10× buffer (Roche). The reaction is stopped by phenol extraction followed by ethanol precipitation. To avoid dephosphorylation, transcription can be conducted in the presence of ApG (5 mM final concentration; Sigma). The 5′-end labeling is performed in 10 μl reactional mixture containing 5 μg 5′-$_{OH}$RNA, 5 U T4 polynucleotide kinase (Fermentas, 10 U/μl), 1 μl 10× commercial buffer (Fermentas), and 5 μl [γ-^{32}P]ATP (3000 Ci/mmol, 150 mCi/ml; Amersham). After 1 h at 37 °C, the reaction is stopped by the addition of 10 μl gel loading buffer. The 5′-labeled RNA is then purified from the excess of ATP by electrophoresis on a denaturing 8% polyacrylamide-8 M urea slab gel.

The 3′-end labeling was performed in the presence of [^{32}P]pCp and T4 RNA ligase according to England and Uhlenbeck (1980). Reaction is done overnight at 4 °C in 10 μl containing 15 pmol RNA, 1 μl 10× commercial buffer (Ambion), 5 μl [^{32}P]pCp (3000 Ci/mmol, 150 mCi/ml; Amersham) and 10 U T4 RNA ligase (Ambion). The sample is then treated as described previously.

7. MAPPING THE RNase III CLEAVAGE SITES *IN VITRO*

S. aureus RNase III hydrolysis is highly dependent of the presence of Mg^{2+}, and the catalysis is inhibited if Mg^{2+} is replaced by Ca^{2+} (see Fig. 16.1). The cleavage patterns of different RNA-RNA complexes were also compared using Sa-RNase III and Ec-RNase III. The experiments have been conducted with end-labeled RNA under limited conditions (see Figs. 16.1 and 16.2). It is essential to adapt the enzymatic hydrolysis to have less than one cut per molecule (i.e., more than 80% of the RNA should not be cleaved). If the reaction is too strong, primary RNase cleavages in the RNA can induce conformational rearrangements of the cleaved RNA that can potentially provide new targets for secondary cleavages. These cleavages can be distinguished from primary cuts by comparing the hydrolysis patterns obtained from the 5′ or 3′-end-labeled RNA because only primary cuts will be detected in both experiments.

S. aureus RNAIII-mRNA complexes or dsRNA are preformed by mixing the pre-renatured cold (50–200 nM final) and labeled RNA (50000 cpm/assay) with 1 μl 10× TMK buffer or 10× TCK buffer (100 mM Tris-HCl, pH 7, 100 mM $CaCl_2$, 1 M KCl) in the presence of 1 mM DTT for 10 min at 37 °C. RNase III hydrolysis is carried out in the presence of 1 μg yeast total tRNA (Sigma) for 2–5 min at 37 °C with different concentrations of enzyme (1–100 nM). In parallel, it is necessary to carry out incubation controls of end-labeled RNA either free or in complex with another RNA in which the RNase III is not added. To verify that observed cleavages resulted specifically from RNase III, the same RNA is digested under the same conditions, except that the reaction buffer contains $CaCl_2$ in place of $MgCl_2$.

The reactions were stopped by phenol-chloroform extraction, followed by ethanol precipitation in the presence of 0.3 M NaAc, pH 6.0. After two washing steps with 200 μl 70% cold ethanol, the pellets were vacuum dried and resuspended in 6 μl gel-loading buffer. It is essential to adjust the volumes of the gel-loading buffer to load the same amount of radioactivity in each sample. The samples were heated at 90 °C for 3 min and 3 μl loaded on a 15% polyacrylamide–8 M urea slab gel (or 12% polyacrylamide–8 M urea for long RNAs > 200 nt). After migration, the 12% polyacrylamide gel is fixed for 5 min in a solution containing 10% ethanol and 7.5% acetic acid, and dried on a Whatman 3 MM paper (30 min at 80 °C). The 15% PAGE was directly transferred without drying on an old autoradiography film and wrapped with a plastic film. Overnight exposure is done at −80 °C using an intensifying screen.

To assign the RNase III cleavages, two ladders (alkaline ladder and RNase T1) were run in parallel. The alkaline ladder induces cleavages at each phosphodiester bond whereas the RNase T1 hydrolysis cleaves at unpaired guanines. For the alkaline ladder, the end-labeled RNA (100,000 cpm) is incubated for 3 min at 90 °C in 5 μl alkaline buffer (0.1 M Na_2CO_3/0.1 M $NaHCO_3$, pH 9) with 1 μg yeast total tRNA. The reaction is stopped by the addition of 5 μl gel-loading buffer. For the RNase T1 digestion, the end-labeled RNA (50,000 cpm) is incubated in 5 μl sequencing buffer (20 mM sodium citrate, pH 4.5, 1 mM of EDTA, 7 M urea, 0.025% xylene cyanol, 0.025% bromophenol blue) for 5 min at 55 °C in the presence of 1 μl total tRNA. The hydrolysis was done at 55 °C for 10 min with 0.5 U RNase T1 (Ambion). The two samples are not heated before loading on polyacrylamide gel electrophoresis.

8. DENATURING AGAROSE GEL ELECTROPHORESIS AND NORTHERN BLOTTING

Total RNAs from *S. aureus* RN6390 and *rnc* mutant were purified using the FastRNA pro blue kit and the FastPrep instrument according to the manufacturer's instructions (Qbiogene). For cell culture, 5 ml BHI (Brain Heart Infusion; AES Laboratoire, France) is inoculated with 1% of an *S. aureus* overnight culture and incubated at 37 °C for 2–6 h to the desired optical density. Cells are pelleted by centrifugation for 10 min at 4000 g at 4 °C, resuspended in 1 ml RNApro solution (Qbiogene), and transferred to a tube containing Lysing Matrix B (Qbiogene). Lysis occurred in a FastPrep instrument during 40 sec at a setting of 6.0. After centrifugation at 15,000*g* for 5 min, the supernatant is transferred to a new tube and incubated for 5 min at 20 °C. Chloroform (300 μl) is then added and incubation continued for another 5 min. After centrifugation at 15,000*g* for 5 min, the upper phase is transferred to a new tube, and the RNA is precipitated by the addition of 500 μl of ice-cold ethanol. All manipulations performed with the pathogen should be carried out in a separate L2 room.

RNA can have different secondary structures that influence the migration behavior on native agarose gel. To avoid this, denaturants are added to the gel. One of the most common products is formaldehyde, which is used at a concentration of 2.2 *M* (Sambrook *et al.*, 1989). However, because of its toxic vapors, formaldehyde can be replaced by guanidine thiocyanate (Goda and Minton, 1995). Instead of MOPS-formaldehyde running buffer, common TBE can be used and the gel can be run in the presence of ethidium bromide. A stock solution of 1 *M* guanidine thiocyanate is freshly prepared and added at a final concentration of 20 m*M* to 1–2% agarose in 0.5× TBE containing 5 μg/ml ethidium bromide. Total RNA (1–10 μg) is diluted 1:1 in formamide loading buffer (95% formamide, 18 m*M* EDTA, 0.025% SDS, 0.025% xylene cyanol, and 0.025% bromophenol blue). Before loading onto the gel, the RNA is denatured for 3 min at 90 °C and quickly cooled on ice. Electrophoresis is run in 0.5× TBE buffer at 20 V/cm. After electrophoresis, the gel can be visualized under ultraviolet (UV) light. The gel is then soaked in a denaturing buffer (50 m*M* NaOH, 1 m*M* NaCl) for 30 min, followed by a washing step in 0.1 *M* Tris-HCl, pH 7.5, for neutralization. This treatment enhances the transfer efficiency of long RNAs (longer than 1 kb) but can be omitted for small RNAs. The transfer of the gel to a Hybond XL membrane (Amersham) is done during 1 h in 1× SSC (150 mM NaCl, 15 mM sodium citrate, pH 7.0) using a vacuum-blot system according to the manufacturer's instructions (Biometra, Germany).

After transfer, the RNA is fixed to the membrane by UV cross-linking (120 mJ) using a UV-Stratalinker (Stratagene). The RNA is then hybridized against a specific oligonucleotide labeled at its 5′-end using T4 polynucleotide kinase and [γ-^{32}P]ATP (see above). We also used a probe to detect 5S rRNA as an internal control. Hybridization is carried out with a protocol derived from Church and Gilbert (1984). The membrane is prehybridized in a buffer for 30 min at a temperature 5 °C below the T$_m$ of the probe (usually between 42 °C and 65 °C). The buffer is then replaced by the radiolabeled probe and hybridization continued overnight. The next day, the membrane is washed twice with washing buffer (40 mM sodium phosphate, pH 7.2, 1 mM EDTA, 5% SDS) for 30 min at the hybridization temperature. The blot is removed from the washing buffer, sealed in a plastic bag, and exposed with an autoradiography (Fuji) overnight at −80 °C.

9. CONCLUDING REMARKS

Our findings indicate that *S. aureus* RNase III is not essential for viability and for ribosomal RNA maturation in cultures grown in rich medium. However, further investigations will be necessary to clarify whether RNase III is required under certain stress conditions (e.g., minimal medium, osmotic and oxidative stress, temperature shift) or in virulence pathways. *S. aureus* RNase III plays an essential role in gene regulation and acts in a coordinated manner with a regulatory RNA to repress the synthesis of a class of virulence factors at the late exponential phase of growth. Although the enzyme is required for irreversible repression, it remains to be addressed whether it is the recognition step or the primary cleavage in the mRNA that is required for efficient control. It will also be of interest to study the function of RNase III in virulence pathways or in response to stress in other pathogens like *Listeria monocytogenes*, in which novel noncoding RNAs have been recently identified and predicted to interact via long base pairings with mRNAs (Mandin *et al.*, 2007). The mechanism of cleavage in several RNA substrates seems to be very similar for *S. aureus* and *E. coli* enzymes. However, it is not known whether the antideterminant sequences play a role in the cleavage site selection of *S. aureus* RNase III. The discovering of more RNA targets in *S. aureus* will certainly help to determine substrate specificity.

Messenger RNA degradation in *B. subtilis* differs in some aspects from that in *E. coli* (Condon, 2007). Strong secondary structures at the 5′-end of mRNA or stalled ribosomes have significant stabilizing effects on the downstream sequences of the mRNA (Huntzinger *et al.*, 2005; Sharp and Bechhofer, 2005). These results have been explained recently with the discovery of the essential RNase J1 that has both endonucleolytic and 5′ to 3′-exonucleolytic activities (Even *et al.*, 2005; Mathy *et al.*, 2007). RNase J1 was recently shown to be involved in the degradation of *B. subtilis glmS* mRNA, which is regulated through a metabolite-induced self-cleavage (Collins *et al.*, 2007). Defining the contribution of the RNA decay machinery components in *S. aureus* would certainly help to more precisely define the links between RNA degradation and gene regulation that might be important for stress adaptation and virulence.

ACKNOWLEDGMENTS

This work was supported by financial support from the Centre National de la Recherche Scientifique (UPR 9002.CNRS), the University Louis Pasteur of Strasbourg, the Ministère de la Recherche (ANR05-MIIME, ANR07-BLANC), and the European Community (FOSRAK, EC005120; BacRNA EC018618).

REFERENCES

Babitzke, P., Granger, L., Olszewski, J., and Kushner, S. R. (1993). Analysis of mRNA decay and rRNA processing in *Escherichia coli* multiple mutants carrying a deletion in RNase III. *J. Bacteriol.* **175**, 229–239.

Bardwell, J. C., Régnier, P., Chen, S. M., Nakamura, Y., Grunberg-Manago, M., and Court, D. L. (1989). Autoregulation of RNase III operon by mRNA processing. *EMBO J.* **8**, 3401–3407.

Benito, Y., Kolb, F. A., Romby, P., Lina, G., Etienne, J., and Vandenesch, F. (2000). Probing the structure of RNAIII, the *Staphylococcus aureus agr* regulatory RNA, and identification of the RNA domain involved in repression of protein A expression. *RNA* **6**, 668–679.

Bernstein, E., Caudy, A. A., Hammond, S. M., and Hannon, G. J. (2001). Role for a bidentate ribonuclease in the initiation step of RNA interference. *Nature* **409**, 363–366.

Blomberg, P., Wagner, E. G. H., and Nordström, K. (1990). Control of replication of plasmid R1: The duplex between the antisense RNA, CopA, and its target, CopT, is processed specifically *in vivo* and *in vitro* by RNase III. *EMBO J.* **9**, 2331–2340.

Boisset, S., Geissmann, T., Huntzinger, E., Fechter, P., Bendridi, N., Possedko, M., Chevalier, C., Helfer, A. C., Benito, Y., Jacquier, A., Gaspin, C., Vandenesch, F., and Romby, P. (2007). *Staphylococcus aureus* RNAIII coordinately represses the synthesis of virulence factors and the transcription regulator Rot by an antisense mechanism. *Genes & Dev.* **21**, 1353–1366.

Brantl, S. (2007). Regulatory mechanisms employed by *cis*-encoded antisense RNAs. *Curr. Opin. Microbiol.* **10**, 102–109.

Calin-Jageman, S., and Nicholson, A. W. (2003). Mutational analysis of an RNA internal loop as a reactivity epitope for *Escherichia coli* ribonuclease III substrates. *Biochemistry* **42**, 5025–5034.

Church, G. M., and Gilbert, W. (1984). Genomic sequencing. *Proc. Natl. Acad. Sci. USA* **81**, 1991–1995.

Collins, J. A., Irnov, I., Baker, S., and Winkler, W. C. (2007). Mechanism of mRNA destabilization by the *glmS* ribozyme. *Genes & Dev* **21**, 3356–3368.

Condon, C. (2007). Maturation and degradation of RNA in bacteria. *Curr. Opin. Microbiol.* **10**, 271–278.

Conrad, C., and Rauhut, R. (2002). Ribonuclease III: New sense from nuisance. *Int. J. Biochem. Cell Biol.* **34**, 116–129.

Conrad, C., Schmitt, J. G., Evguenieva-Hackenberg, E., and Klug, G. (2002). One functional subunit is sufficient for catalytic activity and substrate specificity of *Escherichia coli* endoribonuclease III artificial heterodimers. *FEBS Lett.* **518**, 93–96.

Court, D. (1993). RNA processing and degradation by RNase III. *In* "Control of messenger RNA stability" (J. G. Belasco and G. Brawerman, eds.), pp. 71–116. Academic Press, New York.

Darfeuille, F., Unoson, C., Vogel, J., and Wagner, E. G. H. (2007). An antisense RNA inhibits translation by competing with standby ribosomes. *Mol. Cell* **26**, 381–392.

Deana, A., and Belasco, J. G. (2005). Lost in translation: The influence of ribosomes on bacterial mRNA decay. *Genes & Dev.* **19**, 2526–2533.

Drider, D., and Condon, C. (2004). The continuing story of endoribonuclease III. *J. Mol. Microbiol. Biotechnol.* **8**, 195–200.

England, T. E., Bruce, A. G., and Uhlenbeck, O. C. (1980). Specific labeling of 3′-termini of RNA with T4 RNA ligase. *Methods Enzymol.* **65**, 65–74.

Even, S., Pellegrini, O., Zig, L., Labas, V., Vinh, J., Brechemmier-Baey, D., and Putzer, H. (2005). Ribonucleases J1 and J2: Two novel endoribonucleases in *B. subtilis* with functional homology to *E. coli* RNase E. *Nucleic Acids Res.* **33**, 2141–2152.

Franch, T., Thisted, T., and Gerdes, K. (1999). Ribonuclease III processing of coaxially stacked RNA helices. *J. Biol. Chem.* **274**, 26572–26578.

Gan, J., Shaw, G., Tropea, J. E., Waugh, D. S., Court, D. L., and Ji, X. (2007). A stepwise model for double-stranded RNA processing by ribonuclease III. *Mol. Microbiol.* **67**, 143–154.

Gan, J., Tropea, J. E., Austin, B. P., Court, D. L., Waugh, D. S., and Ji, X. (2006). Structural insight into the mechanism of double-stranded RNA processing by ribonuclease III. *Cell* **124**, 355–366.

Geisinger, E., Adhikari, R. P., Jin, R., Ross, H. F., and Novick, R. P. (2006). Inhibition of *rot* translation by RNAIII, a key feature of *agr* function. *Mol. Microbiol.* **61**, 1038–1048.

Gelhaus, S. L., LaCourse, W. R., Hagan, N. A., Amarasinghe, G. K., and Fabris, D. (2003). Rapid purification of RNA secondary structures. *Nucleic Acids Res.* **31**, e135.

Goda, S. K., and Minton, N. P. (1995). A simple procedure for gel electrophoresis and Northern blotting of RNA. *Nucleic Acids Res.* **23**, 3357–3358.

Graña, N. P., Gardella, T., and Susskind, M. M. (1988). The effects of mutations in the *ant* promoter of phage P22 depend on context. *Genetics* **120**, 319–327.

Herskovitz, M. A., and Bechhofer, D. H. (2000). Endoribonuclease RNase III is essential in *Bacillus subtilis*. *Mol. Microbiol.* **38**, 1027–1033.

Hsieh, H. Y., Tseng, C. W., and Stewart, G. C. (2008). Regulation of Rot expression in *Staphylococcus aureus*. *J. Bacteriol.* **190**, 546–554.

Huntzinger, E., Boisset, S., Saveanu, C., Benito, Y., Geissmann, T., Namane, A., Lina, G., Etienne, J., Ehresmann, B., Ehresmann, C., Jacquier, A., Vandenesch, F., and Romby, P. (2005). *Staphylococcus aureus* RNAIII and the endoribonuclease III coordinately regulate *spa* gene expression. *EMBO J.* **24**, 824–835.

Jahn, M. J., Jahn, D., Kumar, A. M., and Soll, D. (1991). Mono Q chromatography permits recycling of DNA template and purification of RNA transcripts after T7 RNA polymerase reaction. *Nucleic Acids Res.* **19**, 2786.

Jarrige, A. C., Mathy, N., and Portier, C. (2001). PNPase autocontrols its expression by degrading a double-stranded structure in the *pnp* mRNA leader. *EMBO J.* **20**, 6845–6855.

Lanzer, M., and Bujard, H. (1988). Promoters largely determine the efficiency of repressor action. *Proc. Natl. Acad. Sci. USA* **85,** 8973–8977.

Li, H., and Nicholson, A. W. (1996). Defining the enzyme binding domain of a ribonuclease III processing signal. Ethylation interference and hydroxyl radical footprinting using catalytically inactive RNase III mutants. *EMBO J.* **15,** 1421–1433.

Li, H. L., Chelladurai, B. S., Zhang, K., and Nicholson, A. W. (1993). Ribonuclease III cleavage of a bacteriophage T7 processing signal. Divalent cation specificity, and specific anion effects. *Nucleic Acids Res.* **21,** 1919–1925.

Malmgren, C., Wagner, E. G. H., Ehresmann, C., Ehresmann, B., and Romby, P. (1997). Antisense RNA control of plasmid R1 replication: The dominant product of the antisense RNA-mRNA binding is not a full RNA duplex. *J. Biol. Chem.* **272,** 12508–12512.

Mandin, P., Repoila, F., Vergassola, M., Geissmann, T., and Cossart, P. (2007). Identification of new noncoding RNAs in *Listeria monocytogenes* and prediction of mRNA targets. *Nucl. Acids Res.* **35,** 962–974.

Mathy, N., Benard, L., Pellegrini, O., Daou, R., Wen, T., and Condon, C. (2007). 5′ to 3′-exoribonuclease activity in bacteria: Role of RNase J1 in rRNA maturation and 5′-stability of mRNA. *Cell* **129,** 681–692.

Milligan, J. F., Groebe, D. R., Witherell, G. W., and Uhlenbeck, O. C. (1987). Oligoribonucleotide synthesis using T7 RNA polymerase and synthetic DNA templates. *Nucleic Acids Res.* **15,** 8783–8798.

Morfeldt, E., Taylor, D., von Gabain, A., and Arvidson, S. (1995). Activation of α-toxin translation in *Staphylococcus aureus* by the trans-encoded antisense RNA, RNAIII. *EMBO J.* **14,** 4569–4577.

Nicholson, A. W. (1999). Function, mechanism and regulation of bacterial ribonucleases. *FEMS Microbiol. Rev.* **23,** 371–390.

Novick, R. P., Ross, H. F., Projan, S. J., Kornblum, J., Kreiswirth, B., and Moghazeh, S. (1993). Synthesis of staphylococcal virulence factors is controlled by a regulatory RNA molecule. *EMBO J.* **12,** 3967–3975.

Pertzev, A. V., and Nicholson, A. W. (2006). Characterization of RNA sequence determinants and antideterminants of processing reactivity for a minimal substrate of *Escherichia coli* ribonuclease III. *Nucleic Acids Res.* **34,** 3708–3721.

Redko, J., Bechhofer, D. H. and Condon, C. (2008). Mini-III, an unusual member of the RNase III family of enzymes catalyses 23S ribosomal RNA maturation in *B. subtilis*. *Mol. Microbiol.* **68,** 1096–1106.

Robertson, H. D., Webster, R. E., and Zinder, N. D. (1968). Purification and properties of ribonuclease III from *Escherichia coli*. *J. Biol. Chem.* **243,** 82–91.

Romaniuk, P. J., de Stevenson, I. L., and Wong, H. H. (1987). Defining the binding site of *Xenopus* transcription factor IIIA on 5S RNA using truncated and chimeric 5S RNA molecules. *Nucleic Acids Res.* **15,** 2737–2755.

Sambrook, J., Fritsch, E., and Maniatis, T. (1989). *Molecular cloning: A laboratory manual.* Cold Spring Harbor Press. Cold Spring Harbor, NY.

Sharp, J. S., and Bechhofer, D. H. (2005). Effect of 5′-proximal elements on decay of a model mRNA in *Bacillus subtilis*. *Mol. Microbiol.* **57,** 484–495.

Srivastava, A. K., and Schlessinger, D. (1990). Mechanism and regulation of bacterial ribosomal RNA processing. *Annu. Rev. Microbiol.* **44,** 105–129.

Studier, F. W. (1975). Genetic mapping of a mutation that causes ribonucleases III deficiency in *Escherichia coli*. *J. Bacteriol.* **124,** 307–316.

Sun, W., Jun, E., and Nicholson, A. W. (2001). Intrinsic double-stranded-RNA processing activity of *Escherichia coli* ribonuclease III lacking the dsRNA-binding domain. *Biochemistry* **40,** 14976–14984.

Sun, W., Li, G., and Nicholson, A. W. (2004). Mutational analysis of the nuclease domain of *Escherichia coli* ribonuclease III. Identification of conserved acidic residues that are important for catalytic function *in vitro*. *Biochemistry* **43,** 13054–13062.

Wagner, E. G. H., Altuvia, S., and Romby, P. (2002). Antisense RNAs in bacteria and their genetic elements. *Adv. Genet.* **46,** 361–398.

Wang, W., and Bechhofer, D. H. (1997). *Bacillus subtilis* RNase III gene: Cloning, function of the gene in *Escherichia coli*, and construction of *Bacillus subtilis* strains with altered *rnc* loci. *J. Bacteriol.* **179,** 7379–7385.

Zhang, K., and Nicholson, A. W. (1997). Regulation of ribonuclease III processing by double-helical sequence antideterminants. *Proc. Natl. Acad. Sci. USA* **94,** 13437–13441.

CHAPTER SEVENTEEN

STUDYING TMRNA-MEDIATED SURVEILLANCE AND NONSTOP MRNA DECAY

Thomas Sundermeier, Zhiyun Ge, Jamie Richards, Daniel Dulebohn, *and* A. Wali Karzai

Contents

Abstract

In bacteria, ribosomes stalled at the 3′-end of nonstop or defective mRNAs are rescued by the action of a specialized ribonucleoprotein complex composed of tmRNA and SmpB protein in a process known as *trans*-translation; for recent reviews see Dulebohn *et al.* [2007], Keiler [2007], and Moore and Sauer [2007]. tmRNA is a bifunctional RNA that acts as both a tRNA and an mRNA.

Department of Biochemistry and Cell Biology and Center for Infectious Diseases of Stony Brook University, Stony Brook, New York

Methods in Enzymology, Volume 447
ISSN 0076-6879, DOI: 10.1016/S0076-6879(08)02217-9

SmpB-bound tmRNA is charged with alanine by alanyl-tRNA synthetase and recognized by EF-Tu (GTP). The quaternary complex of tmRNA-SmpB-EF-Tu and GTP recognizes stalled ribosomes and transfers the nascent polypeptide to the tRNA-like domain of tmRNA. A specialized reading frame within tmRNA is then engaged as a surrogate mRNA to append a 10 amino acid (ANDENYALAA) tag to the C-terminus of the nascent polypeptide. A stop codon at the end of the tmRNA reading frame then facilitates normal termination and recycling of the translation machinery. Through this surveillance mechanism, stalled ribosomes are rescued, and nascent polypeptides bearing the C-terminal tmRNA-tag are directed for proteolysis. Several proteases (ClpXP, ClpAP, Lon, FtsH, and Tsp) are known to be involved in the degradation of tmRNA-tagged proteins (Choy *et al.*, 2007; Farrell *et al.*, 2005; Gottesman *et al.*, 1998; Herman *et al.*, 1998, 2003; Keiler *et al.*, 1996). In addition to its ribosome rescue and peptide tagging activities, *trans*-translation also facilitates the selective decay of nonstop mRNAs in a process that is dependent on the activities of SmpB protein, tmRNA, and the 3' to 5'-exonuclease, RNase R (Mehta *et al.*, 2006; Richards *et al.*, 2006; Yamamoto *et al.*, 2003). Here, we describe methods and strategies for the purification of tmRNA, SmpB, Lon, and RNase R from *Escherichia coli* that are likely to be applicable to other bacterial species. Protocols for the purification of the Clp proteases, Tsp, and FtsH, as well as EF-Tu and other essential *E. coli* translation factors may be found elsewhere (Joshi *et al.*, 2003; Kihara *et al.*, 1996; Makino *et al.*, 1999; Maurizi *et al.*, 1990; Shotland *et al.*, 2000). In addition, we present biochemical and genetic assays to study the various aspects of the *trans*-translation mechanism.

1. INTRODUCTION

Translation is the process by which genetic information encoded within the nucleotide sequence of an mRNA is decoded for the biosynthesis of a polypeptide. A large ribonucleoprotein machine, the ribosome, catalyzes protein synthesis with transfer RNAs (tRNAs) serving as the adaptor molecules that link mRNA codons to the amino acid sequence of nascent polypeptides. Protein synthesis takes place in three stages: initiation, elongation, and termination and is regulated by stage-specific protein factors. Initiation is the loading of ribosomes onto the mRNA through a specialized tRNA molecule (tRNA^{fMet}) and the activities of three protein initiation factors (IF-1, IF-2, and IF-3). The 70S initiation complex contains the specialized fMet-initiator-tRNA in the ribosomal P-site and the next codon poised to be decoded in the ribosomal A-site. Elongation requires tRNAs, aminoacyl tRNA synthetases, and elongation factors EF-Tu, EF-Ts, and EF-G. Polypeptide synthesis begins by sampling of the A-site codon by a ternary complex of aminoacyl-tRNA, EF-Tu, and GTP. Cognate aminoacyl-tRNAs, capable of complementary base pairing with the awaiting A-site codon, are able to trigger GTP hydrolysis on EF-Tu. Release of EF-Tu-(GDP) from the ribosome permits the accommodation of aminoacyl-tRNA into the ribosomal A-site, followed by peptide bond formation. The resulting peptidyl-tRNA residing in the A-site, is translocated to the P-site through the action of EF-G and hydrolysis of GTP. Translocation positions the next codon in the A-site decoding center to start another round of aminoacyl-tRNA sampling, peptide bond formation, and translocation. Translation elongation continues until the ribosome encounters an in-frame stop codon, which signals the end the elongation phase of translation.

Bacteria have three stop codons (UAA, UAG, and UGA). Two protein release factors recognize these codons (RF-1 recognizes UAA and UAG, and RF-2 recognizes UAA and UGA). These release factors bind in the ribosomal A-site and promote hydrolysis of the polypeptide-tRNA linkage, releasing the newly synthesized polypeptide. RF-3, a GTP binding protein, then binds the A-site promoting release of RF-1 or RF-2. Finally, GTP hydrolysis on RF-3 promotes the dissociation of RF-3 from the ribosome. Peptide release leaves a 70S ribosome complex with mRNA and a deacylated P-site tRNA. Disassembly of this complex into free 50S and 30S ribosomal subunits requires the action of ribosome recycling factor (RRF) and EF-G (GTP) and is necessary to initiate a new round of protein synthesis. It is important to note that the presence of an in-frame termination codon is required to elicit this elaborate process of nascent polypeptide release and ribosome recycling.

Gene mutation, DNA damage, premature transcription termination, mRNA damage, and translational errors could all lead to mRNAs that lack in-frame stop codons. These nonstop mRNAs can still serve as templates for translation initiation and elongation. In the absence of a stop codon, translating ribosomes can decode to the end of an mRNA but are unable to properly terminate, recycle, or release the attached polypeptide (Fig. 17.1). Such translation blockage events could have two potentially hazardous consequences for the cell. First, because an in-frame stop codon is required to recruit the translation termination apparatus, mRNAs that lack an in-frame stop codon lead to ribosome stalling and significant loss of translational efficiency. Second, aberrant protein products translated from incomplete mRNAs may be harmful to cells. How do cells deal with these nonstop mRNAs? Bacteria have evolved a unique quality control mechanism, called *trans*-translation, mediated by the SmpB-tmRNA complex to deal with problematic mRNAs that cause ribosome stalling (Fig. 17.1). tmRNA, the central player in *trans*-translation, is a small stable RNA consisting of an alanyl-tRNA-like domain linked to an mRNA-like domain encoding a short reading frame followed by a stop codon (Fig. 17.2). tmRNA is recruited in a complex with its requisite protein cofactor, SmpB, to stalled ribosomes (Karzai *et al.*, 1999; Sundermeier and Karzai, 2007). The amino

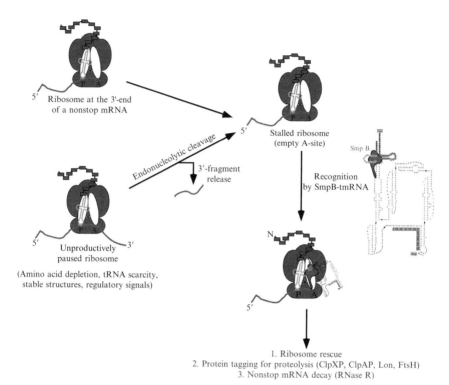

Figure 17.1 *Trans*-translation. Ribosomes stall either at the 3′-end of a nonstop message or at internal positions on the mRNA. Pausing at internal positions is thought to elicit an mRNA cleavage event, generating a nonstop mRNA. Both types of ribosome stalling are alleviated by the action of SmpB protein and tmRNA through a mechanism known as *trans*-translation. *Trans*-translation results in rescue of stalled ribosomes, directed degradation of the aberrant protein product, and facilitated decay of the mRNA.

acylated–tRNA–like domain of tmRNA is recognized by EF-Tu (GTP), enters the ribosomal A-site, and transfers the nascent polypeptide chain to the alanine charge of the tRNA-like domain of tmRNA to initiate the *trans*-translation process (Barends *et al.*, 2001; Rudinger-Thirion *et al.*, 1999). The short reading frame of the mRNA-like domain is then engaged as a surrogate mRNA, providing stalled ribosomes with an alternate means for elongation, termination, and recycling. In addition, the released nascent polypeptide is extended by an 11-amino acid (A-ANDENYALAA) C-terminal tag that is recognized by ClpXP, ClpAP, and Lon proteases and rapidly degraded (Choy *et al.*, 2007; Farrell *et al.*, 2005; Gottesman *et al.*, 1998; Herman *et al.*, 1998, 2003; Keiler *et al.*, 1996).

A number of additional conditions could lead to nonproductive pausing of the translation machinery and elicit *trans*-translation (Fig. 17.1). It is not clear how long such translation blockage events must persist before the

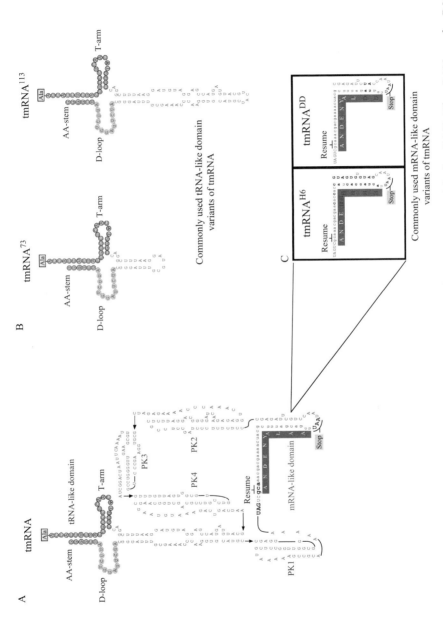

Figure 17.2 Commonly used synthetic variants of tmRNA. (A) Secondary structure model of tmRNA. The tRNA- and mRNA-like domains are highlighted. The four pseudoknots within tmRNA are labeled as PK1, PK2, PK3 and PK4. (B) tmRNA[73] and tmRNA[113] are two

tmRNA surveillance system is activated. It is known that under some circumstances nonproductive ribosome stalling events lead to cleavage of the causative mRNA in the ribosomal A-site (Hayes and Sauer 2003; Hayes *et al.*, 2002a, 2002b; Richards *et al.*, 2006; Sunohara *et al.*, 2002, 2004a, 2004b). However, the endonuclease(s) involved has not yet been identified. Our current understanding is that the SmpB-tmRNA complex recognizes these A-site vacant stalled ribosomes to commence the *trans*-translation process, leading to ribosome rescue, selective nonstop mRNA decay, and protein tagging for directed proteolysis (Fig. 17.1).

The universal presence of the SmpB-tmRNA system in eubacteria suggests that it confers a significant selective advantage. Indeed, recent studies of SmpB and tmRNA mutants in several bacterial species suggest that tmRNA activity is required for normal cellular physiology. In *Escherichia coli, Yersinia pseudotuberculosis,* and *Bacillus subtilis,* tmRNA is required for optimal growth under a variety of nutritional and environmental stress conditions (Karzai *et al.*, 1999; Muto *et al.*, 2000; Okan *et al.*, 2006; Shin and Price 2007). In *Caulobacter crescentus,* tmRNA activity is required for efficient cell-stage–specific developmental transitions (Keiler and Shapiro 2003). In addition, tmRNA activity is shown to be essential for growth of *Neisseria gonorrhoeae* (Huang *et al.*, 2000), *Haemophilus influenzae* (Akerley *et al.*, 2002), and species of *Mycoplasma* (Glass *et al.*, 2006; Hutchison *et al.*, 1999). Most intriguingly, tmRNA mutants in *Salmonella enterica* (Baumler *et al.*, 1994; Julio *et al.*, 2000) and *Yersinia pseudotuberculosis* (Okan *et al.*, 2006) have decreased virulence. In *Yersinia pseudotuberculosis,* loss of SmpB and tmRNA function leads to severe defects in the expression and delivery of virulence effector proteins (Okan *et al.*, 2006). These findings suggest an important regulatory role for the SmpB-tmRNA surveillance system in bacterial fitness and survival under a variety of stress conditions.

2. METHODOLOGY

In recent years, a number of genetic, molecular, and biochemical approaches have been developed to study the tmRNA-mediated surveillence mechanism. We will describe some of the methods and assays used in our research to study tmRNA, the *trans*-translation process, and its

---

aimed at separating the tRNA- and mRNA-like activities of tmRNA. tmRNA truncation variants are also commonly used for binding studies as both SmpB and EF-Tu proteins bind the tRNA-like domain. (C) tmRNA^{H6} and tmRNA^{DD} carry substitutions in the mRNA-like domain of tmRNA, changing the amino acid sequence of the tag peptide. Both variants stabilize tmRNA-tagged proteins against proteolysis, facilitating their detection in cells.

associated protein factors. These methods and assays are based on both modification of existing techniques and development of new approaches. The methods described herein are best suited for evaluating the various aspects of the *trans*-translation process, including SmpB-tmRNA interactions, recognition of stalled ribosomes, selective decay of defective mRNAs, and targeted degradation of tmRNA tagged proteins.

2.1. Purification of SmpB protein

E. coli SmpB protein, overexpressed in the absence of tmRNA, is insoluble. We developed the pET28-BA expression plasmid to coexpress soluble SmpB protein and tmRNA (Karzai and Sauer, 2001). The *smpB* and *ssrA* (the gene encoding tmRNA) genes are present in tandem on the *E. coli* chromosome. We have cloned these genes into the pET28b expression vector (Novagen) under the control of a T7 promoter. The genes were cloned such that the SmpB protein expressed from the construct carries an N-terminal 6-Histidine (6-His) epitope. The resultant plasmid (pET28-BA) is used for purification of SmpB and tmRNA, as well as *in vivo* activity assays. We use a two-step strategy to purify SmpB protein, including both affinity and cation exchange chromatography steps. For SmpB purification, 6 L of BL21(DE3)/pLysS/pET28-BA cells are grown at 37 °C in Luria Bertani (LB) broth supplemented with 50 μg/ml kanamycin and 30 μg/ml chloramphenicol, to an OD OD_{600} that is between 0.5 and 0.7. SmpB and tmRNA expression is induced by addition of isopropyl β-D-1–thiogalactopyranoside (IPTG) to 1 mM final concentration. Cells are grown for an additional 3 h at 37 °C and harvested by centrifugation. Alternately, cells may be grown overnight ($\sim$16 h) in ZYM-5052 autoinducing medium and harvested (Studier, 2005). Cells are resuspended in 60 ml of cold lysis buffer (50 mM HEPES, pH 7.5, 1 M NH$_4$Cl, 5 mM MgCl$_2$, 2 mM β-mercaptoethanol (β-ME), and 10 mM imidazole). The high NH$_4$Cl concentration disrupts SmpB-tmRNA interactions allowing purification of free SmpB. Cells are separated into two 50 ml conical tubes and lysed with three 30 sec sonication pulses. Between each sonication pulse, 100 μl of 100 mM phenylmethanesulfonyl fluoride (PMSF) is added to prevent proteolysis, and the lysate is left on ice for at least 1 min before the next pulse. Cell debris is removed by centrifugation at 30,000g for 1 h, and 1 ml of 50% Ni^{2+}-NTA agarose resin (Qiagen) (preequilibrated in lysis buffer) is added to the supernatant and rotated for 1 h at 4 °C. The cell lysate/Ni^{2+}-NTA slurry is applied to a 35 ml Econo-chromatography column (BioRad), and the resin is allowed to settle. Unbound material is allowed to flow through by gravity and collected. The resin is washed three times with 20 ml of lysis buffer. To remove nonspecifically bound material, it is important to completely resuspend the resin in lysis buffer with each 20 ml wash, allow it to settle, and then allow the dislodged material to flow through. The resin is washed once more with buffer W (50 mM HEPES, pH 7.5, 50 mM

KCl, 5 mM MgCl$_2$, 2 mM β-ME, and 10 mM imidazole) to remove excess salt and prepare the resin for the elution step. SmpB is eluted in three steps with 5 ml of elution buffer (50 mM HEPES, pH 7.5, 50 mM KCl, 5 mM MgCl$_2$, 2 mM β-ME, and 200 mM imidazole). With each 5 ml elution step, the resin is resuspended in elution buffer, allowed to settle for 5 min, and the eluted material is collected. Protein-containing fractions are pooled, and the volume is raised to 50 ml with buffer A (50 mM HEPES, pH 7.5, 50 mM KCl, 5 mM MgCl$_2$, and 2 mM β-ME). It is important to remove any insoluble material present in the eluted sample by centrifugation at 30,000g for 30 min.

The second step in the SmpB purification process is a cation exchange step performed by fast protein liquid chromatography (FPLC) with a MonoS (HR10/10) column (GE Healthcare). The FPLC column is washed and equilibrated in buffer A (see earlier). The cleared SmpB containing sample is applied to the column, and unbound material is removed by washing with 5 bed volumes of FPLC buffer A. SmpB is eluted with a linear gradient of 0 to 75% of buffer B (50 mM HEPES, pH 7.5, 1 M KCl, 5 mM MgCl$_2$, and 2 mM β-ME). SmpB elutes as a single peak at approximately 450 mM KCl. Purified SmpB is concentrated and exchanged into storage buffer (50 mM Tris-HCl, pH 7.5, 50 mM KCl, 5 mM MgCl$_2$, 2 mM β-ME, and 5% [v/v] glycerol) with a 15 ml Amicon Ultra centrifugal filter device with a 5-kDa molecular weight cutoff (Millipore). The product is then aliquoted, flash frozen using liquid nitrogen, and stored at $-80\,^\circ$C. The typical yield from this procedure is approximately 500 nmol of SmpB protein at greater than 95% purity.

2.2. Purification of tmRNA

Purified preparations of tmRNA may be generated either by *in vitro* transcription and purification or by purifying from cells overexpressing tmRNA. tmRNA contains several modified nucleotides in its tRNA-like domain (Felden *et al.*, 1998). tmRNA purified from cells should possess these modified bases, making cell purification the preferred method for tmRNA isolation. We generally purify full-length tmRNA from cells but use *in vitro* transcription to generate tmRNA truncation variants (Fig. 17.2) for binding studies or other functional assays. Because tmRNA undergoes extensive posttranscriptional modification (5′- and 3′-end processing and modification of the D-loop and T-arm bases), care must be taken in choosing the method of tmRNA expression to isolate a pure preparation of mature tmRNA. Full induction of tmRNA expression may lead to a mixture of the pre-tmRNA transcript and the mature form.

For purifying tmRNA from cells, we first express and purify the SmpB-tmRNA complex. This approach has two advantages. First, SmpB stabilizes tmRNA within cells. Second, this approach allows for an initial Ni^{2+}-NTA affinity step, purifying the SmpB-tmRNA complex on the basis of the 6-His epitope attached to SmpB. Six liters of BL21(DE3)/pLysS/pET28-BA cells are grown to OD_{600} of ~0.5 in LB broth supplemented with 10 μM IPTG, 50 μg/ml kanamycin and 30 μg/ml chloramphenicol. Ni^{2+}-NTA purification is accomplished as described previously except that the NH_4Cl concentration in the lysis buffer is reduced from 1 M to 100 mM, and a total of 500 μl of 50% Ni^{2+}-NTA resin slurry is used. The Ni^{2+}-NTA eluate fractions are combined and concentrated down to 5 ml with an Amicon Ultra centrifugal filter device with a 30-kDa MW cutoff (Millipore). tmRNA is separated from SmpB by RNA extraction with 15 ml of TriReagent LS (Molecular Research Center) following the manufacturer's instructions. RNA is then resuspended in 10 ml of FPLC buffer A (see earlier) and loaded onto an FPLC MonoQ (HR10/10) anion exchange column. tmRNA is eluted with a linear gradient of 30 to 100% FPLC buffer B (see earlier) over 30 bed volumes. tmRNA elutes at approximately 650 mM KCl. Additional smaller peaks of contaminating material elute at higher salt concentrations. tmRNA containing fractions are pooled, a 1/10 volume of 3 M sodium acetate is added, and tmRNA is precipitated with isopropanol. tmRNA-containing pellets are washed with 75% ethanol and resuspended in TE buffer. The typical yield of tmRNA is approximately 10 nmol.

Wild-type tmRNA or tmRNA variants may also be generated by *in vitro* runoff transcription. We transcribe tmRNA with T7 RNA polymerase using dsDNA PCR products as the transcription template. The following protocol describes the production of tmRNA[113], a tmRNA truncation variant containing the tRNA-like domain of tmRNA but lacking the three pseudoknots and the mRNA-like reading frame (Fig. 17.2B). Two primers representing the 5'- and 3'-ends of the tmRNA[113] coding sequence (with overlapping complementary sequence) are extended by PCR with Pfu-Turbo DNA polymerase (Stratagene). This PCR extension product is used as template in a second PCR reaction using a 5'-primer designed to add the T7 promoter consensus sequence immediately 5'of the first transcribed base. This PCR product is gel purified and used as template for *in vitro* transcription. We generally use the template DNA derived from one 100 μl PCR reaction as template for one transcription reaction. However, increasing the DNA template concentration may, in some cases, improve the RNA yield of the transcription reaction. *In vitro* transcription reactions are performed in an *in vitro* transcription buffer (40 mM Tris-HCl, pH 8, 26 mM $MgCl_2$, 1 mM spermidine, 0.01% [v/v] Triton X-100, and 1 mM DTT). The reaction mixture (100 μl) contains the template DNA, 2 mM ATP, 2 mM CTP, 2 mM GTP, 2 mM UTP, 2.5 mM GMP, and 100 units of T7 RNA polymerase (USB). RNase inhibitors may also be added to the transcription reaction to control any potential low-level RNase contamination. Reaction mixtures are incubated at 37 °C for 3 h. Template DNA is removed by adding 10 μl of RNase-free DNase I (GE Healthcare) and incubating at 37 °C for an additional 30 min. Transcription reaction products (tmRNA[113]) are purified by electrophoresis in denaturing poly-acrylamide gels (SequaGel, National Diagnostics). An equal volume (100 μl) of gel loading dye (8 M urea, 10% [v/v] glycerol, 50 mM Tris-HCl, pH7.5, 0.05% [w/v] bromophenol blue) is added to the transcription reaction product, loaded onto a 12% sequagel, and run at 200 V until the bromophenol blue front reaches the bottom of the gel (3 to 4 h). UV shadowing, aided by fluorophore-coated TLC plates (Ambion), is used to visualize RNA bands. The fraction of the gel containing tmRNA[113] is cut out of the gel with a clean razor blade and cut into small sections (approximately 0.5 cm^2). Each gel section is then placed into a 1.5-ml microcentrifuge tube and crushed into smaller pieces with a pipette tip; 600 μl of gel elution buffer (50 mM Tris-HCl, pH 7.5, 50 mM HEPES, pH 7.5, and 300 mM sodium acetate) is added to each tube and gently mixed overnight on a rotator at 4 °C. The gel slices are pelleted by centrifugation at 18,000g in a microcentrifuge for 5 min. The supernatant is transferred to a new tube and precipitated by adding 2.5 volumes of cold 100% ethanol and incubating for 3 h at −20 °C. tmRNA samples are resuspended in TE buffer and stored at −80 °C. *In vitro* transcription reaction yields are often variable and depend on the activity of the specific lot of T7 RNA polymerase used.

2.3. Purification of RNase R

We use a three-step strategy to purify RNase R, including affinity, cation exchange, and size exclusion chromatography steps. For RNase R purification, 3 L of BL21(DE3) *rnb*−/plysS/pET15b-rnr cells are grown at 37 °C in Luria Bertani (LB) broth supplemented with 100 μg/ml ampicillin and 30 μg/ml chloramphenicol to an OD OD$_{600}$ that is between 0.5 and 0.7. RNase R expression is induced by addition of IPTG to 1 mM final concentration. Cells are grown for an additional 3 h at 30 °C. Cultures are harvested by centrifugation at 3000g for 60 min at 4 °C, and the resulting cell pellet stored at −80 °C. The frozen cell pellet is thawed on ice and resuspended in 60 ml of a lysis buffer (20 mM HEPES, pH 7.5, 300 mM KCl, 2 mM MgCl$_2$, 0.5 mM EDTA, 2 mM β-ME). Cells are separated into two 50 ml conical tubes and lysed with three 30 sec sonication pulses. After each sonication pulse, 100 μl of 100 mM PMSF is added to prevent proteolysis, and the lysate is left on ice for at least 1 min before the next pulse. The lysate is centrifuged at 30,000g for 1 h at 4 °C. The supernatant is collected and filtered through a 0.45 μm syringe filter.

2.3.1. Affinity chromatography
The resulting cleared lysate is then applied to a preequilibrated 5 ml Hi-Trap Blue HP FPLC column (GE Healthcare). Unbound material is removed by washing with 10 bed volumes of buffer A (20 mM HEPES, pH 7.5, 2 mM MgCl$_2$, 0.5 mM EDTA, 2 mM β-ME, 150 mM KCl).

Bound RNase R is eluted with a linear gradient of 0 to 75% buffer B (20 mM HEPES, pH 7.5, 2 mM MgCl$_2$, 0.5 mM EDTA, 2 mM β-ME, 1 M KCl). A broad RNase R peak is observed between 300 and 500 mM KCl. Fractions containing the purest RNase R sample are then pooled and diluted to 300 mM KCl with dilution buffer R (20 mM HEPES, pH 7.5, 100 mM KCl, 2 mM MgCl$_2$, 0.5 mM EDTA, 2 mM β-ME). With this approach, we routinely observe a strong absorbance peak at 260 nm for this sample, suggesting that RNase R is associated with cellular RNA. The protein–RNA complex is incubated at 37 °C for 1 h to degrade the associated RNA.

2.3.2. Cation exchange chromatography

After incubation, the RNase R sample is cooled to 4 °C, filtered through a 0.45 μm syringe filter, and applied to a preequilibrated MonoS HR10/10 FPLC column (GE Healthcare). Unbound material is removed by washing with 5 bed volumes of buffer A (20 mM HEPES, pH 7.5, 2 mM MgCl$_2$, 0.5 mM EDTA, 2 mM β-ME, 100 mM KCl). RNA-free RNase R is eluted at approximately 500 to 550 mM KCl on application of a linear gradient of 0 to 70% buffer B (20 mM HEPES, pH 7.5, 2 mM MgCl$_2$, 0.5 mM EDTA, 2 mM β-ME, 1 M KCl).

2.3.3. Gel filtration chromatography

RNase R–containing fractions are pooled and concentrated with a 15 ml Amicon Ultra centrifugal filter device with a 30-kDa molecular weight cutoff (Millipore). The concentrated protein is applied to a Sephacryl S-200 gel filtration column (GE Healthcare), preequilibrated with RNase R storage buffer (20 mM Tris-HCl, pH 8.0, 300 mM KCl, 0.25 mM MgCl$_2$, 2 mM β-ME, 10% [v/v] glycerol). RNase R, with greater than 95% purity, elutes as a single peak from this column. Purified RNase R is concentrated with a 15 ml Amicon Ultra centrifugal filter device with a 30-kDa molecular weight cutoff (Millipore). The purified protein is aliquoted, flash frozen in liquid nitrogen, and stored at −80 °C. The typical yield from this procedure is approximately 300 nmol of purified RNase R protein.

2.3.4. RNase R purification: Alternate procedure for obtaining RNA-free protein

The affinity chromatography and cation exchange chromatography steps are the same as described previously, except the 37 °C incubation, after MonoS purification, is replaced with a dialysis step. Concentrated RNase R containing sample is placed in a 3.5-kDa molecular weight cutoff Slide-A-Lyzer dialysis cassette (Pierce) and dialyzed against 4 L of dialysis buffer (20 mM Tris-HCl, pH 8.0, 300 mM KCl, 2 mM MgCl$_2$, 2 mM β-ME, 10% [v/v] glycerol) at room temperature for 24 h. The rationale for this step is that RNase R will degrade the associated RNA during dialysis, and the released nucleotides can diffuse away. Indeed, the absorbance spectrum of the postdialysis RNase R sample does not contain an absorbance peak at 260 nm. This step is followed by the gel filtration chromatography step described previously.

2.4. Purification of Lon

We use C-terminally 6-His–tagged Lon protease for most of our studies. In our experience, placing a 6-His-tag on the N-terminus of Lon protease from *E. coli*, and several other bacterial species, leads to inactivation of the protein. In contrast, a C-terminal 6-His tag has no significant effect on Lon activity. We purify C-terminally 6-His–tagged Lon from *E. coli* strain CH1019 (Hayes and Sauer 2003) carrying a Lon expression plasmid by Ni^{2+}-NTA affinity and MonoQ (GE Healthcare) anion exchange chromatography steps. Three liters of CH1019/pET21b-Lon-6-His cells are grown at 37 °C in LB broth supplemented with 100 μg/ml ampicillin to OD_{600} between 0.5 and 0.7. Lon expression is induced with 1 mM IPTG for 2 h at 37 °C. Cell pellets are resuspended in lysis buffer (1 M NH_4Cl, 20 mM KPO_4, pH 7.4, 1 mM EDTA, 2 mM β-ME, 20 mM imidazole) and lysed by sonication. Cell debris is removed by centrifugation at 30,000g for 1 h, and 1 ml of 50% Ni^{2+}-NTA agarose resin (Qiagen) (preequilibrated in lysis buffer) is added to the supernatant and rotated for 1 h at 4 °C. The cell lysate/Ni^{2+}-NTA slurry is applied to a 35 ml Econo chromatography column (BioRad), and the resin is allowed to settle. Unbound material is allowed to flow through by gravity. Resin is washed three times with a total of 60 ml of lysis buffer. The resin is washed once more with buffer W (50 mM HEPES, pH 7.5, 50 mM KCl, 2 mM β-ME, and 20 mM imidazole) to remove excess salt and prepare the resin for the elution step. Lon is eluted in three steps with 5 ml of elution buffer (50 mM HEPES, pH 7.5, 50 mM KCl, 2 mM β-ME, and 200 mM imidazole). With each 5 ml elution step, the resin is resuspended in elution buffer, allowed to settle for 5 min, and the eluted material is collected. Lon-6-His-containing fractions are pooled, clarified by centrifugation, and applied onto a MonoQ column in buffer A (20 mM Tris-HCl, pH 7.6, 50 mM KCl, 2 mM EDTA, 2 mM β-ME, 10% [v/v] glycerol). Lon protease is eluted from the column on application of a linear gradient of 0 to 60% buffer B (20 mM Tris-HCl, pH 7.6, 1 M KCl, 2 mM EDTA, 2 mM β-ME, 10% [v/v] glycerol). The absence of contaminating proteins is verified by SDS-PAGE followed by Coomassie blue staining. Lon-containing fractions are concentrated and exchanged into storage buffer (50 mM Tris-HCl, pH 8, 50 mM KCl, 1 mM EDTA, 2 mM β-ME, and 10% [v/v] glycerol) using a 15mL Amicon Ultra centrifugal filter device with 30-kDa molecular weight cutoff (Millipore). Concentrated protein samples are flash frozen using liquid nitrogen and stored at −80 °C. The typical yield from this procedure is approximately 400 nmol of purified Lon protein.

3. Activity Assays

We have developed several assays to study different aspects of the *trans*-translation mechanism. SmpB and tmRNA are required for lytic development of two phages, namely *λimm*P22 *dis c2*–5 hybrid phage and a temperature-sensitive variant of phage Mu (Mu$_{CTS}$) (Karzai *et al.*, 1999; Ranquet *et al.*, 2001; Retallack *et al.*, 1994; Withey and Friedman, 1999). Analysis of the ability of *E. coli* mutants to support induction of these phages represents a simple assay for *trans*-translation activity that is amenable to high-throughput screening. *In vivo trans*-translation activity may also be directly assayed by looking at its protein products under normal growth conditions with the tmRNA^{H6} endogenous tagging assay. Association of components of the tmRNA quaternary complex with ribosomes may be assessed through a ribosome copurification assay. Binding of quaternary complex components to tmRNA is assayed with electrophoretic mobility shift assays. We have developed protein and mRNA stability assays to study the proteolysis and mRNA decay activities of *trans*-translation. Finally, we have developed a stalled ribosome enrichment protocol to identify additional components of the SmpB-tmRNA–stalled ribosome complex.

3.1. Phage induction assays

Mu$_{Cts}$ is a temperature-sensitive variant of phage Mu that requires SmpB and tmRNA for its lytic development (Karzai *et al.*, 1999; Ranquet *et al.*, 2001). Mu$_{Cts}$ induction assays may be used to monitor activity of the *trans*-translation system. Mu lysogens are grown at 30 °C to OD$_{600}$ of 0.4 to 0.5 then transferred to 42 °C. Induction of lytic development is assayed by monitoring the OD$_{600}$ of the cultures at 20 min time intervals for 2 h. Clearing of the culture caused by efficient Mu$_{Cts}$ induction suggests an intact *trans*-translation system, whereas cell survival indicates that *trans*-translation function might be impaired (Karzai *et al.*, 1999).

*λimm*P22 *dis c2*–5 hybrid phage induction assays provide a more quantitative assessment of *trans*-translation system function; 3 ml of *E. coli* culture is grown at 37 °C to saturation in LB supplemented with 5 m*M* MgCl$_2$ and 10 m*M* CaCl$_2$. Four separate titrations of phage are prepared at dilutions between 10^{-2} and 10^{-9}, depending on phage titer, in LB with 5 m*M* MgCl$_2$ and 10 m*M* CaCl$_2$; 100 μl of each phage dilution is added to a separate 5 ml culture tube, and 500 μl of bacterial culture is added to each tube. Samples are mixed gently and incubated for 30 min at 37 °C, with tubes being gently mixed every 10 min; 4 ml of 0.7% top agar in LB (supplemented with 5 m*M* MgCl$_2$ and 10 m*M* CaCl$_2$ immediately before use) is added to each tube, mixed, and plated on LB agar plates, and cells are grown at 37 °C overnight. Phage induction results in the formation of plaques in the bacterial lawn within the top agar. Counting of plaques formed at each phage dilution gives a quantitative assessment of phage-induced cell lysis. Lysis suggests that the *trans*-translation system is active.

The *trans*-translation dependence of $\lambda immP22$ *dis c2*–5 hybrid phage induction may also be used for high-throughput screening with a phage cross-streaking assay (Fig. 17.3). This method is less quantitative than that described previously but more amenable to assessing *trans*-translation system activity in a large number of mutant strains (Choy *et al.*, 2007). SmpB and tmRNA functions are required for lytic development of the $\lambda immP22$ *dis c2*–5 hybrid phage but not for lytic development of the related $\lambda immP22$ *dis c1*–7 phage (Choy *et al.*, 2007; Karzai *et al.*, 1999; Retallack *et al.*, 1994; Withey and Friedman, 1999). Analysis of phage induction phenotypes of mutant strains with both hybrid phages allows for elimination of candidates with phage biology defects. Phage-containing solutions (at appropriate

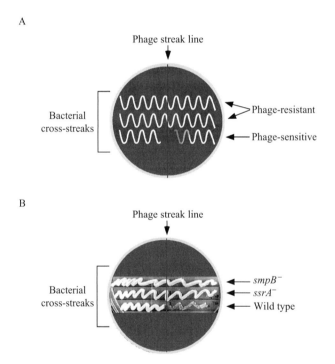

Figure 17.3 $\lambda immP22$ *dis c2*–5 hybrid phage cross-streak assay. (A) Schematic representation of the $\lambda immP22$ *dis c2*–5 hybrid phage cross-streak assay. Phage is spotted at the center of each lane on an agar plate, and cells are streaked across the spot. If the strain is sensitive to hybrid phage induction, cell lysis is observed in the area around the phage spot. Strains resistant to the phage exhibit continuous growth along the line of the streak. (B) A representative phage cross-streak assay that uses wild-type, *smpB⁻*, and *ssrA⁻ E. coli* strains is shown to illustrate the phage phenotypes of these strains.

dilutions depending on phage titer) are spotted at the center of each lane on an LB agar plate. Mutant strains are then streaked across the lane, and the plates are incubated overnight at 37 °C. The absence of bacterial cell growth in the area surrounding the spotted phage is indicative of cell lysis by phage induction (Fig. 17.3). Strains that are lysed by the $\lambda immP22\ dis\ c1$-7 phage but resistant to $\lambda immP22\ dis\ c2$-5 phage are likely to exhibit defects in *trans*-translation system activity.

3.2. Endogenous tmRNA^{H6} tagging assay

Trans-translation system activity may also be studied directly *in vivo* by monitoring the accumulation of tmRNA-tagged protein products by use of the endogenous tmRNA^{H6} tagging assay. The endogenous tagging assay uses a variant of tmRNA (tmRNA^{H6}) with mRNA-like sequence mutated such that the last six amino acids of the tag peptide are histidines (Fig. 17.2C). Cotranslational addition of the tmRNA^{H6} tag to the C-termini of target proteins renders them more resistant to proteolysis and simpler to purify by Ni^{2+}-NTA affinity chromatography (Dulebohn *et al.*, 2006; Sundermeier *et al.*, 2005). We have mutated the *ssrA* gene in pET28-BA plasmid to code for the tmRNA^{H6} variant (pET28-BA^{H6}). We use the endogenous tmRNA^{H6} tagging activity to test the activity of SmpB protein variants, but the technique may be extended to study the effect of mutations to any component of *trans*-translation; 50 ml cultures of $\Delta smpB$(DE3) cells carrying the pET28-BA^{H6} construct, exhibiting various mutations to *smpB,* are grown at 37 °C to late log phase (OD$_{600}$ = 0.8 to 1.0) under continuous low level (5 μM) IPTG induction. Cells are harvested by centrifugation and resuspended in 1 ml of ET buffer (8 M urea, 100 mM KPO$_4$, pH 8, 10 mM Tris-HCl, pH 8, 5 mM β-ME). Resuspended cells are lysed by rocking at room temperature for 1 h, and cellular debris is removed by centrifugation at 30,000g for 1 h. The supernatant is added to 200 μl of 50% Ni^{2+}-NTA agarose resin (Qiagen) preequilibrated in ET buffer. The soluble lysate/Ni^{2+}-NTA agarose slurry is incubated for 1 h at room temperature. Slurries are applied to a Micro-Bio-Spin chromatography column (BioRad) and allowed to flow through by gravity. The resin is washed four times with 1 ml of ET buffer. After the final wash, the column is placed into a 1.5 ml microcentrifuge tube and centrifuged for 1 min at 3000g to remove excess ET buffer. His-tagged proteins are eluted in 200 μl of elution buffer (8 M urea, 100 mM acetic acid, 20 mM β-ME). Resin is resuspended in elution buffer, incubated at room temperature for 2 min, and the eluate is collected by centrifugation at 3000g for 1 min. The eluate pH is neutralized by the addition of 1/10 volume (20 μl) of 2 M Tris-HCl, pH 9.5. tmRNA^{H6}-tagged proteins are resolved by electrophoresis in a 15% Tris-Tricine polyacrylamide gel, transferred to PVDF membrane, and Western blots are developed with an antibody specific to the 6-His epitope. Figure 17.4 shows endogenous tmRNA^{H6} tagging phenotypes of *smpB, relEB,* and *dinJ-yafQ* mutants along-side the otherwise isogenic parental (WT) *E. coli* strain. As expected, no tmRNA tagging activity is observed in cells lacking SmpB protein, whereas tagging activity is observed for all other strains.

3.3. SmpB-tmRNA interactions with electrophoretic mobility shift assay

An essential early step in the *trans*-translation mechanism is association of SmpB with tmRNA and EF-Tu (GTP) to form the quaternary complex that recognizes and binds stalled ribosomes to initiate the ribosome rescue and protein-tagging activities. Association of these components may be studied *in vitro* with an electrophoretic mobility shift assay. The following is a protocol for studying SmpB-tmRNA interactions. Because of the small size of SmpB (18.3 kDa) and the large size of tmRNA (117 kDa), binding of one molecule of SmpB to one molecule of tmRNA produces only a small and difficult-to-resolve shift in electrophoretic mobility. To get around this technical challenge, we have engineered a series of tmRNA truncation variants that possess confirmed and putative SmpB binding sites (Fig. 17.2B). A variety of evidence points to interaction of SmpB protein with bases within the tRNA-like domain (TLD) of tmRNA. As such, we

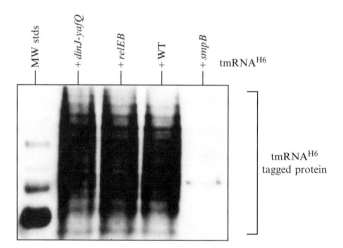

Figure 17.4 Endogenous tagging assay. Western blotting with antibodies against the 6-His epitope reveals the typical pattern of tmRNA^{H6}-tagged proteins under normal growth conditions. Endogenously tagged proteins from a wild-type *E. coli* strain and from three mutant strains are shown. SmpB protein is required for *trans*-translation; hence, no endogenously tagged proteins are detected in Δ*smpB* cells. In contrast, the other mutants show no defect in endogenous tagging activity.

study this interaction with tmRNA[113], a tmRNA variant containing the TLD but lacking all pseudoknots and mRNA sequence (Fig. 17.2B).

3.3.1. tmRNA labeling

We use the terminal tRNA nucleotidyl transferase catalyzed ATP-pyrophosphate exchange reaction to 3′-end label tmRNA and tmRNA variants (Fig. 17.2B) with ^{32}P. The labeling reaction (100 μl) contains 50 mM glycine, pH 9, 10 mM MgCl$_2$, 1 mM sodium pyrophosphate (NaPP$_i$), approximately 100 pmol of tmRNA[113], 10 mM [α^{32}P]-ATP (3000 Ci/mmol), and 1 μg of purified terminal tRNA nucleotidyl transferase (Sundermeier *et al.*, 2005). Reactions are incubated at 37 °C for 15 min. Subsequently, 5 units of yeast pyrophosphatase (Sigma) are added and reactions are incubated at 37 °C for an additional 1 min before quenching with TriReagent LS (MRC). RNA is extracted by use of manufacturer's instructions, resuspended in TE buffer, aliquoted, and stored at −80 °C.

3.3.2. Electrophoretic mobility shift assay

3′-end–labeled tmRNA[113] (10 to 20 pM) is incubated with varying concentrations of purified SmpB in a gel shift buffer containing 50 mM Tris-HCl, pH 7.5, 10 mM MgCl$_2$, 300 mM KCl, 2 mM β-ME, 100 μg/ml BSA, 0.01% (v/v) NP-40, 10% (v/v) glycerol, and 200 nM total-*E. coli* tRNA (included as nonspecific competitor). Reactions are incubated for 30 min at 4 °C then loaded onto native 12% polyacrylamide gels (run in 1/2× TBE, pH 7.5). Gels are dried and exposed to PhosphorImager screens. PhosphorImager data are collected with a Storm840 PhosphorImager and analyzed with ImageQuant Tools (Molecular Diagnostics). Figure 17.5 shows a typical

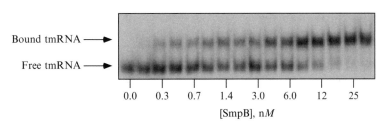

Figure 17.5 Gel mobility-shift assay for the detection of specific interactions between SmpB protein and tmRNA. Gel mobility-shift assay was performed with 3′-end–labeled tmRNA[113]. A typical binding reaction was carried out under high-stringency binding conditions in the presence of 300 mM KCl and 200 nM total-cell *E. coli* tRNA. The reaction buffer contained 50 mM Tris-HCl, pH 7.5, 10 mM MgCl$_2$, 300 mM KCl, 2 mM β-mercaptoethanol, 100 μg/ml BSA, 0.01% (v/v) Nonidet P-40, 10% (v/v) glycerol, and 200 nM total *E. coli* tRNA. Reaction mixtures containing 10 pM tmRNA[113] and various concentrations of SmpB protein were incubated at 4 °C for 30 min, and the bound and free species of tmRNA were resolved by electrophoresis in a native 10% acrylamide gel in 0.5× TBE.

electrophoretic mobility shift assay of SmpB-tmRNA interactions, in which 1 to 2 nM SmpB is required to shift half of the labeled tmRNA[113].

3.4. Ribosome association assay

Another important experimentally separable activity of the *trans*-translation system is binding of the SmpB-tmRNA-EF-Tu (GTP) quaternary complex to stalled ribosomes. Association of components of the quaternary complex with ribosomes may be monitored through ribosome copurification assays. Care must be taken in the choice of stringencies used for ribosome purification, because many ribosome-associated factors, including SmpB, are highly basic proteins. Therefore, ionic stringency at least similar to that encountered in the cytosol must be maintained throughout the ribosome purification process to eliminate nonspecific charge interactions between basic proteins and the ribosome (Sundermeier and Karzai, 2007). A variety of different protocols are available for ribosome purification. Most separate ribosomes from other cellular components on the basis of their high density, by use of density gradient centrifugation or pelleting of ribosomes by centrifugation through a sucrose cushion. The following protocols describe purification of tight-coupled 70S ribosomes, 70S ribosomes, and 50S and 30S ribosomal subunits. Association of protein and RNA components with these ribosome preparations may be assayed by Western and Northern blot analysis, respectively.

   E. coli culture (750 ml) is grown in LB to an OD_{600} of approximately 0.8. Cells are harvested by centrifugation and resuspended in 25 ml of cold ribosome buffer A (20 mM Tris-HCl, pH 7.5, 300 mM NH$_4$Cl, 10 mM MgCl$_2$, 0.5 mM EDTA, 6 mM β-ME). Cells are lysed by French press, with one pass of 2 min duration in a 25 ml cell (Sim-Aminco Spectronic Instruments). Cell debris is removed by centrifugation at 30,000g for 1 h at 4 °C; 20 ml of the supernatant (S30 extract) is then layered onto 20 ml of a 32% sucrose cushion in ribosome buffer A. Ribosomes are pelleted by centrifugation at 120,000g for 22 h at 4 °C. We use a Beckman Type 45Ti rotor in a Beckman-Coulter L100 XP Optima ultracentrifuge, but the protocol may be adapted to many different ultracentrifuges and rotors. The pellet from this centrifugation step is known as tight-coupled 70S ribosomes. The pellet is washed with 5 ml of ice-cold ribosome buffer A. A thin layer of lipid contamination is generally present on top of the clear ribosome pellet. This may be removed by pipetting up and down during the 5 ml wash. Tight-coupled 70S ribosomes are resuspended in 500 μl of buffer A by incubating at 4 °C for 2 to 3 h, with gentle mixing every 30 min.

   Alternately, 70S ribosomes may be isolated by density gradient centrifugation. For this method, S30 extract is prepared as previously. Ribosomes are pelleted by centrifugation at 100,000g for 4 h at 4 °C (without a sucrose cushion). The pellet from this centrifugation step is known as crude ribosomes.

The crude ribosome pellet is washed twice with 5 ml of ribosome buffer B (20 mM Tris-HCl, pH 7.5, 500 mM NH$_4$Cl, 10 mM MgCl$_2$, 0.5 mM EDTA, 6 mM β-ME). The thin layer of lipids on top of the crude ribosome pellet is removed as described previously, although the crude ribosome pellet remains cloudy. Next, the crude ribosomes are resuspended in 500 μl of ribosome buffer A. Approximately 400 A$_{260}$ units (or 10 μmol) of crude ribosomes is layered onto a 38 ml, 10 to 50% sucrose gradient in ribosome buffer A. Gradients are prepared with the Gradient Master-107 gradient maker (Biocomp). The gradient is then subjected to centrifugation at 60,000g for 16 h in a Beckman SW28 swinging bucket ultracentrifuge rotor. The gradients are fractionated drop wise by piercing the bottom of the tube with a 25-gauge needle and collecting fractions of approximately 300 μl. The A$_{260}$ of each gradient fraction is measured with a microplate reader, with UV transparent 96-well microplates. 70S ribosomes migrate as a single sharp peak near the center of the gradient. Fractions representing the top third of the 70S peak are pooled and analyzed for association of SmpB, tmRNA or other quaternary complex components.

Association of *trans*-translation factors with 30S and 50S ribosomal subunits may also be monitored by dissociating tight-coupled 70S ribosomes, or crude ribosomes, into their component subunits. This is accomplished by incubating 70S ribosomes, or a mixed population of ribosomes, in buffer containing a low (1 mM) magnesium concentration. Crude ribosomes are prepared as described previously but washed and resuspended in 500 μl ribosome buffer C (20 mM Tris-HCl, pH 7.5, 300 mM NH$_4$Cl, 1 mM MgCl$_2$, 6 mM β-ME). Four hundred A$_{260}$ units of crude ribosomes are layered onto 38 ml of a 10 to 40% sucrose gradient in ribosome buffer C and subjected to centrifugation at 100,000g for 16 h. The A$_{260}$ of each fraction is measured as described previously. This time the A$_{260}$ profile of fractions resolves into two distinct peaks. A smaller peak localizes in the top third of the gradient volume and corresponds to 30S subunits. A larger peak, corresponding to the large (50S) ribosomal subunit migrates farther down the density gradient. Again, fractions corresponding to the top third of each peak are pooled and analyzed for association of SmpB, tmRNA, or other quaternary complex components.

3.5. Stalled ribosome enrichment assay

We have developed protocols for isolation of pools of ribosomes enriched for those that are stalled on either nonstop or rare codon-containing mRNAs (Sundermeier and Karzai, 2007). This stalled ribosome-enrichment procedure allows us to address what ribosome interactions are specific to *trans*-translation. We have generated plasmids that overexpress nonstop (λ-cI-NS) or rare codon-containing (λ-cI-N-4-AGG) mRNAs but also code for N-terminal 6-His epitopes. Expression of these constructs leads to ribosome stalling, either at the rare codon region or at the end of the nonstop mRNAs. Having already translated the N-terminus of the reporter protein product, the stalled ribosomes then display the 6-His epitope, enabling Ni^{2+}-NTA affinity purification (Fig. 17.6). It should be noted that the product of this purification is not expected to contain only stalled ribosomes, because ribosomes might be captured at any point beyond translation of the N-terminal 6-His epitope. However, because the stalled state is expected to be the most kinetically long-lived state between 6-His epitope translation and peptide release, this pool of ribosomes should be greatly enriched for stalled ribosomes.

Reporter mRNAs are overexpressed in an *E. coli* strain maintaining either pλ-cI-N-AGG or pλ-cI-NS; 1.5 L of culture is grown to an OD_{600} of 1.0, and expression of the reporter mRNA is induced with 1 mM IPTG for 45 min. Cells are harvested by centrifugation and resuspended in 15 ml of enrichment buffer (50 mM Tris-HCl, pH 7.5, 20 mM $MgCl_2$, 2 mM β-ME, 10 mM imidazole, 300 mM NH_4Cl). Cells are lysed by French press (1 pass, 2 min, 25 ml cell). The gentle French press lysis and high magnesium concentrations are both important for keeping stalled ribosome complexes intact. Ribosomes are purified by the tight-coupled 70S ribosome method, pelleted through a 32% sucrose cushion in enrichment buffer as described previously. Ribosomes are washed as described earlier and resuspended in 5 ml of enrichment buffer, to which 200 μl of Ni^{2+}-NTA agarose resin (Qiagen) is added. Ribosome-resin slurries are incubated for 2 h at 4 °C and applied to micro-bio spin columns (Biorad). Unbound material is allowed to flow through by gravity and the resin is washed four times with 1 ml of enrichment buffer. After the final wash, the column is placed into a 1.5 ml microcentrifuge tube and centrifuged for 1 min at 3000g. Stalled ribosomes are eluted with 250 μl of enrichment buffer containing 200 mM imidazole. The eluate is collected by centrifugation at 3000g for 1 min. An increase in factor binding to the stalled ribosome-enriched fraction versus the tight coupled 70S ribosome starting material suggests a functional involvement of that factor in *trans*-translation (Fig. 17.6B).

3.6. *In vitro* RNA degradation assay

There are three $3'$ to $5'$-exoribonucleases in *E. coli*: RNase R, RNase II, and polynucleotide phosphorylase (PNPase). RNase R is distinguished from RNase II and PNPase by its ability to degrade structured RNAs, such as rRNA and RNAs containing repetitive extragenic palindromic sequences, without the aid of a helicase activity or a protein cofactor (Vincent and Deutscher, 2006). RNase R is also distinguished from other exonucleases by its ability to participate in the tmRNA-facilitated degradation of defective mRNAs (Richards *et al.*, 2006). We routinely use an

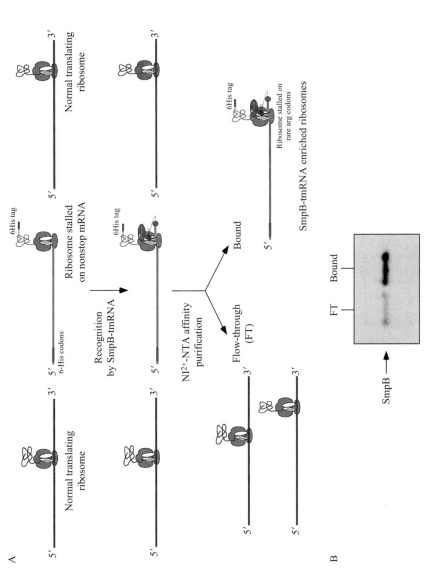

Figure 17.6 Stalled ribosome enrichment assay. (A) Schematic representation of the stalled ribosome enrichment assay. Ribosomes translating normal and nonstop mRNAs are depicted in the upper part of the panel. The SmpB-tmRNA complex recognizes only those ribosomes that have stalled at the 3′-end of a nonstop mRNA. The nonstop mRNA encodes for a polypeptide with an N-terminal 6-His epitope and,

in vitro RNA degradation assay that is designed to distinguish between RNase R and RNase II activities. The assay relies on two complementary oligonucleotides that, when annealed, form a double-stranded (ds) duplex substrate with a 15-nucleotide single-stranded (ss) 3′-poly-A overhang.

3.6.1. Materials required for *in vitro* decay assay

Complementary RNA oligos:

ss-32: 5′-<u>CCCCACCACCAUCACUU</u>AAAAAAAAAAAAAAAA-3′
Complementary ss-17: 5′-<u>AAGUGAUGGUGGUGGGG</u>-3′
The underlined sequences designate the complementary regions of the two oligos.
Purified RNase R: 0.5 μg/μl in storage buffer (20 mM Tris-HCl, pH 8.0, 300 mM KCl, 0.25 mM MgCl$_2$, 2 mM β-ME, 10% [v/v] glycerol).
10× RNase R activity buffer (200 mM Tris-HCl, pH 8.0, 1 M KCl, 2.5 mM MgCl$_2$, 10 mM DTT).
2× Gel loading buffer (95% formamide, 20 mM EDTA, 0.025% [w/v] bromophenol blue, and 0.025% [w/v] xylene cyanol).
20% Polyacrylamide gel containing 8 M urea.
PhosphorImager screen (GE Healthcare).

3.6.2. Preparation of duplex RNA substrate and RNA decay assay

For 5′-labeling of the single-stranded ss-32 substrate, we use [γ-^{32}P]-ATP and T4 polynucleotide kinase, in accordance with the manufacturer's instructions (New England BioLabs). Briefly, 50 pmol of ss-32 RNA oligo is mixed with 20 units of T4 polynucleotide kinase and 50 pmol of [γ-^{32}P]-ATP in 50 μl reaction volume of 1× T4 polynucleotide kinase buffer. Radioactive ATP is added last to start the reaction. The reaction mixture is incubated at 37 °C for 30 min, followed by denaturation of the kinase at 70 °C for 20 min. Unincorporated ATP is removed by applying the kinase reaction mixture to a Sephadex G-25 quick spin column (Roche). Mixing the 5′-^{32}P-labeled ss-32 with the nonradioactive complementary ss-17 in a 1:1.2 molar ratio generates dsRNA-duplex RNase R substrate. The mixture is heated to 95 °C for 5 min and allowed to cool slowly to room temperature. Nonradioactive dsRNA-duplex is generated

---

therefore, can be captured by affinity chromatography on a Ni^{2+}-NTA column. The affinity capture of the 6-His–carrying nascent polypeptide also captures the associated stalled ribosomes and the SmpB-tmRNA complex. The unbound flow-through material from the Ni^{2+}-NTA affinity chromatography step does not contain a significant portion of stalled ribosomes or the SmpB-tmRNA complex. (B) Western blot of the flow-through (FT) and bound material, with antibodies against native SmpB protein, showing the levels of SmpB protein present in these fractions.

by mixing both nonradioactive RNA oligos in a 1:1.2 molar ratio, heating to 95 °C for 5 min and slowly cooling to room temperature.

The typical degradation assay is carried out in 30 μl reaction mixtures containing 3 μl of 10× RNase R activity buffer, 5 μM nonradioactive dsRNA substrate supplemented with radioactive dsRNA (to make the total radioactivity of each reaction ∼50,000 cpm), and 0.5 μg of purified RNase R. The volume of each reaction mixture is brought up to 30 μl with Milli-Q water. RNase R is added last to initiate the degradation assay. Analytical samples (5 μl) are taken at desired time points and the reaction is terminated by the addition of equal volume of 2× gel loading buffer (95% formamide, 20 mM EDTA, 0.025% [v/v] bromophenol blue, and 0.025% [v/v] xylene cyanol). Reaction products are resolved by electrophoresis in a 20% polyacrylamide–8 M urea gel and visualized by exposing the dried-gel to a PhosphorImager screen (GE Healthcare). Figure 17.7 shows a typical RNase degradation assay with this substrate.

3.7. *In vivo* nonstop mRNA (*λ-cI-N-trpAt*) stability and decay assay

We routinely use the rifampicin–chase assay to determine the role of various *trans*-translation related factors, such as RNase R, in nonstop mRNA decay. We overexpress a nonstop reporter mRNA, block transcription with rifampicin, and measure the half-life of the mRNA in the presence or absence of putative tmRNA-dependent mRNA decay factors. For this assay, cultures of MG1655-*rnr*$^-$ and the isogenic wild-type cells harboring pPW500, the *λ-cI-N* nonstop mRNA expression plasmid, are grown to an OD_{600} of 0.5 to 0.6. Expression of the nonstop reporter transcript is induced by the addition of IPTG for 15 min. At this point, rifampicin is added to a final concentration of 150 μg/ml to inhibit *de novo* synthesis of RNA. Analytical samples (3 ml) are taken at 0, 2, 4, 8, 16, and 32 min after the addition of rifampicin and harvested by centrifugation at 18,000g for 1 min. Whole cell pellets are resuspended in equal volumes of sterile water, and OD_{600} is measured to enumerate cell numbers. Total *E. coli* RNA from equal numbers of cells (corresponding to equal OD_{600}) is then extracted.

3.7.1. RNA extraction

Place 1.5 ml of bacterial culture in a screw cap tube and harvest the cells by centrifugation at 18,000g for 1 min.

Resuspend the cell pellet in 250 μl sterile Milli Q H_2O.

Add 750 μl Tri-Reagent LS (Molecular Research Center), vortex well.

Heat the samples in a heating block at 70 °C for 15 min.

Add 150 μl of chloroform, vortex well, and incubate at room temperature for 5 min.

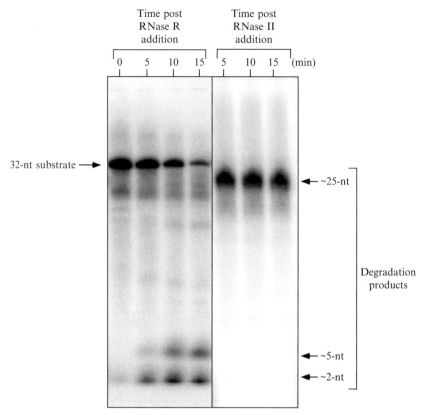

Figure 17.7 Difference in the ability of RNase R and RNase II to degrade a duplex RNA substrate with a 3′-overhang. RNA degradation reaction mixtures (30 μl), containing a ^{32}P-labeled duplex RNA substrate, were prepared as described in the *in vitro* RNA degradation assay. The reaction was initiated by the addition of 0.5 μg of purified RNase R or RNase II, and samples (5 μl) were taken for analysis at 0, 5, 10, and 15 min after enzyme addition. The degradation products were resolved by electrophoresis in a 20% denaturing polyacrylamide gel. RNase R is fully capable of degrading the double-stranded region of the substrate to produce 2 to 5 nucleotide degradation products. In contrast, RNase II is unable to degrade the structured portion of the substrate and generates a product of ∼25 nucleotides, suggesting that it stalls 6 to 8 nucleotides before the secondary structure.

Centrifuge for 15 min at 18,000g at 4 °C.

Carefully transfer the RNA containing supernatant to a fresh tube, taking care not to transfer any of the Tri-Reagent or interface layer.

Add 375 μl of isopropanol to the RNA containing supernatant and vortex well.

Incubate at −20 °C for 15 to 60 min.

Centrifuge for 15 min at 18,000g at 4 °C and discard the supernatant.

Add 1 ml of room temperature 75% ethanol to the pellet.

Centrifuge for 15 min at 18,000g at 4 °C and discard the supernatant.

Leave the open tubes in a fume hood for 15 to 30 min to air dry (it is important to not let the pellets overdry). A small white pellet should be visible.

If the pellets are not completely dry, the remaining supernatant can be removed with a pipette.

Resuspend the pellet in 25 to 50 μl of sterile dH$_2$O.

Vortex well and heat at 70 °C for 10 min.

Vortex again and quantify by A$_{260}$.

The yield of RNA from 1.5 ml of E. coli, at an OD$_{600}$ of ~0.800, is normally between 300 and 1000 ng/μl.

3.7.2. Northern blot analysis

Equal amounts of RNA for each sample are resolved by electrophoresis in a denaturing 1.5% (v/v) formaldehyde agarose gel; 15 μl RNA samples are mixed with 10 μl of denaturing solution (6.5 μl formamide, 2 μl formaldehyde, 1.5 μl 10 × MOPS buffer) and heated at 65 °C for 10 min; 5 μl of sample dye is added (2% bromophenol blue, 10 mM EDTA, 50% [v/v] glycerol) to each sample before loading on the gel. RNA is resolved by electrophoresis in a 1.5% formaldehyde-agarose gel. Resolved RNA is transferred to Hybond N$^+$ membranes (GE Healthcare) by overnight capillary transfer in 20 × SSC buffer (175.3 g/L sodium chloride, 88.2 g/L sodium citrate). Blots are developed with alkaline phosphatase–conjugated streptavidin and chemiluminescent substrate (Biotin Luminescent Detection Kit, Roche). Membranes are exposed to film and the resulting bands quantified with Image-J software (NIH, http://rsb.info.nih.gov/ij/). Figure 17.8 shows the experimental analysis of the nonstop reporter mRNA with the previously described assay and detection approach.

3.8. tmRNA-tagged protein stability assay

We routinely use two tmRNA-tagged reporter proteins (λ-CI-N-ssrA and GFP-ssrA) as substrates for in vivo and in vitro proteolysis assays by Lon protease. In addition, we use untagged versions of these proteins as controls to demonstrate the specificity of Lon for tmRNA-tagged protein (Choy et al., 2007). We will first describe our in vivo proteolysis assays followed by purification of substrate proteins and our standard in vitro proteolysis assays.

3.8.1. In vivo GFP-ssrA protein stability assay

Strains carrying pBAD-GFP or pBAD-GFP-ssrA (a plasmid encoding a GFP protein variant carrying a tmRNA tag at its C-terminus) are cultivated in LB containing 30 μg/ml chloramphenicol at 37 °C until the culture

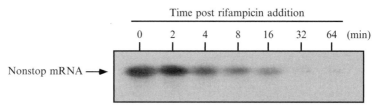

Figure 17.8 *In vivo* analysis of nonstop mRNA decay with the rifampicin–chase assay. The RNA used in this Northern blot was purified from *E. coli* MG1655 cells harboring a plasmid encoding a nonstop mRNA under the control of an IPTG inducible promoter. The promoter was induced in logarithmically growing cells with an IPTG at a final concentration of 0.5 mM for 15 min. At this point, rifampicin was added to a final concentration of 150 μg/ml and 1.5 ml samples of the culture were taken at 0, 2, 4, 8, 16, and 32 min. Equal amounts of RNA from these samples were resolved with electrophoresis in a 1.5% denaturing agarose gel followed by Northern blot analysis. A biotinylated DNA oligonucleotide probe complementary to the nonstop RNA was used to detect the nonstop mRNA.

OD_{600} reaches ~0.45. Expression of GFP or GFP-ssrA is induced by addition of arabinose to a final concentration of 0.01%. After 1 h of induction, cells are gently harvested and washed once with warmed LB, resuspended in one culture volume of warmed LB containing 30 μg/ml chloramphenicol and 200 μg/ml spectinomycin, and placed back at 37 °C. Analytical samples are obtained from the cultures at 0, 10, 20, 40, 60, 90, and 120 min time points. Equal numbers of cells from each sample are harvested. Cell pellets are resuspended and lysed in 1 × SDS sample buffer. Total-cell proteins are resolved by electrophoresis in a 15% Tris-Tricine gel and analyzed by Western blotting with HRP-conjugated rabbit polyclonal anti-GFP-serum (Santa Cruz Biotechnology).

3.8.2. *In vivo* λ-CI-N protein stability assay

E. coli cells carrying the pPW500 plasmid, which harbors the λ-CI-N reporter construct under the control of an IPTG inducible promoter (Choy *et al.*, 2007; Keiler *et al.*, 1996), are cultivated in LB containing 100 μg/ml ampicillin at 32 °C until culture OD_{600} reaches ~0.45. Expression of λ-cI-N-trpAt nonstop mRNA is induced by addition of IPTG to a final concentration of 1 mM. After a 30 min induction, cells are gently harvested and washed once with warmed LB, resuspended in one culture volume of warmed LB containing 100 μg/ml ampicillin and 200 μg/ml spectinomycin, and placed back at 32 °C. Analytical samples are obtained from the cultures at 0, 5, 10, 15, 20, and 30 min time points. Equal numbers of cells from each sample are harvested. Cell pellets are resuspended and lysed in 1 × SDS sample buffer. Total-cell protein is resolved by electrophoresis in a 15% Tris-Tricine gel and analyzed by Western blotting with mouse monoclonal anti-FLAG (Scientific Imaging Systems) and goat anti-mouse-HRP (Santa Cruz Biotechnology).

3.8.3. *In vitro* proteolysis assay

In vitro Lon proteolysis assays require purified tmRNA tagged and untagged proteins. The following sections describe the purification of tagged and untagged λ-CI-N and GFP proteins and their use in the *in vitro* proteolysis assay.

3.8.3.1. *λ-CI-N reporter protein purification* The λ-CI-N reporter protein (Choy *et al.*, 2007) has internal FLAG and His6 epitopes that are used for protein detection and purification by Ni^{2+}-NTA chromatography, respectively. tmRNA-tagged and untagged species of λ-CI-N protein are simultaneously purified from an *E. coli clpP clpX lon* triple mutant carrying the pPW500 plasmid. Three liters of *E. coli* clpX clpP lon/pPW500 cells are grown at 37 °C in LB containing 100 μg/ml ampicillin to an OD_{600} of ~0.5. Reporter protein expression is induced with 1 mM IPTG at 37 °C for 3 h. The high-level expression of the reporter protein results in production of cotranslationally tmRNA-tagged and untagged version of the protein. Cells are harvested and resuspended in lysis buffer (50 mM NaPO$_4$, pH 8.0, 300 mM NaCl, 10 mM imidazole), and lysed by sonication. Cell debris is removed by centrifugation at 30,000g for 1 h, and 1 ml of 50% Ni^{2+}-NTA agarose resin is added to the supernatant and rotated for 1 h at 4 °C. The Ni^{2+}-NTA bound material is washed extensively with lysis buffer containing 20 mM imidazole. Both forms of λ-CI-N protein are eluted with elution buffer (50 mM NaPO$_4$, pH 8.0, 300 mM NaCl, 250 mM imidazole). The absence of contaminating proteins is verified by Tris-Tricine-PAGE followed by Coomassie blue staining. Protein samples are concentrated, exchanged into storage buffer (50 mM Tris-HCl, pH 8, 50 mM KCl, 5 mM MgCl$_2$, 2 mM βME, 10% [v/v] glycerol), aliquoted, flash frozen using liquid nitrogen, and stored at -80 °C.

3.8.3.2. *GFP and GFP-ssrA purification* GFP and GFP-ssrA are purified as described (Choy *et al.*, 2007). GFP is purified from *E. coli* strain JM109 expressing pBAD-GFP. GFP-ssrA (a GFP protein variant carrying a tmRNA tag at its C-terminus) is purified from *E. coli clpP clpX lon* triple mutant expressing pBAD-GFP-ssrA. This strain permits the purification of GFP-ssrA with intact tmRNA tag. The absence of contaminating proteins in the final preparation is verified by SDS-PAGE followed by Coomassie blue staining. Protein samples are concentrated and stored at -80 °C.

3.8.3.3. Proteolysis of tmRNA-tagged proteins

In vitro Lon proteolysis assays are performed with a minimal activity buffer (50 mM Tris-HCl, pH 8.0, 10 mM MgCl$_2$, 1 mM DTT). Complete assays contain 1 μM Lon-6-His, 10 μM of tmRNA-tagged or untagged substrate, an ATP regeneration system (50 mM creatine phosphate, 80 μg/ml creatine kinase [Roche]), and 4 mM ATP. Complete reaction mixtures are incubated at 37 °C, and analytical samples are taken at various time points. The levels of untagged and tmRNA-tagged proteins at selected time points are analyzed by electrophoresis in a 15% Tris-Tricine gel and quantified with a GS-710 imaging densitometer and Quantity One software (BioRad). In addition, the levels of GFP and GFP-ssrA can be easily assessed through fluorescence measurements of analytical samples with a SpectraMax M2 microplate reader (Molecular Devices) that is configured with empirically determined excitation (476 nm) and emission wavelengths and (519 nm).

ACKNOWLEDGMENTS

We thank all current and former members of the Karzai laboratory for their contributions in developing methods and assays described in this chapter. We owe special thanks to Jennifer Choy, Andrew Michaels, Nihal Okan, and Latt Latt Aung for their many contributions over the years. We are grateful to Dr. Jorge Benach and members of The Center for Infectious Diseases for their continued support. This work was supported in part by National Institutes of Health Grants GM65319 and AI055621 (to A. W. K.) and The Pew Scholars Program.

REFERENCES

Akerley, B. J., Rubin, E. J., Novick, V. L., Amaya, K., Judson, N., *et al.* (2002). A genome-scale analysis for identification of genes required for growth or survival of *Haemophilus influenzae*. *Proc. Natl. Acad. Sci. USA* **99**(2), 966–971.

Barends, S., Karzai, A. W., Sauer, R. T., Wower, J., and Kraal, B. (2001). Simultaneous and functional binding of SmpB and EF-Tu-TP to the alanyl acceptor arm of tmRNA. *J. Mol. Biol.* **314**(1), 9–21.

Baumler, A. J., Kusters, J. G., Stojiljkovic, I., and Heffron, F. (1994). *Salmonella typhimurium* loci involved in survival within macrophages. *Infect. Immun.* **62**(5), 1623–1630.

Choy, J. S., Aung, L. L., and Karzai, A. W. (2007). Lon protease degrades transfer-messenger RNA-tagged proteins. *J. Bacteriol.* **189**(18), 6564–6571.

Dulebohn, D., Choy, J., Sundermeier, T., Okan, N., and Karzai, A. W. (2007). Trans-Translation: The tmRNA-mediated surveillance mechanism for ribosome rescue, directed protein degradation, and nonstop mRNA decay. *Biochemistry.* **46**(16), 4681–4693.

Dulebohn, D. P., Cho, H. J., and Karzai, A. W. (2006). Role of conserved surface amino acids in binding of SmpB protein to SsrA RNA. *J. Biol. Chem.* **281**(39), 28536–28545.

Farrell, C. M., Grossman, A. D., and Sauer, R. T. (2005). Cytoplasmic degradation of ssrA-tagged proteins. *Mol. Microbiol.* **57**(6), 1750–1761.

Felden, B., Hanawa, K., Atkins, J. F., Himeno, H., Muto, A., Gesteland, R. F., McCloskey, J.-A., and Crain, P. F. (1998). Presence and location of modified nucleotides in *Escherichia coli* tmRNA: structural mimicry with tRNA acceptor branches. *EMBO J.* **17**(11), 3188–3196.

Glass, J. I., Assad-Garcia, N., Alperovich, N., Yooseph, S., Lewis, M. R., *et al.* (2006). Essential genes of a minimal bacterium. *Proc. Natl. Acad. Sci. USA* **103**(2), 425–430.

Gottesman, S., Roche, E., Zhou, Y., and Sauer, R. T. (1998). The ClpXP and ClpAP proteases degrade proteins with carboxy-terminal peptide tails added by the SsrA-tagging system. *Genes Dev.* **2**(9), 1338–1347.

Hayes, C. S., and Sauer, R. T. (2003). Cleavage of the A Site mRNA codon during ribosome pausing provides a mechanism for translational quality control. *Mol. Cell.* **12** (4), 903–911.

Hayes, C. S., Bose, B., and Sauer, R. T. (2002a). Proline residues at the C terminus of nascent chains induce SsrA tagging during translation termination. *J. Biol. Chem.* **277**(37), 33825–33832.

Hayes, C. S., Bose, B., and Sauer, R. T. (2002b). Stop codons preceded by rare arginine codons are efficient determinants of SsrA tagging in *Escherichia coli*. *Proc. Natl. Acad. Sci. USA* **99**(6), 3440–3445.

Herman, C., Thevenet, D., Bouloc, P., Walker, G. C., and D'Ari, R. (1998). Degradation of carboxy-terminal-tagged cytoplasmic proteins by the *Escherichia coli* protease HflB (FtsH). *Genes Dev.* **12**(9), 1348–1355.

Herman, C., Prakash, S., Lu, C. Z., Matouschek, A., and Gross, C. A. (2003). Lack of a robust unfoldase activity confers a unique level of substrate specificity to the universal AAA protease FtsH. *Mol. Cell.* **11**(3), 659–669.

Huang, C., Wolfgang, M. C., Withey, J., Koomey, M., and Friedman, D. I. (2000). Charged tmRNA but not tmRNA-mediated proteolysis is essential for *Neisseria gonorrhoeae* viability. *EMBO J.* **19**(5), 1098–1107.

Hutchison, C. A., Peterson, S. N., Gill, S. R., Cline, R. T., White, O., *et al.* (1999). Global transposon mutagenesis and a minimal Mycoplasma genome. *Science.* **286**(5447), 2165–2169.

Joshi, S. A., Baker, T. A., and Sauer, R. T. (2003). C-terminal domain mutations in ClpX uncouple substrate binding from an engagement step required for unfolding. *Mol. Microbiol.* **48**(1), 67–76.

Julio, S. M., Heithoff, D. M., and Mahan, M. J. (2000). ssrA (tmRNA) plays a role in *Salmonella enterica serovar typhimurium* pathogenesis. *J. Bacteriol.* **182**(6), 1558–1563.

Karzai, A. W., and Sauer, R. T. (2001). Protein factors associated with the SsrA.SmpB tagging and ribosome rescue complex. *Proc. Natl. Acad. Sci. USA* **98**(6), 3040–3044.

Karzai, A. W., Susskind, M. M., and Sauer, R. T. (1999). SmpB, a unique RNA-binding protein essential for the peptide-tagging activity of SsrA (tmRNA). *EMBO J.* **18**(13), 3793–3799.

Keiler, K. C. (2007). Physiology of tmRNA: what gets tagged and why? *Curr. Opin. Microbiol.* **10**(2), 169–175.

Keiler, K. C., and Shapiro, L. (2003). TmRNA is required for correct timing of DNA replication in *Caulobacter crescentus*. *J. Bacteriol.* **185**(2), 573–580.

Keiler, K. C., Waller, P. R., and Sauer, R. T. (1996). Role of a peptide tagging system in degradation of proteins synthesized from damaged messenger RNA. *Science.* **271**(5251), 990–993.

Kihara, A., Akiyama, Y., and Ito, K. (1996). A protease complex in the *Escherichia coli* plasma membrane: HflKC (HflA) forms a complex with FtsH (HflB), regulating its proteolytic activity against SecY. *EMBO J.* **15**(22), 6122–6131.

Makino, S., Makino, T., Abe, K., Hashimoto, J., Tatsuta, T., *et al.* (1999). Second transmembrane segment of FtsH plays a role in its proteolytic activity and homo-oligomerization. *FEBS Lett.* **460**(3), 554–558.

Maurizi, M. R., Clark, W. P., Kim, S. H., and Gottesman, S. (1990). Clp P represents a unique family of serine proteases. *J. Biol. Chem.* **265**(21), 12546–12552.

Mehta, P., Richards, J., and Karzai, A. W. (2006). tmRNA determinants required for facilitating nonstop mRNA decay. *RNA.* **12**(12), 2187–2198.

Moore, S. D., and Sauer, R. T. (2007). The tmRNA system for translational surveillance and ribosome rescue. *Annu. Rev. Biochem.* **76,** 101–124.

Muto, A., Fujihara, A., Ito, K. I., Matsuno, J., Ushida, C., *et al.* (2000). Requirement of transfer-messenger RNA for the growth of *Bacillus subtilis* under stresses. *Genes Cells.* **5**(8), 627–635.

Okan, N. A., Bliska, J. B., and Karzai, A. W. (2006). A role for the SmpB-SsrA system in *Yersinia pseudotuberculosis* pathogenesis. *PLoS Pathog.* **2**(1), 50–62e6.

Ranquet, C., Geiselmann, J., and Toussaint, A. (2001). The tRNA function of SsrA contributes to controlling repression of bacteriophage Mu prophage. *Proc. Natl. Acad. Sci. USA* **98**(18), 10220–10225.

Retallack, D. M., Johnson, L. L., and Friedman, D. I. (1994). Role for 10Sa RNA in the growth of lambda-P22 hybrid phage. *J. Bacteriol.* **176**(7), 2082–2089.

Richards, J., Mehta, P., and Karzai, A. W. (2006). RNase R degrades non-stop mRNAs selectively in an SmpB-tmRNA-dependent manner. *Mol. Microbiol.* **62**(62), 1700–1712.

Rudinger-Thirion, J., Giegâ, E. R., and Felden, B. (1999). Aminoacylated tmRNA from *Escherichia coli* interacts with prokaryotic elongation factor Tu. *RNA.* **5**(8), 989–992.

Shin, J. H., and Price, C. W. (2007). The SsrA-SmpB ribosome rescue system is important for growth of *Bacillus subtilis* at low and high temperatures. *J. Bacteriol.* **189**(10), 3729–3737.

Shotland, Y., Shifrin, A., Ziv, T., Teff, D., Koby, S., *et al.* (2000). Proteolysis of bacteriophage lambda CII by *Escherichia coli* FtsH (HflB). *J. Bacteriol.* **182**(11), 3111–3116.

Studier, F. W. (2005). Protein production by auto-induction in high density shaking cultures. *Protein Expression Purification.* **41**(1), 207–234.

Sundermeier, T. R., and Karzai, A. W. (2007). Functional SmpB-ribosome interactions require tmRNA. *J. Biol. Chem.* **282**(48), 34779–34786.

Sundermeier, T. R., Dulebohn, D. P., Cho, H. J., and Karzai, A. W. (2005). A previously uncharacterized role for small protein B (SmpB) in transfer messenger RNA-mediated trans-translation. *Proc. Natl. Acad. Sci. USA* **102**(7), 2316–2321.

Sunohara, T., Abo, T., Inada, T., and Aiba, H. (2002). The C-terminal amino acid sequence of nascent peptide is a major determinant of SsrA tagging at all three stop codons. *RNA.* **8**(11), 1416–1427.

Sunohara, T., Jojima, K., Yamamoto, Y., Inada, T., and Aiba, H. (2004a). Nascent-peptide-mediated ribosome stalling at a stop codon induces mRNA cleavage resulting in nonstop mRNA that is recognized by tmRNA. *RNA.* **10**(3), 378–386.

Sunohara, T., Jojima, K., Tagami, H., Inada, T., and Aiba, H. (2004b). Ribosome stalling during translation elongation induces cleavage of mRNA being translated in *Escherichia coli. J. Biol. Chem.* **279**(15), 15368–15375.

Vincent, H. A., and Deutscher, M. P. (2006). Substrate recognition and catalysis by the exoribonuclease RNase R. *J. Biol. Chem.* **281**(40), 29769–29775.

Withey, J., and Friedman, D. (1999). Analysis of the role of trans-translation in the requirement of tmRNA for lambdaimmP22 growth in *Escherichia coli. J. Bacteriol.* **181**(7), 2148–2157.

Yamamoto, Y., Sunohara, T., Jojima, K., Inada, T., and Aiba, H. (2003). SsrA-mediated trans-translation plays a role in mRNA quality control by facilitating degradation of truncated mRNAs. *RNA.* **9**(4), 408–418.

CHAPTER EIGHTEEN

Analyses of mRNA Destabilization and Translational Inhibition Mediated by Hfq-Binding Small RNAs

Teppei Morita, Kimika Maki, Mieko Yagi, *and* Hiroji Aiba

Contents

Division of Biological Science, Graduate School of Science, Nagoya University, Chikusa, Nagoya, Japan

Methods in Enzymology, Volume 447
ISSN 0076-6879, DOI: 10.1016/S0076-6879(08)02218-0

Abstract

A major class of bacterial small RNAs binds to an RNA chaperone Hfq and acts via imperfect base pairing to regulate the translation and stability of target mRNAs under specific physiological conditions. SgrS, an example for this class of small RNAs, is induced in response to the accumulation of glucose phosphates and downregulates the *ptsG* mRNA, which encodes the glucose transporter IICB^{Glc} in *Escherichia coli*. SgrS forms a specific ribonucleoprotein complex with RNase E through Hfq. The regulatory outcomes of SgrS are the inhibition of translation and RNase E–dependent degradation of *ptsG* mRNA. Translational inhibition is the primary event for gene silencing. The crucial base pairs for the action of SgrS are confined to the 6-nt region overlapping the Shine–Dalgarno sequence of the target mRNA. Hfq accelerates the rate of duplex formation between SgrS and the target mRNA. Membrane localization of the target mRNA contributes to efficient SgrS action by competing with ribosome loading. Here, we describe major experimental methods and results used to study functions of Hfq-binding small RNAs in our laboratory. These are illustrated using the regulation of *ptsG* mRNA by SgrS is used as an example.

1. OVERVIEW

The regulation of gene expression by small RNAs (sRNAs) has been known for many years in *Escherichia coli* since the serendipitous discovery of MicF, which is involved in the downregulation of the expression of *ompF* gene, which encodes a major outer membrane porin (Mizuno *et al.*, 1984). During the 1990s, several additional sRNAs such as DsrA (Sledjeski and Gottesman, 1995) and OxyS (Altuvia *et al.*, 1997) were found also fortuitously to be involved in the regulation of gene expression, primarily at posttranscriptional levels. Over the past several years, an increasing number of bacterial sRNAs have been identified in *E. coli* and other organisms (Argaman *et al.*, 2001; Chen *et al.*, 2002; Kawano *et al.*, 2005; Vogel *et al.*, 2003; Wassarman *et al.*, 2001; Zhang *et al.*, 2003), but the functions of many of them remain to be explained. A major class of *E. coli* sRNAs binds to an RNA chaperone Hfq and acts by imperfect base pairing to regulate the translation and stability of target mRNAs under specific physiological conditions (Gottesman, 2004, 2005; Storz and Gottesman, 2006; Storz *et al.*, 2004, 2005). In particular, studies of two sRNAs, SgrS and RyhB, have uncovered several important features regarding the mechanisms of action of Hfq-binding sRNAs (Aiba, 2007). SgrS downregulates *ptsG* mRNA, which encodes the glucose transporter IICB^{Glc}, in response to the accumulation of glucose phosphates (Vanderpool and Gottesman, 2004), whereas RyhB RNA downregulates several genes encoding iron-binding proteins, including *sodB*, which encodes superoxide dismutase, in response to iron limitation (Masse and Gottesman, 2002). Both RNAs destabilize their target mRNAs in an RNase E–dependent manner (Kawamoto *et al.*, 2005; Masse *et al.*, 2003; Morita *et al.*, 2005). They form a specific ribonucleoprotein complex with RNase E through Hfq, resulting in translational repression and the rapid degradation of the target mRNAs (Morita *et al.*, 2005). The inhibition of translation rather than accelerated degradation is the primary event for gene silencing (Morita *et al.*, 2006). The degradation of sRNAs is coupled to the degradation of target mRNAs in an RNase E–dependent manner (Masse *et al.*, 2003). In the case of SgrS, membrane localization of the target mRNA supports the action of SgrS presumably by affecting the level of competition between the sRNA and ribosomes for binding of *pts*G mRNA (Kawamoto *et al.*, 2005). The crucial base pairs for SgrS action are confined to the 6-nt region overlapping the Shine–Dalgarno (SD) sequence of the target mRNA (Kawamoto *et al.*, 2006). Hfq stimulates base pairing between SgrS and the target mRNA by accelerating the rate of duplex formation (Kawamoto *et al.*, 2006). Figure 18.1 provides a summary of the pathway and the mechanism by which the *pts*G gene is silenced through the actions of SgrS, Hfq, and RNase E. The physiological role of SgrS-mediated gene silencing of *pts*G is to stop the production of IICB^{Glc} thereby limiting the accumulation of toxic glucose phosphates. In addition to the

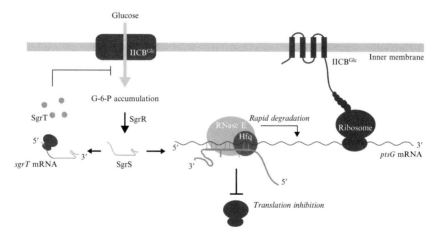

Figure 18.1   Model for *pts*G gene silencing by SgrS RNA. In *E. coli*, external glucose is transported into the cells coupled to phosphorylation by the phosphoenolpyruvate-dependent sugar phosphotransferase system (PTS). IICB^{Glc} encoded by the *pts*G gene is the membrane component of glucose-specific PTS. When G6P accumulates abnormally, transcription of the *sgr*S gene is induced, depending on a specific transcription factor, SgrR. SgrS forms a ribonucleoprotein complex with RNase E via Hfq. The complex acts on the ribosome-binding site of the *pts*G mRNA by by means imperfect base pairing. This results in translational inhibition and rapid RNase E–dependent degradation of the message. As shown by a recent study of Wadler and Vanderpool (2007), SgrS also acts as an mRNA template for a small functional protein SgrT to prevent glucose uptake, presumably by inhibiting the transport activity of IICB^{Glc}. This figure was redrawn from Morita and Aiba, 2007.

riboregulation, SgrS has been discovered to act as an mRNA template for a small protein SgrT that prevents glucose uptake (Wadler and Vanderpool, 2007). Here, we describe major experimental methods and results that were used to analyze the gene silencing of *pts*G by SgrS. Most of the procedures should be useful for the analysis of the action of other Hfq-binding sRNAs in bacteria. Chapter XXX (Regnier and Hajnsdorf) also focuses on HFQ and describes methods to characterize the role of Hfq in poly(A) metabolism.

2. THE *PTSG* mRNA IS DESTABILIZED IN RESPONSE TO GLUCOSE PHOSPHATE STRESS

In *E. coli*, external glucose is transported into the cells coupled to phosphorylation by the phosphoenolpyruvate: sugar phosphotransferase system (PTS) (Postma, 1996). The PTS consists of two common cytoplasmic proteins and a series of sugar-specific enzyme II complexes. IICB^{Glc} encoded by the *pts*G gene is the membrane component of glucose-specific EII. During a study of glucose repression, we that mutation of either the *pgi* or *pfkA*, which encode phosphoglucose isomerase or phosphofructokinase respectively, causes a strong reduction in IICB^{Glc} expression. This was associated with RNase E–dependent destabilization of the *pts*G mRNA in the presence of glucose (Kimata *et al.*, 2001). Moreover, the accumulation of glucose-6-phosphate (G6P) or α-methylglucoside 6-phosphate (αMG6P) was linked to the RNase E–mediated destabilization of the *pts*G mRNA (Morita *et al.*, 2003). The physiological relevance of *pts*G downregulation is probably avoid excessive accumulation of glucose phosphates, which could be toxic to the cell. The critical experiments that led to these conclusions were the analyses of production of *pts*G mRNA and IICB^{Glc} under various physiological and genetic conditions. Example experiments for the analysis of *pts*G mRNA are provided in Fig. 18.2. The *pts*G mRNA is destabilized in Δ*pgi* cells or when wild-type cells are exposed to α-methylglucoside (αMG) (Fig. 18.2A). The destabilization of *pts*G mRNA in *pgi* cells no longer occurred at 42 °C when cells carried the temperature-sensitive *ams1* allele of the gene encoding RNase E (Fig. 18.2B). The folowing describes procedures for the isolation of total-cell RNA and Northern blot analysis.

2.1. Isolation of total-cell RNA

2.1.1. Procedure

1. Inoculate 5 ml of LB medium with a single bacterial colony. Incubate at 37 °C overnight with vigorous shaking.

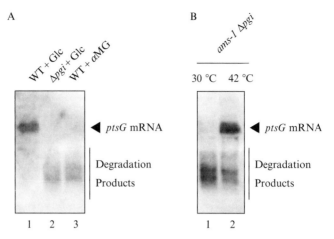

Figure 18.2 The *ptsG* mRNA is destabilized in response to the accumulation of glucose phosphate in an RNase E–dependent manner. (A) IT1568 (wild-type) and TM162 (Δ*pgi::cat*) cells were grown in LB medium. At $OD_{600} = 0.5$, 0.4% glucose (Glc) or α-methylglucoside (αMG) was added, and the incubation was continued for 10 min. Total-cell RNAs were prepared, and 10 μg of each RNA was subjected to Northern blot analysis with a 305-bp, DIG-labeled DNA probe corresponding the 5′-region of *ptsG* mRNA. (B) TM442 (*ams1* Δ*pgi::cat*) cells were grown in LB medium supplemented with 0.4% glucose at 30 °C. At $OD_{600} = 0.5$, the temperature was shifted to 42 °C or kept at 30 °C, and the incubation was continued for 10 min. Total-cell RNA was prepared, and 10 μg of each RNA sample was subjected to Northern blotting. Essentially, the same results were published previously (Kimata *et al.*, 2001; Morita *et al.*, 2003).

2. Inoculate 10 to 20 ml of appropriate medium in 50-ml Falcon tube with 0.1 to 0.2 ml of the overnight culture. Incubate at 37 °C with vigorous shaking until the culture reaches the mid-log phase ($OD_{600} = 0.4$ to 0.5).

3. Chill the culture and harvest cells by centrifugation at 3,500 rpm for 5 min at 4 °C. Discard the supernatant.

4. Resuspend the cell pellet in 400 μl of RNA buffer (0.02 M sodium acetate, pH 5.5, 0.5% SDS, and 1 mM EDTA).

5. Add 400 μl of phenol equilibrated with RNA buffer and shake gently at 65 °C for 5 min.

6. Transfer the mixture to an Eppendorf tube and centrifuge at 12,000 rpm for 5 min.

7. Transfer the aqueous phase into a new Eppendorf tube and add 1 ml of absolute ethanol. Mix well and store at −20 °C for 10 min.

8. Centrifuge at 12,000 rpm for 10 min at 4 °C. Discard the supernatant and wash the pellet with 70% ethanol at room temperature.

9. Dissolve the RNA pellet in 400 μl of RNA buffer and add 1 ml of ethanol. Mix well and store at −20 °C for 10 min.

10. Repeat steps 8 and 9 twice.

11. Discard as much ethanol as possible, dry the RNA pellet briefly, and dissolve the pellet in 50 to 100 μl of RNA buffer.
12. Determine the RNA concentration by measuring the absorbance at 260 nm.

2.1.2. Comments

The procedure is a modified version of our original SDS/hot phenol method (Aiba *et al.*, 1981), which was developed for the isolation of total-cell RNA from *E. coli*. This method has been successfully used for isolation of total-cell RNA from other bacteria. Appropriate antibiotics should be added in the culture medium depending on strains at step 2. When the temperature-sensitive strains are used, cells can be grown initially at a lower temperature, for example, at 30 °C. At steps 7 and 9, the cooling at −20 °C for 10 min can be omitted, because the precipitation of RNA occurs efficiently just after the addition of ethanol at room temperature. Do not dry the RNA pellet completely at step 11, because it becomes difficult to dissolve the RNA in the buffer. Use appropriately diluted samples with distilled water for A_{260} determination at step 12. The RNA sample can be stored stably at −20 °C for at least several weeks.

2.2. Northern blot analysis

2.2.1. Procedure

1. Prepare samples for electrophoresis by mixing 6 μl of RNA (0.1 to 20 μg), 2.5 μl of 10× running buffer (0.2 M MOPS, pH 7.0, 50 mM sodium acetate, 10 mM EDTA), 12.5 μl of formamide, and 4 μl of formaldehyde.
2. Incubate the samples at 65 °C for 5 min and then at 0 °C for 5 min.
3. Add 3 μl of loading buffer (40% glycerol, 1 mM EDTA, 0.25% bromophenol blue, and 0.25% xylene cyanol).
4. Load the samples into the lanes of 1.2% or 1.5% agarose gel in running buffer containing 3.7% formaldehyde and run the gel.
5. Transfer the RNA from the gel to a Hybond-N$^+$ membrane (Amersham Pharmacia Biosciences) overnight by the capillary elution method in 20× SSC buffer (3 M NaCl, 0.3 M sodium citrate).
6. Fix the RNA by putting the membrane (transferred surface should be upside) on a 3MM paper soaked with 0.05 M NaOH for 5 min. Wash the membrane gently with 2× SSC buffer.
7. Place the membrane between two pieces of 3MM paper for 15 min.
8. For hybridization and detection of a specific RNA, we use the digoxigenin (DIG) reagents and kits for the nonradioactive labeling and detection of nucleic acid (see following) (Roche Diagnostics).

2.2.2. Comments

This procedure, a modified version of the standard Northern blot hybridization method (Sambrook and Maniatis, 1989), works well, at least for the analysis of *E. coli* RNAs. See the original reference for detailed procedures, reagents, and conditions for RNA electrophoresis and blotting. The procedures for preparation of the DIG-labeled DNA probe, hybridization, and detection are given by the manufacturer. If necessary, dilute or concentrate the stock RNA solution before the sample preparation at step 1. The double-stranded DIG-labeled DNA probes of 100 to 300 bp corresponding to individual genes are prepared by PCR with DIG-labeled dUTP at step 8. Denhardt's solution (Sambrook and Maniatis, 1989) can work also well as the hybridization buffer at step 8.

3. REQUIREMENT OF C-TERMINAL SCAFFOLD REGION OF RNASE E, HFQ, AND SGRS

RNase E forms, through its C-terminal scaffold region, a multiprotein complex called the RNA degradosome. The other major components are the 3′ exonuclease polynucleotide phosphorylase (PNPase), the RhlB RNA helicase, and the glycolytic enzyme enolase (Carpousis, 2002) (see also Chapters XXX regarding methods for the studying the degradosome and other multiprotein complexes in this volume also chapters in RNA Turnover, Part B, ed. Maquat and Kiledjian). We found that the C-terminal scaffold region of RNase E is required for the rapid degradation of *pts*G mRNA under stress condition (Morita *et al.*, 2004). Vanderpool and Gotteman discovered that the overproduction of RyaA, an uncharacterized Hfq-binding sRNA, inhibited cell growth on glucose (Vanderpool and Gottesman, 2004). They found that the synthesis of RyaA is induced in response to glucose phosphate stress and noticed that a region of around 30 nt within RyaA is partially complementary to the translation initiation region of the *pts*G mRNA. These observations led them to propose that RyaA acts on the *pts*G mRNA through base pairing. RyaA was renamed SgrS (*sugar* transport-related *s*RNA). Another gene encoding a putative transcription factor protein, SgrR, is required for the induction of SgrS under the stress condition. The destabilization of *pts*G mRNA that is induced by stress no longer occurs in cells lacking an RNA chaperone Hfq. This indicates that SgrS acts on the *pts*G mRNA in an Hfq-dependent manner (Kawamoto *et al.*, 2005). Hfq is also involved in the stabilization of SgrS RNA. Figure 18.3 summarizes experiments that demonstrate the requirement for the C-terminal scaffold region of RNase E, Hfq, and SgrS in the destabilization of the *pts*G mRNA. When wild-type cells are exposed to a non–metabolizable glucose analogue, methylglucoside (αMG), SgrS is induced, resulting in destabilization of the *pts*G mRNA (lane 2). The destabilization of the *pts*G mRNA

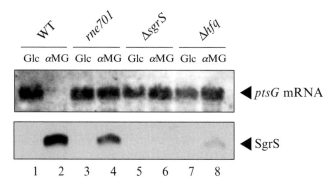

Figure 18.3 Effects of C-terminal truncation of RNase E, and the lack of SgrS or Hfq on the destabilization of the *ptsG* mRNA. IT1568 (wild-type), TM528 (*rne701-FLAG-cat*), TM587 (Δ*hfq::cat*), and TM540 (Δ*sgrS-sgrR::cat*) cells were grown in LB medium. At $OD_{600} = 0.5$, 0.4 % Glc or αMG was added, and the incubation was continued for 10 min. Total-cell RNA was prepared, and 10 μg and 5 μg of each RNA sample were subjected to Northern blot analysis with, respectively, the *ptsG* and 230-bp *sgrS* DIG-labeled DNA the probes. Essentially the same results were published previously (Kawamoto *et al.*, 2005; Morita *et al.*, 2004).

by αMG no longer occurred in cells expressing C-terminally truncated RNase E (lane 4) or lacking Hfq (lane 6) or SgrS (lane 8).

4. IDENTIFICATION OF RIBONUCLEOPROTEIN COMPLEX CONTAINING SGRS, HFQ, AND RNASE E

An important discovery was that Hfq associates with RNase E through the C-terminal scaffold region of the latter, and SgrS associates with RNase E through Hfq (Morita *et al.*, 2005). Thus, RNase E forms a ribonucleoprotein complex with SgrS through Hfq. The RNase E–based ribonucleoprotein complex containing SgrS, Hfq, and RNase E is distinct from the conventional RNA degradosome. The physical association of SgrS with RNase E nicely explains how the functional cooperation of SgrS/Hfq and RNase E is achieved. Example experiments to show the formation of a specific ribonucleoprotein complex containing SgrS, Hfq, and RNase E are provided in Fig. 18.4. Hfq and SgrS can be copurified with RNase E-FLAG (lane 6) but not with RNase E701-FLAG (lane 7). The following is the protocol for these experiments.

4.1. Pull-down assay with strains expressing a Flag-tagged protein

4.1.1. Procedure

1. Inoculate 200 ml of LB medium with 1 ml of overnight culture of a strain carrying a FLAG-tagged version of the protein of interest

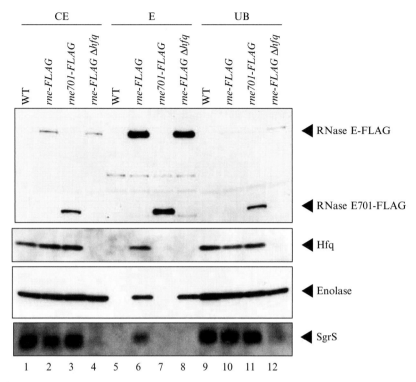

Figure 18.4 Physical interaction between RNase E and SgrS RNA. IT1568 (WT), TM338 (*rne-FLAG-cat*), TM528 (*rne701-FLAG-cat*), and TM618 (*rne-FLAG-cat Δhfq*) cells were grown in LB medium. At $A_{600} = 0.6$, 0.2% αMG was added to each culture, and incubation was continued for 10 min. To analyze proteins associated with RNase E, crude extracts (CE) were prepared and subjected to the pull-down assay with anti-FLAG agarose. The crude extracts, unbound fractions (UB), and bound fractions (E) were analyzed by Western blotting with anti-FLAG, anti-Hfq, and anti-enolase antibodies. To analyze RNAs associated with RNase E analysis, 5 μl of deproteinized crude extracts, unbound fractions, and bound fractions were subjected to Northern blotting with the *sgrS* DNA probe. Essentially the same results were published previously (Morita *et al.*, 2005).

(for example, RNase E-FLAG). Incubate at 37 °C with vigorous shaking until the culture reaches late-log phase (OD$_{600}$ = ~0.6).

2. Add 0.2% αMG to induce SgrS and continue incubation for 10 min.

3. Harvest cells by centrifugation at 4,500 rpm for 5 min at 4 °C. Discard the supernatant.

4. Wash the cells with 10 ml STE buffer (100 mM NaCl, 10 mM Tris-HCl, pH 8.0, 1 mM EDTA), and resuspend in 10 ml of ice-cold IP buffer 2 (20 mM Tris-HCl, pH 8.0, 0.2 M KCl, 5 mM MgCl$_2$, 10% glycerol, 0.1% Tween 20) containing Complete Mini, a broad spectrum of serine and cystein proteases (Roche).

5. Sonicate the cell suspension and centrifuge at 10,000 rpm for 1 h at 4 °C.

6. Transfer the supernatant (crude extract) into a new tube. Add 50 μl of anti-FLAG M2-agarose suspension (Sigma) equilibrated with IP buffer 2 and incubate gently for 1 h at 4 °C.

7. Filtrate the mixture with a mini-chromatography column (Bio-Rad). Save the filtrate as the unbound fraction. Wash the agarose beads three times with 10 ml of IP buffer 2.

8. Add 50 μl of IP buffer 2 containing 0.4 mg/ml FLAG peptide (Sigma) to the beads on the column (the bottom must be sealed with the cap) and incubate for 10 min at 4 °C with shaking.

9. Collect the eluate by centrifugation at 1,000 rpm for 1 min at 4 °C. The eluate corresponds to the bound fraction.

10. Use the eluate at step 9 (bound fraction) and the filtrate at step 7 (unbound fraction) for protein analysis (SDS-PAGE followed by Western blotting and/or Coomassie brilliant blue staining).

11. To analyze RNAs associated with the FLAG-tagged protein, treat appropriate volumes of samples (crude extract, unbound fraction, and bound fraction) with equal volumes of phenol that has been equilibrated with RNA buffer, precipitate the aqueous phase with ethanol, and wash the RNA pellet with 70% ethanol. Dissolve the RNA precipitate in 10 μl of distilled water (or RNA buffer) and use for Northern blotting.

4.1.2. Comments

This method works well for pull-down assays with strains expressing a FLAG-tagged protein from the chromosomal modified gene such as *rne-FLAG*, *hfq-FLAG*, and *rhlB-FLAG*. This procedure allows one to identify proteins and/or RNAs associated with a particular protein in its natural context rather than in artificial conditions (overproduction). We constructed strains in which a specific gene is manipulated to encode a FLAG-tagged protein by the modified Datsenko–Wanner protocol (Datsenko and Wanner, 2000) with pSU313 harboring the FLAG-tag sequence (Uzzau *et al.*, 2001).

5. TRANSLATIONAL REPRESSION IS THE PRIMARY EVENT FOR SILENCING OF *PTSG* MRNA BY SGRS/HFQ/RNASE E

SgrS and RyhB RNAs are involved in the RNase E–dependent degradation and translational repression of target mRNAs. Pulse labeling and immunoprecipitation (IP) experiments revealed that the synthesis of IICB^{Glc} is markedly reduced in wild-type cells, but not in cells lacking SgrS (Morita *et al.*, 2006). Importantly, the synthesis of IICB^{Glc} is still strongly inhibited in cells expressing the truncated RNase E in which the destabilization the *ptsG* mRNA no longer occurred. Efficient translational inhibition is also observed when temperature-sensitive RNase E is thermally inactivated. Thus, SgrS appears to downregulate IICB^{Glc} expression primarily by inhibiting translation directly. The accelerated degradation of *ptsG* mRNA is considered a consequence rather than the primary effector of gene silencing. Example experiments are provided in Fig. 18.5. The translation of the *ptsG* mRNA is markedly inhibited in both wild-type (lanes 7 and 8) and *rne701* (lanes 9 and 10) cells in response to αMG exposure. The following is the procedure for these experiments.

5.1. Pulse labeling and IP experiments

5.1.1. Procedure

1. Introduce a low-copy-number plasmid carrying the *ptsG-FLAG* gene in three isogenic strains (*rne-HA*, *rne701-HA*, and *rne-HA* Δ*sgrS*).
2. Grow cells in M9-0.4% glycerol medium supplemented with all the amino acids with the exception of methionine (50 μg/ml each). Also add ampillicin ampicillin to select for the plasmid.
3. Add 0.01% glucose or α-methyl glucoside at OD$_{600}$ = 0.4 and continue incubation for 10 min.

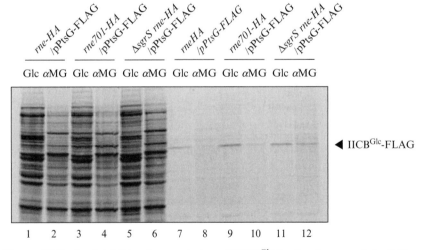

Figure 18.5 SgrS inhibits the translation of IICB^{Glc} in the absence of the RNase E–dependent destabilization of the *ptsG* mRNA. TM641 (*rne-HA*), TM642 (*rne701-HA*), and TM666 (*rne-HA* Δ*sgrS*) harboring pPtsG-FLAG (plasmid carrying the *ptsG-FLAG* allele) were grown until A$_{600}$ = 0.4 in M9-0.4% glycerol medium without methionine. To each culture, 0.01% Glc or αMG was added, and incubation was continued for 10 min. Each culture was exposed to ^{35}S-methionine for 1 min and subjected to immunoprecipitation experiments. The crude extract (total, lanes 1 to 6) and bound fraction (IP, lanes 7 to 12), corresponding to 0.04 A$_{600}$ unit or to 0.12 A$_{600}$ unit, respectively, were analyzed by 12% SDS-PAGE followed by autoradiography. Data were published previously (Morita *et al.*, 2006).

4. Add 0.1 nmol (37 kBq) of [^{35}S]methionine to 400 μl of culture and incubate for 1 min.
5. Add trichloroacetic acid to a final concentration of 5% and store for 10 min on ice.
6. Centrifuge at 14,000 rpm for 10 min at 4 °C, wash the precipitate with acetone, and dissolve in 40 μl SDS-EDTA buffer (1% SDS, 20 mM Tris-HCl, pH 8.0, 0.1 mM EDTA) by heating the sample at 45 °C. This can be used as a total-cell protein fraction.
7. Take 20 μl of total-cell protein fraction and add 0.5 ml IP buffer 1 (20 mM Tris-HCl, pH 8.0, 0.1 M KCl, 5 mM MgCl$_2$, 10% glycerol, 0.1 % Tween 20) containing Complete Mini (Roche Diagnostics) for IP experiments.
8. Add 20 μl of anti-FLAG M2-agarose suspension (Sigma) and incubate the mixture for 1 h at 4 °C.
9. Centrifuge at 15,000 rpm for 1 min, wash the pellet beads two times with 1 ml of IP buffer 1 by centrifugation.
10. Elute the proteins bound to the beads with 15 μl of SDS sample buffer at 95 °C for 5 min and use as IP sample.
11. Analyze the total-cell protein and IP samples by SDS-PAGE followed by autoradiography.

6. BASE PAIRING NEAR TRANSLATION INITIATION REGION IS CRUCIAL FOR SGRS ACTION

A systematic mutational study of the base pairing required for sRNA action has been carried out in the SgrS-*ptsG* mRNA system (Kawamoto *et al.*, 2006). Mutations in *ptsG* mRNA or SgrS that affect the action of SgrS are confined to the 6-nt region overlapping the *ptsG* SD sequence. In particular, two single mutations that disrupt a G–C base pair within this short sequence almost completely eliminate SgrS function and compensatory mutations restore it. Duplex formation *in vitro* was also eliminated by a single mutation disrupting the C-G base pair within the critical region and restored by a compensatory mutation. Thus, the 6 base pairs overlapping the *ptsG* SD sequence among the predicted 23 base pairs are crucial for SgrS action. The requirment for base pairing between SgrS and *ptsG* mRNA at a short sequence overlapping the *ptsG* SD sequence is consistent with a view that SgrS competes with and prevents the binding of 16S rRNA of the ribosome resulting in translational inhibition. Example experiments that support these conclusions are provided in Fig. 18.6. The *sgrS* gene was cloned on a plasmid in which its expression was under the control of an IPTG-inducible promoter. The *ptsG* gene was cloned on a low-copy-number plasmid. Point mutations at individual nucleotides in the *ptsG* or

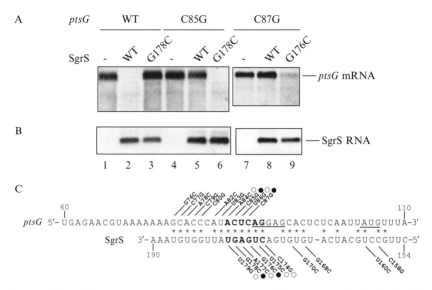

Figure 18.6 Effect of compensatory mutations on *ptsG* silencing by SgrS. (A, B) AS23 (△*ptsG* △*sgrS-sgrR*) cells harboring pMG001 (wild-type *ptsG*) plus pQE80L (vector) (lane 1), pMG001 plus pQESgrS1 (wild-type *sgrS*) (lane 2), pMG001 plus pQESgrS2 (*sgrS*G178C) (lane 3), pMG309 (*ptsG*C85G) plus pQE80L (lane 4), pMG309 plus pQESgrS1 (lane 5), pMG309 plus pQESgrS2 (lane 6), pMG311 (*ptsG*C87G) plus pQE80L (lane 7), pMG311 plus pQESgrS1 (lane 8), pMG311 plus pQESgrS3 (*sgrS*G176C) (lane 9) were grown to A$_{600}$ = 0.5; IPTG was added at a final concentration of 0.1 m*M*, and incubation was continued for 20 min. Total-cell RNA was prepared and 3 μg (*ptsG* probe; A) or 1 μg (*sgrS* probe; B) of each RNA sample were subjected to Northern blot analysis. (C) Nucleotide sequences of *ptsG* mRNA and SgrS RNA around the predicted base-pairing region. RNA the sequences corresponding to nucleotide positions 60–110 of *ptsG* mRNA and 154–190 of SgrS RNA are shown. The predicted base pairs between *ptsG* mRNA and SgrS RNA (Vanderpool and Gottesman, 2004) are shown by asterisks. The point mutations in Shine-Dalgarno sequence and translation initiation codon of *ptsG* mRNA are underlined. The point mutations in *ptsG* and *sgrS* that were generated are indicated above and below the sequences with thin lines, respectively. The closed and open circles indicate mutations that affect the SgrS action strongly and weakly, respectively. Data were published previously (Kawamoto *et al.*, 2006).

sgrS gene that are predicted to base pair were introduced. The SgrS- and *ptsG*-containing plasmids were introduced into cells lacking the endogenous *ptsG* gene. Expression of *ptsG* and SgrS RNAs from the cloned genes was examined in the presence and absence of IPTG by Northern blot analysis. The wild-type *ptsG* mRNA was destabilized by IPTG-induced wild-type SgrS (lane 2) but not by SgrS$_{G178C}$ (lane 3). The *ptsG*$_{C85G}$ and *ptsG*$_{C85G}$ mRNAs were no longer destabilized by wild-type SgrS (lanes 5 and 8). The *ptsG*$_{C85G}$ and *ptsG*$_{C85G}$ mRNAs were destabilized when a compensatory mutation was introduced in SgrS (lanes 6 and 9).

7. BASE PAIRING BETWEEN SGRS AND *PTSG* MRNA *IN VITRO* AND ACCELERATION OF DUPLEX FORMATION BY HFQ

Base pairing between SgrS and *pts*G mRNA was also tested *in vitro* by gel mobility shift assay (Kawamoto *et al.*, 2006). The wild-type *pts*G but not *pts*G$_{C85G}$ RNA formed a stable RNA-RNA duplex with wild-type SgrS RNA. SgrS$_{G178C}$ RNA formed a duplex with wild-type *pts*G RNA as expected. These results are consistent with the *in vivo* data indicating that SgrS RNA does indeed interact directly with *pts*G RNA through base pairing at a short sequence overlapping the *pts*G SD sequence to down-regulate *pts*G mRNA. Hfq, originally identified as a host factor required for the *in vitro* replication of the RNA phage Qβ in *E. coli* (Franze de Fernandez *et al.*, 1968), is an RNA-binding protein extensively involved in the regulation of RNA metabolism (Valentin-Hansen *et al.*, 2004). Chapter XXX (Regnier and Hajnsdorf) also focuses on Hfq and describes methods to characterize the role of Hfq in poly(A) metabolism.

Hfq was shown to bind to and be necessary for the regulatory activity of a number of chromosomally encoded antisense sRNAs (Valentin-Hansen *et al.*, 2004). The protein resembles the Sm and Sm-like proteins involved in splicing and mRNA degradation in eukaryotic cells. It has a mass of 11 kDa and forms a doughnut-shaped hexameric structure (Moller *et al.*, 2002; Zhang *et al.*, 2002). For several examples, Hfq has been shown to stimulate base pairing between a given sRNA and its target mRNA *in vitro*. This is achieved either by acting as a chaperone that changes RA changing the RNA structure(i.e. Hfq acts as an RNA chaperone), thereby allowing accessibility of two complementary RNAs, or by binding to a given sRNA and its target mRNA simultaneously, thereby increasing the local concentrations of two RNAs (Storz *et al.*, 2004). Another role of Hfq is to stabilize sRNAs, thereby to support appropriate expression levels of these molecules. We have shown also that Hfq dramatically enhances the rate of duplex formation between the *pts*G and SgrS RNAs (Kawamoto *et al.*, 2006). The rapid association of SgrS with *pts*G mRNA in the presence of Hfq is thought to prevent the efficient ribosome binding to *pts*G mRNA. Example experiments to show base pairing between SgrS and *pts*G mRNA *in vitro* and the acceleration of duplex formation by Hfq are provided in Fig. 18.7A and 7B, respectively.

7.1. Gel mobility shift assay for detection of base pairing

7.1.1. Procedure

1. Amplify DNA fragments corresponding to the 5′-region (1 to 120) of wild–type and mutant C85G) *pts*G by PCR and clone into pBluescript II

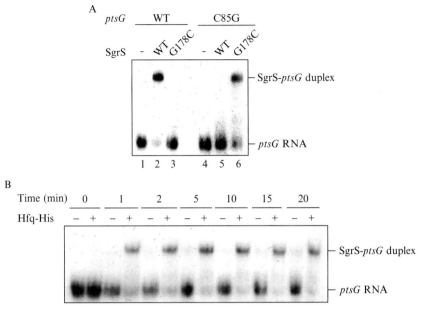

Figure 18.7 Duplex formation between *ptsG* mRNA and SgrS RNA. (A) Effect of mutations on *ptsG* mRNA -SgrS complex formation. ^{32}P-labeled wild-type or *ptsG*$_{C85G}$ RNA (10 nM) was incubated with 500 nM of unlabeled wild-type or SgrS$_{G178C}$ RNA at 37 °C for 30 min. The complex formation was monitored by a gel mobility shift assay using a native polyacrylamide gel. (B) Effect of Hfq on duplex formation. ^{32}P-labeled *ptsG* mRNA (10 nM) was incubated with 20 nM of unlabeled SgrS RNA in the presence and absence of 20 nM Hfq-His$_6$ hexamer at 37 °C. At the indicated times, the samples were extracted with phenol and loaded onto a native gel. Data were published previously (Kawamoto *et al.*, 2006).

SK(−). Amplify DNA fragments corresponding to wild-type and G178C *sgrS* and clone into pBluescript II SK(−).

2. Linearize the plasmids by restriction digestion, and purify the linearized plasmids through agarose gel electrophoresis.

3. Transcribe the linearized plasmids with CUGA®T7 *in vitro* transcription kit (Nippon Genetech) according to the manufacturer's instructions. Add [α-^{32}P]-UTP (Amersham Biosciences) to the reaction mixture to generate ^{32}P-labeled transcripts from the *ptsG* DNA fragment.

4. Separate the reaction mixtures by electrophoresis in a 8% polyacrylamide gel, cut out the RNA bands, elute in buffer containing 20 mM Tris-HCl (pH 7.5), 0.5 M NH$_4$OAc, 10 mM Mg(OAc)$_2$, 1 mM EDTA, and 0.1% SDS at 37 °C for overnight, followed by ethanol precipitation.

5. Mix ^{32}P-labeled *ptsG* RNA (10 nM) and varying concentrations of SgrS RNA in 20 μl of binding buffer (10 mM Tris-HCl, pH 8.0, 50 mM NaCl, 50 mM KCl, 10 mM MgCl$_2$, 200 ng/μl yeast tRNA).

6. Incubate the reaction mixture at 37 °C for the indicated times.

7. Add 2 μl of loading buffer (50 % glycerol, 0.1 % bromophenol blue) and analyze by gel electrophoresis in a 4% polyacrylamide gel in 0.5× TBE bufer containing 5% glycerol at 4 °C.
8. After electrophoresis, dry the gel and subject to autoradiography.
9. To examine the effect of Hfq, incubate 10 nM ^{32}P-labeled *ptsG* RNA with 50 nM SgrS RNA in the presence and absence of 50 nM Hfq-His$_6$ in 20 of μl binding buffer at 37 °C. Treat the reaction mixture with phenol at the indicated times. After centrifugation, take the aqueous phase, add 2 μl of loading buffer, and subject to electrophoresis in a 4% polyacrylamide gel at 4 °C. After electrophoresis, dry the gel and subject to autoradiography.

7.2. Purification of His-tagged Hfq

7.2.1. Procedure

1. Grow cells harboring pQE-Hfq-His expressing the C-terminally His-tagged Hfq under the IPTG-inducible promoter in 200 ml of LB medium at 37 °C.
2. Add 1 mM of IPTG when $A_{600} = 0.2$ and continue incubation for 90 min.
3. Harvest cells by centrifugation at 4,500 rpm for 5 min at 4 °C. Discard the supernatant.
4. Wash the cells with 10 ml STE and resuspend in 2 ml of 50 mM NaH$_2$PO$_4$, 300 mM NaCl, 10 mM imidazole, pH 8.0.
5. Add lysozyme (2.5 mg/ml) and incubate for 10 min at 0 °C.
6. Sonicate cell suspension and centrifuge at 10,000 rpm for 10 min at 4 °C.
7. Add RNase A (0.1 μg/ml) to the supernatant, incubate for 10 min at 0 °C, and then for 10 min at 80 °C.
8. Centrifuge at 10,000 rpm for 10 min at 4 °C.
9. Add 100 μl of Ni^{2+}-NTA agarose resin (Qiagen) to the supernatant and incubate for 15 min with gentle shaking at 4 °C.
10. Purify Hfq-His$_6$ protein according to the manufacturer's instructions.
11. Check the concentration and purity of protein by SDS-PAGE followed by Coomassie brilliant blue staining.

8. MEMBRANE LOCALIZATION OF *PTSG* MRNA IS REQUIRED FOR SGRS ACTION

IICB^{Glc} consists of the N-terminal IIC domain containing eight transmembrane segments and the C-terminal IIB domain (Buhr and Erni, 1993). During our study of the downregulation of *ptsG* mRNA by SgrS, we found that the region corresponding to the first two transmembrane stretches of

IICB^{Glc} is required for the rapid degradation of *pts*G mRNA (Kawamoto *et al.*, 2005). When the transmembrane region was replaced with that of another protein, the mRNA was still destabilized by SgrS, suggesting that the membrane-targeting property is important for the destabilization of mRNA mediated by SgrS. In addition, the cytoplasmic target mRNA is destabilized by SgrS when its translation is reduced by mutations. On the basis of these observations, the following model has been proposed. The cotranslational membrane insertion of nascent peptide of IICB^{Glc} brings the *pts*G mRNA near the membrane. This membrane localization of *pts*G mRNA may decrease the efficiency of ribosome loading, thereby allowing SgrS RNA to base pair more efficiently with *pts*G mRNA. In turn this would result in accelerated RNase E–dependent degradation of the *pts*G mRNA. This suggests that sRNA action on target mRNAs could be modulated by translational status (ribosome loading) in some cases.

9. CONCLUSIONS

Studies on SgrS and RyhB RNAs in the past few years have demonstrated that both RNAs share common features with other Hfq-binding sRNAs involved in the downregulation of target mRNAs. In addition, studies of SgrS and RyhB have uncovered several novel mechanistic features of sRNA action. Nevertheless, several important questions remain concerning the mechanisms of action of Hfq-binding sRNAs. For example, an important question is how Hfq stimulates the base-pairing between a specific sRNA and its target mRNA. We also need to learn more about the mechanisms and roles of RNase E–dependent degradation of both target mRNAs and sRNAs. The turnover of sRNAs was shown to be coupled to and dependent on pairing with target mRNAs (Masse *et al.*, 2003). The coupled degradation of sRNAs and their target mRNAs provides an elegant way by which cells escape from regulation by certain sRNAs when stress signals are have no longer present. However, this view must be tested experimentally. Furthermore, little is known about how a specific stress induces transcription of a given sRNA that modulates the activity of the cognate transcription factor. Finally, eukaryotic microRNAs also act on target mRNAs through base pairing by forming ribonucleoprotein complexes (Bartel, 2004), and their regulatory outcomes are translational inhibition and mRNA destabilization (Bagga *et al.*, 2005; Lim *et al.*, 2005). see also chapters in RNA Turnover, Part B, ed. Maquat and Kiledjian. Thus, bacterial sRNAs apparently share common features with eukaryotic counterparts. It is certainly worthwhile investigating further general similarities and differences between both systems.

ACKNOWLEDGMENTS

This work was supported by Grants-in-Aid from the Ministry of Education, Culture, Sports, Science and Technology of Japan and by Ajinomoto Co., Inc.

REFERENCES

Aiba, H., Adhya, S., and de Crombrugghe, B. (1981). Evidence for two functional gal promoters in intact *Escherichia coli* cells. *J. Biol. Chem.* **256,** 11905–11910.

Aiba, H. (2007). Mechanism of RNA silencing by Hfq-binding small RNAs. *Curr. Opin. Microbiol.* **10,** 134–139.

Altuvia, S., Weinstein-Fischer, D., Zhang, A., Postow, L., and Storz, G. (1997). A small, stable RNA induced by oxidative stress: Role as a pleiotropic regulator and antimutator. *Cell* **90,** 43–53.

Argaman, L., Hershberg, R., Vogel, J., Bejerano, G., Wagner, E. G., Margalit, H., and Altuvia, S. (2001). Novel small RNA-encoding genes in the intergenic regions of *Escherichia coli. Curr. Biol.* **11,** 941–950.

Bagga, S., Bracht, J., Hunter, S., Massirer, K., Holtz, J., Eachus, R., and Pasquinelli, A. E. (2005). Regulation by let-7 and lin-4 miRNAs results in target mRNA degradation. *Cell* **122,** 553–563.

Bartel, D. P. (2004). MicroRNAs: Genomics, biogenesis, mechanism, and function. *Cell* **116,** 281–297.

Buhr, A., and Erni, B. (1993). Membrane topology of the glucose transporter of *Escherichia coli. J. Biol. Chem.* **268,** 11599–11603.

Carpousis, A. J. (2002). The *Escherichia coli* RNA degradosome: structure, function and relationship in other ribonucleolytic multienzyme complexes. *Biochem. Soc. Trans.* **30,** 150–155.

Chen, S., Lesnik, E. A., Hall, T. A., Sampath, R., Griffey, R. H., Ecker, D. J., and Blyn, L. B. (2002). A bioinformatics based approach to discover small RNA genes in the *Escherichia coli* genome. *Biosystems* **65,** 157–177.

Datsenko, K. A., and Wanner, B. L. (2000). One-step inactivation of chromosomal genes in *Escherichia coli* K-12 using PCR products. *Proc. Natl. Acad. Sci. USA* **97,** 6640–6645.

Franze de Fernandez, M. T., Eoyang, L., and August, J. T. (1968). Factor fraction required for the synthesis of bacteriophage Qbeta-RNA. *Nature* **219,** 588–590.

Gottesman, S. (2004). The small RNA regulators of *Escherichia coli*: Roles and mechanisms. *Annu. Rev. Microbiol.* **58,** 303–328.

Gottesman, S. (2005). Micros for microbes: Non-coding regulatory RNAs in bacteria. *Trends Genet* **21,** 399–404.

Kawamoto, H., Morita, T., Shimizu, A., Inada, T., and Aiba, H. (2005). Implication of membrane localization of target mRNA in the action of a small RNA: Mechanism of post-transcriptional regulation of glucose transporter in *Escherichia coli. Genes Dev.* **19,** 328–338.

Kawamoto, H., Koide, Y., Morita, T., and Aiba, H. (2006). Base-pairing requirement for RNA silencing by a bacterial small RNA and acceleration of duplex formation by Hfq. *Mol. Microbiol.* **61,** 1013–1022.

Kawano, M., Reynolds, A. A., Miranda-Rios, J., and Storz, G. (2005). Detection of 5′- and 3′ UTR-derived small RNAs and cis-encoded antisense RNAs in *Escherichia coli. Nucleic Acids Res.* **33,** 1040–1050.

Kimata, K., Tanaka, Y., Inada, T., and Aiba, H. (2001). Expression of the glucose transporter gene, ptsG, is regulated at the mRNA degradation step in response to glycolytic flux in *Escherichia coli. EMBO J.* **20**, 3587–3595.

Lim, L. P., Lau, N. C., Garrett-Engele, P., Grimson, A., Schelter, J. M., Castle, J., Bartel, D. P., Linsley, P. S., and Johnson, J. M. (2005). Microarray analysis shows that some microRNAs downregulate large numbers of target mRNAs. *Nature* **433**, 769–773.

Masse, E., and Gottesman, S. (2002). A small RNA regulates the expression of genes involved in iron metabolism in *Escherichia coli. Proc. Natl. Acad. Sci. USA* **99**, 4620–4625.

Masse, E., Escorcia, F. E., and Gottesman, S. (2003). Coupled degradation of a small regulatory RNA and its mRNA targets in *Escherichia coli. Genes Dev.* **17**, 2374–2383.

Mizuno, T., Chou, M. Y., and Inouye, M. (1984). A unique mechanism regulating gene expression: Translational inhibition by a complementary RNA transcript (micRNA). *Proc. Natl. Acad. Sci. USA* **81**, 1966–1970.

Moller, T., Franch, T., Hojrup, P., Keene, D. R., Bachinger, H. P., Brennan, R. G., and Valentin-Hansen, P. (2002). Hfq: A bacterial Sm-like protein that mediates RNA-RNA interaction. *Mol. Cell* **9**, 23–30.

Morita, T., El-Kazzaz, W., Tanaka, Y., Inada, T., and Aiba, H. (2003). Accumulation of glucose 6-phosphate or fructose 6-phosphate is responsible for destabilization of glucose transporter mRNA in *Escherichia coli. J. Biol. Chem.* **278**, 15608–15614.

Morita, T., Kawamoto, H., Mizota, T., Inada, T., and Aiba, H. (2004). Enolase in the RNA degradosome plays a crucial role in the rapid decay of glucose transporter mRNA in the response to phosphosugar stress in *Escherichia coli. Mol. Microbiol.* **54**, 1063–1075.

Morita, T., Maki, K., and Aiba, H. (2005). RNase E-based ribonucleoprotein complexes: Mechanical basis of mRNA destabilization mediated by bacterial noncoding RNAs. *Genes Dev.* **19**, 2176–2186.

Morita, T., Mochizuki, Y., and Aiba, H. (2006). Translational repression is sufficient for gene silencing by bacterial small noncoding RNAs in the absence of mRNA destruction. *Proc. Natl. Acad. Sci. USA* **103**, 4858–4863.

Morita, T., and Aiba, H. (2007). Small RNAs making a small protein. *Proc. Natl. Acad. Sci. USA* (Electronic: published December 11, 2007).

Postma, P. W., Lengeler, J. W., and Jacobson, G. R., Ed. (1996). "Phosphoenolpyruvate: Carbohydrate Phosphotransferase Systems." ASM Press, Washington, DC.

Sambrook, J. F. E., and Maniatis, T. (1989). "Molecular Cloning: A Laboratory Manual," 2nd ed. Cold Spring Harbor Laboratory Press. Cold Spring Harbor, NY.

Sledjeski, D., and Gottesman, S. (1995). A small RNA acts as an antisilencer of the H-NS-silenced rcsA gene of *Escherichia coli. Proc. Natl. Acad. Sci. USA* **92**, 2003–2007.

Storz, G., Opdyke, J. A., and Zhang, A. (2004). Controlling mRNA stability and translation with small, noncoding RNAs. *Curr. Opin Microbiol.* **7**, 140–144.

Storz, G., Altuvia, S., and Wassarman, K. M. (2005). An abundance of RNA regulators. *Annu. Rev. Biochem.* **74**, 199–217.

Storz, G., and Gottesman, S. (2006). "Versatile Roles of Small RNA Regulators in Bacteria," Cold Spring Harbor Laboratory Press, Cold Spring Harbor, New York.

Uzzau, S., Figueroa-Bossi, N., Rubino, S., and Bossi, L. (2001). Epitope tagging of chromosomal genes in *Salmonella. Proc. Natl. Acad. Sci. USA* **98**, 15264–15269.

Valentin-Hansen, P., Eriksen, M., and Udesen, C. (2004). The bacterial Sm-like protein Hfq: A key player in RNA transactions. *Mol. Microbiol.* **51**, 1525–1533.

Vanderpool, C. K., and Gottesman, S. (2004). Involvement of a novel transcriptional activator and small RNA in post-transcriptional regulation of the glucose phosphoenol-pyruvate phosphotransferase system. *Mol. Microbiol.* **54**, 1076–1089.

Vogel, J., Bartels, V., Tang, T. H., Churakov, G., Slagter-Jager, J. G., Huttenhofer, A., and Wagner, E. G. (2003). RNomics in *Escherichia coli* detects new sRNA species and indicates parallel transcriptional output in bacteria. *Nucleic Acids Res.* **31**, 6435–6443.

Wadler, C. S., and Vanderpool, C. K. (2007). A dual function for a bacterial small RNA: SgrS performs base pairing-dependent regulation and encodes a functional polypeptide. *Proc. Natl. Acad. Sci. USA* Electronic: published November 27, 2007.

Wassarman, K. M., Repoila, F., Rosenow, C., Storz, G., and Gottesman, S. (2001). Identification of novel small RNAs using comparative genomics and microarrays. *Genes Dev.* **15,** 1637–1651.

Zhang, A., Wassarman, K. M., Ortega, J., Steven, A. C., and Storz, G. (2002). The Sm-like Hfq protein increases OxyS RNA interaction with target mRNAs. *Mol. Cell* **9,** 11–22.

Zhang, A., Wassarman, K. M., Rosenow, C., Tjaden, B. C., Storz, G., and Gottesman, S. (2003). Global analysis of small RNA and mRNA targets of Hfq. *Mol. Microbiol.* **50,** 1111–1124.

ARCHAEA

CHAPTER NINETEEN

IN VIVO AND *IN VITRO* STUDIES OF RNA DEGRADING ACTIVITIES IN ARCHAEA

Elena Evguenieva-Hackenberg, Steffen Wagner, *and* Gabriele Klug

Contents

Institut für Mikrobiologie und Molekularbiologie, University of Giessen, Giessen, Germany

Methods in Enzymology, Volume 447
ISSN 0076-6879, DOI: 10.1016/S0076-6879(08)02219-2

Abstract

Controlled degradation of RNA is important for the regulation of gene expression in Bacteria and Eukarya, but information about these processes is limited in the domain of Archaea. To address this, we studied the half-life of different mRNAs in halophilic Archaea after blocking transcription with actinomycin D. We found that the stability of mRNAs of the *gvp* operons in *Haloferax mediterranei* varies under different growth conditions. To understand regulated mRNA decay in Archaea, we need to identify stability determinants within mRNAs and proteins, mainly ribonucleases (RNases), which recognize these determinants. First, we wanted to identify archaeal RNases independently of their sequence similarity to known RNases from Bacteria and Eukarya. To this end we performed fractionation of proteins from *Halobacterium salinarum* and tested the fractions for RNase activity with an internally labeled *in vitro*–synthesized mRNA. After three purification steps, we isolated an endoribonucleolytically active protein with similarities to the eukaryotic initiation factor 5A. Further characterization was performed with recombinant halobacterial IF-5A, which was purified from *H. salinarum* or *Escherichia coli*. Mutational analysis confirmed unambiguously its RNase activity. In another study, we aimed to purify a double-strand–specific endoribonuclease from *Sulfolobus solfataricus*. Seven purification steps led to the isolation of two different dehydrogenases with RNase properties. Interestingly, their RNase activity resembled that of aIF-5A and of highly diluted RNase A. RNA was cleaved preferentially between C and A nucleotides in single-stranded regions, and the activity was inhibited at $MgCl_2$ concentrations >5 mM and at KCl concentrations >200 mM. However, it was possible to distinguish the activity of the archaeal proteins from the activity of RNase A. In a different approach, we used a bioinformatics prediction of the archaeal exosome to purify this protein complex from *S. solfataricus*. Isolation by coimmunoprecipitation revealed the presence of four orthologs of eukaryotic exosomal subunits and at least one archaea-specific subunit. We characterized the *S. solfataricus* exosome as a major enzyme involved in phosphorolytic RNA degradation and in RNA polyadenylation. Here we describe in detail the techniques used to achieve these results.

1. INTRODUCTION

Different types of RNA are often transcribed as precursors, which are processed by enzymes, mostly ribonucleases (RNases), into mature functional molecules. These mature RNAs can be very stable (transfer RNA, ribosomal RNA) or they can be a subject of fast turnover (most messenger RNAs) (Deutscher, 2006; Houseley et al., 2006). Small noncoding RNAs have different functions and stabilities (Matera et al., 2007; Vogel et al., 2003). The half-life of eukaryotic messenger RNAs (mRNAs) ranges from tens of minutes to hours (Raghavan et al., 2002). Bacterial mRNAs are much less stable; their half-lives are in the range of seconds to minutes, enabling quick responses to changing environmental conditions (Bernstein et al., 2002; Rauhut and Klug, 1999). Nothing was known about the half-lives of archaeal mRNAs at the beginning of our studies.

It is well documented that in bacteria the half-lives of specific mRNA molecules significantly vary under different growth conditions—a phenomenon known to control gene expression at the posttranscriptional level. The half-life of an mRNA depends on the presence of destabilizing elements, which are recognized by degrading enzymes, and stabilizing elements, which interfere with RNase activities (Condon, 2007; Rauhut and Klug, 1999). Known stabilizing elements include the triphosphate at the mRNA $5'$-end and hairpin loops at the mRNA $5'$-end or $3'$-end. Cleavage sites for the endoribonucleases RNase E and RNase III, and short poly(A)-tails at the $3'$-end, which serve as a loading platform for the $3'$ to $5'$-exoribonucleases PNPase, RNase II, and RNase R, destabilize RNA. RNases are often found in complexes with other RNases and with RNA helicases like RhlB, which help to unwind secondary structures. For example, PNPase is found in E. coli together with RNase E, RhlB, and other proteins in a protein complex called the degradosome (see chapters by Carpousis et al., Mauria and Dehò and Mackie et al.), but it also forms an independent complex only with RhlB (Condon, 2007; Lin-Chao et al., 2007). Bacterial enzymes involved in mRNA degradation also participate in the processing and degradation of stable RNA molecules like rRNA and tRNA (Deutscher, 2006). The preceding mechanisms were well studied in the model organism Escherichia coli, and similar decay pathways seem to exist in other gram-negative bacteria. Different mechanisms operate in Bacillus subtilis, a gram-positive bacterium that harbors a very recently characterized $5'$ to $3'$-exoribonuclease (Condon, 2007; see chapters by Bechhofer et al. and Condon et al.).

Control of gene expression on the posttranscriptional level is seemingly more complex in Eukarya (Moore, 2005; Zamore and Haley, 2005). Eukaryotic mRNA stabilizing elements are the $5'$-methylated guanosine cap at the $5'$-end and the long poly(A)-tail at the $3'$-end. The removal of

poly(A) and decapping are followed by exoribonucleolytic degradation in a
5′ to 3′-direction by enzymes like Xrn1 or in a 3′ to 5′-direction by the
exosome, a protein complex with structural similarity to PNPase. Very
recently, short destabilizing poly(A)-tails were also described in Eukarya
(LaCava *et al.*, 2005; Vanacova *et al.*, 2005). This very short and incomplete
depiction of important details concerning mechanisms for mRNA degradation in Eukarya (see chapters in Maquat and Kiledjian, RNA Turnover,
Part B) already outlines differences and similarities to bacterial decay pathways. Generally, the molecular mechanisms operating in Archaea show
more similarities to the eukaryotic ones, despite Archaea having a more
prokaryotic-like than eukaryotic-like cellular structure. Very little is known
about RNA degradation in Archaea. At the begin of our studies, proteins
with homology to RNase III were not found in the sequenced archaeal
genomes, the existence of an RNase E–like activity in archaeal extracts was
reported, and orthologs of bacterial and eukaryotic exoribonucleases were
predicted (Anantaraman *et al.*, 2002; Franzetti *et al.*, 1998; Koonin *et al.*,
2001; Zuo and Deutscher, 2001). The structural features of mRNA that
would be important for mRNA stability were assumed to resemble those of
bacteria, because archaeal mRNAs are not capped and do not carry long
poly(A)-tails (Brown and Reeve, 1985). Our work focuses on RNA stability in Archaea and on the identification of archaeal ribonucleases.

2. *IN VIVO* STUDIES ON RNA DEGRADING ACTIVITIES IN ARCHAEA

One aim of our studies was to analyze whether there is posttranscriptional regulation of gene expression in Archaea. Therefore, we studied the
half-lives of selected mRNAs in the halophilic archaeon *Haloferax mediterranei* under different growth conditions. Furthermore, we wanted to obtain
general insight into mRNA stability in Archaea. For this purpose, we used
the halophilic archaeon *Halobacterium salinarum* NRC-1, the complete
genome sequence of which was available, in microarray studies.

2.1. *In vivo* labeling of RNA and inhibition of transcription with actinomycin D

To understand the role of degradation of certain mRNAs in gene expression, it is necessary to follow mRNA decay *in vivo* under different conditions. The amount of individual RNA species in total-cell RNA represents
a steady-state level, which is determined by the rates of RNA synthesis and
decay. One method to follow RNA decay is to determine RNA abundance

at different time points after efficiently blocking transcription. In Bacteria, transcription is usually blocked by the addition of rifampicin, which directly interacts with RNA polymerase, to cell cultures. The archaeal RNA polymerase shares similarity with the eukaryotic RNA polymerase and is different from the bacterial enzyme. Accordingly, the archaeal RNA polymerase is insensitive to rifampicin, like its eukaryotic counterpart (Langer et al., 1995; Prangishvilli et al., 1982).

We decided to use actinomycin D, which is known to interact with DNA and, thereby, block RNA transcription in Bacteria and in Eukarya (Goldberg et al., 1962; Reich et al., 1961) to block transcription in Haloferax and Halobacterium. To measure the efficiency of the block, it is necessary to have a method to detect RNA synthesis in vivo. Incorporation of radiolabeled uracil or uridine into TCA (trichloroacetic acid) precipitable material can generally be used to determine the in vivo rate of total-cell RNA synthesis (Pato et al., 1973). To monitor the inhibition of transcription by the addition of chemicals, cultures of Haloarchaea were grown to early exponential phase ($OD_{600nm} = 0.45$) or stationary phase ($OD_{600nm} = 2.8$). Radiolabeled uridine ([5,6-^{3}H], 32 Ci/mmol; Amersham Biosciences) was added at time point 0 to a final concentration of 100 μCi/ml when exponentially growing cultures were used or 300 μCi/ml when stationary-phase cultures were used. Samples of 20 μl were collected at several time points thereafter and mixed with 500 μl of unlabeled cells. Cold TCA was added to a final concentration of 11.5% (w/v). After incubation on ice for 5 min, samples were centrifuged, and TCA-precipitated radioactivity was quantified with a scintillation counter (LS 6500 Multi Purpose Scintillation Counter, Beckman). Optimal results were obtained when 500 μl of Rotiszint Ecoplus (Roth, Germany) cocktail was used to dissolve the pellet (Hundt et al., 2007; Jäger et al., 2002).

To test the effect of chemicals on transcription, the agent was added 12 min after the addition of uridine. Actinomycin D was dissolved overnight at 4 °C in water to generate a stock of 5 mg/ml that was used the next day. In H. mediterranei and H. salinarum NRC1, a final concentration of 100 μg/ml of actinomycin D efficiently blocks RNA synthesis (Jäger et al., 2002). It was shown by Bini et al. (2002) that actinomycin D is stable enough at high temperatures and acidic pH to be used to block transcription in Sulfolobus solfataricus. A final actinomycin D concentration of 10 μg/ml is sufficient to block RNA synthesis efficiently in this archaeon; the stock solution of (1 mg/ml) was prepared in 50% ethanol. In S. solfataricus, uridine is not incorporated to significant levels but uracil ([5,6-^{3}H], 33.1 Ci/mmol) is well incorporated. For in vivo labeling, S. solfataricus was grown to early exponential phase ($OD_{540nm} = 0.17$). Time linear synthesis was observed when 189 μCi/10^8 cells were used (Bini et al., 2002). In another study, Andersson et al. (2006) used actinomycin D, which was dissolved in DMSO to a final concentration of 15 μg/ml, to block transcription in S. solfataricus and Sulfolobus acidocaldarius.

2.2. RNA isolation from Archaea and Northern blot analysis

Total-cell RNA from haloarchaea for use in Northern blot analysis can be isolated by one of three methods.

In the standard hot phenol method, a 50-ml centrifuge tube for each sample was filled with 20 ml of ice-cold culture medium and stored on ice; 7 to 10 ml of cultured cells was added to this tube, and cells were sedimented in a Sorval centrifuge SA600 rotor (6000 rpm, 4 °C) for 10 min. The cell pellet was resuspended in 0.5 ml of ice-cold culture medium and transferred to a clean, nonautoclaved 1.5-ml tube. Cells were pelleted in a tabletop centrifuge (8000 rpm, 4 °C) for 5 min and used immediately for RNA isolation. Alternately, they were frozen in liquid nitrogen and stored at −80 °C.

Cells were resuspended in 125 μl of cold resuspension buffer (0.3 M saccharose, 0.01 M sodium acetate, pH 4.5), mixed with 125 μl 2% SDS in 0.01 M sodium acetate, pH 4.5, and incubated at 65 °C with a water bath or a heat block for 90 sec. After this, 250 μl of phenol water (e.g., Roth ready-to-use solution is the name of the product (Phenol saturated with water) preheated to 65 °C) was added. Samples were thoroughly mixed, incubated at 65 °C for 3 min, and then frozen in liquid nitrogen for at least 30 sec. After incubation for 2 min at room temperature (to warm the plastic ware, which may crack when centrifuged at very low temperatures), centrifugation was performed at 13,000 rpm for 10 min at room temperature. The aqueous phase was transferred into a new 1.5-ml tube. Phenol treatment was repeated two times; 40 μl 3 M sodium acetate, pH 4.5, was added followed by ethanol precipitation. After centrifugation (13,000 rpm, 4 °C) for 20 min, the pellet was washed once with 80% ethanol, air dried, and resuspended in 180 μl of storage buffer (20 mM NaPO$_4$, pH 6.5, 1 mM EDTA); 20 μl of 10× DNase buffer (200 mM sodium acetate, pH 4.5, 100 mM MgCl$_2$, 100 mM NaCl) were added together with 1 μl DNase I (e.g., Life Technologies, diluted 1:5 in 1× DNase buffer), and samples were incubated at room temperature for 30 min; 20 μl of 250 mM EDTA, pH 7, was added, and the DNase was removed by two phenol chloroform extractions. After final ethanol precipitation, the RNA was resuspended in a suitable volume of storage buffer (usually 50 to 100 μl).

The second method follows the protocol of Nieuwlandt *et al.* (1995) followed by two phenol/chloroform extraction steps.

Cells were pelleted and washed as described for the hot phenol method. Then 1.25 ml lysis buffer (2% SDS in 0.01 M sodium acetate, pH 4.5) and 35 μl DEPC were added, and the cells were resuspended by shaking and the use of a vortex mixer. Samples were incubated for 15 min at 37 °C and then transferred to ice for an additional 15 min; 625 μl NaCl-saturated water was added to the lysate, and samples were gently mixed several times. Do not use a vortex mixer. Samples were incubated on ice for 15 min. The SDS-DNA-protein-aggregate was sedimented in a tabletop centrifuge (13,000 rpm,

4 °C or room temperature) for 30 min; 1 ml of the supernatant was carefully transferred into a nonautoclaved 2-ml tube. Two phenol/chloroform extractions and ethanol precipitation followed.

Third, total RNA was alternately prepared from harvested cells by use of the RNeasy Midi kit (Qiagen, Hilden, Germany) including a DNase I treatment according to the manufacturer's instructions. RNA isolated by this method also proved useful for microarray analyses on mRNA decay in Haloarchaea (Hundt et al., 2007).

For studies of mRNA half-lives in S. solfataricus (Bini et al., 2002), RNA was isolated by acid guanidinium thiocyanate–phenol chloroform extraction (Chomczinski and Sacchi, 1987).

For Northern blot analyses the RNA was dissolved in storage buffer, 7 to 10 μg of RNA per lane are run on a 1% (w/v) agarose, 2.2 M formaldehyde gel, and transferred to nylon membrane (e.g., Pall Biodyne B) by vacuum pressure blotting according to the manufacturer's recommendations. Specific DNA fragments were radiolabeled with α^{32}P-dCTP with nick translation (e.g., Nick translation kit, Amersham Biosciences) and purified on microcolumns (e.g., Probe Quant G-50, Amersham Biosciences). Aliquots of 2×10 cpm are used per hybridization reaction. The signals were detected and quantified with a BioRad molecular imager and the Quantity One (BioRad) software.

With the preceding techniques, we analyzed transcripts of the mc-*gvp* operons for gas vesicle formation in *Haloferax mediterranei*. The determined half-lives were between 4 and 80 min. The stability of *gvpA* mRNA or a 0.45-kb transcript population derived from the 5′-part of *gvpD* differed when the cultures were grown at different salt concentrations or to different densities (Jäger et al., 2002). This strongly suggests that, like in Bacteria and Eukarya, mRNA processing contributes to regulated gene expression in Archaea.

We also studied mRNA degradation on a global level with DNA microarrays and RNA prepared from exponentially growing cultures of *Halobacterium salinarum* NRC-1 using the RNaesy Midi Kit (Qiagen). The technical details for performing microarrays and for the data analysis are not given here. The half-lives of the transcripts ranged from 5 min to more than 18 min, with an overall mean half-life of 10 min. We found some relationship between gene function and transcript stability but no correlation between transcript length and its stability, providing initial insights into mRNA turnover in a euryarchaeon (Hundt et al., 2007).

Stability of mRNA was also studied in the crenarchaeota S. *solfataricus* and S. *acidocaldarius* by other research groups. In a genome-wide study, the median mRNA half-life in the two species was determined to be 5 min (Andersson et al., 2006). In contrast, Bini et al. (2002) determined an average half-life of 54 min with selected mRNAs and Northern blot analysis.

3. *IN VITRO* STUDIES ON NOVEL RNA DEGRADING ACTIVITIES IN ARCHAEA

We aimed to identify and to characterize ribonucleases (RNases) from Archaea to understand the mechanisms for RNA degradation. In two independent attempts to purify novel RNases from the halophilic archaeon *H. salinarum* and from the thermoacidophilic archaeon *S. solfataricus*, very different proteins with similar endoribonucleolytic activity were purified. In an alternate approach, a protein complex with predicted RNase function—the exosome—was purified from *S. solfataricus* and was subsequently characterized. For the detection of RNase activity in protein fractions and for the characterization of recombinant proteins as RNases, tests for ribonuclease activity must be performed and evaluated. For this, it is necessary to prepare suitable substrates, to incubate them with the proteins of interest, to separate the reaction products from the substrate, and to detect them. The techniques used by our group for these purposes are described in the following.

3.1. Detection of RNA degrading activity *in vitro* and determination of cleavage sites

3.1.1. RNA substrates

The following types of radioactively labeled RNA substrates were used in our studies: synthetic poly(A) oligoribonucleotide (30-mer, CureVac GmbH, Tübingen, Germany) labeled at the 5′-end by T4 polynucleotide kinase, the same oligoribonucleotide labeled at the 3′-end by polyadenylation with poly (A)-polymerase I, and internally labeled transcripts of different lengths produced *in vitro* with T7 RNA polymerase (Conrad *et al.*, 1998; Evguenieva-Hackenberg *et al.*, 2002; Wagner and Klug, 2007; Walter *et al.*, 2006).

3.1.2. Labeling and purification of substrates

For 5′-end labeling, 10 pmol of the poly(A) oligoribonucleotide was incubated with T4 polynucleotide kinase (NEB or Fermentas) and 30 μCi [γ-^{32}P] ATP (3000 Ci/mmol) in a 10-μl reaction mixture containing buffer supplied by the manufacturer for 30 to 60 min at 37 °C. The reaction was stopped by adding 50 μl STE buffer (100 mM NaCl, 10 mM Tris-HCl, pH 8.0, 1 mM EDTA), and the unincorporated nucleotides were removed with ProbeQuant G-25 Micro columns (Amersham Pharmacia Biotech) or MicroSpin G-25 columns (GE Healthcare). For labeling at the 3′-end with *in vitro* polyadenylation, 50 pmol poly(A) oligoribonucleotide was incubated with 30 μCi [α-^{32}P]- ATP (3000 Ci/mmol) and poly(A) poly-merase I (Invitrogen) in PAPI-buffer (100 mM Tris, pH 8, 200 mM NaCl, 20 mM MgCl$_2$, 2 mM EDTA, 2 mM DTT, and 2.5 mM MnCl$_2$) in a 20-μl reaction mixture for 30 min at 37 °C. After addition of 30 μl STE buffer, the unincorporated nucleotides were removed as just described for the 5′-end labeling. The labeled substrates were stored at −20 °C.

In vitro transcription with T7 RNA polymerase in the presence of [α-^{32}P]-UTP, and purification of the resulting labeled transcripts in denaturing gels was performed as previously described with some modifications (Conrad et al., 1998; Milligan and Uhlenbeck, 1989). When cloned DNA was the substrate for in vitro transcription, the plasmid was linearized with a suitable restriction endonuclease to enable runoff transcription, thus determining the 3′-end of the transcript. When we used pUC18 for cloning, the primer matching the sequence of the RNA substrate (sense primer) was extended by the T7 promoter sequence at the 5′-end. Alternately, we used vectors containing the T7 promoter (for example, pDrive, Qiagen). When the T7 promoter was present in the sense primer, PCR products were also directly used for in vitro transcription. In addition, we used oligonucleotides with an annealed 18-mer promoter oligonucleotide for in vitro transcription of very short RNAs, such as the N26 transcript (Conrad et al., 1998; Evguenieva-Hackenberg and Klug, 2000; Schweisguth et al., 1994; Wagner and Klug, 2007). When T7 RNA polymerase and the supplied buffer from NEB were used, reaction mixtures contained ribonucleoside triphosphates (rNTPs), where the concentration of rUTP was lower than that of the other rNTPs (final concentrations: 0.5 mM rATP, 0.5 mM rGTP, 0.5 mM rCTP, 0.1 mM rUTP) and 20 μCi [α-^{32}P]-UTP (3000 Ci/mmol). Alternately, when the T7-MEGAshortscript High Yield Transcription Kit (Ambion) was used according to the recommendations of the manufacturer, we added 20 μCi [α-^{32}P]-UTP (3000 Ci/mmol) in a final volume of 20 μl. Each in vitro transcription reaction contained 1 to 3 μg of plasmid DNA, approximately 200 ng of PCR product or 5 pmol template oligonucleotide with annealed T7 promoter oligonucleotide (25 pmol template oligonucleotide and 30 pmol T7-oligonucleotide were heated for 5 min at 70 °C and incubated for 30 min at 25 °C in 50 μl 10 mM Tris-HCl, pH 8). The in vitro transcription was performed for 2 to 4 h at 37 °C.

Internally labeled in vitro transcripts were purified with denaturing gel electrophoresis in a 6 to 12% polyacrylamide gel (25 cm × 25 cm × 0.5 mm) containing 8 M urea and Tris-borate-EDTA (TBE) buffer as the running buffer. Samples were amended with formamide containing dye (80% v/v deionized formamide, 6 M urea, 1× TBE buffer, 0.1% (w/v) xylene cyanol, and 0.1% (w/v) bromphenol blue), heated for 10 min at 65 °C, placed on ice, and loaded after excess urea was removed from the slots by flushing them with a syringe and TBE. If radiolabeled RNA substrate does not enter the gel because of high protein content, samples can be treated with proteinase K (Roche Diagnostics, PCR Grad) before adding the formamide loading dye. Usually, the electrophoresis was performed for 90 min at 500 V. After removing one glass plate, the gel was wrapped with cling film and exposed together with the other glass plate on an X-ray film for 1 to 2 min; illumination with red light was performed during this time to visualize the slots and to enable the localization of the transcript in the gel thereafter (shadowing). The desired bands were cut out and the substrates eluted from the crushed gel pieces for 1 to 2 h at room temperature in 100 μl RNA elution buffer (0.5 M sodium acetate, pH 5.0, 1 mM EDTA, pH 8.0, 2.5% v/v phenol). Freeze–thaw cycles and repeated elution maximized the yield of extraction, which was followed in a scintillation counter. After phenol/chloroform extraction, RNA was precipitated with isopropanol, dissolved in RNase-free water, and stored at $-20\,^{\circ}$C. Incorporated radioactivity was determined in a scintillation counter.

3.1.3. RNA degradation assays

A typical 10-μl degradation assay was performed with 1000- to 5000-cpm substrate. The used enzymes or protein fractions, substrates, buffers, incubation times, and temperatures varied and were published; many examples are given in the corresponding text that follows. In general, small reaction tubes (0.5 ml) help to minimize evaporation of water. Relatively short incubations of larger reaction volumes at low temperatures (up to 37 $^{\circ}$C) can be done in a heating block or water bath. Long incubations of small volumes and incubations at high temperatures (60 $^{\circ}$C to 80 $^{\circ}$C) were done in a heating oven or in a thermocycler.

3.1.4. Thin-layer chromatography

For separation of the reaction products, denaturing gel electrophoresis in TBE buffer (see 3.1.2.) or thin-layer chromatography (TLC) was performed. TLC was used to resolve nucleoside monophosphates and nucleoside diphosphates that are produced by the hydrolytic and phosphorolytic activity of RNases, respectively. The nucleosides were separated with 0.1 mm cellulose MN 300 polyethylenimine (Polygram®CEL 300 PEI TLC plate; Machery & Nagel) in 0.9 M guanidinium hydrochloride, pH 6.3; 3 to 6 μl of each reaction mixture (without any loading dye) was spotted 5 to 10 mm apart. After drying, the thin layer plate was soaked for 20 min in methanol and completely dried before performing the chromatography in a closed chamber. The signals were detected and quantified with a BioRad molecular imager and Quantity one (BioRad) software.

To obtain UDP and UMP or AMP and ADP standards, we incubated the substrate with a PNPase-containing fraction from *E. coli* and with an RNase R–containing fraction from *Pseudomonas syringae*, respectively, as previously described (Purusharth *et al.*, 2005). Briefly, 20 fmol substrate (1000 to 3000 cpm) was incubated with highly active degradosome fractions containing approximately 5 ng PNPase or RNase R as determined by silver staining. The reaction mixture of 10 μl contained in addition 25 mM Tris-HCl, pH 8.0, 5 mM MgCl$_2$, 60 mM KCl, 100 mM NH$_4$Cl, 0.5 mM dithiothreitol, 5% glycerol, and 10 units RNasin (Promega). In addition,

10 mM inorganic phosphate was present in the reaction buffer for RNA degradation by PNPase. The incubation of the samples was performed for 10 min at 22 °C.

3.1.5. Primer extension analyses

The exact cleavage sites of an RNase were identified by determination of 5′-ends of the distal cleavage products by primer extension analysis. The unlabeled RNA substrate was incubated with the protein of interest under suitable conditions. Generally, these cleavage tests were performed in a way that intact substrate and cleavage products were detectable. After phenol-chloroform extraction and ethanol precipitation, RNA was dissolved in water and primer extension was performed. RNA was incubated with 1 to 2 × 10^5 cpm 5′-^{32}P–labeled primer in a total volume of 8 μl in 5 mM Tris-HCl, pH 7.5, 1 mM EDTA for 5 min at 70 °C. Within 10 min, the mixture was cooled down to 50 °C, kept for 5 min on ice, and brought back to room temperature. Thereafter, 1.5 μl of 40 mM sodium pyrophosphate, AMV reverse transcriptase reaction buffer (1 × final concentration), 8 units of AMV reverse transcriptase, 10 units of RNasin (Promega) and 1 mM of each deoxyribonucleotide triphosphate (dNTP) were added in a final volume of 20 μl. Reverse transcription was carried out for 10 min at 37 °C, 1 h at 42 °C, and 50 min at 50 °C. The reaction was stopped by adding 30 μl of formamide loading buffer. Radioactively labeled (^{T7}Sequencing kit, USB) reactions of the cloned DNA template were loaded on the same gel to map the positions of the cleavage sites.

3.2. Isolation of RNA degrading activities by fractionation of *S. solfataricus* cell lysates

In 1999, we started our attempts to isolate an RNase III–like activity from the thermoacidophilic archaeon *Sulfolobus solfataricus* P2. In Bacteria and Eukarya, RNase III and the proteins belonging to the RNase III family are important double-strand (ds)–specific RNases that are involved in RNA processing and degradation (Gegenheimer and Apirion, 1981; Huntzinger *et al.*, 2005). Archaea do not harbor proteins with sequence similarities to RNase III (Anantharaman *et al.*, 2002), but this does not exclude the possibility that novel archaeal RNases exist, which are specific for dsRNA and are different from the already known intron endonuclease (Kleman-Leyer *et al.*, 1997). To identify endoribonucleases, which cleave dsRNA, from *S. solfataricus*, we monitored protein fractions for such activity with N26, a well-characterized small substrate of RNase III from *E. coli* derived from phage T7 R1.1 (Conrad *et al.*, 1998; Schweisguth *et al.*, 1994). This substrate is a mostly double-stranded RNA consisting of 46 bases.

 S. solfataricus was grown at 70 °C in a fermenter with air supply of 10 L/min^{-1} for 5 days in the following medium: 1 g yeast extract, 1 g casamino acid, 3.1 g KH_2PO_4, 2.5 g $(NH_4)_2SO_4$, 0.2 g $MgSO_4 \times 7\ H_2O$, 0.25 g $CaCl_2 \times 2H_2O$, 0.1 ml of the following solutions: 1.8% $MnCl_2 \times 4\ H_2O$ and 4.5% $Na_2B_4O_7 \times 10\ H_2O$, and 10 μl of each of the following solutions: 2.2% $ZnSO_4 \times 7\ H_2O$, 0.5% $CuCl_2 \times 2\ H_2O$, 0.3% $NaMoO_4 \times 2\ H_2O$, 0.15% $CoCl_2 \times 6\ H_2O$ per 1 L. The medium was adjusted to pH of 4.2 to 4.4, autoclaved, and 10 μl sterile filtered 0.3% $VOSO_4 \times 2\ H_2O$ per 1 L was added. *S. solfataricus* cells (60 g) were resuspended in 120 ml cold extraction buffer (Rauhut *et al.*, 1995; 50 mM Tris-HCl, pH 7.9, 0.25 M KCl, 2 mM EDTA, 1 mM β-mercaptoethanol, 0.1 mM dithiothreitol [DTT], and 0.5 mM phenylmethylsulfonyl fluoride [PMSF]) and opened by sonication. After 45 min of ultracentrifugation at 4 °C and 100,000g, 2 μl of the cytoplasmic fraction was tested for endonucleolytic activity with 3000 cpm of the radioactively labeled *in vitro* transcript N26 as substrate in a final volume of 10 μl in the reaction buffer TMKG (30 mM Tris-HCl, pH 7.5, 10 mM $MgCl_2$, 130 mM KCl, and 5% glycerol). These buffer conditions were used in our laboratory for specific *in vitro* processing of dsRNA by RNase III (Conrad *et al.*, 1998). The 47-nt N26 substrate is a double-stranded RNA with a central bulge comprising of one major scissile bond; the cleavage products have sizes of 38 nt and 9 nt. The RNase III binding site in this minimal RNase III substrate is located between the bulge and the terminal loop (Schweisguth *et al.*, 1994; Zhang and Nicholson, 1997). We analyzed the N26 substrate by *mfold* (Mathews *et al.*, 1999). The same secondary structures was predicted at temperatures between 20 °C and 50 °C; 11 nucleotides at the 5′-end and 12 nucleotides at the 3′-end were proposed to form single-stranded regions at higher temperatures. Thus, at higher temperatures only the part representing the functional RNase III–binding site is expected to remain double stranded.

We performed the RNA degradation assays for 5 min at 45 °C and in parallel at 65 °C. Substrate and degradation products were analyzed on 10% denaturing polyacrylamide gels (2.1). We observed two degradation products of approximately 35 nt and 19 nt. The 35-nt product may nearly arise by cleavage at the RNase III cleavage site; the product of 19 nt may be due to a cleavage in the region remaining double stranded even at high temperatures. An identical degradation pattern was observed when different incubation temperatures between 30 °C and 90 °C were applied, and the highest activity was detected at 80 °C. Further assays were performed at 70 °C, and the endoribonucleolytic activity that produced this cleavage pattern was followed in the subsequent fractionation experiments.

The cytoplasmic fraction was subjected to precipitation with ammonium sulfate at 25%, 35%, 50%, 65%, 80%, and 100% saturation. The precipitated proteins were pelleted and dissolved in TEG buffer (40 mM Tris, pH 8.0, 0.1 mM EDTA, 5% glycerol), dialyzed against the same buffer, and tested for activity. The endoribonucleolytic activity precipitated between final ammonium sulfate concentrations from 35 to 65%.

These fractions were pooled and used for further purification by chromatography after dialysis against TEG buffer. The chromatography steps were performed with the FPLC System (Pharmacia). The following columns were used in this order: HiTrap heparin column (Pharmacia), cation exchanger (BioRad Econo-Pac S-cartridge), anion exchanger (BioRad Econo-Pac Q-cartridge), and HiLoad™ Superdex 200 size exclusion chromatography column (Pharmacia). The columns were used according to the instructions of the manufacturer. The elution fractions were dialyzed if necessary and were tested for activity. In parallel, their protein content was visualized in silver-stained 10% SDS-polyacrylamide gels. The endoribonuclease activity was associated with a major peak in the elution profiles and, consequently, the corresponding protein fractions contained many proteins even after the application of the four different columns. The apparent native molecular weight of the endoribonuclease-containing fractions corresponded to a 90-kDa protein, as determined by gel filtration. After this step, the endonuclease-containing pool was applied on preparative 8% native PAGE, 0.25 cm slices were cut, and the proteins recovered by diffusion. Two major proteins were detected in the fractions with the highest RNase activity. They were transferred to Immobilon-PVDF membrane (Millipore) and identified by N-terminal sequencing. The 50-kDa band corresponded to an aspartate semialdehyde dehydrogenase (NCBI accession number AAK41162), whereas the 44-kDa band was found to represent an acyl-CoA dehydrogenase (NCBI accession number AAK42872; Evguenieva-Hackenberg *et al.*, 2002).

3.3. Characterization of the recombinant RNA degrading dehydrogenases from *S. solfataricus*

On the basis of the knowledge that eukaryotic dehydrogenases can bind RNA and that certain RNases share homology with dehydrogenases (Baker *et al.*, 1998; Nagy and Rigby, 1995), we characterized the RNase properties of the two *S. solfataricus* dehydrogenases after their purification as His-tagged proteins from *E. coli*. We used the vector pQE30, and the genes of interest were cloned in frame from the second codon to the stop codon. The recombinant proteins could not cleave the N26 transcript in the TMKG buffer, but they produced a different cleavage pattern in the same buffer without $MgCl_2$ (TKG buffer). This third type of N26-cleavage pattern (the approximate sizes of the detected products were 42 nt and 36 nt) was also produced by the original *S. solfataricus* proteins when $MgCl_2$ was omitted from the reaction mixture. This type of cleavage reaction was possible at $MgCl_2$ concentrations up to 5 mM; the optimum was achieved at 60 °C, pH between 6.0 and 7.5, and at 50 mM KCl. In addition to the N26 substrate, other structured transcripts derived from bacterial pre-rRNAs were also endonucleolytically cleaved in TKG buffer by the recombinant *S. solfataricus* proteins and by the original *S. solfataricus* fractions. Further assays were routinely performed at 60 °C. During characterization of the recombinant proteins, we observed that the RNA cleavage reaction was blocked in presence of RNasin, an RNase A inhibitor (Promega). Moreover, we found that RNase A in femtomolar concentrations produces similar RNA cleavage patterns as the recombinant dehydrogenases in nanomolar concentrations. Similar cleavage patterns were also produced by some bacterial and eukaryotic glyceraldehyde-3-phosphate dehydrogenase (GAPDH) proteins. Primer extension analyses demonstrated that the dehydrogenase-containing fractions and RNase A cleave the substrates in loops and bulges between pyrimidine and A, preferentially between C and A. Thus, it was necessary to clarify whether the RNase activity of the dehydrogenases was due to contamination with RNase A or, indeed, is their intrinsic property.

We observed that the activity of RNase A can be distinguished from the RNA degradation by dehydrogenases by the addition of 50 ng μl^{-1} tRNA or single-stranded DNA (ssDNA). Whereas the presence of tRNA or ssDNA inhibits the degradation of the N26 transcript by the dehydrogenases, the RNA cleavage by RNase A was enhanced. To exclude the possibility that the excess of dehydrogenase polypeptides changes the behavior of contaminant RNase A in the reaction mixture, control reactions were performed with pure RNase A in femtomolar concentration together with dehydrogenases in nanomolar concentration. In the presence of ssDNA, RNase A still efficiently cleaved the radioactively labeled RNA substrate, strongly suggesting that the RNase activity of the dehydrogenases is similar to, but distinct from, that of RNase A. In addition, we subcloned and purified His-tagged deletion derivatives of the aspartate semialdehyde dehydrogenase from *S. solfataricus*. A polypeptide consisting of the 205 N-terminal amino acids, which contains a predicted Rossmann fold, showed RNase activity, whereas a polypeptide consisting of the 1440 C-terminal amino acids was not active. Further subcloning revealed that the RNase active site was located in the first 73 N-terminal amino acids, which comprise the first mononucleotide binding site of the putative Rossmann fold (Evguenieva-Hackenberg *et al.*, 2002).

3.4. Purification of a protein with RNase activity from *H. salinarum* NRC-1 and its identification as aIF-5A

To identify enzymes involved in RNA processing and degradation in *Halobacterium*, we performed a biochemical screen for such activities. As a first purification step, we chose a heparin column (Amersham Biosciences) to enrich nucleic acid binding proteins.

H. salinarum NRC-1 was cultivated in ATCC medium #2185 at aerobic conditions and 37 °C or 42 °C. Cells from 1 L culture (10 to 15 g cells, wet weight) were harvested at early stationary growth phase ($OD_{600nm} = \sim 2.0$)

by centrifugation (20 min, 5000g, room temperature), washed two times with basal salt solution (4.3 M NaCl, 80 mM MgSO$_4$, 10 mM trisodium citrate, 27 mM KCl, pH 7.2), and resuspended in 10 to 15 ml volumes of the same solution. The suspension was diluted fivefold in lysis buffer (50 mM Tris-HCl, pH 7.5), and the cells were opened by sonication. After ultra-centrifugation for 1 h at 40,000 rpm (45Ti, Beckman) and 4 °C, proteins of the supernatant were precipitated with ammonium sulfate (85% saturation) and dissolved in 50 mM Tris-HCl, pH 7.5. Dialysis against the same buffer, containing additionally 100 mM KCl and 1 mM EDTA, was performed to remove the ammonium sulfate. The proteins were then loaded onto a 5-ml heparin column (Pharmacia, HiTrap), and bound proteins were eluted with an increasing KCl gradient. RNA degrading activity of fractions was ana-lyzed by incubating protein samples with the internally radiolabeled RNA substrate "pZBP" in "low salt" reaction buffer (200 mM KCl, 50 mM Tris-HCl, pH7.2) for 3 h at 37 °C. The reaction products were separated on denaturing TBE gels containing 8 M urea and 10% polyacrylamide. This long incubation time and maximal KCl concentrations of 200 mM were necessary for detection of RNase activity. We found that RNA remains stable for hours when incubated at 37 °C or 42 °C with cell-free extract from *Halobacterium* under physiologic salt conditions in buffer containing 10 mM MgCl$_2$. Endoribonucleolytic activity was observed at KCl or NaCl concentrations <200 mM without the addition of MgCl$_2$. This is consistent with the data of Schierling *et al.* (2002), who have shown that *in vitro*, the optimal reaction conditions for RNase Z from the halophilic archaeon *Haloferax volcanii* are at 5 mM KCl.

The chosen substrate, the *in vitro* transcript "pZBP," is derived from mRNA of the *puf*-operon of *Rhodobacter capsulatus* and contains single-stranded regions with RNase E cleavage sites (Fritsch *et al.*, 1995). An RNase-containing peak eluted at approximately 200 mM KCl. The cleav-age pattern was complex and could be explained by multiple endonucleo-lytic cleavages. Structured *in vitro* transcripts derived from bacterial pre-rRNA were also cleaved. The fractions corresponding to this activity were pooled and brought to Mono Q binding buffer (50 mM Tris-HCl, pH 7.75) by centrifugal filtration (Vivaspin, Sartorius; or Centricon®, Millipore). After loading onto a 5-ml Mono Q column (Pharmacia), bound proteins were eluted with an increasing KCl gradient and RNA degrading activity of fractions was analyzed. Fractions with RNase activity were pooled and further purified by size exclusion chromatography (gel filtration; Superdex 75 16/60 column with 1.5 ml/min of 200 mM KCl, 50 mM Tris-HCl, pH 7.5). The RNase activity was in fractions containing a pure protein that was identified by mass spectroscopy as a homolog of the eukaryotic translation initiation factor 5A (Wagner and Klug, 2007). The purification procedure was optimized as follows: the Mono Q chromatography was omitted, and the gel filtration was performed at a flow rate of 1.2 ml/min in a buffer consisting of 1 M KCl, 50 mM Tris-HCl, 2 mM EDTA, pH 7.5, before the highly purified protein aIF-5A was finally concentrated by centrifugal filtration. The elution volume of aIF-5A of 75 to 81 ml corresponds to a molecular size of 23.8 to 29.4 kDa, whereas the calculated monomeric mass of aIF-5A is 14.3 kDa. In addition, we observed a protein complex corresponding to a higher molecular weight of 41 to 46.5 kDa eluting at 62 to 67 ml.

The eukaryotic protein eIF-5A carries a unique posttranslational modification, hypusination, which is restricted to Eukarya and Archaea. The enzyme that carries out this modification, deoxyhypusine synthase, is also only found in these two domains of life. Both the genes for eIF-5A and deoxyhypusine synthase are essential in yeast. It was shown that eIF-5A efficiently binds structured RNA containing certain motifs and that this interaction with RNA is hypusine dependent (Sasaki et al., 1996; Schnier et al., 1991; Xu and Chen, 2001; Xu et al., 2004). Studies on temperature-sensitive mutants indicated involvement of this protein in mRNA turnover (Valentini et al., 2002; Zuk and Jacobson, 1998). This prompted us to investigate in vitro the aIF-5A from Halobacterium in regard to its interaction with RNA.

3.5. Purification of recombinant haloarchaeal aIF-5A protein from E. coli

Large amounts of recombinant haloarchaeal proteins can easily be obtained by heterologous expression of these proteins in E. coli. For this purpose, we use E. coli M15 (REP4) cells and the pQE vector system (Qiagen) as specified by the manufacturer. Recombinant proteins expressed with the pQE vector system carry a hexahistidine (His$_6$) tag and can rapidly be purified by immobilized metal affinity chromatography with Ni-NTA agarose (Qiagen) according to the manufacturer's recommendation.

E. coli M15 cells, containing a pQE30 plasmid for expression of recombinant His$_6$~aIF-5A from H. salinarum NRC-1, were grown (at 37 °C and 180 rpm on a shaker) in 150 ml Standard I medium (Difco) in a 500-ml flask with ampicillin and kanamycin concentrations as specified by QIAexpressionist (Qiagen). When cell density reached an OD$_{600nm}$ of 0.6 to 1.0, protein expression was induced by adding isopropyl-beta-d-thiogalactoside (IPTG) to a final concentration of 0.8 to 1.0 mM. After 3 to 5 h (at 37 °C and 180 rpm on a shaker), cells were harvested by centrifugation (15 min, 5000g at 4 °C), resuspended in 4 ml lysis buffer (50 mM NaH$_2$PO$_4$, 1 M NaCl, 10 mM β-mercaptoethanol, 5% [v/v] glycerol, 10 mM imidazole, pH 8.0) with the addition of 1 mg/ml lysozyme, and incubated for 15 min on ice. Cells were lysed 3 to 5 times by sonication on ice, each time with increasing intensity for 1 min (stepwise from 20 to 70%) followed by cooling for 1 min. After centrifugation (15 min, 13,000g at 4 °C) the supernatant was incubated for 30 to 60 min at 4 °C with 1 ml (50% slurry)

Ni–NTA resin (Qiagen), equilibrated in lysis buffer, and packed into a gravity flow column. The resin was washed with 20 ml lysis buffer and 10 to 15 ml wash buffer (50 mM NaH$_2$PO$_4$, 1 M NaCl, 5 mM β-mercaptoethanol, 5% [v/v] glycerol, 15 mM imidazole, pH 8.0). Elution of His$_6$~aIF-5A was done with buffer containing 80 mM imidazole (50 mM NaH$_2$PO$_4$, 1 M NaCl, 5 mM β-mercaptoethanol, 5% [v/v] glycerol, 80 mM imidazole, pH 8.0). If necessary, His$_6$~aIF-5A–containing fractions were pooled and concentrated by centrifugal filtration. Further purification of His$_6$~aIF-5A was done by gel filtration (Superdex 75 16/60 column, 1.5 ml/min flow rate, 50 mM sodium phosphate buffer, pH 7.0, containing 1 M NaCl, 5 mM β-mercaptoethanol, and 2 mM EDTA). The elution volume of His$_6$~aIF-5A was 74.5 to 80.5 ml, and purity of the protein yield of approximately 50 mg was almost complete.

3.6. Purification of the recombinant haloarchaeal aIF-5A protein from *H. salinarum* NRC-1

Protein expression in *E. coli* usually yields good quantities of biologically active proteins, but for proteins from extreme halophilic Archaea, this is often not the case. Lack of the high intracellular ion concentrations present in halophilic Archaea may result in unproper protein folding. The lack of specific posttranslational modifications such as hypusination might suppress enzymatic activity of the recombinant protein as well. To overcome these problems, we used the pBPH-M vector (Patenge *et al.*, 2000) for homologous protein expression directly in *H. salinarum* NRC-1 with a C-terminal heptahistidine (His$_7$) tag. This plasmid carries an ampicillin resistance for cloning purposes in *E. coli* and a mevinoline resistance as a selection marker for *Halobacterium*. The gene to be expressed was ligated into a cloning site located in front of seven histidine codons and downstream of the *bop*-promoter, which is induced at low oxygen tension and a certain degree of DNA supercoiling (Gropp *et al.*, 1995; Yang *et al.*, 1996). The cloning site is flanked by upstream and downstream parts of the gene coding for sensory rhodopsin I for stable integration of the plasmid into the haloarchaeal genome. Transformation of *H. salinarum* NRC-1 was performed as previously described (Cline *et al.*, 1989), with the modification that regeneration of the cells after transformation was done for 1.5 to 2 days at 42 °C on a shaker instead of only overnight. The prodrug for mevinolin was a gift from Merck Sharp and Dohme Ltd. and was used in a concentration of 20 mg/L, after conversion into the active form according to the manufacturer's recommendation. We had a good experience when *H. salinarum* NRC-1 cells containing pBPH-M for expression of recombinant proteins were harvested at room temperature (5000g, 20 min) after shifting exponentially grown cultures from aerobic to anaerobic conditions overnight and back to aerobic conditions.

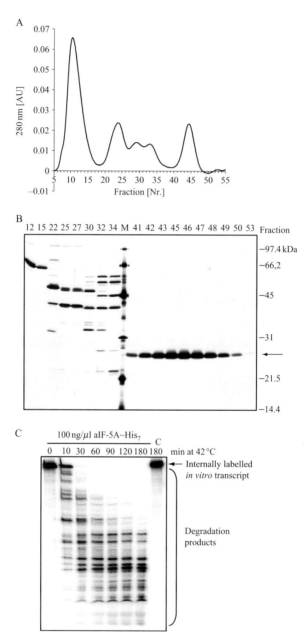

Figure 19.1 Purification of ribonucleolytically active aIF-5A∼His₇ from *H. salinarum* NRC-1. Recombinant protein was expressed in *H. salinarum* NRC-1 and partial purification was performed with Ni-NTA chromatography and high salt the concentration was 3 M. Eluted fractions were pooled, concentrated with a Centricon 10 (Millipore), and subjected to size-exclusion chromatography (for details, see the text). (A) Absorption profile of the gel filtration. (B) Silver-stained 12% SDS-PAGE showing the proteins in

A culture of *H. salinarum* NRC-1, containing a pBPH-M–derived plasmid for expression of recombinant aIF-5A~His$_7$ integrated within the chromosome, was kept growing under aerobic conditions (500 ml ATCC medium #2185 with 10 to 20 mg/L mevinolin in a 2-L flask, on a shaker at 180 rpm) at 42 °C until cell density reached an OD$_{600nm}$ of 0.5 to 1.0. After shifting the exponentially grown cultures from aerobic to anaerobic conditions overnight (by incubating the culture volume in completely filled flasks, at 42 °C without shaking) and again at aerobic conditions for 6 to 8 h, cells were harvested by centrifugation for 20 min at 5000g and room temperature. Cell material, corresponding to 1 L of culture volume, was resuspended in 7 ml (or as less as possible) of basal salt solution, filled up with high salt lysis buffer for *Halobacterium* (3 M KCl, 10 mM β-mercaptoethanol, 5% [v/v] glycerol, 100 mM sodium phosphate buffer, pH 7.5) to 35 ml and mixed with care. While sonicating the suspension, two times for 2 min with increasing intensity (stepwise from 20 to 70%), the temperature was checked, and the suspension was chilled on ice if the mixture was warming up above room temperature. The lysate was centrifuged at room temperature for 15 min at 13,000g. Thereafter the supernatant was incubated for 30 to 60 min at 4 °C with 1 ml (50% slurry) Ni-NTA resin (Qiagen), equilibrated in high salt lysis buffer for *Halobacterium,* and packed into a gravity flow column. The resin was washed with 20 ml high salt lysis buffer and 10 to 20 ml high salt wash buffer (3 M NaCl, 1 M KCl, 2 mM β-mercaptoethanol, 5 mM imidazole, 100 mM sodium phosphate buffer, pH 7.5). Elution was done with the same buffer containing 227.5 mM imidazole (high salt elution buffer). aIF-5A~His$_7$ containing fractions were pooled, concentrated by centrifugal filtration, and subjected to size exclusion chromatography. The gel filtration was performed with a Superdex 200 16/60 column at a flow rate of 1.0 ml/min with 100 mM sodium phosphate buffer, pH 7.5, containing in addition 3 M NaCl, 1 M KCl, 2 mM β-mercaptoethanol, and 2 mM EDTA. aIF-5A~His$_7$ was purified to homogeneity and was shown to exhibit RNase activity (Fig. 19.1).

---

20 μl of each 1.5-ml fraction, which are numbered above the panel. M specifies protein markers; the sizes of which are indicated to the right side of the panel. Contaminating proteins (fractions 12 to 34) were successfully separated from aIF-5A~His$_7$ (marked with an arrow), which was purified to homogeneity (fractions 41 to 50). (C) Phosphor image of a 8% denaturing polyacrylamide gel (for details see 2.1) showing RNA degradation assays with recombinant aIF-5A~His$_7$ from fraction 43 as shown in (B). The internally labeled, 308-nt sub2.(OE2664) mRNA was used as substrate (Wagner and Klug, 2007). Each reaction mixture contained 120 mM KCl, 50 mM sodium phosphate, pH 7.0, 2 mM β-mercaptoethanol, 2 mM EDTA, approximately 100 ng/μl aIF-5A~His$_7$, and 3000 cpm substrate. The samples were incubated at 42 °C for different time periods as indicated. The intact substrate band is marked with an arrow. Lane C Serves as a negative control; its substrate was incubated for the specified time in the reaction buffer in the absence of added protein.

3.7. Characterization of the endoribonucleolytic activity of the haloarchaeal aIF-5A

The RNase activity of the nonrecombinant aIF-5A was tested with different *in vitro* transcripts containing double-stranded and single-stranded regions (Wagner and Klug, 2007) and was compared with the activity of the recombinant proteins after homologous or heterologous expression. All protein preparations led to identical RNA cleavage patterns, showing that hypusination is not necessary for RNA cleavage *in vitro*. The proteins were ribonucleolytically active in presence of 2 mM EDTA, and, thus, divalent cations are not needed for cleavage. MgCl$_2$ concentrations above 5 mM and salt concentrations (KCl or NaCl) above 300 mM inhibited cleavage; maximal activity was observed at 120 mM KCl. We also tested the effect of polyamines on RNA cleavage, because polyamines are known to bind RNA, and the hypusine is a modified lysine with structural contribution from the polyamine spermidine (summarized in Wolff *et al.*, 2007). At low concentrations (1 mM for putrescine, 0.5 mM for spermidine, 10 mM for ornithine), the polyamines increased RNase activity by 20 to 40%, whereas at higher concentrations (20 mM, 4 mM, and 120 mM, respectively), they inhibited RNase activity almost completely. Primer extension analysis revealed that major cleavage sites were located in single-stranded regions between C and A (Wagner and Klug, 2007). These characteristics strongly resemble those of RNase A and the two archaeal dehydrogenases with RNase properties (see 3.2 and 3.3). Similar to the RNase activity of the dehydrogenases, the addition of tRNA at concentrations above 25 ng/μl inhibited RNase activity of aIF-5A (Wagner and Klug, unpublished results), whereas RNase A activity was increased in the presence of tRNA (Evguenieva-Hackenberg *et al.*, 2002).

Additional experiments were performed to demonstrate that the RNase activity is an intrinsic property of the aIF-5A protein and is not due to contamination with RNases or to copurification of highly active RNases, which cannot be detected in the silver-stained polyacrylamide gels used to monitor the purity of the isolated aIF-5A. The following results show that aIF-5A itself can cleave RNA *in vitro*: (1) the two separated domains of the protein did not show any RNase activity; (2) site-directed mutagenesis resulting in the exchange of one (glutamic acid at position 117 was replaced by alanine) or four (arginine-lysine at positions 72 and 73 were exchanged to glycine-alanine and arginine-lysine at positions 122 and 123 were exchanged to alanine-glycine), amino acid residues drastically reduced RNase activity. The assays were performed as described previously, but the incubation was at 42 °C for 1 h. Further characterization, which is not described here in detail, revealed that aIF-5A forms oligomers with RNase activity (results from refolding experiments, gel filtration, and RNase activity assays and from pull-down assays) and that hypusination stabilizes the interaction between aIF–5A and RNA (results from gel retardation experiments). Moreover, in the course of our experiments, we found out that in addition to its RNA-binding properties, the eukaryotic eIF–5A also shows hypusine-dependent RNase activity (Wagner and Klug, 2007).

Our attempts to purify novel RNase activities from Archaea resulted in the isolation of very different proteins (aIF–5A from *Halobacterium* and two dehydrogenases from *Sulfolobus*) with rather similar endoribonucleolytic properties. We think that many different proteins show *in vitro* this type of RNase activity, which may be described as "strong destabilizing activity." A purification scheme that starts with a heparin column and in which the assays are performed at low MgCl$_2$ concentrations without the RNase inhibitor RNasin is likely to result in the isolation of different proteins showing this type of endoribonucleolytic activity with unknown physiologic relevance.

4. ISOLATION AND CHARACTERIZATION OF A PROTEIN COMPLEX WITH PREDICTED RIBONUCLEASE ACTIVITY FROM *SULFOLOBUS SOLFATARICUS*: THE ARCHAEAL EXOSOME

The availability of several completed archaeal genomes in the data bank enabled the straightforward investigation of ribonucleases in Archaea, which share structural similarities to already characterized RNases in Eukarya or Bacteria. Koonin *et al.* (2001) predicted the existence of the archaeal exosome on the basis of the finding that three orthologs of eukaryotic exosomal subunits are encoded in an array in a highly conserved archaeal superoperon. Two of the orthologs (Rrp41 and Rrp42) contain RNase PH-like domains, suggesting that the archaeal exosome exhibits phosphorolytic exoribonuclease activity; the third predicted subunit, Rrp4, contains the RNA binding domains S1 and KH. A fourth ortholog of a eukaryotic exosome subunit, Csl4, was found to be encoded by another operon in Archaea. It contains the RNA binding domains S1 and Zn-ribbon. We decided to prove the existence of the archaeal exosome with the thermoacidophilic archaeon *Sulfolobus solfataricus* as a model organism. To do this, we followed the procedures for detection and purification of the eukaryotic exosome with immunologic techniques as described by Mitchell *et al.* (1997), with some modifications. We found out that the archaeal exosome consists of SsoRrp4, SsoRrp41, SsoRrp42, SsoCsl4, and SsoDnaG. Two proteins with chaperone properties, the Cdc48 homolog and Cpn, were also isolated together with the exosome. We analyzed the structure and the function of the exosome in more detail with cell-free extracts and reconstituted protein complexes.

4.1. Production of recombinant proteins and polyclonal antibodies

On the basis of the preceding information, we expected that the predicted subunits of the archaeal exosome are present in a high molecular weight protein complex in the cell, and we aimed to isolate this protein complex from *S. solfataricus* cell-free lysate by coimmunoprecipitation. To this end, the genes for SsoRrp4, SsoRrp41, SsoRrp42, and SsoCsl4 were cloned from the second codon to the stop codon in pQE30. The recombinant proteins with a hexahistidine (His$_6$) tag at the N-terminus were overexpressed in *E. coli* and purified under denaturing conditions with Ni-NTA agarose according to the instructions of the manufacturer. They were used for production of polyclonal antibodies in rabbits (Eurogentec, Belgium and Biogenes, Berlin, Germany). To check the specificity of the antibodies, 200 ng of purified His-tagged proteins was separated on 12% SDS-PAGE, and Western blot analysis was performed separately with each antiserum. The polypeptides were transferred to Protran nitrocellulose membranes (Schleicher & Schuell), and the transfer was controlled by Ponceau staining. The stain was removed by washing in 0.1 *N* NaOH, H$_2$O, and 1× TBS, and the membranes were blocked in 1× TBS containing 5% nonfat dry milk overnight at 4 °C. Signals were detected with primary antibodies (antisera or purified IgG of suitable dilution, usually 1:1000) and secondary antibodies (anti-rabbit IgG, diluted 1:50,000) conjugated with alkaline phosphatase (Sigma).

When the anti-Rrp41 antiserum was diluted 1:1000 and used in Western blot analysis, SsoRrp41 was detected as a very strong signal. Weak signals were obtained with the other recombinant proteins, most probably because of the His$_6$-tag. In accordance with this interpretation, SsoRrp41 that consists of 248 amino acid residues was detected in the cell-free *S. solfataricus* lysate as a single band migrating as a 27-kDa protein, whereas other proteins did not react with the antiserum. We concluded that this antiserum is suitable for specific detection of SsoRrp41 and for immunoprecipitation. The sensitivity of the antisera against SsoRrp4 and SsoRrp42 was not high enough, and IgG purification was performed to detect the respective polypeptides in Western blots. IgG was purified from the antiserum with a protein A-column (Pharmacia) according to the suggestion of the manufacturer. The antibody raised against SsoCsl4 was of low sensitivity. After the purification of the *S. solfataricus* exosome and the identification of SsoDnaG as a subunit of the archaeal exosome (see following), this protein was also produced in recombinant form and used for production of polyclonal antibodies. The sensitivity and specificity of the antiserum against SsoDnaG was comparable to that of the anti-Rrp41 antiserum.

4.2. Detection of putative subunits of the *S. solfataricus* exosome in high molecular weight fractions

In the next experiment, we analyzed whether SsoRrp41 is present in a high molecular weight complex in the cell. For this, cell-free lysate of *S. solfataricus* was fractionated in different glycerol gradients by ultracentrifugation. The cell-free lysate was prepared by sonication of 0.5 g cells (wet weight), which were grown to a $OD_{600} = 0.5$ and were resuspended in 2 ml of TMN buffer (10 mM Tris, pH 7.5, 150 mM NaCl, 5 mM MgCl$_2$, 10% glycerol, 0.1% NP-40; Allmang *et al.*, 1999) supplemented with 1 mM PMSF; cell debris was removed by centrifugation (18,000g, 20 min at 4 °C), and 600 μl of the supernatant was layered on an 11-ml 10 to 30% (w/v) glycerol gradient containing TMN-PMSF buffer. Centrifugation was performed in a TH-641 rotor at 4 °C for 21.5 h at 26,600 rpm (121, 262 g). For calibration purposes, ovalbumin (45 kDa), BSA (66 kDa), IgG (150 kDa), catalase (240 kDa), and urease (trimer of 272 kDa, hexamer of 544 kDa) were treated under identical conditions. Fractions of 630 μl were collected, 30 μl of each fraction was used for 12% SDS-PAGE. The *S. solfataricus* fractions were analyzed further by Western blot with anti-SsoRrp41 antiserum, whereas the marker proteins were detected by silver staining.

SsoRrp41 was not detected in monomeric form. Most of SsoRrp41 was in the pellet, the rest remained in complexes of an approximate molecular weight of 230 to 400 kDa. To analyze the pellet fraction of the 10 to 30% glycerol gradient, 10 to 60% glycerol gradient was used under the same conditions. SsoRrp41 was detected in two peaks that correspond to 240 to 270 kDa and 30S to 50S, suggesting an association with ribosomal subunits. According to the intensity of the signal, approximately two thirds of the cellular SsoRrp41 was present in the 30S to 50S fractions. In a control, 5 to 20% glycerol gradient containing 500 mM KCl SsoRrp41 was detected in fractions corresponding to 240 kDa and in the pellet, showing that the presence of SsoRrp41 in the 30S to 50S fractions is not due to unspecific interactions because of low salt and high glycerol concentrations. Thus, after separation in a glycerol gradient by ultracentrifugation, SsoRrp41can be detected as a part of high molecular weight complexes of 240 kDa and 30S to 50S (Evguenieva-Hackenberg *et al.*, 2003; Walter *et al.*, 2006).

The fractions of the 10 to 60% glycerol gradient were also analyzed with antiserum against SsoDnaG (diluted 1:1000) and with SsoRrp4-specific IgG. SsoDnaG and SsoRrp4 were detected only in the 30S to 50S fractions, but the corresponding signals were weaker than the signals obtained in the Western blot with the anti-Rrp41 antiserum. This suggests that the exosome in the 240-kDa fractions of the glycerol gradient was under the limit of detection in Western blots with antibodies against SsoDnaG and SsoRrp4. Indeed, the two proteins were present in exosomes that were immunoprecipitated from those glycerol gradient fractions (Walter *et al.*, 2006).

4.3. Purification of the exosome from *S. solfataricus* by coimmunoprecipitation and identification of its subunits

The next step was to purify the SsoRrp41-containing complexes by coimmunoprecipitation. The SsoRrp41-specific antibodies were directly coupled to protein A–Sepharose beads (Harlow and Lane, 1988). Briefly, 200 mg dry weight protein A–Sepharose CL-4B beads (Amersham) were pretreated overnight in 1 ml phosphate-buffered saline (PBS), mixed with 1 ml antiserum, incubated for 1 h at 20 °C, washed with the 10-fold volume of 0.2 M sodium borate, pH 9.0, resuspended in the 10-fold volume of sodium borate, and treated with 20 mM dimethyl pimelinidate (DMP, Sigma) for 30 min at 20 °C. The coupling reaction was interrupted by washing the beads once with 0.2 M ethanolamine, pH 8.0, and further incubation in the ethanolamine solution for 2 h at 20 °C was performed. The antibodies coupled to protein A-Sepharose were stored up to 1 year at 4 °C in PBS.

Coimmunoprecipitation was performed as follows: 0.05 g protein A-Sepharose with coupled antiserum was washed three times in 1 ml TMN buffer, resuspended in 150 μl of the same buffer, mixed with 500 μl cell-free extract or glycerol gradient fractions, incubated for 2 h at 4 °C on a tumbler, washed 10-fold with 1 ml TMN buffer, and eluted 10 times with 50 μl 0.1 M glycine, pH 1.8. The elution fractions were neutralized with 1/10 volume of 1 M potassium phosphate buffer (pH 8.0) and were analyzed on SDS-PAGE. Coimmunoprecipitated polypeptides were detected by silver staining and were identified by mass spectrometry (Evguenieva-Hackenberg *et al.*, 2003). The following proteins were immunoprecipitated together with SsoRrp41: SsoRrp4, SsoRrp42, SsoCsl4, the archaeal DnaG protein, the Cdc48 homolog, and Cpn. Cdc48 was copurified only from 30S to 50S fractions of the glycerol gradient and was absent when unfractionated cell-free extract was used for immunoprecipitation (Walter *et al.*, 2006). In a control experiment, pre-immune serum which was coupled to protein A-Sepharose was used. In this case, only very small amounts of Cpn (thermosome) were isolated from *S. solfataricus*, confirming the specific coimmunoprecipitation of the exosome with SsoRrp41-directed antibodies. In an other control experiment, the washed samples were treated with 15 μg/μl RNase A and 20 μg/μl RNase T1 in a final volume of 100 μl, washed four times with 1 ml TMN, and eluted. The RNase treatment had no influence on the composition of the immunoprecipitated exosome with one exception: the copurified Cpn was then at the limit of detection. This last control shows that the immunoprecipitation of SsoRrp4, SsoRrp42, SsoCsl4, SsoDnaG, and SsoCdc48 together with SsoRrp41 is due to protein–protein interactions and not due to independent binding to native RNA substrates.

The archaeal orthologs of eukaryotic Rrp4, Rrp41, Rrp42, and Csl4 proteins were predicted subunits of the archaeal exosome. We were surprised to identify the archaeal DnaG protein, Cpn, and the Cdc48 homolog among the coimmunoprecipitated proteins instead of other proteins that were predicted to be subunits of the exosome because of their location on the chromosome (Koonin et al., 2001). Cpn and the Cdc48 homolog are chaperones, and it is still not clear whether they are exosomal subunits; the significance of their copurification with the exosome was previously discussed (Evguenieva-Hackenberg et al., 2003; Walter et al., 2006). Here we describe the experiments showing that SsoDnaG is, indeed, a subunit of the S. solfataricus exosome.

First, we performed coimmunoprecipitation with SsoDnaG-specific antiserum and compared the protein profile of the isolated complex with the profile of the complex that was purified with the SsoRrp41-specific antiserum. The eluted proteins were analyzed by 12% SDS-PAGE and silver staining. The controls with the preimmune sera and the RNase treatment were also repeated. As judged from their migration behavior, the same proteins were coimmunoprecipitated in very similar relative amounts with the two different antisera, strongly suggesting that SsoDnaG is a component of the exosome comparable to the orthologs of yeast Rrp4, Rrp41, Rrp42, and Csl4. In a second experiment, we removed the exosome from the cell-free lysate by three rounds of coimmunoprecipitation with the SsoRrp41-specific antiserum and tested the depleted extract by Western blotting for presence of SsoRrp41 and SsoDnaG. Washed protein A–Sepharose (0.05 g) beads with coupled antiserum were incubated with 500 μl cell-free extract for 2 h at 4 °C on a tumbler and briefly centrifuged to pellet the beads. The supernatant was incubated again with the same amount of fresh beads for 2 h at 4 °C, the beads with the bound exosomes were removed, and the last step was repeated once again; 30 μl of the depleted extract (the supernatant of the last step) was analyzed by Western blotting. As a negative control, the cell-free extract was subjected to three rounds of incubation with preimmune serum coupled to protein A-Sepharose and analyzed in parallel. Western blotting with the SsoRrp41-specific antiserum revealed that 94% of the SsoRrp41 protein was specifically removed by the depletion. The detection was performed with anti-rabbit IgG conjugated with peroxidase (Sigma) and Lumi-Light Western Blotting Substrate (Roche). SsoDnaG was well detectable in the nondepleted extract and in the negative control, but it was under the limit of detection in the depleted extract. These results demonstrated that SsoRrp41-depletion (exosome-depletion) is paralleled by comparable SsoDnaG-depletion. We concluded that the S. solfataricus protein, which is annotated as DnaG because of its topoisomerase/primase domain, is an exosomal subunit involved in RNA metabolism. The possible role of SsoDnaG in the exosome was previously discussed (Evguenieva-Hackenberg et al., 2003; Walter et al., 2006).

SsoDnaG is associated very tightly with the exosome. When after coimmunoprecipitation with anti-SsoRrp41 antiserum, the exosome is eluted with pH 1.8, all polypeptides elute together and the highest protein concentrations are detectable in the second to fourth elution fraction. Elution with glycine at pH 3.0 led to slower elution, but the relative amount of the *S. solfataricus* proteins Rrp4, Rrp41, Rrp42, Csl4, and DnaG in the particular elution fractions remained the same (Fig. 19.2). Similar results were obtained when the elution was performed with increasing $MgCl_2$ concentrations from 0.2 to 1.8 *M*. Treatment with 2 *M* NaCl does not lead to elution of the exosome or some of its subunits (unpublished results).

4.4. Analysis of protein–protein interactions *in vitro* and reconstitution of the *S. solfataricus* exosome by refolding

After identification of the subunits of the immunoprecipitated archaeal exosome, we wanted to characterize it in respect to structure and function. At that time, nothing was known about the interactions between its

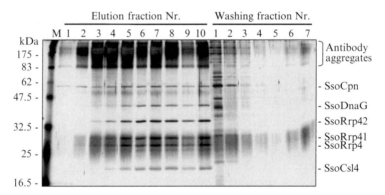

Figure 19.2 Isolation of the *S. solfataricus* exosome by coimmunoprecipitation with antiserum directed against His_6-SsoRrp41. Silver-stained 12% SDS-PAGE showing the cellular proteins that copurify with SsoRrp41 (for details see the text). The antiserum was covalently coupled to protein A–Sepharose, incubated with *S. solfataricus* cell-free extract (for details see the text), washed seven times with 1 ml of TMN buffer, and eluted 10 times with 50 μl of 0.1 *M* glycine, pH 3.0; 30 μl of each wash and elution fraction was used for analysis. M denotes protein marker, the size of which are indicated to the left of the panel. Fractions are numbered at the top of the panel. On the right side, coimmunoprecipitated proteins are marked. In addition, antibody chains and aggregated antibodies, the degradation product of Rrp4 was visible on the original gel, but not in the pdf version we printed out, were detected. The reuse of the protein A–Sepharose beads with coupled antibodies minimized the amount of antibodies in the elution fractions (Walter *et al.*, 2006).

particular subunits, but it was predicted that the RPD-containing subunits Rrp41 and Rrp42 should form a hexameric ring, and the S1-containing subunits Rrp4 and Csl4 should bind to the ring, resembling the structure of the bacterial PNPase (Aloy *et al.*, 2002; Raijmakers *et al.*, 2002; Symmons *et al.*, 2002). To analyze which subunits of the *S. solfataricus* exosome are in direct contact, we refolded them together in different combinations. The refolding was performed by decreasing the concentration of the denaturing reagent and separation of correctly folded thermostable proteins and protein complexes from incorrectly folded ones by heat precipitation as described by Werner and Weinzierl (2002). The His_6-tag proteins were isolated from *E. coli* M15 cells by Ni-NTA chromatography under denaturing conditions. We used 50 μg/ml of SsoRrp41 and SsoRrp42 (corresponds approximately to 1.8 nmol/μl of each protein), as well as 100 μg/ml of SsoRrp4, SsoCsl4, and SsoDnaG (corresponds to approximately 3.3 nmol SsoRrp4 per ml, 4.8 nmol SsoCsl4 per ml, and 2.2 nmol SsoDnaG per ml) in a total volume of 10 ml. The sample was dialyzed in denaturing buffer P_8 (10 mM Tris, pH 7, 5 mM $MgCl_2$, 0.5 mM EDTA, 200 mM NaCl, 5% glycerol, 0.05% Tween 20, 0.2 mM DTT, and 8 M urea) and then renatured by seven consecutive dialysis steps at 20 °C into buffers P_6 to P_0 with lower concentrations of urea (6, 4, 3, 2, 1, 0.5, and 0 mM, respectively) and otherwise identical composition. After incubation at 70 °C for 20 min, the precipitated proteins were separated by centrifugation at 13,000 rpm. The protein content of the pellet and the supernatant (soluble fraction) was analyzed on SDS-PAGE.

It was not possible to efficiently refold SsoRrp41 or SsoDnaG to monomers. After refolding, only 5% of the Rrp41 polypeptides remained soluble. Half of the SsoDnaG polypeptides were in the soluble fraction after refolding, but they formed aggregates when stored for a few hours at 4 °C or at 20 °C, as analyzed by gel filtration. In contrast, SsoRrp4, SsoRrp42, and SsoCsl4 were successfully refolded in monomeric form. SsoRrp41 remained soluble after refolding in complex with SsoRrp42 or SsoRrp4. The protein–protein interactions between SsoRrp41 and SsoRrp42, as well as between SsoRrp41 and SsoRrp4, were demonstrated by coimmunoprecipitation of the respective protein complex with the anti-Rrp41 antiserum from the soluble fraction after refolding. To do this, the soluble fraction was mixed with 0.05 g beads in 150 μl PBS and treated as described in 4.3. We could not detect other protein–protein interactions by refolding of two different exosomal subunits together. The same question was addressed by Far Western blot analysis. The purified polypeptides to be tested were run in separate lanes on SDS-PAGE, transferred on a membrane, fixed by Ponceau staining, washed as described previously, and renatured in buffer P_0 with 1% nonfat dry milk. The membrane was then incubated for 1 h at 20 °C in the same buffer containing 5% milk. The next step was the incubation overnight at 4 °C with the putative interaction partner in monomeric form (SsoRrp4, SsoRrp42, or SsoCsl4). We used 0.5 to 1 mg of protein in buffer

P_0 with 1% milk. Thereafter, the membrane was washed three times in the same buffer and treated further with primary antibodies directed against the putative interaction partner. Then, secondary antibodies were applied and the signals were detected as described for Western blotting. By this analysis, we confirmed the interactions between SsoRrp41 and SsoRrp42 as well as between SsoRrp41 and SsoRrp4; other interactions were not detected.

Because of the proposed similarities to bacterial PNPase and RNase PH, we assumed that the SsoRrp41–SsoRrp42 complex is the interaction partner for SsoCsl4, and that this or a higher order complex is required for binding of SsoDnaG to the archaeal exosome. To test this assumption, we refolded exosomes of different composition, concentrated them on solid polyethylene glycol (PEG) in dialysis tubes, in Centricon 30 (Millipore) or in Vivaspin 500 (Sartorius) tubes, and analyzed them by size-exclusion chromatography. We used a HiLoad^{TM} Superdex 200 size exclusion chromatography column (Amersham Biosciences) and the FPLC System (Pharmacia). The gel filtration was performed at a flow rate of 1.0 ml/min in a buffer containing 10 mM Tris, pH 7.5, 150 mM NaCl, 5 mM MgCl$_2$, and 0.5 mM EDTA. The column was calibrated with marker proteins (aldolase, 158 kDa; catalase, 232 kDa; ferritin, 440 kDa; thyroglobulin, 669 kDa) and blue dextran (to determine the void volume) from Amersham. The apparent molecular weight of the refolded SsoRrp41–SsoRrp42 complex was 230 kDa, although it was found that it is, indeed, a hexamer of three SsoRrp41 and three SsoRrp42 subunits, each of approximately 28 kDa (Lorentzen *et al.*, 2005; Lorentzen and Conti, this volume). Most probably it elutes faster from the column because of its planar shape. We refolded and purified by gel filtration also the SsoRrp41-SsoRrp42-SsoRrp4 complex (270 kDa), the SsoRrp41-SsoRrp42-SsoCsl4 complex (240 kDa), and a mixture of these complexes probably containing also SsoRrp41-SsoRrp42-SsoRrp4-SsoCsl4 complexes (the peak at 280 nm corresponded to 270 kDa) when all four subunits were used. In general, our results are in agreement with the crystal structures of the exosomes from the hyperthermophilic Archaea *S. solfataricus* and *Archaeoglobus fulgidus* (Büttner *et al.*, 2005; Lorentzen and Conti, this volume; Lorentzen *et al.*, 2005, 2007).

By use of the five subunits of the archaeal exosome (SsoRrp41, SsoRrp42, SsoRrp4, SsoCsl4, SsoDnaG), complexes containing SsoDnaG were also refolded. These exosomes were well detectable by coimmunoprecipitation with SsoRrp41-directed antiserum. However, very low amounts of SsoDnaG-containing exosomes were detected by size-exclusion chromatography, suggesting that these complexes were not stable (Walter *et al.*, 2006). Therefore, we were not able to analyze the function of the archaeal DnaG protein. The other exosomal complexes were analyzed in respect to their activity.

4.5. Functional characterization of reconstituted exosome complexes *in vitro*

On the basis of the domain composition of the subunits of the *S. solfataricus* exosome (Koonin *et al.*, 2001) and on previous characterization of individual recombinant subunits of eukaryotic exosomes *in vitro* (Chekanova *et al.*, 2002; Mitchell *et al.*, 1997), we expected that SsoRrp41 or SsoRrp4 exhibits 3′ to 5′-exoribonuclease activity *in vitro*. To be sure that this activity is not impaired by secondary structures of the substrate, we decided to perform the assays with a 5′-end-labeled 30-meric poly(A)-oligoribonucleotide (CureVac GmbH, Tübingen, Germany). However, we did not detect any RNase activity with individual recombinant subunits of the *S. solfataricus* exosome even when the proteins were applied in a 1000-fold molar overexcess in respect to the substrate; the proteins used were purified under native conditions or refolded separately. We assumed that complex formation might be necessary for activity. Indeed, refolded Rrp41–Rrp42 complexes showed RNase activity. Unfortunately, some preparations yielded highly active exosomes, whereas the activity of independently refolded complexes was very low. Therefore, the assays for functional characterization of the recombinant *S. solfataricus* exosome were performed with proteins and protein complexes prepared under native conditions by Esben Lorentzen (this volume). The proteins were frozen in aliquots in liquid nitrogen and stored at −80 °C. Freeze–thaw cycles strongly decrease the activity of the exosome and should be avoided. By use of the native protein preparations, we found that the ribonuclease activity of the SsoRrp41–SsoRrp42 complex is increased in presence of SsoRrp4 and SsoCsl4 (Fig. 19.3; Walter *et al.*, 2006).

In a typical assay for ribonuclease activity 6 fmol poly(A)-substrate and 1 pmol protein complex were used in a final volume of 10 μl. The reaction buffer contained of 20 mM HEPES, pH 7.9, 60 mM KCl, 5 mM MgCl$_2$, 0.1 mM EDTA, 2 mM DTT, 12% glycerol, 375 mM trehalose, and 10 mM KH$_2$PO$_4$. In this buffer, we obtained the most consistent results with refolded proteins, and, therefore, it was used for the initial characterization of the exosome. Further experiments revealed that in assays with native protein preparations, trehalose and glycerol could be omitted. The RNase activity was very low on ice and at 20 °C and increased with increasing temperature. The optimum *in vitro* was at 70 °C; somewhat lower activity was observed at 60 °C and at 80 °C. We achieved the best time course results when a master mix reaction was prepared on ice, and 10-μl aliquots were then incubated for different times (2, 5, 10, 15 min) in a thermocycler at 60 °C. The RNase activity of the Rrp41–Rrp42 complex was lost at concentrations <0.03 pmol/μl, whereas Rrp4 and Csl4-containing complexes were still active. The protein complex was present in excess

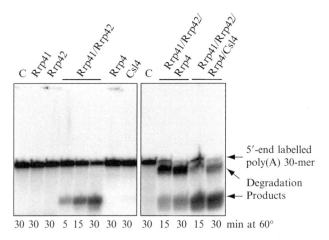

30 30 30 5 15 30 30 30   30 15 30 15 30 min at 60°

Figure 19.3 RNA degradation assays that use recombinant complexes of different composition and with individual subunits of the *S. solfataricus* exosome. Phosphor image of a 10% denaturing polyacrylamide gel, in which the 5′-end-labeled substrate and the detectable reaction products were resolved. The exosome subunits assayed are indicated above the panels. Lane C is a negative control and consists of substrate incubated in reaction buffer in the absence of added protein. The reaction buffer contained 20 mM HEPES, pH 7.9, 60 mM KCl, 5 mM MgCl$_2$, 0.1 mM EDTA, 2 mM DTT, 12% glycerol, 375 mM trehalose, and 10 mM KH$_2$PO$_4$. Assays were performed at 60 °C for the indicated time in a final volume of 10 μl with 6 fmol of RNA substrate. Approximately 200 ng of each polypeptide was used when individual subunits were tested. When complexes containing SsoRrp41 and SsoRrp42 or SsoRrp41, SsoRrp42, and Rrp4 were tested, approximately 100 ng of each polypeptide was present. When SsoRrp41, SsoRrp42, Rrp4, and Csl4 were used, approximately 100 ng SsoRrp41, 100 ng SsoRrp42, 50 ng Rrp4, and 50 ng Csl4 were present.

compared with the substrate in the assay mixtures, because the results of the degradation experiments implied that only part of the protein complex is active. Alternately, we may miss an important buffer or protein component for full activity *in vitro*.

We detected RNase activity even without addition of KH$_2$PO$_4$ in the reaction buffer. This raised the question whether RNA was degraded hydrolytically. To analyze the mode of action of the different exosome complexes, we performed thin-layer chromatography (TLC) analysis of the released products (see 3.1.4). In these assays, 20 fmol substrate was used per 10-μl reaction mixture. The results demonstrated that the reconstituted exosomal complexes degrade RNA and release nucleoside diphosphates. Thus, the SsoRrp41–SsoRrp42 complex has phosphorolytic exoribonuclease activity. The presence of SsoRrp4 and SsoCsl4 leads to faster phosphorolytic degradation because of better substrate binding or because of conformational changes or stabilization of the hexameric ring. The degradation of the substrate without the extra addition of inorganic phosphate to the reaction mixture is most probably due to the presence of phosphate ions in low concentrations in our buffer substances or in the protein fractions used.

The degradation activity of the archaeal exosome was demonstrated so far exclusively on poly(A) RNA (Büttner *et al.*, 2005; Lorentzen *et al.*, 2005; Ramos *et al.*, 2006; Walter *et al.*, 2006). In assays with internally labeled transcript derived from a *Rhodobacter capsulatus* mRNA (Fritsch *et al.*, 1995), very low phosphorolytic activity was detected (Walter *et al.*, 2006). The 3′-end of this transcript was not characterized experimentally so far, and we do not know whether it is accessible or whether secondary structures impairing the RNA degradation are present near the 3′-end. We performed assays with other *in vitro* transcripts and found that the hexameric ring of the *S. solfataricus* exosome can degrade short RNA consisting of all four nucleotides (Verena Roppelt, unpublished results).

At low concentrations of inorganic phosphate and sufficient high concentration of nucleoside diphosphates, phosphorolytic exoribonucleases like PNPase can synthesize RNA in a reaction which is reverse to phosphorolysis. PNPase is known to produce heterogeneous adenine-rich RNA tails *in vivo* that are important for degradation of RNA by exoribonucleases in Bacteria and organelles (Gadi Schuster, this volume; Rott *et al.*, 2003; Yehudai-Resheff *et al.*, 2001). The presence of adenine-rich RNA tails in Archaea correlates with the presence of genes for orthologs of Rrp4, Rrp41, Rrp42, and Csl4 in the archaeal genomes (Portnoy and Schuster, 2006; Portnoy *et al.*, 2005). This suggested a role of the exosome in RNA polyadenylation. Indeed, reconstituted exosomes can polyadenylate the 30-meric poly(A) oligoribonucleotide in the presence of 20 m*M* ADP and without addition of inorganic phosphate under the conditions otherwise identical to the degradation assay conditions. The synthesized tails were longer than 500 nt (Portnoy *et al.*, 2005; Lorentzen *et al.*, 2005). The polyadenylation activity was very low at 4 °C and at 20 °C. High polyadenylation activity was observed at temperatures between 60 °C and 80 °C.

4.6. *In vitro* characterization of the native *S. solfataricus* exosome

To characterize the native exosome in cell-free extracts from *S. solfataricus*, we compared the exoribonucleolytic activity and the polyadenylation activity of depleted and nondepleted extracts. As described in 4.3, the exosome depletion was achieved by three rounds of coimmunoprecipitation resulting in the removal of approximately 94% of SsoRrp41. Consistent with the assumption that the exosome is the major degradative exoribonuclease and the major polyadenylation enzyme in *S. solfataricus*, the degradation and the polyadenylation activity of the extract were strongly reduced after depletion. A control treatment of the extract with preimmune serum coupled to protein A–Sepharose beads had nearly no effect on activity. The assays were performed as described in 4.5, but instead of 30 pmol of pure exosome, 100 ng of total *S. solfataricus* proteins was added to the reaction mixtures (Portnoy *et al.*, 2005; Walter *et al.*, 2006). A TLC analysis of the released nucleosides revealed the presence of adenosine diphosphates (ADP) and adenosine monophosphates (AMP).

We assume that AMP arises because of dephosphorylation of ADP by another enzyme in the extract and not directly during RNA degradation. To test this assumption, 10,000 cpm of poly(A) RNA, which was freshly labeled by a tailing reaction with the poly(A) polymerase I, was first degraded with the degradosome of *E. coli* as described in 3.1.4. The detectable product of this RNA degradation reaction is ADP. Then, 1000 cpm containing aliquots of this reaction were used as a substrate in a typical RNA degradation assay for 10 min at 60 °C with extracts from *S. solfataricus* depleted or nondepleted for the exosome, and the reaction mixtures were subjected to TLC analysis. ADP remained stable in the negative controls, which were incubated without addition of extract. In contrast, more than 90% of the ADP was converted to AMP by the depleted and the nondepleted extract. Thus, the RNA-degrading activity of the extract, which was decreased after depletion (the activity of the exosome), was different from the activity converting ADP to AMP. This result confirms that the exosome of *S. solfataricus* exhibits only phosphorolytic activity (Walter *et al.*, 2006).

The tests just described characterize the exosome in cell-free extracts and pure recombinant complexes containing orthologs of eukaryotic exosomal subunits. We also wanted to test pure native exosome that contains SsoDnaG. For this purpose we used exosome purified by coimmunoprecipitation with anti-Rrp41 antiserum as described previously, with 0.05 g beads and 500 μl extract. Coimmunoprecipitation with 500 μl 30S to 50S fraction of a glycerol density gradient (see 4.2.) was also performed to obtain an exosome that contains the Cdc48 homolog in addition. Because no activity was detected when elution fractions were used, we performed the assays with the protein complex still bound to antibodies coupled to protein A–Sepharose beads. After 10 washing steps with 1 ml TMN buffer, the beads with the bound exosome were resuspended in 50 μl TMN buffer, and 3 μl of this suspension was used in a typical assay for RNA degradation or polyadenylation as described in 4.5. The RNA substrate remained intact in negative control reactions containing antiserum coupled to the beads. The native exosome showed phosphorolytic RNase activity and polyadenylation activity under the respective reaction conditions. The activity of these native preparations was lost after few hours' incubation at 4 °C, and it was strongly decreased after one thawing. These difficulties in the handling of native exosomes hampered the investigation of the role of the Archaea-specific subunits of the *S. solfataricus* exosome.

ACKNOWLEDGMENTS

We thank Verena Roppelt and Pamela Walter for sharing unpublished results. Our work was supported by the Deutsche Forschungsgemeinschaft, the Justus-Liebig-Universität, Giessen, and the Fonds der Chemischen Industrie.

REFERENCES

Allmang, C., Petfalski, E., Podtelejnikov, A., Mann, M., Tollervey, D., and Mitchell, P. (1999). The yeast exosome and human PM-Scl are related complexes of 3′ to 5′-exonucleases. *Genes Dev.* **13**, 2148–2158.

Aloy, P., Ciccarelli, F. D., Leutwein, C., Gavin, A. C., Superti-Furga, G., Bork, P., Bottcher, B., and Russell, R. B. (2002). A complex prediction: Three-dimensional model of the yeast exosome. *EMBO Rep.* **3**, 628–635.

Anantharaman, V., Koonin, E. V., and Aravind, L. (2002). Comparative genomics and evolution of proteins involved in RNA metabolism. *Nucleic Acids Res.* **30**, 1427–1464.

Andersson, A. F., Lundgren, M., Eriksson, S., Rosenlund, M., Bernander, R., and Nilsson, P. (2006). Global analysis of mRNA stability in the archaeon *Sulfolobus*. *Genome Biol.* **7**, R99.

Baker, M. E., Grundy, W. N., and Elkan, C. (1998). Spinach CSP41, an mRNA-binding protein and ribonuclease, is homologous to nucleotide-sugar epimerases and hydroxy-steroid dehydrogenases. *Biochem. Biophys. Res. Commun.* **248**, 250–254.

Bernstein, J. A., Khodursky, A. B., Lin, P. H., Lin-Chao, S., and Cohen, S. N. (2002). Global analysis of mRNA decay and abundance in *Escherichia coli* at single-gene resolution using two-color fluorescent DNA microarrays. *Proc. Natl. Acad. Sci. USA* **99**, 9697–9702.

Bini, E., Dikshit, V., Dirksen, K., Drozda, M., and Blum, P. (2002). Stability of mRNA in the hyperthermophilic archaeon *Sulfolobus solfataricus*. *RNA* **9**, 1129–1136.

Brown, J. W., and Reeve, J. N. (1985). Polyadenylated, noncapped RNA from the archaebacterium *Methanococcus vannielii*. *J. Bacteriol.* **162**, 909–917.

Büttner, K., Wenig, K., and Höpfner, K. P. (2005). Structural framework for the mechanism of archaeal exosomes in RNA processing. *Mol. Cell.* **20**, 461–471.

Chekanova, J. A., Dutko, J. A., Mian, I. S., and Belostotsky, D. A. (2002). *Arabidopsis thaliana* exosome subunit AtRrp4p is a hydrolytic 3′ to 5′-exonuclease containing S1 and KH RNA-binding domains. *Nucleic Acids Res.* **30**, 695–700.

Chomczinski, P., and Sacchi, N. (1987). Single-step method of RNA isolation by acid guanidinium thiocyanate-phenol-chloroform extraction. *Anal. Biochem.* **162**, 156–159.

Cline, S. W., Lam, W. L., Charlebois, R. L., Schalkwyk, L. C., and Doolittle, W. F. (1989). Transformation methods for halophilic archaebacteria. *Can. J. Microbiol.* **35**, 148–152.

Condon, C. (2007). Maturation and degradation of RNA in bacteria. *Curr. Opin. Microbiol.* **10**, 271–278.

Conrad, C., Rauhut, R., and Klug, G. (1998). Different cleavage specificities of RNases III from *Rhodobacter capsulatus* and *Escherichia coli*. *Nucleic Acids Res.* **26**, 4446–4453.

Deutscher, M. (2006). Degradation of RNA in bacteria: Comparison of mRNA and stable RNA. *Nucleic Acids Res.* **34**, 659–666.

Evguenieva-Hackenberg, E., and Klug, G. (2000). RNase III processing of intervening sequences found in helix 9 of 23S rRNA in the alpha subclass of *Proteobacteria*. *J. Bacteriol.* **182**, 4719–4729.

Evguenieva-Hackenberg, E., Schiltz, E., and Klug, G. (2002). Dehydrogenases from all three domains of life cleave RNA. *J. Biol. Chem.* **277**, 46145–46150.

Evguenieva-Hackenberg, E., Walter, P., Hochleitner, E., Lottspeich, F., and Klug, G. (2003). An exosome-like complex in *Sulfolobus solfataricus. EMBO Rep.* **4,** 889–893.

Franzetti, B., Sohlberg, B., Zaccai, G., and von Gabain, A. (1998). Biochemical and serological evidence for an RNase E-like activity in halophilic Archaea. *J. Bacteriol.* **179,** 1180–1185.

Fritsch, J., Rothfuchs, R., Rauhut, R., and Klug, G. (1995). Identification of an mRNA element promoting rate-limiting cleavage of the polycistronic *puf* mRNA in *Rhodobacter capsulatus* by an enzyme similar to RNase E. *Mol. Microbiol.* **15,** 1017–1029.

Gegenheimer, P., and Apirion, D. (1981). Processing of procaryotic ribonucleic acid. *Microbiol. Rev.* **45,** 502–541.

Goldberg, I. H., Rabinowitz, M., and Reich, E. (1962). Basis of actinomycin action. I. DNA binding and inhibition of RNA-polymerase synthetic reactions by actinomycin. *Proc. Natl. Acad. Sci. USA* **48,** 2094–2101.

Gropp, F., Gropp, R., and Betlach, M. C. (1995). Effects of upstream deletions on light- and oxygen-regulated bacterio-opsin gene expression in *Halobacterium halobium. Mol. Microbiol.* **16,** 357–364.

Harlow, E., and Lane, D. (1988). "Antibodies. A Laboratory Manual," pp. 522–523. CSH Press, Cold Spring Harbor, NY.

Houseley, J., LaCava, J., and Tollervey, D. (2006). RNA-quality control by the exosome. *Nat. Rev. Mol. Cell. Biol.* **7,** 529–539.

Hundt, S., Zaigler, A., Lange, C., Soppa, J., and Klug, G. (2007). Global analysis of mRNA decay in *Halobacterium salinarum* NRC-1 at single-gene resolution using DNA microarrays. *J. Bacteriol.* **189,** 6936–6944.

Huntzinger, E., Boisset, S., Saveanu, C., Benito, Y., Geissmann, T., Namane, A., Lina, G., Etienne, J., Ehresmann, B., Ehresmann, C., Jacquier, A., Vandenesch, F., *et al.* (2005). *Staphylococcus aureus* RNAIII and the endoribonuclease III coordinately regulate *spa* gene expression. *EMBO J.* **24,** 824–835.

Jäger, A., Samorski, R., Pfeifer, F., and Klug, G. (2002). Individual gvp transcript segments in *Haloferax mediterranei* exhibit varying half-lives, which are differentially affected by salt concentration and growth phase. *Nucleic Acids Res.* **30,** 5436–5443.

Kleman-Leyer, K., Armbruster, D. W., and Daniels, C. J. (1997). Properties of *H. volcanii* tRNA intron endonuclease reveal a relationship between the archaeal and eucaryal tRNA intron processing systems. *Cell* **89,** 839–847.

Koonin, E. V., Wolf, Y. I., and Aravind, L. (2001). Prediction of the archaeal exosome and its connections with the proteasome and the translation and transcription machineries by a comparative-genomic approach. *Genome Res.* **11,** 240–252.

LaCava, J., Houseley, J., Saveanu, C., Petfalski, E., Thompson, E., Jacquier, A., and Tollervey, D. (2005). RNA degradation by the exosome is promoted by a nuclear polyadenylation complex. *Cell* **121,** 713–724.

Langer, D., Hain, J., Thuriaux, P., and Zillig, W. (1995). Transcription in archaea: Similarity to that in eucarya. *Proc. Natl. Acad. Sci. USA* **92,** 5768–5772.

Lin-Chao, S., Chiou, N. T., and Schuster, G. (2007). The PNPase, exosome and RNA helicases as the building components of evolutionarily-conserved RNA degradation machines. *Biomed. Sci.* **14,** 523–532.

Lorentzen, E., Walter, P., Fribourg, S., Evguenieva-Hackenberg, E., Klug, G., and Conti, E. (2005). The archaeal exosome core is a hexameric ring structure with three catalytic subunits. *Nat. Struct. Mol. Biol.* **12,** 575–581.

Lorentzen, E., Dziembowski, A., Lindner, D., Seraphin, B., and Conti, E. (2007). RNA channelling by the archaeal exosome. *EMBO Rep.* **8,** 470–476.

Matera, A. G., Terns, R. M., and Terns, M. P. (2007). Non-coding RNAs: Lessons from the small nuclear and small nucleolar RNAs. *Nat. Rev. Mol. Cell. Biol.* **8,** 209–220.

Mathews, D. H., Sabina, J., Zuker, M., and Turner, D. H. (1999). Expanded sequence dependence of thermodynamic parameters provides robust prediction of RNA secondary structure. *J. Mol. Biol.* **288,** 911–940.

Milligan, J. F., and Uhlenbeck, O. C. (1989). Synthesis of small RNAs using T7 RNA polymerase. *Methods Enzymol.* **180,** 51–62.

Mitchell, P., Petfalski, E., Shevchenko, A., Mann, M., and Tollervey, D. (1997). The exosome: A conserved eukaryotic RNA processing complex containing multiple 3′ to 5′-exoribonucleases. *Cell* **91,** 457–466.

Moore, M. J. (2005). From birth to death: The complex lives of eukaryotic mRNAs. *Science* **309,** 1514–1518.

Nagy, E., and Rigby, W. F. (1995). Glyceraldehyde-3-phosphate dehydrogenase selectively binds AU-rich RNA in the NAD(+)-binding region (Rossmann fold). *J. Biol. Chem.* **270,** 2755–2763.

Nieuwlandt, D. T., Palmer, J. R., Armbruster, D. T., Kuo, Y.-P., Oda, W., and Daniels, C. J. (1995). A rapid procedure for the isolation of RNA from *Haloferax volcanii.* *In* "Archaea; A Laboratory Manual," (F. T. Robb and A. R. Place, eds.), pp. 161–162. Cold Spring Harbor Laboratory Press, New York.

Patenge, N., Haase, A., Bolhuis, H., and Oesterhelt, D. (2000). The gene for a halophilic beta-galactosidase (bgaH) of *Haloferax alicantei* as a reporter gene for promoter analyses in *Halobacterium salinarum. Mol. Microbiol.* **36,** 105–113.

Pato, M. L., Bennett, P. M., and von Meyenburg, K. (1973). Messenger ribonucleic acid synthesis and degradation in *Escherichia coli* during inhibition of translation. *J. Bacteriol.* **116,** 710–718.

Portnoy, V., Evguenieva-Hackenberg, E., Klein, F., Walter, P., Lorentzen, E., Klug, G., and Schuster, G. (2005). RNA polyadenylation in Archaea: Not observed in *Haloferax* while the exosome polynucleotidylates RNA in *Sulfolobus. EMBO Rep.* **6,** 1188–1193.

Portnoy, V., and Schuster, G. (2006). RNA polyadenylation and degradation in different Archaea; roles of the exosome and RNase R. *Nucleic Acids Res.* **34,** 5923–5931.

Prangishvilli, D., Zillig, W., Gierl, A., Biesert, L., and Holz, I. (1982). DNA-dependent RNA polymerase of thermoacidophilic archaebacteria. *Eur. J. Biochem.* **122,** 471–477.

Purusharth, R. I., Klein, F., Sulthana, S., Jäger, S., Jagannadham, M. V., Evgenieva-Hackenberg, E., Ray, M. K., and Klug, G. (2005). Exoribonuclease R interacts with endoribonuclease E and an RNA-helicase in the psychrotrophic bacterium *Pseudomonas syringe* Lz4W. *J. Biol. Chem.* **280,** 14572–14578.

Raghavan, A., Ogilvie, R. L., Reilly, C., Abelson, M. L., Raghavan, S., Vasdewani, J., Krathwohl, M., and Bohjanen, P. R. (2002). Genome-wide analysis of mRNA decay in resting and activated primary human T lymphocytes. *Nucleic Acids Res.* **30,** 5529–5538.

Raijmakers, R., Egberts, W. V., van Venrooij, W. J., and Pruijn, G. J. (2002). Protein-protein interactions between human exosome components support the assembly of RNase PH-type subunits into a six-membered PNPase-like ring. *J. Mol. Biol.* **323,** 653–663.

Ramos, C. R., Oliveira, C. L., Torriani, I. L., and Oliveira, C. C. (2006). The *Pyrococcus* exosome complex: Structural and functional characterisation. *J. Biol. Chem.* **281,** 6751–6759.

Rauhut, R., and Klug, G. (1999). mRNA degradation in bacteria. *FEMS Microbiol. Rev.* **23,** 353–370.

Reich, E., Franklin, R. M., Shatkin, A. J., and Tatum, E. L. (1961). Effect of actinomycin D on cellular nucleic acid synthesis and virus production. *Science* **134,** 556–557.

Rott, R., Zipor, G., Portnoy, V., Liveanu, V., and Schuster, G. (2003). RNA polyadenylation and degradation in cyanobacteria are similar to the chloroplast but different from *Escherichia coli. J. Biol. Chem.* **278,** 15771–15777.

Sasaki, K., Abid, M. R., and Miyazaki, M. (1996). Deoxyhypusine synthase gene is essential for cell viability in the yeast *Saccharomyces cerevisiae. FEBS Lett.* **384,** 151–154.

Schierling, K., Rösch, S., Rupprecht, R., Schiffer, S., and Marchfelder, A. (2002). tRNA 3'-end maturation in archaea has eukaryotic features: The RNase Z from *Haloferax volcanii. J. Mol. Biol.* **316,** 895–902.

Schnier, J., Schwelberger, H. G., Smit-McBride, Z., Kang, H. A., and Hershey, J. W. (1991). Translation initiation factor 5A and its hypusine modification are essential for cell viability in the yeast *Saccharomyces cerevisiae. Mol. Cell. Biol.* **11,** 3105–3114.

Schweisguth, D. C., Chelladurai, B. S., Nicholson, A. W., and Moore, P. B. (1994). Structural characterization of a ribonuclease III processing signal. *Nucleic Acids Res.* **22,** 604–612.

Symmons, M. F., Williams, M. G., Luisi, B. F., Jones, G. H., and Carpousis, A. J. (2002). Running rings around RNA: A superfamily of phosphate-dependent RNases. *Trends Biochem. Sci.* **27,** 11–18.

Valentini, S. R., Casolari, J. M., Oliveira, C. C., Silver, P. A., and McBride, A. E. (2002). Genetic interactions of yeast eukaryotic translation initiation factor 5A (eIF5A) reveal connections to poly(A)-binding protein and protein kinase C signaling. *Genetics* **160,** 393–405.

Vanácová, S., Wolf, J., Martin, G., Blank, D., Dettwiler, S., Friedlein, A., Langen, H., Keith, G., and Keller, W. (2005). A new yeast poly(A) polymerase complex involved in RNA quality control. *PLoS Biol.* **3,** 189.

Vogel, J., Bartels, V., Tang, T. H., Churakov, G., Slagter-Jäger, J. G., Hüttenhofer, A., and Wagner, E. G. (2003). RNomics in *Escherichia coli* detects new sRNA species and indicates parallel transcriptional output in bacteria. *Nucleic Acids Res.* **31,** 6435–6443.

Wagner, S., and Klug, G. (2007). An archaeal protein with homology to the eukaryotic translation initiation factor 5A shows ribonucleolytic activity. *J. Biol. Chem.* **282,** 13966–13976.

Walter, P., Klein, F., Lorentzen, E., Ilchmann, A., Klug, G., and Evguenieva-Hackenberg, E. (2006). Characterisation of native and reconstituted exosome complexes from the hyperthermophilic archaeon *Sulfolobus solfataricus. Mol. Microbiol.* **62,** 1076–1089.

Werner, F., and Weinzierl, R. O. (2002). A recombinant RNA polymerase II-like enzyme capable of promoter-specific transcription. *Mol. Cell.* **10,** 635–646.

Wolff, E. C., Kang, K. R., Kim, Y. S., and Park, M. H. (2007). Posttranslational synthesis of hypusine: Evolutionary progression and specificity of the hypusine modification. *Amino Acids* **33,** 341–350.

Xu, A., and Chen, K. Y. (2001). Hypusine is required for a sequence-specific interaction of eukaryotic initiation factor 5A with postsystematic evolution of ligands by exponential enrichment RNA. *J. Biol. Chem.* **276,** 2555–2561.

Xu, A., Jao, D. L., and Chen, K. Y. (2004). Identification of mRNA that binds to eukaryotic initiation factor 5A by affinity co-purification and differential display. *Biochem. J.* **384,** 585–590.

Yang, C. F., Kim, J. M., Molinari, E., and DasSarma, S. (1996). Genetic and topological analyses of the bop promoter of *Halobacterium halobium*: Stimulation by DNA supercoiling and non-B-DNA structure. *J. Bacteriol.* **178,** 840–845.

Yehudai-Resheff, S., Hirsh, M., and Schuster, G. (2001). Polynucleotide phosphorylase functions as both an exonuclease and a poly(A) polymerase in spinach chloroplasts. *Mol. Cell. Biol.* **21,** 5408–5416.

Zamore, P. D., and Haley, B. (2005). Ribo-gnome: The big world of small RNAs. *Science* **309,** 1519–1524.

Zhang, K., and Nicholson, A. W. (1997). Regulation of ribonuclease III processing by double-helical sequence antideterminants. *Proc. Natl. Acad. Sci. USA* **94,** 13437–13441.

Zuk, D., and Jacobson, A. (1998). A single amino acid substitution in yeast eIF-5A results in mRNA stabilization. *EMBO J.* **17,** 2914–2925.

Zuo, Y., and Deutscher, M. P. (2001). Exoribonuclease superfamilies: Structural analysis and phylogenetic distribution. *Nucleic Acids Res.* **29,** 1017–1026.

CHAPTER TWENTY

EXPRESSION, RECONSTITUTION, AND STRUCTURE OF AN ARCHAEAL RNA DEGRADING EXOSOME

Esben Lorentzen* *and* Elena Conti[†]

Contents

Abstract

The exosome is a protein complex that participates in a wide variety of RNA processing, degradation, and quality-control pathways. The exosome is conserved in all eukaryotes studied to date and is also present in many archaeal organisms, albeit in a simpler form. To gain insights into the architecture of the

* Birkbeck College London, Institute of Structural Molecular Biology, London, United Kingdom
† Max-Planck Institute of Biochemistry, Martinsried, Germany

Methods in Enzymology, Volume 447   © 2008 Elsevier Inc.
ISSN 0076-6879, DOI: 10.1016/S0076-6879(08)02220-9   All rights reserved.

exosome complex, we have chosen the hyperthermophilic archaeum *Sulfolobus solfataricus* as a model system. Here we describe the coexpression, purification, and crystal structure determination of archaeal exosome complexes. To understand how the archaeal exosome binds and degrades RNA, we designed RNA substrates for degradation experiments exploiting the knowledge of the geometric constraints of the exosome structure. Furthermore, we describe several crystal structures in which RNA substrates were diffused into crystals and how anomalous scattering from 5-iodo-uridine–modified RNA was used to locate low-occupancy RNA binding sites.

1. INTRODUCTION

The RNA degrading exosome was initially discovered in yeast strains with incorrectly processed 5.8 ribosomal RNA (rRNA) (Mitchell *et al.*, 1997; see chapters in Maquat and Kiledjian, RNA Turnover, Part B). Since then, the exosome has been shown to play a number of important roles in RNA processing, degradation, and quality-control pathways (reviewed in Houseley *et al.*, 2006). The eukaryotic exosome consists of a conserved core complex of 10 subunits present in both the nucleus and the cytoplasm. The nuclear form of the eukaryotic exosome participates in the maturation of stable RNA precursors and in the rapid decay of defective messenger RNA (mRNA) precursors (Allmang *et al.*, 1999; Bousquet-Antonelli *et al.*, 2000; Briggs *et al.*, 1998; Mitchell *et al.*, 1997; van Hoof *et al.*, 2000). In addition to the 10 core subunits, the nuclear exosome is known to associate with the hydrolytic RNase Rrp6 (Briggs *et al.*, 1998) and to require the TRAMP complex for stimulation of its nuclease activity (LaCava *et al.*, 2005; Vanacova *et al.*, 2005; Wyers *et al.*, 2005). The TRAMP complex is a macromolecular assembly containing a poly(A) polymerase (Trf4), a putative zinc-knuckle RNA-binding protein (Air1 or Air2), and a putative helicase (Mtr4).

Besides the nuclear functions described previously, several roles have been reported for the cytoplasmic exosome. These include 3′ to 5′-directed turnover of mRNA (Anderson and Parker, 1998) and quality control functions such as the removal of aberrant mRNAs that either lack a normal stop codon or contain a premature stop codon (Conti and Izaurralde, 2005; van Hoof *et al.*, 2002). Both normal turnover of mRNA and the quality-control pathways in the cytoplasm rely on the SKI complex as an activator of the exosome (reviewed in Houseley *et al.*, 2006). The SKI complex also contains a putative helicase (Ski2) but is otherwise not homologous to the activating TRAMP complex found in the nucleus. Additional proteins of the SKI complex include the proteins Ski3 and Ski8 that are thought to mediate protein–protein interactions (Wang *et al.*, 2005). In *S. cerevisiae*, a protein (Ski7) with homology to translational elongation factors bridges the interaction between the exosome and the other SKI components (Araki *et al.*, 2001).

Ten different proteins of equal stoichiometry associate to form the yeast exosome core. Six of these subunits (ribosomal RNA–processing protein (Rrp) 41, Rrp42, Rrp43, Rrp45, Rrp46, and Mtr3) show significant sequence identity to RNase PH, a bacterial protein that cleaves transfer RNA (tRNA) precursors in a phosphate-dependent manner. As predicted from the structure of bacterial polynucleotide phosphorylase (PNPase) (Symmons *et al.*, 2000), the six RNase PH–like exosome proteins form a hexameric ring (reviewed in Buttner *et al.*, 2006; Lorentzen and Conti, 2006; Vanacova and Stefl, 2007). Three additional proteins with putative RNA-binding domains (Rrp4, Rrp40, and Csl4) form a trimeric cap that associates with the hexamer. The tenth core subunit, Rrp44, is a hydrolytic RNase with sequence homology to members of the RNase II superfamily (Frazao *et al.*, 2006; Zuo *et al.*, 2006). Recent research has shown that only one subunit in either yeast or human core exosomes, the Rrp44 protein, has RNase activity (Dziembowski *et al.*, 2007; Liu *et al.*, 2006). Rrp44 hydrolyzes RNA substrates processively from the 3′-end and can degrade extensive regions of secondary structure provided that a short 3′-end overhang of single-stranded RNA is present (Dziembowski *et al.*, 2007, Lorentzen *et al.*, 2008).

The exosome is not only conserved in eukaryotes but also in many archaeal organisms, although with a simpler subunit composition. Genomic analysis identified the four gene products Rrp41, Rrp42, Rrp4, and Csl4 as likely constituents of the archaeal exosome (Koonin *et al.*, 2001). The archaeal exosome does not seem to require association with an Rrp44 homolog or with activating components similar to the TRAMP or SKI complexes for activity. The archaeal Rrp41 and Rrp42 proteins also show homology to the RNase PH protein. Archaeal Rrp4 is homologous to the eukaryotic Rrp4 and Rrp40 subunits, whereas the archaeal Csl4 subunit displays homology to eukaryotic Csl4. In the hyperthermophilic archaeum *Sulfolobus solfataricus*, it was shown that these proteins indeed form a 300-kDa complex *in vivo* that is endowed with phosphorolytic RNase activity (Evguenieva-Hackenberg *et al.*, 2003; Walter *et al.*, 2006). The high molecular weight of the archaeal exosome and the clear homology to the eukaryotic exosome components suggested that the two complexes could have similar overall architectures (Koonin *et al.*, 2001).

In an endeavor to characterize the structure of the exosome and its mechanism of RNA degradation, we chose to study the exosome from the *S. solfataricus*. The rationale behind this choice is the simplicity of its subunit composition (advantageous in reconstituting a functional complex) and its thermostability (well known to increase the chances of obtaining high-resolution diffracting crystals). These factors, combined with the homology to eukaryotic exosomes, makes it feasible to use structure-based sequence alignments to extrapolate the mechanistic insights obtained from the archaeal exosome structure to eukaryotes. Here, we describe the details of the methods leading to the atomic-level characterization of the architecture and RNA-binding sites of the *S. solfataricus* archaeal exosome.

2. CLONING, EXPRESSION, PURIFICATION, AND RECONSTITUTION OF AN ARCHAEAL EXOSOME

This section describes the production of abundant (milligram) quantities of pure *S. solfataricus* exosome complex suitable for biochemical and structural characterization. The complex was first reconstituted by mixing the subunits that had been individually overexpressed in bacterial cells and purified. The method has then been optimized to increase the speed of the purification protocol and the amount of material obtained by use of coexpression in *E. coli*.

2.1. Cloning of the *S. solfataricus* exosome genes

The four individual genes encoding the *S. solfataricus* exosome core (Rrp4, Rrp41, Rrp42, and Csl4) were originally cloned from genomic DNA into the pQE30 vector by E. Evguenieva-Hackenberg and coworkers and purified under denaturing conditions followed by refolding (described by Hackenberg *et al.*, chapter 19 in this volume). For the purpose of crystallization, we subcloned the genes into pET-11b (which encodes an N-terminal hexa-His tag and ampicillin resistance), pET-28b (which encodes no tag and kanamycin resistance), and different pET-MCN vectors (which encode N-terminal His and/or GST tags and ampicillin resistance) by use of *Nde*I and *Xho*I restriction sites (Fribourg *et al.*, 2001; Romier *et al.*, 2006). The vectors encoding a His and/or GST tag all have an engineered tobacco etch virus (TEV) protease cleavage site that allows removal of the tag after affinity purification (Carrington and Dougherty, 1988). This is routinely implemented for structural studies, because flexible termini of proteins can prevent successful crystallization. In addition, the pET-MCN vector facilitates easy construction of polycistronic expression vectors. The bicistronic Rrp41/Rrp42 construct was cloned into the pET-MCN vector by cutting out the insert of Rrp41 with *Spe*I and *Nhe*I restriction enzymes and ligating the inserts into a pET-MCN vector containing Rrp42 and cut with *Spe*I. The tricistronic Rrp4/Rrp41/Rrp42 vector was then engineered by cutting out the Rrp4 insert with *Spe*I/*Nhe*I and ligating it into the Rrp41/Rrp42 bicistronic pET-MCN vector cut with *Spe*I. Each gene cloned into the polycistronic vector has its own ribosome-binding site and encodes an N-terminal His tag. All vectors used for subsequent protein expression were sequenced with T7 polymerase primers.

2.2. Overexpression of the archaeal exosome components

Before initiating large-scale cultures, overexpression of the individual gene products was tested in small scale. Electrocompetent Bl21 (DE3) *E. coli* cells (50 μl) were mixed with 0.5 μl of plasmid. After transformation by electroporation, 500 μl of sterile Luria–Bertani (LB) medium without antibiotics was added, and the Eppendorf tubes were then left to shake at 200 rounds per minutes (rpm) at 37 °C for 45 min. The solution was plated onto agar containing the antibiotic whose resistance is encoded by the transfected plasmid and left overnight at 37 °C. On the following morning, single colonies were picked and transferred into 10 ml LB medium and grown to an OD_{600} of 0.8 at 37 °C. Cultures were then left to cool down to 18 °C for 2 h before inducing 3 to 4 h (or overnight) with 0.5 mM isopropyl β-D-1-thiogalactopyranoside (IPTG). Before adding IPTG, 0.4 ml of the culture was spun down and resuspended in 40 μl of sodium dodecyl sulfate (SDS) loading buffer for later use (uninduced sample). Experience in our laboratory suggests that many recombinantly expressed proteins give a higher yield of soluble protein when expressed at room temperature rather than at 37 °C, and we, therefore, carry out low-temperature expression routinely.

After induction, the cells were centrifuged for 10 min at 4 °C and 4000 rpm and resuspended in 1 ml of buffer A. Cells were lysed by sonication for 20 sec, and 40 μl of the lysate was removed and mixed with 40 μl of SDS loading buffer and stored for later use (induced sample). The rest of the lysate was spun for 10 min at 16,000g at 4 °C. The supernatant was then mixed with 50 μl of Ni^{2+}-nitrilotriacetic acid (Ni–NTA) beads (preincubated with 1 ml of buffer A for 30 min) and left on a rolling table for 30 min at 4 °C. The beads were then spun down at 2000 rpm and washed three times with buffer A before adding 40 μl of SDS loading buffer. Expression levels of the proteins were evaluated by running 5 μl of the crude extracts (uninduced and induced cells) and 10 μl of the Ni–TNA beads on a 12% SDS polyacrylamide gel (BioRad mini-gel system). Before loading, the samples were heated to 95 °C for 2 min and spun at 16000g for 10 min.

After small–scale expression, large-scale cultures (2 to 12 L) were grown to obtain milligram amounts of proteins for complex reconstitution and crystallization. Freshly transformed BL21 (DE3) *E. coli* cells were plated out with the appropriate antibiotic and left to grow overnight at 37 °C. Grown bacteria were directly transferred into 3-L flasks containing 0.5 L of LB medium with sterile pipette tips. Usually, one plate contained enough bacteria to inoculate a total of six flasks. The cells were grown to an OD_{600} of 0.8 at 37 °C with 200 rpm shaking. The temperature was then adjusted to 18 °C and the flasks left to cool down while still shaking for 3 h before induction with 0.5 mM IPTG. The bacteria were left to shake for 12 to 16 h (overnight) at room temperature.

2.3. Purification of archaeal exosome components

One of the advantages of working with proteins from hyperthermophilic organisms is the possibility of purifying the overexpressed thermostable protein by heat precipitation of the nonthermostable *E. coli* proteins. Cells were first lysed in buffer A (10 ml buffer per liter of grown culture) by four rounds of sonication, each 1 min long. After centrifugation at 25,000 rpm for 45 min to remove insoluble material such as membranes, the solution was heated to 75 °C for 30 min in a waterbath and again centrifuged at 25,000 rpm for 45 min to remove the heat-precipitated *E. coli* proteins. After heat precipitation, the overexpressed proteins tended to be 70 to 90% pure. The protein or protein complex was then incubated in batch with 5 ml Ni-NTA resin on a rolling table for 30 min at 4 °C, the resin spun down at 2000 rpm, resuspended in buffer A, and used to pack a column. After extensive washing with buffer A followed by buffer B (10 to 20 column volumes) a linear 0 to 200 mM imidazole gradient at 3 ml/min was carried out with an AKTA Explorer system (GE Healthcare) to elute the protein. The wash with buffer B containing 1 M NaCl was carried out to remove impurities, in particular nucleic acids from *E. coli* that might otherwise remain bound. Protein-containing fractions, as judged by A_{280} absorption, were further analyzed by sodium dodecyl sulfate polyacrylamide gel electrophoresis (SDS-PAGE).

Fractions containing protein of the highest purity were combined and dialyzed overnight against at least a 20-fold (vol/vol) excess of buffer C. TEV protease was added to the dialysis membrane in a ratio of 1 mg protease per 50 mg of His-tagged protein. In general, adding TEV protease at the dialysis step reduces the purification time by cleaving the tag and removing the imidazole at the same time. The completeness of the cleavage was evaluated by SDS-PAGE, and the solution was passed over 1 ml Ni-NTA resin in a disposable column to remove the His-tagged TEV protease along with remaining traces of uncleaved protein. The protein was then concentrated with a Centricon (Millipore) with an appropriate molecular weight cutoff and loaded onto a Superdex 200 size exclusion chromatography column (GE Healthcare) pre-equilibrated in buffer C. The column was run at 0.5 ml/min, and protein-containing fractions were evaluated with SDS-PAGE. The four individual subunits all eluted from the column with molecular weights close to that predicted for a monomer. Pure fractions were combined and concentrated with a Centricon to 5 to 20 mg/ml for reconstitution.

Buffer A: 50 m*M* Tris, pH 7.6, 150 m*M* NaCl
Buffer B: 50 m*M* Tris, pH 7.6, 1.0 *M* NaCl
Buffer C: 10 m*M* Tris, pH 7.6, 150 m*M* NaCl

2.4. *In vitro* reconstitution of archaeal exosome complexes

With all the four individual proteins purified, different subcomplexes could
be reconstituted. Rrp41 and Rrp42 were mixed at a 1:1.5 ratio for 30 min
at 20 °C (protein concentrations between 1 and 10 mg/ml have been
tested). Excess of Rrp42 was easily removed from the Rrp41/Rrp42
complex by passing the reconstitution mix over a Superdex 200 size-
exclusion chromatography column. The Rrp41/Rrp42 complex eluted as
a 200-kDa complex and appeared to contain an equal number of Rrp41 and
Rrp42 subunits as judged from SDS-PAGE (Fig. 20.1). Dynamic light-
scattering experiments revealed a molecular mass of 170 kDa for the
Rrp41/Rrp42 complex, consistent with a hexameric structure containing
three Rrp41 and three Rrp42 subunits (the expected molecular weight of a
hexameric complex is 174 kDa). To verify that the reconstituted exosome
complex was active, RNA degradation experiments were performed
(Lorentzen *et al.*, 2005; Portnoy *et al.*, 2005; Walter *et al.*, 2006).
These experiments were carried out by E. Evguenieva-Hackenberg and
coworkers and are described in chapter 19 of this volume.

The trimeric Rrp4/Rrp41/Rrp42 complex was reconstituted by mixing
purified Rrp41/Rrp42 complex with a fourfold molar excess of Rrp4 for
30 min at 20 °C. Excess Rrp4 was removed by size-exclusion chromatography. The trimeric complex eluted with a molecular weight of 250 kDa and has
a 1:1:1 stoichiometry as judged by SDS-PAGE, suggesting that the complex
contained three of each subunit. Similarly, a trimeric Rrp41/Rrp42/Csl4
complex containing three copies of each subunit could be reconstituted and
purified. Attempts to purify a tetrameric complex resulted in substoichiometric

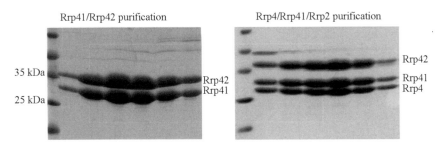

Figure 20.1 SDS gels of purified Rrp41/Rrp42 (left) and Rrp4/Rrp41/Rrp42 (right)
exosome complexes used for structural analysis. The first lane contains the molecular
weight markers, and the following lanes are fractions from size-exclusion chromatography
on a Superdex 200 column.

amounts of Rrp4 and Csl4. The degree of the substoichiometry depended on the relative amounts and the mixing order of subunits, suggesting that Rrp4 and Csl4 associate with the Rrp41/Rrp42 subcomplex through an overlapping binding site. This is consistent with published biochemical and structural analysis of the *Archaeoglobus fulgidus* exosome (Buttner *et al.*, 2005). It is likely that different populations of the archaeal exosome containing different numbers of Rrp4 and Csl4 subunits exist *in vivo,* although the functional implications of this, if any, are currently not known.

2.5. Coexpression of archaeal exosome complexes

Although all four proteins of the *S. solfataricus* exosome could be individually expressed, yields were relatively low, particularly for Rrp41 (Table 20.1). In addition, Rrp41, Rrp4, and Csl4 all had a tendency to aggregate on purification. To overcome these limitations, we coexpressed the subunits so as to purify the exosome complexes directly. Coexpression of the Rrp41/Rrp42 subcomplex was carried out by using the same protocol as described previously for the individual components. Remarkably, the coexpression led to a fivefold increase in the soluble expression of Rrp41. This is likely to be due to a stabilization of the Rrp41 subunit by associating with Rrp42, which prevents degradation and/or aggregation of Rrp41. Although Rrp42 was present in approximately 50% excess compared with Rrp41 on Ni–NTA affinity chromatography, excess Rrp42 was readily removed by size exclusion chromatography. Similarly, the Rrp4/Rrp41/Rrp42 exosome complex was coexpressed and resulted in a more than two-fold increase in the expression of the limiting subunit, Rrp41. The large amounts of pure Rrp41/Rrp42 and Rrp4/Rrp41/Rrp42 obtained by coexpression were used for a detailed structural characterization of the complexes.

Table 20.1 Over-expression rates for the arcaheal exosome subunits

Protein	Final yield in mg of protein per L of culture
Rrp4	4
Rrp41	1
Rrp42	20
Csl4	2
Rrp41/Rrp42	11
Rrp41/Rrp42/Rrp4	7

3. ARCHAEAL EXOSOME STRUCTURE

Structure determination of macromolecular complexes often gives valuable insights into function and mechanism. In this section, we describe the structure determination of the archaeal exosome by X-ray crystallography and the use of chemically modified RNA to identify substrate-binding sites.

3.1. Crystallization of apo *S. solfataricus* Rrp41/Rrp42 and Rrp4/Rrp41/Rrp42

To obtain crystals of the archaeal exosome suitable for diffraction studies, we first screened approximately 900 crystallization conditions from commercial kits (Qiagen, Hampton Research, and Emerald Biostructure) in small scale (with a drop size of 0.2 μl). The crystallization screening was carried out with a Mosquito crystallization Robot (TTP Labtech), which can dispense accurate drop sizes of 0.1 to 0.2 μl into 96-well Greiner crystallization plates (Hampton research). For crystallization, the complexes were concentrated to 10 mg/ml after size exclusion chromatography (as judged by A_{280} measurements on a ND-1000 Spectrophotometer, Nano-Drop). The Rrp41/Rrp42 complex yielded crystals in different screening conditions, generally characterized by high salt contents (sodium malonate or sodium citrate). The crystals grew at 18 °C within 2 to 4 days. We then refined the conditions yielding the most promising crystals (as judged by their size and morphologic regularity) in larger format with 1 μl protein drops in 24-well sitting-drop plates (VDX plates, Hampton Research), with the same equilibrating technique (sitting-drop vapor diffusion). The conditions were refined by modifying the salt concentration and the pH value. The best crystals were obtained by mixing the Rrp41/Rrp42 complex at a concentration of 10 mg/ml with an equal volume of 50 mM MES and 2.0 to 2.4 M sodium malonate, pH 6.5 to 7.0, and equilibrating against a 500-μl reservoir at 18 °C. The high salt concentration also served as a cryoprotectant for X-ray data collection, allowing the crystals to be cooled directly in the nitrogen stream without any ice formation.

The Rrp4/Rrp41/Rrp42 complex was concentrated to 10 mg/ml and screened for crystallization conditions in the same way as the Rrp41/Rrp42 complex. Optimization was achieved by testing the effect of the buffer type and pH, as well as the type of PEG (several different molecular weights) and its concentration. Crystals of the highest quality were obtained by mixing the protein complex with an equal amount of 40% PEG 400, 50 mM Tris-HCl at pH 8.0, and equilibrated by sitting-drop vapor diffusion against a well containing 500 μl of the precipitant solution. Crystals grew to a maximum dimension of 0.2 mm within 2 days at 18 °C.

Both the Rrp41/Rrp42 and Rrp4/Rrp41/Rrp42 complexes yielded crystals with freshly prepared sample, as well as sample stored at −80 °C after flash freezing in liquid nitrogen (without the addition of any glycerol). Crystals were reproducibly obtained with different batches of protein with samples either purified from coexpression or reconstituted after individual purification. Complexes of S. solfataricus Rrp41/Rrp42/Csl4 and Rrp4/Rrp41/Rrp42/Csl4 were also extensively screened for crystallization conditions at different concentrations (between 5 and 15 mg/ml) at both 4 °C and 18 °C with freshly prepared complexes from several different purification batches. No crystals of diffraction quality were obtained for S. solfataricus Rrp41/Rrp42/Csl4 or Rrp4/Rrp41/Rrp42/Csl4.

3.2. Structure determination of the Rrp41-Rrp42 complex

Crystals of S. solfataricus Rrp41/Rrp42 reproducibly diffracted X-rays to a maximum resolution of 2.8 Å at the Swiss Light Source (SLS) X06SA synchrotron beam line (Villigen, Switzerland) (Lorentzen et al., 2005). Because of a large unit cell with the longest cell dimension of 440 Å, the diffraction data were collected with the MarCCD 225 detector at SLS and fine slicing with a $\Delta\varphi$ of 0.2° to avoid overlaps between reflections. The crystals are of space group $C222_1$ and appeared to contain a large number of Rrp41/Rrp42 copies (8 to 16 as judged from the predicted solvent content of the crystals). Indexing and scaling of the diffraction data with the program XDS (Kabsch, 1993) gave superior results to other processing programs, probably because of a better handling of the finely sliced data.

The structure of the Rrp41/Rrp42 complex was solved by molecular replacement with the previously solved bacterial RNase PH (pdb code 1YOR) as a search model in the program PHASER (Storoni et al., 2004). RNase PH is a homohexamer (trimer of dimers) that phosphorolytically cleaves the 3′-end of tRNA precursors (Harlow et al., 2004). Despite the relatively low sequence identity of 19% between Rrp42 and bacterial RNase PH (28% for Rrp41), a clear solution with four hexamers was found. Data were only included up to 6 Å for the rotation and translation functions to speed up the procedure, while the final refinement and phasing were done with all available data. The log likelihood function in PHASER increased considerably with each hexamer found until a total of four hexamers were included and without further gain for a potential fifth hexamer. The four hexamers in the asymmetric unit pack into a tetrahedron giving a crystal solvent content of 60%.

Despite the clear molecular replacement solution, the resulting electron density maps suffered heavily from model bias, making rebuilding rather ambiguous (Fig. 20.2). Maps were significantly improved by carrying out density modification with noncrystallographic symmetry (NCS) averaging of the twelve identical copies of Rrp41/Rrp42 in the asymmetric unit with

No NCS   12-fold NCS

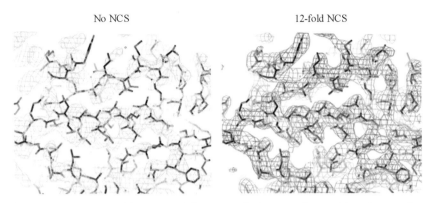

Figure 20.2   Effects of density modification techniques on the resulting electron density maps. The map shown in the left panel was calculated with phases from molecular replacement with the 25% sequence identity model of RNase PH solvent flattened in the program DM. The right panel shows the same part of a map calculated with phases in which 12-fold noncrystallographic symmetry (NCS) averaging was used in combination with solvent flattening. The final model of the Rrp41/Rrp42 complex is shown for both maps.

the CCP4 program DM (Collaborative Computational Project, 1994). Similarly to previous crystallographic work on high-symmetry particles such as viruses, phases were only included to 5 Å. Beyond 5 Å the F_c and F_o showed a very poor correlation. F_c represents the calculated structure factors from the molecular replacement model and F_o the observed structure factors for the Rrp41/Rrp42 complex. Phases were extended from 5 Å to 2.8 Å in 1000 cycles of density modification by applying NCS matrices extracted from superposing dimers from the molecular replacement solution, leading to a greatly improved electron density map (see Fig. 20.2). Whereas bacterial RNase PH contains just one monomer, Rrp41 and Rrp42 adopt the same overall fold but share only 20% sequence identity. In the procedure, it was thus important to correctly average the same subunit among the four hexamers (Rrp41 with Rrp41 and Rrp42 with Rrp42). In total there are 8 (2^3) different ways to perform the averaging among the four different hexamers, with only one way being correct. Fortuitously, the four hexamers in the asymmetric unit obeyed point group symmetry, allowing averaging to be carried out unambiguously. If this were not the case, it would have been necessary to test each of the eight possible averaging protocols and compare the resulting electron density maps. The Rrp41/Rrp42 heterodimer was manually built into the maps with the RNase PH protein as guidance in the programs O (Jones *et al.*, 1991) and COOT (Emsley and Cowtan, 2004). Refinement of the atomic coordinates against the diffraction data was carried out in the program REFMAC (Murshudov *et al.*, 1997) with tight NCS restraints between equivalent Rrp41/Rrp42 dimers.

3.3. Structure determination of the Rrp4/Rrp41/Rrp42 complex

The Rrp4/Rrp41/Rrp42 crystals were of high quality and diffracted X-rays to a maximum resolution of 1.6 Å at the Swiss light source (SLS) synchrotron (Villigen, Switzerland), beam line X06SA (Lorentzen *et al.*, 2007). The crystals belong to the cubic space group *P*213, contain one Rrp4/Rrp41/Rrp42 heterotrimer in the asymmetric unit, and have a solvent content of 47%. The biologically relevant oligomer is a trimer of heterotrimers that is created by a three-fold symmetry axis in the crystal. The previously determined Rrp41/Rrp42 crystal structure (see earlier) was used as a molecular replacement model for initial phasing in the program PHASER. The maps based on model phases from the resulting solution showed poor and disconnected electron density for Rrp4. Attempts to place the previously determined structure of *A. fulgidus* Rrp4 (Buttner *et al.*, 2005) did not yield a promising solution. However, solutions could be obtained in a separate search with the N-terminal region of Rrp4 and with the C-terminal S1/KH region. With hindsight, the reason why the search with the full-length Rrp4 protein failed is because of a conformational change characterized by an 8 Å domain movement in Rrp4 of the *A. fulgidus* exosome relative to the *S. solfataricus* exosome. The resulting map showed clear electron density for Rrp4 and allowed model building of this subunit in the program COOT. The structure was refined with REFMAC. The Rrp4 subunit displays weaker electron density than the hexameric Rrp41/Rrp42 core even in the fully refined structure. Analysis of the relative temperature factors in the structure of Rrp4/Rrp41/Rrp42 showed that the Rrp4 protein has an average temperature factor twice as high as that of the Rrp41/Rrp42 core, suggesting it has more pronounced flexibility.

The overall structure of the *S. solfataricus* exosome displays a barrel-like architecture with the Rrp4 protein forming a trimeric cap atop of the Rrp41/Rrp42 hexameric ring. The central channel of the Rrp4 cap forms a continuous pore with the channel of the hexamer. This architecture is consistent with that observed for the *A. fulgidus* and *P. abyssi* exosomes (Buttner *et al.*, 2005, Navarro *et al.*, 2008), for the human exosome core (Liu *et al.*, 2006), and for bacterial PNPase (Symmons *et al.*, 2000) and seems to be evolutionarily conserved (Lorentzen and Conti, 2006).

4. RNA Binding Sites in the Archaeal Exosome

The structures of the archaeal exosome revealed the overall architecture but did not show how the complex recruits and degrades RNA. This section describes the methods we used to visualize RNA substrates bound to the exosome. To achieve binding without degradation, we diffused RNA molecules either into crystals of the wild-type exosome in the absence of phosphate or in crystals of a catalytically inactive mutant. To assess the access route of the RNA substrates reaching the exosome active site, we character-ized the substrate requirement biochemically, and we made use of modified RNA to unambiguously model the substrate in the electron density.

4.1. Structures of oligoribonucleotides bound at the archaeal exosome active site

To identify the mode of RNA binding at the active site of the hexameric RNase PH–like ring, crystals of Rrp41/Rrp42 were incubated in solutions containing short 5 to 10 nucleotide (nt) RNA molecules (Lorentzen and Conti, 2005). Because of the phosphorolytic nature of RNA degradation by the archaeal exosome (Lorentzen *et al.*, 2005), these experiments were initially carried out with the native protein complex with the buffer solution devoid of inorganic phosphate. Equilibrated drops containing Rrp41/Rrp42 crystals with an approximate volume of 1 μl were mixed with 0.2 μl of the following solutions adjusted to a pH value of 7.0 with MES buffer: 10 mM of either A_5, U_8, or A_{10} RNA (Dhamacon, 2′-deprotected and desalted but not PAGE-purified) or 100 mM of ADP. Soaking time before flash cooling was 5 min for the ADP and A_5 molecules and 10 min for the larger U_8 and A_{10} molecules. Despite being relative short, the soaking time allowed the RNA ligands to diffuse into the crystals. This is probably because of the large solvent channels in the Rrp41/Rrp42 crystals that have 60% solvent content allowing the ligands to readily diffuse through the crystal.

As typical for soaked crystals, the diffraction for the complexes was of somewhat lower quality than for the apo structures. The visualization of RNA substrates and the ADP was, however, clear in the 2.8 to 3.3 Å resolution data sets. Electron density maps were calculated by use of model phases after rigid-body refinement and refinement of the atomic coordinates and B factors in REFMAC. The maps showed the well-ordered binding of 4 nt of RNA at the active site. The nucleotides interact mainly by means of phosphate-backbone interactions with conserved arginine residues located in a binding groove between the Rrp41 and Rrp42 subunits (Lorentzen and Conti, 2005). Somewhat surprisingly, the high salt concentration of the crystallization condition (2.0 to 2.4 M sodium malonate) did not interfere with the electrostatic RNA-protein interactions.

4.2. RNA degradation experiments to probe the RNA access route

The structures of the wild-type RNase PH core bound to short oligoribonu-cleotides provided a snapshot of the catalytic site and suggested that recruitment of substrates to the active site occurs by means of the central channel of the exosome complex. However, these structures did not reveal from which side of the central channel substrate recruitment occurs. To probe this biochemically, we designed RNA substrates on the basis of the geometry of the exosome structure. The design was based on the knowledge that the archaeal exosome stalls on encountering RNA substrates with stable secondary structure. The structure of the Rrp41-Rrp42 complex shows a barrel-like architecture with a central channel that is open at both ends. The active site of the complex is located at the end of a solvent-shielded grove, 4 nt in length and perpendicular to the central channel (Fig. 20.3). One end of the central channel of the barrel-like structure has a wide opening (25 Å), whereas the other end has more restricted access with an opening of only 10 Å. From the crystal structure it is possible to measure distances and predict that an RNA substrate approaching from the 25 Å wide opening would require at least 4 nt of single-stranded overhang to reach the site of catalysis. However, if the substrate approached from the other 10 Å wide pore, it would require at least 9 nt of single-stranded RNA. To experimentally test these two possibilities, degradation experiments were carried out with substrates with a stable 8 base-pair stem loop at the 5′-end and 5, 10, 15, or 25 nucleotides of single-stranded poly(A) overhang at the 3′-end: 5′-CCCCCCCCGAAAG-GGGGGGGGAAAAA-3′.

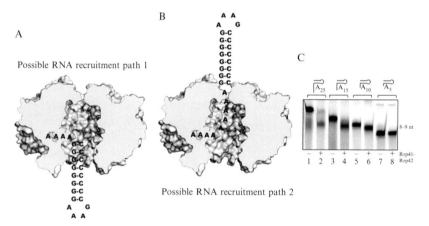

Figure 20.3 Surface representation of the Rrp41/Rrp42 exosome structure sliced in half showing the central channel of the complex. The architecture suggests that there are two different possible paths for RNA recruitment. (A) RNA can be recruited from the 25-Å wide end in which case only 4 nt of single-stranded RNA overhang is required to reach the active site. (B) Alternately, RNA substrates can be recruited from the narrower 10 Å pore located farther from the active site. In this case 9 nt of single-stranded RNA is required to reach the active site. (C) RNA degradation experiment with stem loop–containing RNA molecules having different lengths of single-stranded overhang. The experiment shows a final product with 8–9 nt of single-stranded overhang, indicating that substrate recruitment by the archaeal exosome occurs from the more restricted 10 Å wide pore (Lorentzen and Conti, 2006).

The RNA degradation experiment was carried out as follows: 5 μg (29 pmol) of Rrp41/Rrp42 wild-type complex was incubated with 5 μg (1.5 nmol) RNA in 40 μl total volume of buffer containing 20 mM Tris, pH 7.6, 60 mM KCl, 10 mM MgCl$_2$, 0.1 mM EDTA, 10% glycerol, 1 U/μl RNasin, 10 mM Na$_2$HPO$_4$, and 2 mM DTT at 70 °C for 20 min. The reaction was stopped by adding 1 volume of loading buffer containing 8 M urea, 10% glycerol, 1× TBE (Tris-borate-EDTA) and bromophenol blue, and placing the tubes on ice. Reaction products were resolved on a 20% polyacrylamide gel containing 8 M urea and 1× TBE. The gel was prerun for 15 min in 1× TBE buffer and unpolymerized acrylamide was removed from the wells before loading of the sample. Products of the RNA degradation experiments were visualized directly by staining for 5 min in 0.1% toluidine followed by destaining in water. One advantage of carrying out the assay at high temperature, at which the thermostable exosome complex is very active, is that any trace contaminants of RNases are likely to be heat inactivated.

The results from these experiments showed that the archaeal Rrp41–Rrp42 complex requires at least 8 to 9 nt of single-stranded RNA for activity (see Fig. 20.3). The smaller RNA molecule with only 5 nt of single-stranded overhang was not degraded at all, whereas the RNA molecules with longer tails (10 to 25 nt) were all partly degraded giving the same product with 8 to 9 nt of overhang. This is consistent with a model where RNA substrates are recruited by way of the more restricted 10 Å opening and channeled through the central pore to the active sites.

4.3. Structures with 5-iodo-uridine RNA substrates to visualize the RNA access route

To obtain the structure of the entire exosome with a substrate bound both at the active site and at the access route, we diffused Rrp4/Rrp41/Rrp42 crystals with larger RNA molecules containing stable secondary structure (Lorentzen *et al.*, 2007). With the knowledge from the structural and biochemical experiments described in the previous sections, we designed RNA molecules to probe for additional binding sites. Because our biochemical experiments showed that the archaeal exosome recruits RNA through a pore 10 Å wide at its narrowest point, we expected an RNA molecule with a stable stem loop to stall at and reveal the exact position of substrate entrance. Soaking was done with the following RNA molecules at a final concentration of 0.3 mM added to the mother liquor:

5′-CCCCCCCCGAAAGGGGGGGGAAAAAAAAAAAAAAA-3′
5′-CCCCCGAAAGGGGGU$_I$U$_I$U$_I$U$_I$U$_I$U$_I$U$_I$U$_I$U$_I$U$_I$U$_I$U$_I$U$_I$U$_I$U$_I$-3′

The molecules contain a stable GC stem loop at the 5′-end followed by 15 nucleotides of single-stranded RNA tail. The rationale behind the use of such a substrate is that although the tail can enter the exosome complex and bind at the active site, the stem-loop structure should stall at the entrance site. For the second substrate, U_I indicates that the uracil base has been chemically modified with iodide (5-iodo-uridine, commercially available at Dharmacon). The idea behind the use of such a modified substrate is that iodide gives a strong anomalous signal at longer X-ray wavelengths and can thus be used to identify even weak RNA binding sites. The relatively low concentration of 0.3 mM of the RNA molecules was used because of their low solubility in the mother liquor containing a high concentration of PEG. Short soaking times of 10 to 60 min resulted in empty RNA binding sites. This was probably because of the larger size and lower concentration of the RNA substrates, the high viscosity of the 40% PEG 400 condition, as well as the smaller size of the solvent channels in the Rrp4/Rrp41/Rrp42 crystals compared with the Rrp41/Rrp42 crystals. To overcome these problems, a longer soaking time of 16 h was used.

To ensure stability of the RNA during the long soaking time, the experiments were carried out with a catalytically dead exosome complex containing a point mutation (D182A) in the Rrp41 subunit (Lorentzen et al., 2005). Crystals soaked with RNA molecule 1 diffracted at best to a resolution of 2.3 Å. The resulting electron density maps revealed an additional RNA-binding site located at of the entrance of central channel at the side where the Rrp4 trimeric cap associates with the Rrp41/Rrp42 hexamer. One additional nucleotide with electron density of sufficient quality to be modeled was found at the neck of the complex where the channel is the most constricted with a diameter of 10 Å, only wide enough to accommodate one RNA chain at a time. The archaeal exosome has three-fold symmetry, and the crystallographic structure, which is an average of all the exosome complexes in the crystal, thus has three identical binding sites, each with an occupancy of one third. In addition to the one nucleotide identified in the channel, four nucleotides of RNA could be traced at the active site. The modeled RNA was given occupancies of either one third or 1 in separate refinements in REFMAC. The temperature factors of this nucleotide refined to values similar to those of the surrounding protein residues only with an occupancy of one third, confirming that only one RNA chain can enter the archaeal exosome at any given time. These results are in agreement with isothermal titration calorimetry (ITC) experiments that also showed a stoichiometry of 1:3 (Oddone et al., 2007). The relatively good resolution of the data allowed us to model the RNA despite the low occupancy and resulting weaker density. To confirm the binding site, anomalous data were collected on RNA molecule 2 with a long (1.55 Å) X-ray wavelength where iodide has a considerable anomalous signal. Although the diffraction data extended to 2.5 Å resolution, the anomalous signal was detectable only up to 4 Å. Anomalous maps were calculated from the anomalous differences in the resolution range of 50 to 4 Å with phases calculated from the Rrp4/Rrp41/Rrp42 model after rigid body refinement and restrained refinement of atomic coordinates and temperature factors against all the data extending to 2.5 Å resolution. The resulting maps clearly allowed the identification of the iodide-labeled uracil base confirming the binding site found with substrate 1. The data further suggest that uracil and adenine bases bind in similar ways in the central channel of the archaeal exosome.

The results described in this section demonstrate the synergy of biochemical and structural approaches. The results of the RNA degradation experiments described in Section 4.2 could not have been mechanistically interpreted in the absence of the 3D structure of the exosome. In turn, the information from the RNA degradation experiments was used in the rational design of RNA substrates for further structural studies.

ACKNOWLEDGMENTS

We thank Doris Lindner for excellent technical assistance and the staff at SLS for excellent assistance with data collection. We also thank Atlanta Cook for critical reading of the manuscript and Peter Brick, Elena Evguenieva-Hackenberg, and Pamela Walter for many valuable discussions.

REFERENCES

Allmang, C., Kufel, J., Chanfreau, G., Mitchell, P., Petfalski, E., and Tollervey, D. (1999). Functions of the exosome in rRNA, snoRNA and snRNA synthesis. *EMBO J.* **18,** 5399–5410.

Anderson, J. S., and Parker, R. P. (1998). The 3′ to 5′-degradation of yeast mRNAs is a general mechanism for mRNA turnover that requires the SKI2 DEVH box protein and 3′ to 5′-exonucleases of the exosome complex. *EMBO J.* **17,** 1497–1506.

Araki, Y., Takahashi, S., Kobayashi, T., Kajiho, H., Hoshino, S., and Katada, T. (2001). Ski7p G protein interacts with the exosome and the Ski complex for 3′ to 5′-mRNA decay in yeast. *EMBO J.* **20,** 4684–4693.

Bousquet-Antonelli, C., Presutti, C., and Tollervey, D. (2000). Identification of a regulated pathway for nuclear pre-mRNA turnover. *Cell* **102,** 765–775.

Briggs, M. W., Burkard, K. T., and Butler, J. S. (1998). Rrp6p, the yeast homologue of the human PM-Scl 100-kDa autoantigen, is essential for efficient 5.8 S rRNA 3′-end formation. *J. Biol. Chem.* **273,** 13255–13263.

Buttner, K., Wenig, K., and Hopfner, K. P. (2005). Structural framework for the mechanism of archaeal exosomes in RNA processing. *Mol. Cell* **20,** 461–471.

Buttner, K., Wenig, K., and Hopfner, K. P. (2006). The exosome: A macromolecular cage for controlled RNA degradation. *Mol. Microbiol.* **61,** 1372–1379.

Carrington, J. C., and Dougherty, W. G. (1988). A viral cleavage site cassette: Identification of amino acid sequences required for tobacco etch virus polyprotein processing. *Proc. Natl. Acad. Sci. USA* **85,** 3391–3395.

Collaborative Computational Project, N. (1994). The CCP4 Suite: Programs for protein crystallography. *Acta Crystallogr. Sec. D* **50,** 760–763.

Conti, E., and Izaurralde, E. (2005). Nonsense-mediated mRNA decay: Molecular insights and mechanistic variations across species. *Curr. Opin. Cell Biol.* **17,** 316–325.

Dziembowski, A., Lorentzen, E., Conti, E., and Seraphin, B. (2007). A single subunit, Dis3, is essentially responsible for yeast exosome core activity. *Nat. Struct. Mol. Biol.* **14,** 15–22.

Emsley, P., and Cowtan, K. (2004). Coot: Model-building tools for molecular graphics. *Acta Crystallogr. D. Biol. Crystallogr.* **60,** 2126–2132.

Evguenieva-Hackenberg, E., Walter, P., Hochleitner, E., Lottspeich, F., and Klug, G. (2003). An exosome-like complex in *Sulfolobus solfataricus. EMBO Rep.* **4,** 889–893.

Frazao, C., McVey, C. E., Amblar, M., Barbas, A., Vonrhein, C., Arraiano, C. M., and Carrondo, M. A. (2006). Unravelling the dynamics of RNA degradation by ribonuclease II and its RNA-bound complex. *Nature* **443,** 110–114.

Fribourg, S., Romier, C., Werten, S., Gangloff, Y. G., Poterszman, A., and Moras, D. (2001). Dissecting the interaction network of multiprotein complexes by pairwise coexpression of subunits in *E. coli. J. Mol. Biol.* **306,** 363–373.

Harlow, L. S., Kadziola, A., Jensen, K. F., and Larsen, S. (2004). Crystal structure of the phosphorolytic exoribonuclease RNase PH from *Bacillus subtilis* and implications for its quaternary structure and tRNA binding. *Protein Sci.* **13,** 668–677.

Houseley, J., LaCava, J., and Tollervey, D. (2006). RNA-quality control by the exosome. *Nat. Rev. Mol. Cell. Biol.* **7,** 529–539.

Jones, T. A., Zou, J. Y., Cowan, S. W., and Kjeldgaard, S. W. (1991). Improved methods for building protein models in electron density maps and the location of errors in these models. *Acta Crystallogr.* **A47**(Pt 2), 110–119.

Kabsch, W. (1993). Automatic processing of rotation diffraction data from crystals of initially unknown symmetry and cell constants. *J. Appl. Crystallogr.* **26,** 795–800.

Koonin, E. V., Wolf, Y. I., and Aravind, L. (2001). Prediction of the archaeal exosome and its connections with the proteasome and the translation and transcription machineries by a comparative-genomic approach. *Genome Res.* **11,** 240–252.

LaCava, J., Houseley, J., Saveanu, C., Petfalski, E., Thompson, E., Jacquier, A., and Tollervey, D. (2005). RNA degradation by the exosome is promoted by a nuclear polyadenylation complex. *Cell* **121,** 713–724.

Liu, Q., Greimann, J. C., and Lima, C. D. (2006). Reconstitution, activities, and structure of the eukaryotic RNA exosome. *Cell* **127,** 1223–1237.

Lorentzen, E., Basquin, J., Tomecki, R., Dziembwski, A., and Conti, E. (2008). Structure of the active subunit of the yeast exosome core, Rrp44: diverse modes of substrate recruitment in the RNase II nuclease family'. *Mol. Cell* **29,** 717–728.

Lorentzen, E., Dziembowski, A., Lindner, D., Seraphin, B., and Conti, E. (2007). RNA channelling by the archaeal exosome. *EMBO Rep.* **8,** 470–476.

Lorentzen, E., and Conti, E. (2006). The exosome and the proteasome: Nano-compartments for degradation. *Cell* **125,** 651–654.

Lorentzen, E., and Conti, E. (2005). Structural basis of 3′-end RNA recognition and exoribonucleolytic cleavage by an exosome RNase PH core. *Mol. Cell* **20,** 473–481.

Lorentzen, E., Walter, P., Fribourg, S., Evguenieva-Hackenberg, E., Klug, G., and Conti, E. (2005). The archaeal exosome core is a hexameric ring structure with three catalytic subunits. *Nat. Struct. Mol. Biol.* **12,** 575–581.

Mitchell, P., Petfalski, E., Shevchenko, A., Mann, M., and Tollervey, D. (1997). The exosome: A conserved eukaryotic RNA processing complex containing multiple 3′ to 5′-exoribonucleases. *Cell* **91,** 457–466.

Murshudov, G. N., Vagin, A. A., and Dodson, E. J. (1997). Refinement macromolecular structures by the maximum-likelihood method. *Acta Cryst* D**53,** 240–255.

Navarro, M. V. A. S., Oliveira, C. C., Zanchin, N. I. T., and Guimaraes, B. G. (2008). Insights into the mechanism of progressive RNA degradation by the archaeal exosome. *J. Biol. Chem.* **283,** 14120–14131.

Oddone, A., Lorentzen, E., Basquin, J., Gasch, A., Rybin, V., Conti, E., and Sattler, M. (2007). Structural and biochemical characterization of the yeast exosome component Rrp40. *EMBO Rep.* **8,** 63–69.

Portnoy, V., Evguenieva-Hackenberg, E., Klein, F., Walter, P., Lorentzen, E., Klug, G., and Schuster, G. (2005). RNA polyadenylation in Archaea: not observed in Haloferax while the exosome polynucleotidylates RNA in Sulfolobus. *EMBO Rep.* **6,** 1188–1193.

Romier, C., Ben Jelloul, M., Albeck, S., Buchwald, G., Busso, D., Celie, P. H., Christodoulou, E., De Marco, V., van Gerwen, S., Knipscheer, P., Lebbink, J. H., Notenboom, V., *et al.* (2006). Co-expression of protein complexes in prokaryotic and eukaryotic hosts: Experimental procedures, database tracking and case studies. *Acta Crystallogr. D. Biol. Crystallogr.* **62,** 1232–1242.

Storoni, L. C., McCoy, A. J., and Read, R. J. (2004). Likelihood-enhanced fast rotation functions. *Acta Crystallogr. D. Biol. Crystallogr.* **60,** 432–438.

Symmons, M. F., Jones, G. H., and Luisi, B. F. (2000). A duplicated fold is the structural basis for polynucleotide phosphorylase catalytic activity, processivity, and regulation. *Structure Fold Des.* **8,** 1215–1226.

van Hoof, A., Frischmeyer, P. A., Dietz, H. C., and Parker, R. (2002). Exosome-mediated recognition and degradation of mRNAs lacking a termination codon. *Science* **295,** 2262–2264.

van Hoof, A., Lennertz, P., and Parker, R. (2000). Yeast exosome mutants accumulate 3'-extended polyadenylated forms of U4 small nuclear RNA and small nucleolar RNAs. *Mol. Cell Biol.* **20,** 441–452.

Vanacova, S., and Stefl, R. (2007). The exosome and RNA quality control in the nucleus. *EMBO Rep.* **8,** 651–657.

Vanacova, S., Wolf, J., Martin, G., Blank, D., Dettwiler, S., Friedlein, A., Langen, H., Keith, G., and Keller, W. (2005). A new yeast poly(A) polymerase complex involved in RNA quality control. *PLoS Biol.* **3,** e189.

Walter, P., Klein, F., Lorentzen, E., Klug, G., and Evguenieva-Hackenberg, E. (2006). Characterization of native and reconstituted exosome complexes from the hypothermophilic archaeon *Sulfolobus solfataricus*. *Mol. Microbiol.* **62,** 1076–1089.

Wang, L., Lewis, M. S., and Johnson, A. W. (2005). Domain interactions within the Ski2/3/8 complex and between the Ski complex and Ski7p. *RNA* **11,** 1291–1302.

Wyers, F., Rougemaille, M., Badis, G., Rousselle, J. C., Dufour, M. E., Boulay, J., Regnault, B., Devaux, F., Namane, A., Seraphin, B., Libri, D., and Jacquier, A. (2005). Cryptic pol II transcripts are degraded by a nuclear quality control pathway involving a new poly(A) polymerase. *Cell* **121,** 7.

Zuo, Y., Vincent, H. A., Zhang, J., Wang, Y., Duetscher, M. P., and Malhotra, A. (2006). Structural basis for processivity and single-stranded specificity of RNase II. *Mol. Cell* **24,** 149–156.

ORGANELLES

POLYADENYLATION-MEDIATED RNA DEGRADATION IN PLANT MITOCHONDRIA

Sarah Holec, Heike Lange, André Dietrich, *and* Dominique Gagliardi

Contents

Abstract

In plant mitochondria, polyadenylation-mediated RNA degradation is involved in several key aspects of genome expression, including RNA maturation, RNA turnover, and RNA surveillance. We describe here a combination of *in vivo*, *in vitro*, and *in organello* methods that have been developed or optimized to characterize this RNA degradation pathway. These approaches include several PCR-based methods designed to identify polyadenylated RNA substrates, as well as *in vitro* and *in organello* systems, to study functional aspects of the RNA degradation processes. Taken together, identification of RNA substrates combined with information from degradation assays are invaluable tools to dissect the mechanisms and roles of RNA degradation in plant mitochondrial genome expression.

Institut de Biologie Moléculaire des Plantes, Centre National de la Recherche Scientifique, Unité Propre de Recherche 2357, Conventionné avec l'Université Louis Pasteur, Strasbourg, France

Methods in Enzymology, Volume 447
ISSN 0076-6879, DOI: 10.1016/S0076-6879(08)02221-0

1. INTRODUCTION

The destabilizing role of polyadenylation was discovered in prokaryotes and subsequently shown in chloroplasts (Bollenbach *et al.*, 2004; Dreyfus and Regnier, 2002). Polyadenylation also triggers degradation of noncoding transcripts in the nucleus of the yeasts *Saccharomyces cerevisiae* and *Schizosaccharomyces pombe* and likely in higher eukaryotes such as human and plants (Doma and Parker, 2007; Houseley *et al.*, 2006; Slomovic *et al.*, 2006; Vanacova and Stefl, 2007; West *et al.*, 2006). In mitochondria, the roles of polyadenylation are quite varied in different organisms: polyadenylation is absent from mitochondria of *S. cerevisiae* and *S. pombe* (Schafer, 2005); in both human and *Trypanosoma* mitochondria, polyadenylation is constitutive and plays complex roles in stabilizing RNAs or possibly promoting translation on one side, and in inducing RNA degradation on the other side; in plant mitochondria, polyadenylation promotes RNA degradation (Gagliardi *et al.*, 2004).

Plant mitochondrial polyadenylated RNAs are degraded by a polynucleotide phosphorylase (PNPase). Actually, plant genomes encode two PNPases that are targeted to chloroplasts and mitochondria, respectively (Hayes *et al.*, 1996; Perrin *et al.*, 2004b). Roles of both PNPases have been investigated *in vivo* by reverse genetic studies. Upon downregulation of the chloroplast PNPase (cpPNPase), mRNA levels are unaffected, which suggests that cpPNPase could be insufficient to mediate transcript degradation *in vivo* (Walter *et al.*, 2002). Furthermore, cpPNPase is not essential for cell viability. However, cpPNPase is required for maturing the 3′-ends of mRNA and 23S rRNA; However, incompletely processed mRNAs are apparently translated into functional proteins. Similarly, 23S rRNA with incompletely processed 3′-ends can still become incorporated into ribosomes without severely impairing translation (Walter *et al.*, 2002). In contrast to the cpPNPase gene, the gene At5g14580 encoding mitochondrial PNPase (mtPNPase) is essential in *Arabidopsis thaliana*. Downregulation of mtPNPase results in the accumulation of mRNA and rRNA precursors, rRNA fragments, tRNA and rRNA maturation by-products, antisense RNAs, and intergenic transcripts (Holec *et al.*, 2006; Perrin *et al.*, 2004a,b).

We describe here a combination of techniques to study polyadenylation-mediated RNA degradation in plant mitochondria. These techniques include PCR-based approaches that are necessary for the molecular characterization of polyadenylated RNA substrates. Identification of these substrates constitutes an essential step toward understanding RNA degradation in plant mitochondria. We also describe an *in vitro* RNA degradation system that reproduces the preferential degradation of polyadenylated RNA in a mitochondrial lysate and, therefore, facilitates the biochemical characterization of the degradation process. Last, we detail a method for *in organello* expression of engineered genes after DNA uptake into plant mitochondria (Koulintchenko *et al.*, 2003). In the absence of the long-awaited method for *in vivo* transformation of plant mitochondria, *in organello* expression systems are precious tools to study the degradation of defective or modified transcripts.

2. IDENTIFICATION OF POLYADENYLATED SUBSTRATES DESTINED FOR RNA DEGRADATION

2.1. Mutant plants affected in RNA degradation

Because poly(A) tails trigger degradation of mitochondrial RNA, polyadenylated RNAs are found at extremely low levels in wild-type plant mitochondria. The low amount of polyadenylated RNAs precludes their analysis by Northern blotting and results in unacceptable levels of background for some of the PCR-based methods (see later). Therefore, it is advisable to interfere with the degradation pathway to accumulate otherwise unstable polyadenylated RNA. Specific inhibition of mitochondrial RNA degradation by chemicals would be interesting, because this strategy could be applied to any plant investigated. However, to our knowledge such a protocol has not been described yet. Inhibition of the degradation process must, therefore, be achieved by genetic means. This approach is facilitated by the relative simplicity of the plant mitochondrial degradation process: the mitochondrial PNPase is required to degrade polyadenylated RNA. In other words, the absence of mtPNPase cannot be compensated by alternate RNases for the substrates investigated to date.

Because mtPNPase is encoded by an essential gene in *A. thaliana*, stable insertion mutants are not viable (Perrin *et al.*, 2004b). Therefore, downregulation of mtPNPase must be achieved by posttranscriptional gene silencing. Whether PNPase is essential in all plant species is unknown, but an insertion mutant collection will anyway not be available for any plant species which implies that in most cases PNPase must be down-regulated by posttranscriptional gene silencing. Several methods have been developed to induce RNA interference in plants, including the expression of long hairpin RNAs, of engineered viral RNAs (VIGS, virus-induced gene silencing) or of artificial microRNAs (amiRNAs) (Small, 2007). The choice of method will depend on several parameters such as transformability of the plant species investigated or its susceptibility to infection by virus vectors. All these approaches require prior knowledge of the sequence to be down-regulated and ideally of other related sequences to avoid or at least monitor off-target gene extinction.

Although any of the aforementioned methods can be tested, we have achieved efficient silencing of the mtPNPase in *Arabidopsis* by cosuppression.

Cosuppression is a process, specific to plants, in which an endogenous gene is silenced by the overexpression of the gene in *trans*. We have observed cosuppression of mtPNPase by expressing PNPase fused at its C-terminus to different fusion tags. The presence of a C-terminal fusion tag is useful for two reasons. First, the fusion tag can include epitopes and is used to monitor the expression of the fusion protein. Second, a C-terminal fusion tag inactivates mtPNPase, thereby avoiding any RNA degradation by the ectopically expressed PNPase. Typically, we subclone the PNPase cDNA fused to the sequence encoding the C-terminal tag into a binary vector, such as pBinPlus (van Engelen *et al.*, 1995), containing the 35S CaMV promoter and terminator sequences. The construct is then transferred into *Agrobacterium tumefaciens*, strain GV3101 (pMP90), which is then used to transform *Arabidopsis* by the floral dip method (Bent, 2006). In independent experiments, we observed that high expression of the transgene results in cosuppression in up to several percent of the plants. Cosuppressed plants (referred to as PNP⁻) exhibit round, downward curled leaves, fail to set up new leaves, and do not resume growth once this initial phenotype is observed. Downregulation of mtPNPase can be assessed by testing the accumulation of some of its RNA substrates, such as the 5′-external transcribed sequence (ETS) of the 18S rRNA primary transcript (Perrin *et al.*, 2004a). The main disadvantage of the cosuppression strategy is that plants have to be characterized individually. However, it remains to date an efficient way to downregulate mtPNPase in *Arabidopsis*, as, for instance, silencing of mtPNPase expression by inducible miRNAs was not successful in our hands.

Polyadenylated RNA substrates of mtPNPase can accumulate to high levels in PNP⁻ plants (Holec *et al.*, 2006; Perrin *et al.*, 2004a,b). This accumulation facilitates their identification and allows their analysis by conventional methods such as Northern blotting, primer extension, and cleavage by RNase H.

2.2. Mapping polyadenylation sites by 3′-RACE

The identification of polyadenylation sites is a prerequisite to study the polyadenylation-mediated RNA degradation pathway in plant mitochondria. The easiest and quickest way to map polyadenylation sites is by use of 3′-RACE (rapid amplification of cDNA ends).

2.2.1. RNA extraction

RNA is classically extracted from plant tissues with one of the numerous versions of the protocol developed by Chomczynski and Sacchi (2006). Disruption of plant tissues in a solution of phenol and guanidine isothiocyanate ensures high-quality RNA. Both commercial (TRIZOL™ Invitrogen, Tri reagent™ MRC) and home-made (38% v/v phenol, 0.8 *M* guanidine isothiocyanate, 0.4 M ammonium isothiocyanate, 0.1 M sodium acetate, 5% w/v glycerol) RNA extraction solutions were used without any noticeable differences. For small-scale RNA extraction, plant tissues (50 mg) are introduced into screw-cap 2-ml tubes containing 500 μl of commercial or home-made RNA extraction solution and 1-mm-diameter glass beads (corresponding to a volume of approximately 150 μl). Tissues are disrupted by precession in a Precellys^{TM} grinder/homogenizer (2 times 30 sec at 5500 rpm). Larger amounts of plant tissues are ground with a mortar and pestle in liquid nitrogen, and the resulting frozen powder is added to RNA extraction solutions (approximately 1 g of tissue for 10 ml). After a 5-min incubation at room temperature, samples are spun at 20,000g for 5 min. Chloroform (0.2 volume) is added to the supernatant, and samples are mixed with vortex, incubated at room temperature for 2 to 3 min, and centrifuged at 16,000g for 5 min at 4 °C. After addition of isopropanol (0.4 volume) and a 10-min incubation at room temperature, samples are centrifuged at 16,000g for 5 min at 4 °C, washed with 70% ethanol, briefly air-dried, and resuspended in water. Concentration and purity are determined by spectrophotometry. An additional phenol-chloroform extraction followed by ethanol precipitation is performed in case of unsatisfactory purity.

2.2.2. cDNA synthesis

A flowchart of the procedure for 3′-RACE is shown in Fig. 21.1. Typically, 10 μg of total RNA is incubated in the presence of 1 U of DNase I (Fermentas) for 1 h at 37 °C. This step eliminates possible residual DNA contamination. After phenol-chloroform extraction and ethanol precipitation, RNA is dissolved in 25 μl of water containing 4 pmol of oligo (dT)-anchor primer, heated at 65 °C for 3 min, and quenched on ice. The sequence of the oligo(dT)-anchor primer is 5′-GAATTCATGTC-GACGGTCTCA(T)$_{12}$-3′. To the sample are added 8 μl of 5× RT buffer (250 mM Tris-HCl, pH 8.3, 375 mM KCl, 15 mM MgCl$_2$), 2 μl 100 mM DTT, 2 μl 10 mM dNTP mix (i.e., each dNTP is at 10 mM), and 2 μl (80 U) RNase inhibitor (e.g., RNaseOUT^{TM}, Invitrogen). A 19-μl aliquot is taken and complemented with 200 U of reverse transcriptase (200 U/μl) (e.g., SuperScript II^{TM}, Invitrogen); the remaining 20 μl constitutes the minus RT control in subsequent PCR analysis. Both samples are incubated for 1 h at 42 °C. If needed, some reverse transcriptases (e.g., ImProm^{TM}, Promega) can operate at 50 °C, which can reduce mispriming of the oligo (dT)-anchor to A-rich regions.

2.2.3. PCR amplification

The cDNA samples are then used to map polyadenylation sites with PCR amplification and a primer corresponding to the anchor sequence (5′-GAATTCATGTCGACGGTCTCA-3′) in combination with a gene-specific primer (Fig. 21.1). The optimal amount of cDNA used in PCR

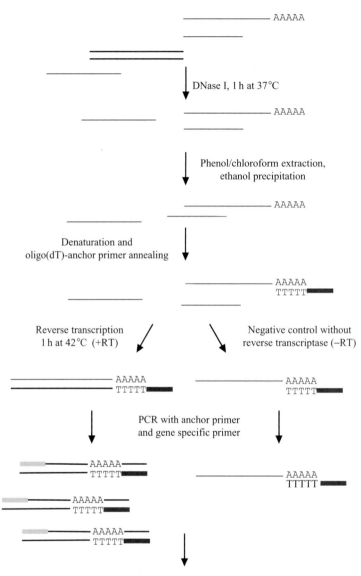

Figure 21.1 Mapping polyadenylation sites with 3′-RACE. RNA and DNA strands are shown as thin and bold lines, respectively. The anchor and gene-specific primers are represented by dark and light grey rectangles, respectively.

reactions has to be determined empirically, because it will depend on several parameters such as the level of target in the cDNA population or the melting temperature of the gene-specific oligonucleotide. Typically, we test 1/20 and 1/200 of the cDNA reaction in a 20-μl PCR reaction. We usually start with the following amplification parameters: 94 °C for 30 sec, 30 cycles of 94 °C for 30 sec, 50 °C for 30 sec, 72 °C for 30 sec, and a final step of 72 °C for 1 min. The annealing temperature is optimized if needed. PCR products are analyzed with agarose gel electrophoresis and cloned in a T/A cloning vector before sequence analysis.

2.2.4. Comments on 3′-RACE

In the most favorable cases, polyadenylation sites are unambiguous, because the poly(A) tails are found at sites devoid of As and/or the poly(A) tails are larger than the oligo(dT)$_{12}$. However, in other cases, results obtained in mapping polyadenylation sites with 3′-RACE have to be interpreted with caution, because not all PCR products will correspond to true polyadenylated transcripts. In fact, false-positive results are frequently caused by annealing of the oligo(dT) primer in A-rich regions encoded by the genome. Therefore, the localization of the gene-specific primer has to be carefully considered; it should not be upstream of A-rich regions where the oligo(dT)-anchor primer could have annealed during cDNA synthesis. Up to 3 to 4 consecutive As can be tolerated, but there is no general rule, because the success of the experiment relies on the relative ratio between true polyadenylated RNA versus templates that can give rise to artifactual amplification. It is worth trying to eliminate false-positive results caused by artifactual priming of oligo(dT) by one of the two following procedures. First, downregulation of mtPNPase should result in the accumulation of the polyadenylated transcripts studied. For instance, while investigating polyadenylation sites in the 18S-5S rRNA cotranscript, only a few percent of clones corresponded to true polyadenylated RNA in wild-type plants, because most amplicons were due to oligo(dT) priming an A-rich region that derived from the body of the transcript. In contrast, this artifact was negligible when PNP$^-$ plants were used because of the significant accumulation of truly polyadenylated transcripts from the rRNA locus (Perrin et al., 2004a). Second, the A-rich region can be eliminated before cDNA synthesis by RNase H cleavage with a reverse primer annealing upstream of the A-rich region. Such a strategy was successfully used to map low-abundant poly(A) sites upstream of the constitutive poly(A) tails of human mitochondrial transcripts (Slomovic et al., 2005).

If needed, sensitivity of the 3′-RACE method can be increased by running semi-nested PCR (i.e., two successive rounds of amplification that use consecutive gene-specific primers in combination with the anchor primer) or by use of mitochondrial RNA instead of total RNA when purification of mitochondria is possible from the plant material investigated.

2.3. Analysis of polyadenylation sites by circular RT-PCR

The basic principle of circular RT-PCR (cRT-PCR) is that the RNAs are circularized (or ligated in concatemers) before cDNA synthesis and PCR amplification of the region spanning joined 5′- and 3′-ends (Fig. 21.2). Sequence analysis reveals the precise 5′- and 3′-junction and hence allows mapping of both extremities at the nucleotide level. More importantly, the presence of nonencoded nucleotides such as poly(A) tails can be determined.

2.3.1. RNA circularization

In a typical experiment, 1 to 10 μg of total RNA is incubated for 1 h at 37 °C in a total volume of 20 μl containing 50 mM Tris-HCl, pH 7.8, 10 mM MgCl$_2$, 1 mM ATP, 10 mM DTT, 40 U RNase inhibitor, 1 U DNase I, and 40 U T4 RNA ligase (e.g., from NEB). This step allows the simultaneous elimination of possible residual DNA contamination and the circularization or intermolecular ligation of RNA. T4 RNA ligase ligates a 3′-hydroxyl extremity to a 5′-monophosphoryl extremity. Therefore, treatment of the RNA sample with tobacco acid pyrophosphatase (Epicentre^{TM} Biotechnologies) according to the manufacturer's instructions is recommended if primary transcripts, which harbor triphosphoryl 5′-extremities, are investigated. However, these triphosphates are relatively unstable, and a proportion of primary transcripts will have monophosphate 5′-ends anyway.

2.3.2. cDNA synthesis and PCR amplification

After phenol-chloroform extraction and ethanol precipitation, cDNA synthesis is performed as previously described for 3′-RACE, except that 2 pmol of a gene-specific reverse primer is used instead of oligo(dT)-anchor oligonucleotide. The gene-specific reverse primer should preferably anneal approximately 200 to 500 nt downstream of the 5′-extremity (primer 1 in Fig. 21.2). The region spanning the joined ends is then amplified with PCR and two additional gene-specific primers; the forward primer must anneal upstream of the 3′-extremity (primer 2 in Fig. 21.2), and the reverse primer must anneal between the primer used for cDNA synthesis and the 5′-extremity (primer 3 in Fig. 21.2). PCR products are analyzed with agarose gel electrophoresis and cloned in a T/A cloning vector before sequence analysis.

2.3.3. Comments on cRT-PCR

Circular RT-PCR is a technique complementary to 3′-RACE to map polyadenylation sites. cRT-PCR brings additional information, such as the ratio of polyadenylated versus nonpolyadenylated extremities for a given transcript. Furthermore, it allows a preliminary molecular characterization of the poly(A) tails in terms of size and composition, with the caveat that short tails will be preferentially observed. Because no oligo(dT) primer is used,

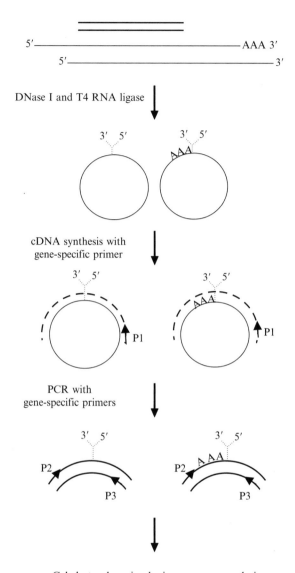

Figure 21.2 Mapping of polyadenylation sites with cRT-PCR. RNA is shown as thin lines, cDNA as broken bold lines, and other DNA as bold lines. Primer P1 is used for cDNA synthesis, and primers P2 and P3 are used for PCR amplification. Positions corresponding to initial 5'- and 3'-extremities of the RNA are indicated throughout by dotted lines.

cRT-PCR allows unambiguous demonstration of polyadenylation, even if it occurs in A-rich regions.

Multiple transcripts can be investigated simultaneously by use of different gene-specific reverse primers in one cDNA synthesis reaction but running separate PCR amplifications. Also, a second round of PCR with nested primers can be undertaken to improve specificity or sensitivity, although we have rarely seen the need to do so. The main disadvantage of this technique is that it requires that at least a few percent of a target RNA be polyadenylated to avoid sequencing hundreds of clones. Therefore, downregulating mtPNPase is strongly recommended.

2.4. A cDNA library to identify polyadenylated substrates

Both previous methods require the design of gene-specific primers, and, therefore, the analysis is restricted to a chosen target RNA. However, it might be interesting to determine RNA substrates of a degradation pathway without any preconceived idea about their identity. Obviously, analysis of mutants by genome-wide, strand-specific DNA tiling arrays is ideally suited for this purpose, but these arrays are costly and currently only available for *Arabidopsis*. We have, therefore, developed a protocol to identify RNA substrates of the polyadenylation-mediated RNA degradation pathway (Holec *et al.*, 2006). The main steps are indicated in Fig. 21.3 and consist in the size selection of RNA and a PCR-based method to construct a library of polyadenylated transcripts representing mitochondrial RNA degradation tags. In theory, this protocol can be used in any species, but it probably requires downregulating PNPase to accumulate sufficient levels of polyadenylated RNAs.

The size-selection step allows the use of total RNA (i.e., without purification of mitochondria) by limiting the contamination by polyadenylated nuclear mRNAs. Full-length cDNAs are then synthesized with the BD SMART^{TM} PCR cDNA synthesis kit (BD Biosciences Clontech). Conditions are slightly modified compared with the manufacturers' instructions and are detailed later. SMART cDNA synthesis uses the template switching ability of RNase H$^-$ point mutants of MMLV RT (Moloney murine leukemia virus reverse transcriptase). cDNA synthesis is initiated by an oligo(dT)-anchor, and when the RT reaches the 5′-extremity of the template, its terminal transferase activity adds a short deoxycytidine (dC) sequence. This sequence anneals to an oligo(G) present at the 3′-end of the BD SMART^{TM} II A oligonucleotide (BD Biosciences Clontech). The RT switches templates to transcribe the 5′-part of the BD SMART^{TM} II A oligonucleotide. The resulting cDNA has three important features. First, different sequence tags are incorporated both at the 3′- and 5′-ends (the oligo(dT)-anchor and SMART primers, respectively). Second, only full-length cDNAs will contain both tags. Third, only polyadenylated RNAs are converted to cDNA. The resulting single-stranded cDNA is then subjected to a few PCR cycles with the anchor and the SMART PCR primers, PCR products are cloned into a T/A cloning vector, and the resulting library is sequenced. All these steps are summarized in Fig. 21.3.

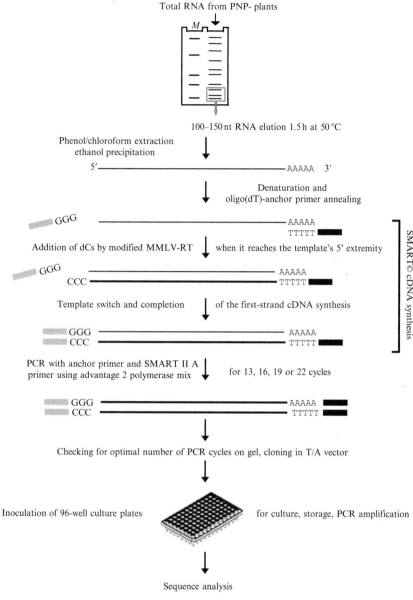

Figure 21.3 cDNA library construction with the BD SMART™ PCR cDNA synthesis kit (BD Biosciences Clontech). RNA and DNA strands are shown as thin and bold lines, respectively. The anchor and SMART primers are represented by dark and light grey rectangles, respectively. M, RNA size marker; MMLV-RT, Moloney murine leukemia virus reverse transcriptase.

2.4.1. Size selection of RNA

For the size-selection step, 5 μg of total RNA from *Arabidopsis* PNP$^-$ plants are dissolved in 8 μl of 60% v/v formamide, 15% w/v glycerol spiked with bromophenol blue and xylene cyanol, heated to 70 °C for 5 min, and electrophoresed in a 6% w/v acrylamide-bisacrylamide (ratio 19:1), 7 M urea, 1× Tris-borate-EDTA gel at 200 V until the bromophenol blue has migrated approximately 7 cm. An RNA size marker is loaded alongside the sample. The RNA size marker lane is cut, stained briefly with ethidium bromide, and positioned alongside the sample lane that has been rinsed three times with water. Rinsing the gel reduces contamination by larger RNAs that migrate between the gel and glass plates. The portion of gel containing RNA from 100 to 150 nt is cut into small pieces that are incubated for 1.5 h at 50 °C in 250 μl of 0.5 M NH$_4$CH$_3$CO$_2$, 0.1% w/v SDS, 1 mM EDTA. After phenol/chloroform extraction and ethanol precipitation, the recovered RNA is dissolved in 3 μl of water.

2.4.2. SMART cDNA library construction

To synthesize cDNA, 1 μl of BD SMART II^{TM} A primer (12 μM, BD Biosciences Clontech) and 1 μl of oligo(dT)-anchor primer (12 μM) are added to the size-selected RNA dissolved in 3 μl of water. The oligo(dT)-anchor primer is identical to the primer used in 3′-RACE experiments: 5′-GAATTCATGTCGACGGTCTCA(T)$_{12}$-3′. The sample is heated at 70 °C for 2 min, quenched on ice, and supplemented with 2 μl of 5× first-strand buffer (250 mM Tris-HCl, pH 8.3, 375 mM KCl, 30 mM MgCl$_2$), 1 μl of 20 mM DTT, 1 μl of 10 mM dNTP mix (i.e., each dNTP is at 10 mM), and 1 μl (100 U/μl) of MMLV RT with reduced RNaseH activity (see addendum PT3980-4 from Clontech). The sample is incubated for 1 h at 42 °C.

The cDNA library is subsequently amplified for a few PCR cycles as detailed below and using the Advantage^{TM} 2 polymerase mix (BD Biosciences Clontech) as follows: 2 μl of the cDNA synthesis reaction medium is added to 80 μl of water, 10 μl of 10× Advantage^{TM} 2 PCR buffer (400 mM Tricine-KOH, pH 8.7, 150 mM KCH$_3$CO$_2$, 35 mM Mg$_2$CH$_3$CO$_2$, 37.5 μg/ml BSA, 0.05% v/v Tween 20, 0.05% v/v Nonidet-P40), 2 μl of 10 mM dNTP mix (i.e., each dNTP is at 10 mM), 4 μl of 5′-PCR primer II A (12 μM, BD Biosciences Clontech), 4 μl of the anchor primer 5′-GAATT-CATGTCGACGGTCTCA-3′-(12 μM), and 2 μl of Advantage^{TM} 2 polymerase mix (BD Biosciences Clontech), which contains TITANIUM *Taq* DNA polymerase, a small amount of proofreading polymerase, and TaqStart antibody. The PCR reaction is divided into four aliquots of 25 μl each. All tubes are subjected to a initial denaturing step. Subsequently, the four aliquots are amplified for 13, 16, 19 and 22 PCR cycles, respectively, with the following conditions: 95 °C for 30 sec, 65 °C for 30 sec, 68 °C for 45 sec. For each tube, 5 to 10 μl are electrophoresed in a 1% w/v agarose gel in 0.5× Tris-acetate-EDTA buffer to determine the optimal number of PCR cycles. This optimal number is defined by the minimal number of PCR cycles that results in amplicons with a size between 100 to 150 bp. A size increase indicates PCR overcycling. PCR products amplified by the optimal number of cycles are cloned into a T/A cloning vector.

The entire transformed cDNA library is plated on agar plates and incubated overnight at 37 °C. Individual colonies are then grown in 96-well culture plates, each well containing 50 μl of LB medium supplemented with the appropriate antibiotic, and incubated overnight at 37 °C. Ordering the library in 96-well culture plates allows all downstream analyses to be performed with multichannel pipettes and 96-well plates (culture, storage, PCR amplification of inserts). For long-term storage of the library, 25 μl of the bacterial cultures are transferred to other 96-well culture plates, each well containing 25 μl of 50% w/v glycerol. Plates are kept at -80 °C for storage and further analysis.

Inserts of the cDNA library are amplified in 96-well plates with PCR and primers flanking the cDNA insertion sites of the vector. Randomly chosen reactions are run in agarose gels to assess the quality and average yield of the PCR step. Inserts amplified with PCR are then directly sequenced without further purification.

2.4.3. Comments on libraries of polyadenylated substrates

Complexity and coverage of this type of library are satisfactory with T/A cloning vectors, but conventional library cloning can be performed if needed. Because the cloning step is not directional in T/A cloning vectors and because poly(A) tails may result in sequencing problems, it is advisable to use the SMART primer (incorporated at the 5′-end of the insert) rather than one of the vector primers for sequencing. Mapping the 5′-extremity of the insert is also useful information that can be determined later after the insert is known from the first round of sequencing. Alternatively, RACE experiments can be performed, because the cDNA has different 5′- and 3′-sequence tags.

When using *Arabidopsis* PNP$^-$ plants, we reported that 70% of such a library correspond to mitochondrial clones (Holec *et al.*, 2006). Discrimination between truly polyadenylated RNAs and artifacts caused by priming of the oligo(dT) primer in A-rich regions is based on the same criteria as those for 3′-RACE experiments. In fact, we obtained very few experimentally introduced artifacts with this method compared with classical 3′-RACE PCR.

The 100- to 150-nt polyadenylated RNAs we identified in a prior study (Holec *et al.*, 2006) correspond to degradation fragments produced from much larger transcripts. Thus, the clones represent *bonafide* degradation intermediates, but they can also be considered as sequence tags corresponding to larger transcripts undergoing degradation. These tags are useful to identify the full-length RNA substrates of degradation. This protocol, therefore, leads to the identification of degradation substrates of any size, despite the initial size-selection step.

 ## 3. Tools to Investigate the Process of Polyadenylation-Mediated RNA Degradation

3.1. An *in vitro* system that mimics polyadenylation-mediated RNA degradation

The characterization of substrates of the polyadenylation-mediated RNA degradation pathway is essential, but a better understanding of this pathway also requires information about the degradation process *per se*. To this end, *in vitro* systems reproducing degradation or maturation of plant mitochondrial RNA have been developed (e.g., Bellaoui *et al.*, 1997; Dombrowski *et al.*, 1997; Gagliardi *et al.*, 2001; Kuhn *et al.*, 2001). These systems allow the study of the fate of mutagenized RNA substrates to determine features of the degradation process such as, for instance, the influence of secondary structures. The directionality of degradation can be studied by use of 5′- or 3′-end-labeled transcripts. Radiolabeled transcripts are also essential to detect degradation intermediates, useful indicators of the nature of the degradation process, or to identify the final products of degradation. Production of nucleoside diphosphate indicates a phosphorolytic pathway, whereas nucleoside monophosphate is produced by a hydrolytic pathway. Finally, *in vitro* degradation systems allow the use of inhibitors (e.g., chemical compounds or antibodies) to assess the involvement of a particular process or factor in the degradation pathway.

3.1.1. Preparation of mitochondrial lysate

A major problem in establishing an *in vitro* system to mimic mitochondrial RNA degradation is contamination by RNases from other cell compartments. This will influence the choice of starting plant material and the protocol used for purification of mitochondria. Also, large amounts of mitochondria are necessary to set up the system. Therefore, our favorite starting plant material is potato tubers, from which large amounts of pure mitochondria can be isolated. We have also occasionally used cauliflower heads. Numerous protocols for mitochondrion isolation have been published and will not be detailed here. Basically, potato and cauliflower mitochondria are isolated by differential centrifugation steps and purification in Percoll gradients (e.g., Koulintchenko *et al.*, 2003; Takenaka and Brennicke, 2007).

A flowchart of the procedure for mitochondrial lysate preparation is shown in Fig. 21.4. Both fresh and frozen mitochondria have been used indifferently. All steps must be performed at 4 °C, because mtPNPase is

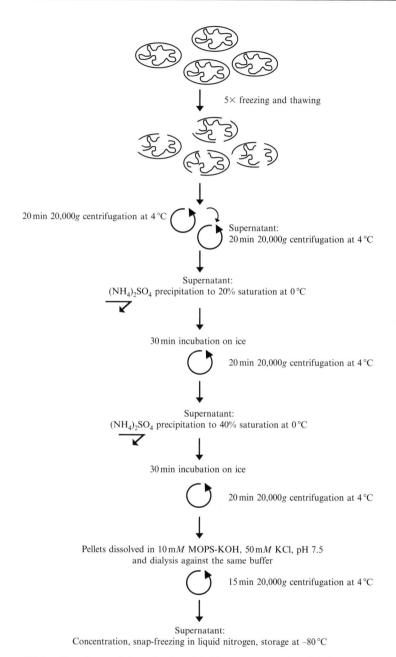

5× freezing and thawing

20 min 20,000g centrifugation at 4 °C

Supernatant:
20 min 20,000g centrifugation at 4 °C

Supernatant:
$(NH_4)_2SO_4$ precipitation to 20% saturation at 0 °C

30 min incubation on ice

20 min 20,000g centrifugation at 4 °C

Supernatant:
$(NH_4)_2SO_4$ precipitation to 40% saturation at 0 °C

30 min incubation on ice

20 min 20,000g centrifugation at 4 °C

Pellets dissolved in 10 mM MOPS-KOH, 50 mM KCl, pH 7.5
and dialysis against the same buffer

15 min 20,000g centrifugation at 4 °C

Supernatant:
Concentration, snap-freezing in liquid nitrogen, storage at –80 °C

Figure 21.4 Flowchart of the mitochondrial lysate preparation for the *in vitro* RNA degradation assay.

particularly prone to proteolytic digestion. Frozen mitochondria (30 mg of protein) are resuspended at a final concentration of 5 mg protein/ml in 50 mM MOPS, 250 mM KCl, 1 mM DTT, pH 7.5 complemented with Complete^{TM} protease inhibitors (Roche). Mitochondria are disrupted by five cycles of freezing and thawing and then centrifuged at 20,000g for 20 min at 4 °C. The supernatant is taken, and the centrifugation step is repeated. The lysate is then fractionated with ammonium sulfate precipitation. The activity responsible for the preferential degradation of polyadenylated RNA is present in the fraction corresponding to 20 to 40% ammonium sulfate saturation at 0 °C. This activity is also in the 40 to 60% fraction, but it is contaminated by an endonucleolytic activity.

The mitochondrial lysate is first brought to 20% saturation. After an incubation of 30 min on ice, the lysate is centrifuged at 20,000g for 20 min at 4 °C. The ammonium sulfate saturation of the resulting supernatant is then brought to 40%. After a further incubation of 30 min on ice, the lysate is again centrifuged at 20,000g for 20 min at 4 °C. The resulting pellets are dissolved in 2.5 ml of 10 mM MOPS-KOH, 50 mM KCl, pH 7.5, and dialyzed against the same buffer at 4 °C. The sample is then centrifuged at 20,000g for 15 min at 4 °C and concentrated to 500 μl with an ultrafiltration device (e.g., Ultrafree^{TM} 0.5 from Millipore). Under these conditions, the protein concentration should be between 3 and 5 μg/μl. Aliquots are snap frozen in liquid nitrogen and stored at −80 °C. Cycles of thawing/ freezing will destroy the activity.

3.1.2. *In vitro* degradation assays

To study the influence of polyadenylation on the degradation of an RNA substrate, both nonpolyadenylated and polyadenylated forms of the substrate are prepared. A convenient substrate size range is 100 to 150 nt. Nonpolyadenylated and polyadenylated substrates are transcribed *in vitro* with PCR products as DNA templates. The DNA template for the nonpolyadenylated substrate is obtained with PCR and a forward primer containing the T7 promoter sequence and a gene-specific sequence in combination with a gene-specific reverse primer. To generate the DNA template for the polyadenylated substrate, the gene-specific reverse primer must have between 15 to 20 dTs at its 5′-end. RNA substrates are synthesized *in vitro* either in the absence of radionucleotide, if subsequent labeling of 5′- or 3′-extremities is required, or in the presence of [α-^{32}P]UTP, to obtain uniformly labeled transcripts.

Typically, 2 fmol of uniformly labeled RNA substrate with a specific activity of approximately 10^4 cpm/fmol is used per 40 μl reaction in 10 mM MOPS-KOH, pH 7.5, 50 mM KCl, 1 mM DTT, 1 mM MgCl$_2$, 1 mM potassium phosphate, pH 7.5, 1 U/μl RNase inhibitor, and 3.5 μg/μl mitochondrial proteins. Three 10-μl aliquots are removed at different time points (usually 0, 30, and 60 min), and the reaction is immediately stopped by extraction with phenol:chloroform:isoamyl alcohol (25:24:1). The aqueous phase (approximately 5 μl) is then frozen with liquid nitrogen until further analysis.

3.1.3. Analysis of degradation assays

Two straightforward methods exist to analyze the results of the degradation assays: gel electrophoresis to assess the preferential degradation of the poly-adenylated substrate and thin-layer chromatography (TLC) to determine the nature of the final degradation products.

For gel analysis, formamide spiked with bromophenol blue and xylene cyanol is added to a final concentration of 60% v/v to the RNA taken at different time points during degradation assays. Usually, 5 μl of formamide is added to 3 μl of the degradation assay aliquots. Samples are denatured for 2 min at 65 °C, quenched on ice, and loaded on a 6% w/v acrylamide-bisacrylamide (19:1 ratio), 7 M urea, 1× Tris-borate-EDTA gel. Acrylamide concentration and gel length obviously depend on the size of RNA substrates. After electrophoresis at 200 V, gels are fixed by two successive 5-min incubations in 10% v/v ethanol/10% v/v acetic acid, dried, and subjected to autoradiography.

The expected result of the degradation assays is the preferential degradation of the polyadenylated RNA substrates (Gagliardi *et al.*, 2001). Secondary structures within the RNA substrate should not impede degradation. However, secondary structures such as a double-stem loop immediately upstream of the poly(A) tail will, to some extent, block degradation. Therefore, a degradation intermediate with the size of the nonpolyadenylated substrate is frequently observed. It is unclear at present whether this intermediate is generated because of the stalling of the mtPNPase or the degradation of the poly(A) tail by the RNase II homolog. This RNase II homolog is present in both mitochondria and chloroplasts (at least in *Arabidopsis*). Although it is able to degrade unstructured RNA, its progression is stopped, even by moderate secondary structures (Perrin *et al.*, 2004b).

To analyze the final degradation products, 1 μl of the degradation assay aliquots is spotted onto polyethylenimine (PEI)-cellulose TLC plates. The chromatography is developed with 0.25 M KH$_2$PO$_4$. Standards (UTP, UDP, UMP) are analyzed alongside the samples. The TLC plate is dried, and the migration of standards is visualized under a 254-nm UV light. The migration of the degradation products is detected with autoradiography. If UTP was the radiolabeled nucleotide, the expected result is the detection of both UDP and UMP/Pi (both UMP and Pi will comigrate under these conditions). The production of UDP indicates phosphorolytic degradation caused by mtPNPase. This production should be higher in the samples with polyadenylated substrates but will not be absent in the samples with nonpolyadenylated substrates, because a few percent of those are degraded in the assay.

3.1.4. Comments on degradation assays

In vitro degradation assays are powerful tools to dissect the degradation process but are quite difficult to master. The two main problems are the contamination by nonmitochondrial RNases and the loss of activity.

Contamination by nonmitochondrial RNases is rather difficult to assess, because there is no substrate specificity (i.e., any RNase will degrade the polyadenylated substrates). Moreover, similar activities exist in other cellular compartments (e.g., PNPase in plastids). Extreme care should, therefore, be taken for purification of mitochondria. Ideally, features determined by the *in vitro* system should be correlated with *in vivo* data; for instance, while studying degradation of potato *atp9* transcripts, we observed neither *in vitro* degradation intermediates ending at the stable stem-loop structure present within the 3′ UTR nor polyadenylation sites at this particular stem-loop *in vivo* (Gagliardi *et al.*, 2001).

In case the preferential degradation of polyadenylated substrates is not observed, the most likely explanation is that mtPNPase has been degraded, because it is rather unstable in mitochondrial lysates. The presence of functional mtPNPase is best tested by TLC analysis of the degradation products. Absence of nucleoside diphosphate production is a clear sign that mtPNPase activity is absent from the lysate. In this case, the experiments should be repeated with increased amounts of protease inhibitors that should be added at all steps. It is essential that all steps in preparing the mitochondrial extract are done at 0 to 4 °C and as quickly as possible. For instance, the use of dialysis cassettes rather than tubing allows considerable shortening of the preparation of the lysate.

3.2. *In organello* expression systems

Studying the *in vivo* decay of engineered RNA substrates seems an obvious step when characterizing degradation processes. However, to date, it is not possible to transform plant mitochondria *in vivo*. Fortunately, two protocols for transfecting purified plant mitochondria have been developed. These *in organello* expression systems are based on DNA electroporation and direct DNA uptake, respectively (Farré and Araya, 2001; Koulintchenko *et al.*, 2003). Expression of electroporated DNA has been reported with mitochondria from wheat embryos, potato tubers, and etiolated seedlings of maize and sorghum (Choury *et al.*, 2005; Farré and Araya, 2001; Staudinger and Kempken, 2003). A detailed protocol for the electrotransfection of plant mitochondria has recently been published and, therefore, is not presented here (Farré *et al.*, 2007). The second *in organello* expression system relies on the competence of plant mitochondria for DNA uptake and has been used with potato mitochondria (Koulintchenko *et al.*, 2003; Placido *et al.*, 2005). Although *in organello* transcription of imported constructs has not been assessed yet in these cases, efficient DNA uptake has also been established for mitochondria from *A. thaliana*, tobacco (*Nicotiana tabacum*), cauliflower (*Brassica oleracea*), turnip (*Brassica rapa*), and maize (*Zea mays*) (Boesch P, Ibrahim N, Nepomnyashchikh D, Lotfi F and Dietrich A, unpublished).

3.2.1. Expression constructs

Usual bacterial cloning vectors are used for the construction of the engineered transgenes. The gene to be expressed must be cloned downstream of a strong mitochondrial promoter active in the species from which the organelles will be isolated. For expression in potato mitochondria, we use the -72 to $+69$ region of the potato 18S rRNA gene (Giese *et al.*, 1996), containing the promoter and the $5'$-extension down to the mature end of the rRNA. This region is part of the sequence with the accession number X98800. The transgene must contain specific sequences allowing discrimination from the corresponding endogenous transcripts in subsequent PCR analysis. Therefore, sequence tags must be incorporated into the transgene, unless the sequence to be expressed is not naturally present in the mitochondria used for the study. For instance, the gfp sequence or larch trnH were expressed in potato mitochondria and are easily discriminated from potato sequences (Koulintchenko *et al.*, 2003; Placido *et al.*, 2005). Detailed examples of DNA constructs successfully expressed by use of this protocol were published previously (Koulintchenko *et al.*, 2003; Placido *et al.*, 2005). In our hands, a minimal DNA fragment strictly restricted to the 18S rRNA promoter region as defined above and the gfp coding sequence did not support significant *in organello* reporter transcript accumulation (Koulintchenko M and Dietrich A, unpublished). In successful experiments, the promoter/transgene construct was flanked by additional sequences in the actual import substrate (i.e., the expression cassette was either inserted into a natural 2.3-kb linear mitochondrial plasmid from *Z. mays* (Koulintchenko *et al.*, 2003) or PCR-amplified with flanks derived from the cloning vector (Placido *et al.*, 2005)).

3.2.2. Import and expression of the DNA construct

Only linear DNA molecules are imported with the DNA uptake protocol. Therefore, the mitochondrial expression cassette is amplified by PCR from the plasmid carrying the construct. Simple linearization of the plasmid by a restriction enzyme is possible but will yield an unnecessarily longer import substrate. This is not recommended, because long DNA molecules are less efficiently imported than shorter ones. However, the sequence context can influence the import efficiency, and we have successfully obtained import of DNA up to 12.5 kb (Ibrahim N, Lotfi F, Handa H and Dietrich A, unpublished). Nucleotides and primers present in PCR reactions are competitors for the import of the DNA of interest. It is thus advisable to purify the PCR products on a Sephadex G-50 spin column or through the protocol of a dedicated purification kit (e.g., Nucleospin Extract or NucleoTraPCR, Macherey-Nagel; QIAquick, Qiagen).

Mitochondria are isolated from potato tubers as described previously (Koulintchenko *et al.*, 2003) by use of differential centrifugation and Percoll gradients. Buffer conditions optimized for *in organello* transcription are not favorable for DNA import, especially because DNA and nucleotides cross-compete for uptake. Import and expression are thus run as distinct successive steps. For each import reaction, mitochondria corresponding to 0.3 to 1 mg of proteins are incubated with 200 to 500 ng of DNA in 200 to 400 μl of 40 mM potassium phosphate and 0.4 M sucrose, pH 7.0 (import buffer). After a 30-min incubation at 25 °C under mild shaking, mitochondria are treated for 20 min with 200 μg of DNase I in the presence of 10 mM MgCl$_2$, washed twice with 1 ml import buffer, and resuspended in 200 to 300 μl of 330 mM sucrose, 90 mM KCl, 10 mM MgCl$_2$, 12 mM Tricine, 5 mM KH$_2$PO$_4$, 1.2 mM EGTA, 1 mM GTP, 2 mM DTT, 2 mM ADP, 10 mM sodium succinate, 0.15 mM CTP, 0.15 mM UTP, pH 7.2 (transcription buffer) essentially according to Farré *et al.* (2007).

DNase treatment can be omitted, but washing should be maintained to eliminate the bulk of nonimported DNA. Expression of the transgene at 25 °C under mild shaking can be carried out for various times, ranging from 1 h to overnight. Two to 3 h is usually a good compromise between significant expression and organelle stability. Mitochondria are finally pelleted and extracted with 150 μl of 10 mM Tris-HCl, 1 mM EDTA, 1% (w/v) SDS, pH 7.5, and 150 μl of phenol. The nucleic acids recovered in the aqueous phase are ethanol-precipitated and treated with RNase-free DNase. The positions of polyadenylation sites are determined by RT-PCR or cRT-PCR as described above. Radioactive transcripts for the hybridization analysis of processing or degradation products can also be generated by including [α^{32}P]UTP (800 Ci/mmol) in the transcription buffer (80 μCi for 200 μl medium). In this case, nonincorporated [α^{32}P]UTP is eliminated from the final nucleic acid fraction with gel filtration over a Sephadex G-50 spin column.

3.2.3. Comments on *in organello* expression systems

Introducing DNA into plant mitochondria by electroporation or direct uptake is reasonably efficient under the reported conditions. *In organello* expression systems should, thus, contribute significantly to a better understanding of polyadenylation-mediated degradation processes in plant mitochondria. However, so far expression of the exogenous DNA is not always observed, hence the need for multiple replicates. This suggests that transcription of imported DNA is still somehow at the edge of what can be expected from isolated mitochondria. Whereas common *in organello* RNA synthesis essentially represents run-on completion of already ongoing transcription, expression of exogenous DNA requires *de novo* recruitment of the transcription initiation and elongation machinery. Moreover, as in yeast and mammalian organelles, the DNA metabolism in plant mitochondria might be located in membrane-associated protein–DNA complexes called nucleoids that would direct replication, repair, and transcription (Dai *et al.*, 2005; Sakai *et al.*, 2004). If this is the case, the imported DNA might have to be recruited into these structures to be efficiently transcribed, a potentially limiting step. Expression of imported DNA may also require species-specific adaptations of reaction conditions. For instance, addition of 1 mg/ml of fatty-acid–free BSA is needed for expression of electroporated DNA in potato mitochondria but dispensable in the case of wheat (Choury *et al.*, 2005). Although in their early stages of development, *in organello* expression systems are particularly well adapted to study the polyadenylation of engineered substrates by RT-PCR or cRT-PCR, which require low amounts of material. By use of direct DNA uptake, we showed previously that a defective tRNA precursor transcribed in isolated plant mitochondria was subjected to polyadenylation-mediated degradation (Placido *et al.*, 2005).

ACKNOWLEDGMENTS

Work in our laboratory is funded by the Centre National de la Recherche Scientifique (CNRS, France), a fellowship of the Deutsche Forschungsgemeinschaft to H. L., and a PhD fellowship from the French Ministère de l'Education Nationale, de l'Enseignement Supérieur et de la Recherche (MENESR, France) to S. H. The work on DNA import and expression in plant mitochondria has been developed in collaboration with the laboratory of Y. Konstantinov (SIFIBR, Irkutsk, Russia).

REFERENCES

Bellaoui, M., Pelletier, G., and Budar, F. (1997). The steady-state level of mRNA from the Ogura cytoplasmic male sterility locus in *Brassica* hybrids is determined post-transcriptionally by its 3′-region. *EMBO J.* **16,** 5057–5068.

Bent, A. (2006). *Arabidopsis thaliana* floral dip transformation method. *Methods Mol. Biol.* **343,** 87–103.

Bollenbach, T. J., Schuster, G., and Stern, D. B. (2004). Cooperation of endo- and exoribonucleases in chloroplast mRNA turnover. *Prog. Nucleic Acid Res. Mol. Biol.* **78,** 305–337.

Chomczynski, P., and Sacchi, N. (2006). The single-step method of RNA isolation by acid guanidinium thiocyanate-phenol-chloroform extraction: Twenty-something years on. *Nat. Protoc.* **1,** 581–585.

Choury, D., Farré, J. C., Jordana, X., and Araya, A. (2005). Gene expression studies in isolated mitochondria: *Solanum tuberosum* rps10 is recognized by cognate potato but not by the transcription, splicing and editing machinery of wheat mitochondria. *Nucleic Acids Res.* **33,** 7058–7065.

Dai, H., Lo, Y. S., Litvinchuk, A., Wang, Y. T., Jane, W. N., Hsiao, L. J., and Chiang, K. S. (2005). Structural and functional characterizations of mung bean mitochondrial nucleoids. *Nucleic Acids Res.* **33,** 4725–4739.

Doma, M. K., and Parker, R. (2007). RNA quality control in eukaryotes. *Cell* **131,** 660–668.

Dombrowski, S., Brennicke, A., and Binder, S. (1997). 3′-Inverted repeats in plant mitochondrial mRNAs are processing signals rather than transcription terminators. *EMBO J.* **16,** 5069–5076.

Dreyfus, M., and Regnier, P. (2002). The poly(A) tail of mRNAs: Bodyguard in eukaryotes, scavenger in bacteria. *Cell* **111,** 611–613.

Farré, J. C., and Araya, A. (2001). Gene expression in isolated plant mitochondria: High fidelity of transcription, splicing and editing of a transgene product in electroporated organelles. *Nucleic Acids Res.* **29,** 2484–2491.

Farré, J. C., Choury, D., and Araya, A. (2007). In organello gene expression and RNA editing studies by electroporation-mediated transformation of isolated plant mitochondria. *Methods Enzymol.* **424,** 483–500.

Gagliardi, D., Perrin, R., Marechal-Drouard, L., Grienenberger, J. M., and Leaver, C. J. (2001). Plant mitochondrial polyadenylated mRNAs are degraded by a 3′ to 5′-exoribonuclease activity, which proceeds unimpeded by stable secondary structures. *J. Biol. Chem.* **276,** 43541–43547.

Gagliardi, D., Stepien, P. P., Temperley, R. J., Lightowlers, R. N., and Chrzanowska-Lightowlers, Z. M. (2004). Messenger RNA stability in mitochondria: Different means to an end. *Trends Genet.* **20,** 260–267.

Giese, A., Thalheim, C., Brennicke, A., and Binder, S. (1996). Correlation of nonanucleotide motifs with transcript initiation of 18S rRNA genes in mitochondria of pea, potato and *Arabidopsis. Mol. Gen. Genet.* **252,** 429–436.

Hayes, R., Kudla, J., Schuster, G., Gabay, L., Maliga, P., and Gruissem, W. (1996). Chloroplast mRNA 3′-end processing by a high molecular weight protein complex is regulated by nuclear encoded RNA binding proteins. *EMBO J.* **15,** 1132–1141.

Holec, S., Lange, H., Kuhn, K., Alioua, M., Borner, T., and Gagliardi, D. (2006). Relaxed transcription in *Arabidopsis* mitochondria is counterbalanced by RNA stability control mediated by polyadenylation and polynucleotide phosphorylase. *Mol. Cell Biol.* **26,** 2869–2876.

Houseley, J., LaCava, J., and Tollervey, D. (2006). RNA-quality control by the exosome. *Nat. Rev. Mol. Cell Biol.* **7,** 529–539.

Koulintchenko, M., Konstantinov, Y., and Dietrich, A. (2003). Plant mitochondria actively import DNA via the permeability transition pore complex. *EMBO J.* **22,** 1245–1254.

Kuhn, J., Tengler, U., and Binder, S. (2001). Transcript lifetime is balanced between stabilizing stem-loop structures and degradation-promoting polyadenylation in plant mitochondria. *Mol. Cell Biol.* **21,** 731–742.

Perrin, R., Lange, H., Grienenberger, J. M., and Gagliardi, D. (2004a). AtmtPNPase is required for multiple aspects of the 18S rRNA metabolism in *Arabidopsis thaliana* mitochondria. *Nucleic Acids Res.* **32,** 5174–5182.

Perrin, R., Meyer, E. H., Zaepfel, M., Kim, Y. J., Mache, R., Grienenberger, J. M., Gualberto, J. M., and Gagliardi, D. (2004b). Two exoribonucleases act sequentially to process mature 3′-ends of atp9 mRNAs in *Arabidopsis* mitochondria. *J. Biol. Chem.* **279,** 25440–25446.

Placido, A., Gagliardi, D., Gallerani, R., Grienenberger, J. M., and Marechal-Drouard, L. (2005). Fate of a larch unedited tRNA precursor expressed in potato mitochondria. *J. Biol. Chem.* **280,** 33573–33579.

Sakai, A., Takano, H., and Kuroiwa, T. (2004). Organelle nuclei in higher plants: Structure, composition, function, and evolution. *Int. Rev. Cytol.* **238,** 59–118.

Schafer, B. (2005). RNA maturation in mitochondria of *S. cerevisiae* and *S. pombe. Gene* **354,** 80–85.

Slomovic, S., Laufer, D., Geiger, D., and Schuster, G. (2005). Polyadenylation and degradation of human mitochondrial RNA: The prokaryotic past leaves its mark. *Mol. Cell Biol.* **25,** 6427–6435.

Slomovic, S., Laufer, D., Geiger, D., and Schuster, G. (2006). Polyadenylation of ribosomal RNA in human cells. *Nucleic Acids Res.* **34,** 2966–2975.

Small, I. (2007). RNAi for revealing and engineering plant gene functions. *Curr. Opin. Biotechnol.* **18,** 148–153.

Staudinger, M., and Kempken, F. (2003). Electroporation of isolated higher-plant mitochondria: Transcripts of an introduced cox2 gene, but not an atp6 gene, are edited in organello. *Mol. Genet. Genomics* **269,** 553–561.

Takenaka, M., and Brennicke, A. (2007). RNA editing in plant mitochondria: Assays and biochemical approaches. *Methods Enzymol.* **424,** 439–458.

van Engelen, F. A., Molthoff, J. W., Conner, A. J., Nap, J. P., Pereira, A., and Stiekema, W. J. (1995). pBINPLUS: An improved plant transformation vector based on pBIN19. *Transgenic Res.* **4,** 288–290.

Vanacova, S., and Stefl, R. (2007). The exosome and RNA quality control in the nucleus. *EMBO Rep.* **8,** 651–657.

Walter, M., Kilian, J., and Kudla, J. (2002). PNPase activity determines the efficiency of mRNA 3′-end processing, the degradation of tRNA and the extent of polyadenylation in chloroplasts. *EMBO J.* **21,** 6905–6914.

West, S., Gromak, N., Norbury, C. J., and Proudfoot, N. J. (2006). Adenylation and exosome-mediated degradation of cotranscriptionally cleaved pre-messenger RNA in human cells. *Mol. Cell* **21,** 437–443.

CHAPTER TWENTY-TWO

IN VIVO AND IN VITRO APPROACHES FOR STUDYING THE YEAST MITOCHONDRIAL RNA DEGRADOSOME COMPLEX

Michal Malecki,* Robert Jedrzejczak,[†] Olga Puchta,*,[‡]
Piotr P. Stepien,*,[‡] and Pawel Golik*,[‡]

Contents

* Department of Genetics and Biotechnology, Warsaw University, Warsaw, Poland
[†] Synchrotron Radiation Research Section, MCL, National Cancer Institute, Argonne National Puchta Laboratory, Argonne, Illinois, USA
[‡] Institute of Biochemistry and Biophysics PAS, Warsaw, Poland

Methods in Enzymology, Volume 447
ISSN 0076-6879, DOI: 10.1016/S0076-6879(08)02222-2

Abstract

The mitochondrial degradosome (mtEXO) of *S. cerevisiae* is the main exoribonuclease of yeast mitochondria. It is involved in many pathways of mitochondrial RNA metabolism, including RNA degradation, surveillance, and processing, and its activity is essential for mitochondrial gene function. The mitochondrial degradosome is a very simple example of a 3′ to 5′-exoribonucleolytic complex. It is composed of only two subunits: Dss1p, which is an RNR (RNase II–like) family exoribonuclease, and Suv3p, which is a DExH/D-box RNA helicase. The two subunits form a tight complex and their activities are highly interdependent, with the RNase activity greatly enhanced in the presence of the helicase subunit, and the helicase activity entirely dependent on the presence of the ribonuclease subunit.

In this chapter, we present methods for studying the function of the yeast mitochondrial degradosome *in vivo*, through the analysis of degradosome-deficient mutant yeast strains, and *in vitro*, through heterologous expression in *E. coli* and purification of the degradosome subunits and reconstitution of a functional complex. We provide the protocols for studying ribonuclease, ATPase, and helicase activities and for measuring the RNA binding capacity of the complex and its subunits.

1. INTRODUCTION

RNA degradation plays an essential role in gene expression, both in assuring regulation through proper mRNA turnover and in controlling RNA quality by generating processing intermediates and destroying processing by-products and/or aberrant transcripts (Gagliardi *et al.*, 2004; Meyer *et al.*, 2004; Mitchell and Tollervey, 2000, 2001; Newbury, 2006). The mechanisms of mitochondrial genome expression are quite divergent among different branches of *Eukaryota*. Although transcriptional control is relatively simple (Amiott and Jaehning, 2006a,b), transcripts undergo extensive processing to yield mature RNAs (Binder and Brennicke, 2003; Fernandez-Silva et al., 2003; Schafer, 2005). The balance between transcription and RNA degradation is thus critical for yeast mitochondria (Rogowska et al., 2006).

Enzymatic complexes performing RNA degradation in different genetic systems of prokaryotic and eukaryotic cells and organelles are very divergent and involve several classes of exoribonucleases (both 3′ to 5′- and 5′ to 3′) and endoribonucleases (Carpousis, 2002; Gagliardi et al., 2004; Meyer et al., 2004; Mitchell et al., 1997; Newbury, 2006; Zuo and Deutscher, 2001). The most common and universal RNA degradation activity is that of a 3′ to 5′-exoribonuclease; it is also the only kind commonly found in mitochondria (Gagliardi et al., 2004). In addition to exoribonucleases and endoribonucleases, RNA degradation enzymes often contain other enzymatic activities, of which RNA helicases are a prominent class (Carpousis, 2002; Cordin et al., 2006; Meyer et al., 2004; Rocak and Linder, 2004). One of the model examples of RNA degrading complexes is the eubacterial degradosome with its three principal activities: endonuclease (RNase E), RNA helicase (RhlB), and phosphorolytic exoribonuclease (PNPase) (Carpousis, 2002).

The yeast mitochondrial degradosome (also known as mtEXO) was the first described mitochondrial RNA degradation system (Dziembowski and Stepien, 2001; Margossian et al., 1996; Min et al., 1993). It is the main exoribonuclease in yeast mitochondria, which, unlike bacterial cells and animal or plant mitochondria, lack the phosphorolytic polynucleotide phosphorylase (PNPase) activity (Gagliardi et al., 2004). The subunit composition of the yeast mitochondrial degradosome is extremely simplified—it is a heterodimer of an RNase subunit, encoded by the DSS1 (MSU1, YMR287C) gene, and an RNA helicase, encoded by the SUV3 (YPL029W) gene (Dmochowska et al., 1995; Dziembowski et al., 1998, 2003; Golik et al., 1995; Malecki et al., 2007; Margossian et al., 1996; Stepien et al., 1992). Both subunits are essential for complex function, and inactivating either subunit leads to incapacitation of the mitochondrial gene expression machinery, with the overaccumulation of excised intronic sequences and high-molecular-weight precursors, depletion of mature transcripts, loss of translation, and, finally, loss of mitochondrial genome stability (Dmochowska et al., 1995; Dziembowski et al., 1998, 2003; Golik et al., 1995; Rogowska et al., 2006; Stepien et al., 1992).

Biochemical analysis of the yeast mitochondrial degradosome revealed remarkably tight functional interdependence of the two subunits (Malecki et al., 2007). The Dss1 exoribonuclease outside the complex does have some basal exoribonuclease activity, which is not ATP dependent. However, in the presence of Suv3p, Dss1 activity is greatly enhanced and becomes entirely ATP dependent, with no activity observed in the absence of ATP. The degradosome complex is, thus, an ATP-dependent exoribonuclease, which is unique among exoribonucleases. The Suv3 protein alone, outside the complex, does not display any detectable RNA-unwinding activity; the

$3'$ to $5'$-directional helicase activity requiring a free $3'$-single-stranded substrate becomes only apparent when Suv3p is in complex with Dss1p. Although the ATPase activity of Suv3p does not depend on the presence of Dss1p, its background activity in the absence of RNA is greatly reduced when the protein is in the complex.

The helicase Suv3p strongly stimulates the ribonuclease activity of the degradosome complex in an ATP-dependent manner even in the presence of short single-stranded substrates that are devoid of any secondary structure. This suggests that the role of RNA helicases in exoribonucleolytic complexes, often thought to be related to unwinding secondary-structure elements in the substrate that impede helicase progress (Coburn *et al.*, 1999; Py *et al.*, 1996), needs to be reconsidered. In the absence of structural information, we hypothesize that the Suv3p helicase acts as a molecular motor feeding the substrate to the catalytic center of the Dss1p ribonuclease (Malecki *et al.*, 2007).

The yeast mitochondrial degradosome can thus serve as a model to study the interplay of different enzymatic activities in shaping the structure and function of RNA turnover enzymes. The methods presented in this chapter were tested in the studies on the mitochondrial degradosome of yeast *S. cerevisiae*, and they can be readily adapted to similar enzymes from other organisms.

2. Genetics of the *S. cerevisiae* Mitochondrial Degradosome

2.1. Working with degradosome-deficient mutants

The principal genetic approach for studying mitochondrial degradosome function is based on inactivating the *SUV3* (YPL029W) and *DSS1* (YMR287C) genes. Because both genes are essential for degradosome function, their inactivation results in a similar phenotype: complete loss of respiratory capability with partial or complete loss of functional mtDNA (discussed in detail later). On the molecular level, the phenotypes include accumulation of RNAs with abnormal $5'$- and $3'$-termini and high-molecular-weight RNA precursors, variations in steady-state levels of mature transcripts, and disruption of mitochondrial translation (Dziembowski *et al.*, 1998, 2003; Rogowska *et al.*, 2006; Stepien *et al.*, 1992).

As it is often the case in studying the phenotypes of mutants deficient in the mitochondrial function, success depends on careful selection of the nuclear and mitochondrial background of the strains used in the project. Common laboratory strains derived from S288C (Brachmann *et al.*, 1998) (including the BY series strains used in systematic deletion projects) are a poor choice for mitochondrial research, because they contain a mutation in the *HAP1* gene that affects their respiratory capability (Gaisne *et al.*, 1999). D273-10B (Sherman, 1963) and W303 (Thomas and Rothstein, 1989) are both good respirers and are often the strains of choice for the study of nuclear–mitochondrial interactions. They also are relatively resistant to catabolite repression, meaning that good yields of mitochondrial preparations (RNA, DNA, protein, etc.) can be obtained from media with glucose as the carbon source. With the exception of some early work (Golik *et al.*, 1995; Stepien *et al.*, 1992, 1995), our studies on the genetics of the *S. cerevisiae* mitochondrial degradosome were performed in the W303 strain. This strain carries the *ade2-1* allele, and supplementing even complete medium with adenine is recommended to avoid slower growth and accumulation of a red dye associated with adenine starvation in *ade2* strains. This red dye will copurify with DNA and RNA preparations and may interfere with enzymatic reactions such as PCR.

All standard yeast techniques (Sherman, 2002) apply to these parental and degradosome-deficient strains with no major modifications, and common LiAc/sssDNA transformation procedures (Gietz and Woods, 2002) work well in the W303 background.

Mitochondrial genotype is of critical importance in the study of degradosome-deficient strains. The severity of certain phenotypes associated with inactivation of *SUV3* or *DSS1* genes depends on the presence of mitochondrial introns. Wild-type and common laboratory strains of *S. cerevisiae* contain up to 13 introns in three of the mitochondrial genes (the *CYTB* gene has 5 introns, the *COX1* gene has 7 introns, and the *LSU-rRNA* gene has 1 intron). Although all *suv3*Δ and *dss1*Δ strains are respiratory deficient, the stability of the mitochondrial genome depends on the presence of introns. In intron-containing strains most, if not all, of the cells are converted to cytoplasmic petites—a mixture of rho⁻ (large deletions in mtDNA) and rho⁰ (lack of mtDNA) mitochondria (Dmochowska *et al.*, 1995; Golik *et al.*, 1995; Rogowska *et al.*, 2006; Stepien *et al.*, 1995). Stability of the mitochondrial genome in degradosome-deficient strains is markedly improved (to approximately 60% rho⁺) when a strain that harbors intronless mtDNA is used. When maintaining intact functional mitochondrial DNA is important, we therefore recommend the use of the intronless variant. The original intronless mitochondrial genome (Seraphin *et al.*, 1987) works well; it was, however, shown to carry a point mutation (G252D) in the *CYTB* gene (Saint-Georges *et al.*, 2002). This mutation slightly impairs respiratory function, particularly at elevated temperatures. An intronless strain with the corrected CYTB sequence is available (Saint-Georges *et al.*, 2002) and should preferably be used in new strain constructions. Strains containing only a few (1 to 3) introns (Golik *et al.*, 1995) can usually maintain rho⁺ mtDNA in degradosome-deficient strains, albeit at a lower frequency. The presence of the ω intron in the *LSU-rRNA* gene particularly affects mtDNA stability, although an *suv3*Δ strain containing this intron alone (Stepien *et al.*, 1995) can maintain a certain proportion of intact mitochondrial genomes. Detailed description of the construction and maintenance of strains containing different mtDNA variants by cytoduction is beyond the scope of this chapter, and the technique has been described elsewhere (Fox *et al.*, 1991).

Another important feature of the strains carrying degradosome deletions is the relatively high frequency of spontaneous nuclear suppressors (or, more correctly, pseudorevertants) that partially restore respiratory capability in strains carrying rho$^+$ mtDNAs devoid of most introns (Golik *et al.*, 2004; Rogowska *et al.*, 2006). The pseudoreversions were mapped to point partial loss-of-function mutations in the genes encoding subunits of the mitochondrial RNA polymerase: *RPO41* and *MTF1* (Rogowska *et al.*, 2006). These mutations decrease the transcription rate in mitochondria, thus restoring balance between transcription and impaired degradation.

Impaired mtDNA stability, on the one hand, and significant spontaneous pseudoreversion, on the other, mean that particular care should be given to maintaining and testing strains used. An intact mitochondrial genome should be verified by crossing each strain to a suitable rho^0 tester strain; and the absence of suppressor mutations is easily detected as the failure to grow on nonfermentable carbon sources. We recommend performing such tests for each starting culture and also setting aside and retesting a small aliquot of each large volume culture used for preparations. Multiple passages and long-term storage of stationary cultures at 4 °C should be avoided. We recommend restreaking from frozen glycerol stocks followed by genotype testing.

2.2. Construction of degradosome-deficient deletant strains

Suv3Δ strains constructed for the original paper (Stepien *et al.*, 1992) and used in early studies (Dmochowska *et al.*, 1995; Dziembowski *et al.*, 2003; Golik *et al.*, 1995; Stepien *et al.*, 1995) harbored partial deletions with the *URA3* marker and removing a portion of the C-terminal sequence of the Suv3 protein. This deletion cassette can be conveniently amplified with primers 9090 (5′-AACTGCGGTTACATGGCCTA) and 9091 (5′-CTCGAAGATGAGAGGTGACC) (Rogowska *et al.*, 2006). More recently, the original suv3::URA3 deletions were replaced by deletion of the entire *SUV3* ORF with the KanMX4 cassette derived from the Euroscarf Y12799 strain (Giaever *et al.*, 2002; Winzeler *et al.*, 1999) by PCR with primers SUV3_A (5′-TCAGAACACAATGTCCTTATTGAAA) and SUV3_D (5′-TATATTTTACTGC-CCTTTGCTCAAC). Both the partial *suv3::URA3* and the complete *suv3::KanMX4* deletions behave identically in all our phenotypic tests and can be considered true *null* alleles of the gene (Rogowska *et al.*, 2006).

A strain bearing a complete deletion of the *DSS1* gene can be constructed with the *KanMX4* cassette obtained by PCR amplification from the

Euroscarf strain Y10873 with primers DSS1_A (5′-GTTTACAAATT GAATCGGATGACTC) and DSS1_D (5′-TTTATAGTGGAGAAGA GAACCATCG). Early studies (Dmochowska *et al.*, 1995; Dziembowski *et al.*, 1998) used an insertion disruption construct, which is not recommended. Presence of the disruption cassette should be verified with PCR and the primers listed previously in addition to testing for the presence of the selection marker.

2.3. Generation of spontaneous nuclear pseudorevertants of the degradosome deficiency phenotype

Plate 100 to 200 μl of an overnight culture of a *suv3Δ* or *dss1Δ* strain carrying rho$^+$ intronless mtDNA on YPGlycerol and incubate at 28 to 30 °C for up to 7 days. Most respiratory-competent colonies that appear will bear recessive hypomorphic mutations in the genes encoding subunits of the mitochondrial RNA polymerase (Rogowska *et al.*, 2006), and the respiratory-competent phenotype will be temperature sensitive (in the W303 background check, the ts phenotype at 35 to 36 °C; higher temperatures may inhibit the wild-type parental strain as well). This can easily be verified by transforming cells with a plasmid-borne wild-type allele of either the *MTF1* gene or the *RPO41* gene and testing for the loss of a respiratory-competent phenotype (Rogowska *et al.*, 2006).

3. RNA METABOLISM IN THE MITOCHONDRIA OF DEGRADOSOME-DEFICIENT STRAINS

Basic tools for assaying the function of the yeast mitochondrial degradosome *in vivo* involve preparation of RNA from mitochondria, determining the steady-state levels of mitochondrial transcripts, and measuring the transcription rate—particularly relevant for the respiring pseudorevertant strains that usually carry partial loss-of-function mutations in genes encoding mitochondrial RNA polymerase (Rogowska *et al.*, 2006). Although successful Northern blot analyses to detect mitochondrial transcripts can be performed with total-cell RNA preparations, some probes work better with RNA prepared from purified mitochondria. Isolation of mitochondria is also required for measuring transcription rates, RNase activity, and many other assays. Mitochondria obtained by differential centrifugation are usually sufficient. However, further purification can be achieved with sucrose or Percoll step gradients if necessary (Boldogh and Pon, 2007).

3.1. Preparation of yeast mitochondria

1. Grow yeast in 1 L of YPGlucose (when strains other than W303 or D273 with strong catabolite repression are used, YPGalactose may give

better results) starting from a tested preculture. At OD_{600} of approximately 1.2 to 1.5 harvest the cells by centrifugation ($1500g$, 5 min, RT), wash with double-distilled (dd) H_2O, and weigh the yeast pellet.

2. Suspend in 15 ml of 0.1 M Tris-H_2SO_4, pH 9.4, 10 mM DTT (add DTT freshly before preparation), and incubate 15 min at 30 °C.

3. Pellet cells ($1500g$, 5 min, 4 °C), wash with 30 ml of zymolyase buffer (1.2 M sorbitol, 20 mM K_2HPO_4/KH_2PO_4 buffer, pH 7.4), pellet cells as previously, and suspend in 2 ml of zymolyase buffer per 1 g of cells.

4. Add 2.5 mg zymolyase per 1 g of cells, incubate at 30 °C for approximately 1 h, monitor digestion by mixing a 5-μl aliquot with an equal volume of 10% SDS and checking lysis by observation under microscope.

5. Pellet the protoplasts ($3000g$, 5 min, 4 °C), and suspend in 5 ml of cold homogenization buffer (0.5 M sorbitol, 60 mM Tris-HCl, pH 7.5, 1 mM EDTA) with BSA added to 0.1% freshly before the procedure.

6. Add cold washed glass beads (0.4 to 0.6 mm) and vortex (30 to 60 sec) at 4 °C, allow to settle, recover supernatant, add 1 to 2 ml of cold homogenization buffer, vortex, and recover supernatant again. Repeat a few times until the supernatant becomes almost clear (the number of washes at this step is a tradeoff between the recovery of mitochondria and the final volume of preparation; adjust to suit your equipment).

7. Centrifuge at $1500g$, 10 min, 4 °C to pellet the cellular debris, recover supernatant, and centrifuge at $12,000g$, 10 min, 4 °C to pellet mitochondria. Resuspend mitochondria in 2 to 5 ml of homogenization buffer (with no BSA added).

8. Repeat step 7 two to three times to get rid of the fluffy debris; the mitochondrial pellet should be hard and slightly translucent (after the high-speed spin, one can remove some of the debris by gentle aspiration with a pipette tip or Pasteur pipette).

9. Suspend mitochondrial preparation in 1 ml of cold homogenization buffer.

10. Determine total protein content of the preparation with the Bradford assay.

11. To purify RNA, take 0.5 or 1 ml of this preparation and extract with an equal volume of TRI-Reagent® (Sigma–Aldrich) according to the manufacturer's instructions. For protein-based assays, aliquot 20 μg of total protein into Eppendorf tubes, centrifuge in a microfuge (4 °C) at maximum speed, remove the supernatant, and freeze the pellets at −80 °C.

3.2. Analysis of the steady-state RNA levels with Northern blotting

Use 1 to 2 μg of mitochondrial RNA or ~10 μg of total-cell RNA per lane. All standard electrophoresis and blotting techniques work well. When total-cell RNA is used, normalize the blot with cytoplasmic rRNAs or any suitable housekeeping RNA control. With mitochondrial RNA preparations, achieving normalization is trickier, because all mitochondrial transcripts are to some extent affected by changes in degradosome function. The total amount of mitochondrial protein in the aliquot used for RNA extraction can provide a rough approximation, but any quantitative results should be treated with caution.

Either synthetic oligonucleotides or cloned or amplified DNA fragments labeled by any suitable method can be used as probes. Examples of suitable probes are given in our published work (Dziembowski *et al.*, 2003; Golik *et al.*, 1995, 2004; Rogowska *et al.*, 2006; Stepien *et al.*, 1995). When choosing hybridization conditions, observe that some of the yeast mitochondrial sequences are extremely AT-rich and adjust stringency conditions accordingly.

3.3. Measuring mitochondrial transcription *in organello*

Measuring mitochondrial transcription rates is important in research on the balance between transcription and RNA degradation with the pseudorevertants obtained in degradosome-deficient strains. The protocol is modified from Krause and Dieckmann (2004).

1. Thaw the mitochondrial preparation aliquot (see earlier) on ice and resuspend in cold 1 M HEPES-KOH, pH 8.0, to 1 to 5 mg/ml (4 to 20 μl for a 20-μg aliquot). Use 1 to 10 μg in a single assay (depending on the transcription rate of the strain, too much preparation will result in the reaction reaching plateau already at the first time point; 5 μg is a good starting point). Adjust the concentration of the mitochondrial preparation so that the chosen amount is added in no more than 1 to 2 μl volume.

2. Prepare a 50-μl reaction containing: 50 mM HEPES-KOH, pH 8.0, 10 mM MgCl$_2$, 25 mM KOAc, 10 mM DTT, 125 μM CTP, ATP, and GTP each, 50 μCi of [α-32P]UTP, 40 U of an RNase inhibitor (we used RiboLock from MBI Fermentas, Hanover, MD, with success), and the chosen amount of mitochondrial preparation.

3. Incubate at 30 °C, at 2, 5, 10, 15, and 20 min, remove 5-μl aliquots, and mix immediately with 5 μl of stop solution (0.5 M Na$_4$P$_2$O$_7$, 1% SDS). Spot 5 μl of this mixture on each of two DEAE–cellulose filter discs (DE81; Whatman, Maidstone, United Kingdom); 30- and 40-min time points may be added if the reaction does not reach plateau at 20 min.

4. Wash one of each pair of filters four times with an excess of 0.5 M Na$_2$HPO$_4$. Allow to air dry.

5. Measure radioactivity of each filter in a scintillation counter with liquid scintillator. The ratio of signal from washed and unwashed filter is the incorporation for a given time point.

4. Expression and Purification of the Yeast Mitochondrial Degradosome Subunits in *E. coli*

Although the native degradosome complex can be isolated from yeast mitochondria with the TAP-TAG strategy (Dziembowski *et al.*, 2003), heterologous expression in bacteria allows us to produce pure proteins in large quantities suitable for biochemical (Malecki *et al.*, 2007) and structural analysis. The proteins are expressed as His$_6$MBP fusions, according to the procedure described by (Tropea *et al.*, 2007).

4.1. Construction of the expression vectors

When designing expression constructs for mitochondrial proteins, it is important to remove the N-terminal leader sequence to avoid problems with the folding and solubility of the expressed protein. In many cases, the length of the sequence and the putative cleavage site can be derived from bioinformatics analysis with algorithms such as SignalP (Bendtsen *et al.*, 2004). For both Suv3p and Dss1p, the first 26 codons of the ORF were omitted from the expression construct. The fusion expression vectors are conveniently constructed with the Gateway Recombinational Cloning (Invitrogen) system, as described (Tropea *et al.*, 2007). Primers N1-27-737ySUV3 (5′-GAGAACCTGTACTTCCAGGGTTACCACAGCGA GCCGCATAG and C-ySUV3 (5′-GGGGACCACTTTGTACAAG AAAGCTGGGTTATTATGTACGCAATCTTCTTCTCGA) are used to amplify codons 27 to 737 of the *SUV3* ORF, whereas primers N1-27-969-yDSS1 (5′-GAGAACCTGTACTTCCAGGGTACCAGAGGCAAA CGACAGCGA) and C-yDSS1 (5′-GGGGACCACTTTGTACAAG AAAGCTGGGTTATTATAGCTTTTCCAACTCTAACATTC) amplify codons 27 to 969 of the *DSS1* ORF. Both primer pairs add a sequence encoding a TEV cleavage site at the 5′-end (primers N1) and an attB2 site at the 3′-end of the amplicon.

An overlapping left primer N2 (5′-GGGGACAAGTTTGTACAAA AAAGCAGGCTCGGAGAACCTGTACTTCCAGGGT) together with right primers C-ySUV3 or C-yDSS1 is used to add the attB1 site to the 5′-terminus of the amplicon. This can be performed in two consecutive PCR reactions, first with primers N1/C and then with primers N2/C with the product of the first reaction as a template. Alternately, a single reaction can be performed with all three primers at the 1:20:20 (N1:C:N2) ratio (Tropea *et al.*, 2007). Use of a high-fidelity thermostable polymerase is recommended.

The final PCR products are cloned first into the pDONR201 (Invitrogen) by recombinational cloning to generate the entry plasmids pDON-SUV3-27-737 and pDON-DSS1-27-969 in the BP reaction according to the manufacturer's manual and subsequently into the pDEST-His$_6$MBP destination vector in the LR reaction, yielding His$_6$MBP-ySUV3-27-737 and His$_6$MBP-DSS1p-27-969, respectively. The two clonase reactions can be performed in a single tube or as two subsequent steps with the isolation of pDONR intermediate constructs, according to the Invitrogen Gateway Cloning System manual. The single-step protocol should perform well; should it fail to yield proper transformants, we recommend the two-step procedure for troubleshooting. Obtained clones should be verified by sequencing.

4.2. Expression, purification, and storage of proteins

Transform the expression vectors encoding the His$_6$MBP fusions into *E. coli* BL21(DE3) CodonPlus-RIL cells (Stratagene). Other *E. coli* strains optimized for eukaryotic protein expression, such as Rosetta (Novagen), can also be used. Freshly obtained transformant colonies are inoculated into Luria broth supplemented with 100 μg/ml ampicillin and 34 μg/ml chloramphenicol and grown overnight at 37 °C. Overnight cultures are diluted 1:100 into the same medium. The volume of the culture depends on the scale of the experiment, 1 to 4 L is typical for biochemical assays (expect ~1 mg protein/1 L of culture). When the entire degradosome complex is to be purified, inoculate 3 L of His$_6$MBP-DSS1p-27-969 culture per every liter of His$_6$MBP-ySUV3-27-737 to compensate for the weaker binding of the Dss1p fusion to the column.

The cultures are grown at 37 °C to OD$_{600}$ between 0.6 and 0.8. At this point, flasks are rapidly cooled down to 20 °C, and IPTG is added to 0.25 m*M* final concentration to induce expression. Cultures are then incubated at 20 °C overnight with vigorous shaking. If purification of the entire complex is desired, cultures expressing His$_6$MBP-DSS1p-27-969 and His$_6$MBP-ySUV3-27-737 (at 3:1 v/v ratio) should be mixed before centrifugation or after resuspension; otherwise, the proteins can be purified separately (always expect a higher yield from the His$_6$MBP-ySUV3-27-737 construct). The buffers and procedure used in the purification (modified from Tropea *et al.*, 2007) are as follows.

4.2.1. Buffers

Buffer A: 0.3 *M* KCl, 25 m*M* imidazole, 10% glycerol, 50 m*M* Tris, 1 m*M* dithiothreitol (DTT), pH adjusted to 8.0. DTT should be added freshly before use or on thawing of the pellet.
Buffer AP: Buffer A + 50 m*M* KH$_2$PO$_4$
Buffer AK: Buffer A + 2 *M* KCl
Buffer AE: Buffer A + 200 m*M* imidazole
Buffer A-i: Buffer A without imidazole
Buffer B: 0.5 *M* NaCl, 10% glycerol, 50 m*M* Tris, 1 m*M* DTT, pH adjusted to 8.0

4.2.2. Protocol

1. Centrifuge cultures (3000*g*, 6 min, 4 °C), resuspend in small volume (∼50 ml for 3 L of culture) of buffer AP. If cultures expressing Suv3p and Dss1p were not mixed before centrifugation, they should be mixed at this stage if the entire complex is to be purified. Freeze the pellet at −80 °C. Even if long-term storage of the cells is not envisioned, do perform the freeze–thaw cycle. Perform the subsequent steps at 4 °C.

2. Thaw the cells, add PMSF to 100 μg/ml, and sonicate.

3. Ultracentrifuge (40,000*g*, 40 min in a fixed-angle rotor).

4. Apply supernatant on a Ni–NTA (Qiagen) column equilibrated with buffer AP. Simple gravity-flow columns work well, although FPLC or HPLC systems can also be used.

5. Wash the column with buffer A, monitor the flow-through for proteins with the Bradford assay until no more protein is detected, and then wash with buffer AK and subsequently with buffer A, monitoring the protein content of the flow-though each time.

6. Elute with buffer AE (approximately twice the column volume), pool the eluted fractions, and add EDTA to 2 mM.

7. Add TEV protease at a ratio of 1 mg/100 mg of protein. We used the autoinactivation-resistant TEV mutant S219V expressed from the pRK793 plasmid (Kapust *et al.*, 2001) (courtesy of Dr. David Waugh), but any commercial TEV protease preparation can also be used. Perform cleavage at 4 °C overnight.

8. Precipitate the proteins with 0.39 g/ml ammonium sulfate (15 min on ice) and centrifuge (10,000*g*, 20 min).

9. Resuspend the pellet in initial volume of the eluate from step 6 of buffer A-i, apply to a column as in step 4, but use ∼10% more resin volume.

10. Wash the column with buffer A-i, and monitor the flow-through for proteins as in step 5.

11. Elute with buffer A, precipitate with 0.39 g/ml ammonium sulfate as in step 8, and suspend in buffer B in a volume determined by the capacity of the size exclusion chromatography column used in the subsequent step.

12. Apply on a size exclusion column and perform chromatography according to the instructions of the instrument's manufacturer. We used a Superdex 200 column in the ÄKTA^{TM} purifier FPLC system (GE Healthcare). Pool the fractions containing the purified complex (or single proteins), optionally concentrate with Amicon filter devices (Millipore), aliquot, and store at −80 °C. Alternately, the proteins can be precipitated with ammonium sulfate (as in steps 8 and 11) and stored as precipitate at 4 °C.

Typical results of purifying the entire degradosome complex from a mixture of Suv3p and Dss1p expressing cultures are shown in Fig. 22.1.

A

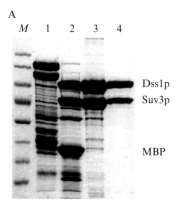

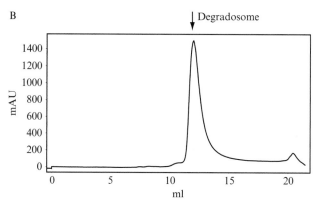

Figure 22.1 Purification of yeast mitochondrial degradosome complex expressed in *E. coli.* (A) SDS-polyacrylamide gel electrophoresis analysis of proteins in various stages of preparation (described in the text). (1) Eluted proteins after the first round of purification with a Ni–NTA column; (2) proteins after cleavage with TEV protease; (3) proteins after the second round of purification with an Ni–NTA column; and (4) protein preparation after the final purification with the Superdex 200 size exclusion column; *M*, molecular weight marker (PageRuler™ Prestained Protein Ladder, Fermentas #SM0671). (B) Elution profile of the purified degradosome complex with a Superdex 200 size exclusion column.

5. Assaying the Ribonuclease Activity of the Degradosome and the Dss1 Protein

The exoribonucleolytic activity of the yeast mitochondrial degradosome is associated with the Dss1p subunit, belonging to the RNR (RNase II-like) superfamily of RNases (Dziembowski *et al.*, 2003; Malecki *et al.*, 2007). On its own, it displays a basal activity (6 ± 0.6 nmol·min^{-1}·mg^{-1}) that is not dependent on ATP, but the activity of the entire complex is significantly higher (55 ± 3 nmol·min^{-1}·mg^{-1}) and, interestingly, becomes entirely ATP-dependent (Malecki *et al.*, 2007).

Both longer polynucleotides and short synthetic oligonucleotides can be used as substrates for the assays of the RNase activity of the yeast mitochondrial degradosome and the Dss1 protein. Longer substrates, prepared by *in vitro* transcription, better reflect the native substrates of the enzyme, whereas short synthetic oligonucleotides can be designed with the presence or absence of particular sequence and structure motifs in mind, giving greater flexibility in studying the reaction mechanisms. Use of end-labeled oligonucleotide substrates also allows visualizing the residual core of approximately four nucleotides left by the enzyme (Malecki *et al.*, 2007).

In all procedures involving RNA substrates, it is important to use RNase-free siliconized test tubes to avoid the adsorption of RNA to the surface of the tube. Addition of RNA grade glycogen ($\sim$1 μg) is recommended in all steps involving RNA precipitation.

5.1. Preparing the polynucleotide substrates

Longer (usually few hundred nucleotides) single-stranded polynucleotide substrates are prepared by *in vitro* transcription with T7 RNA polymerase (T7 Transcription Kit, Fermentas) with α-^{32}P-UTP added to the reaction to produce radioactive RNA, according to the manufacturer's instructions. Templates can be conveniently obtained by PCR with the T7 promoter sequence added as a 5′-overhang to the left primer. An example of a suitable template is the PCR product corresponding to 831 bp of the 3′-sequence of the mature yeast mitochondrial *CYTB* gene obtained with primers Lp (TAATACGACTCACTATAGGGAGATATTACTAATTTATTCT CAG, containing the T7 promoter as a 5′-overhang) and Rp (TTAAGAA TATTATTAAAGTA). PCR amplification of yeast mitochondrial genome sequences is sometimes problematic because of their very high AT content. In some cases, the temperature of the elongation step has to be lowered from 72 °C to approximately 50 °C.

After transcription, the products are separated in a denaturing polyacrylamide/7 *M* urea gel (percentage, size, and electrophoresis conditions are determined by the size of the expected product), the position of the full-length radioactive product is determined by autoradiography, the corresponding band is excised from the gel and isolated by crushing the frozen gel slice, soaking overnight in 0.2 *M* Tris, pH 7.5, 0.3 *M* NaCl, 25 mM EDTA, 2% SDS, and precipitation with ethanol from the supernatant (Peattie, 1979).

5.2. Preparing the oligonucleotide substrates

Commercially purchased synthetic RNA oligonucleotides of any desired sequence can be used as substrates. Substrates approximately 30 nt long were successfully used to demonstrate the RNase activity (Malecki *et al.*, 2007).

RNA oligonucleotides are radiolabeled at the 5′-terminus with ^{32}P with the T4 polynucleotide kinase (New England Biolabs) according to the manufacturer's instructions. Labeled products were purified by electrophoresis in a denaturing polyacrylamide/7 M urea gel (12 to 15%) and isolated as described previously for the polynucleotide substrates.

5.3. The RNase reaction

Assays were carried out in a 20-μl reaction volume containing 10 mM Tris-HCl, pH 8.0, 25 mM KCl, 10 mM MgCl$_2$, and 1 mM DTT and, optionally, 0.5 to 1 mM ATP. Approximately 30 to 40 cps of the radioactive substrate and ~0.1 μg of the purified protein are used per reaction. Reactions proceed for 30 min at 30 °C and are stopped by addition of EDTA to 25 mM.

5.4. Thin-layer chromatography assay

Thin-layer chromatography (TLC) with PEI-cellulose plates is a rapid method of visualizing and measuring ribonucleolytic activity suitable for radioactive polynucleotide substrates uniformly labeled with α-^{32}P-UTP, which can be successfully applied to the study of the mitochondrial degradosome and its subunits (Dziembowski and Stepien, 2001; Malecki et al., 2007). It allows separating nucleoside monophosphates, diphosphates, and triphosphates from the oligonucleotides and polynucleotides.

PEI-cellulose F TLC plates (Merck), cut to approximately 10 cm in height, are prerun in ddH$_2$O. When the entire plate is wet, spot 1 μl of the reaction mixture approximately 1.5 cm from the bottom edge of the plate. Plates are then developed in 1:1 mixture of 0.5 M formic acid and 2 M LiCl. The progress is monitored under UV illumination with UTP, UDP, and UMP as markers. Plates are then dried at room temperature or at 50 °C and visualized by autoradiography (film or PhosphorImager). Radioactive nucleoside monophosphates are the expected reaction product. A typical result is shown in Fig. 22.2A.

5.5. Gel assay

Reaction products can also be separated with standard polyacrylamide denaturing gels (with 7 M urea). This method allows monitoring the completeness of digestion of both polynucleotide and oligonucleotide substrates and is the only method of visualizing the progress of the assays with end-labeled oligonucleotide substrates. The format and percentage of the gel depend on the particular application, with gels ranging in concentration from 10 to 20% being routinely used. Long (sequencing type) 20% denaturing gels, together with a 5′-end-labeled oligonucleotide substrate, allow visualization of the residual core (of approximately 4 nt) left undigested by

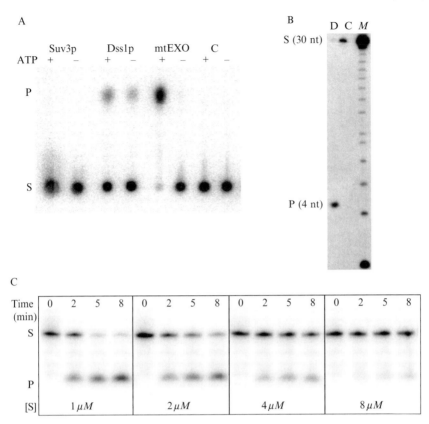

Figure 22.2 RNase activity of the yeast mitochondrial degradosome. All assays were performed as described in the text. (A) A TLC assay showing the ATP-independent basal RNase activity of Dss1p alone, and the much increased and entirely ATP-dependent activity of the degradosome complex (mtEXO). Neither Suv3p nor the BSA negative control (C) shows any detectable RNase activity. Assays were performed with a polynucleotide substrate synthesized and labeled with *in vitro* transcription. (B) A 20% polyacrylamide/7 *M* urea denaturing gel assay showing the degradation of a 30-nt oligonucleotide substrate labeled at the 5′-end by the degradosome complex (D). No sign of RNase activity is visible in the negative control reaction (C). Marker (M) was prepared by limited digestion of the substrate oligonucleotide with the T1 RNase (Fermentas). Positions of the 30-nt substrate and the 4-nt residual core remaining after the digestion are indicated. (C) Typical time-course results obtained with the entire mitochondrial degradosome complex at several substrate (5′-labeled oligonucleotide) concentrations. Reaction mixtures were separated in 10% denaturing polyacrylamide gels, and the positions of substrate (S) and product (P) are indicated. Reaction velocities at different substrate concentrations required for the Michaelis–Menten analysis can be obtained from such assays.

the mitochondrial degradosome (Malecki *et al.*, 2007), which is typical of this family of exoribonucleases. An example of such assay is shown in Fig. 22.2B.

The reactions are performed as described previously and stopped by the addition of the formamide RNA loading buffer. The samples are then denatured by heating to 65 °C for 10 min with cooling on ice and then applied on the gel. As in this assay, the proteins are in large excess; protein-bound RNA may remain in the wells and fail to enter the gel. To alleviate this problem, carrier RNA (0.1 to 0.5 μg/sample of commercially available *E. coli* tRNA or total yeast RNA) can be added to the samples after the reactions (this way it will not affect the assay itself).

5.6. Measuring reaction kinetics

To estimate the kinetics of the RNase reaction of the yeast mitochondrial degradosome, an assay is performed essentially as described previously. A 5′-end-labeled oligonucleotide is used as the substrate, and the products are separated on a 10% polyacrylamide denaturing gel. Assays are performed for a series of substrate concentrations, typically ranging from 0.1 μM to 8 μM with 0.1 μg of the enzyme. ATP is added to the assay at 0.5 mM. The total reaction volume is 20 μl.

The same amount of the radioactive substrate is used in each reaction, and the concentration is adjusted with the unlabeled substrate oligonucleotide. Because the amount of the radioactive substrate is very low (approximately 0.02 pmol in a typical reaction, orders of magnitude less than the required final substrate concentration) it can be omitted from the calculation, and the concentration of the unlabeled substrate can be considered as the final substrate concentration, which greatly simplifies the preparation of the assay. Reaction mixtures without the enzyme are prepared in advance, and the enzyme is added last. Samples (4 μl each) are removed from each reaction at 0, 2, 5, and 8 min and immediately stopped by addition of the formamide-loading buffer with 25 mM EDTA. Samples from each time point from every reaction are separated on a 10% denaturing polyacrylamide gel, which is then dried and exposed in a PhosphorImager and quantified with Image Quant (Molecular Dynamics), and the amounts of product and undigested substrate are calculated from the band intensities. Values from each time point are then plotted, and the Vo value is calculated for each substrate concentration. Examples of time-course measurements at several substrate concentrations are shown in Fig. 22.2C.

To calculate the K_m and V_{max} the obtained values are fitted to the Michaelis–Menten equation by nonlinear regression. Many commercially available software packages can be used to perform the fit; a good multi-platform open-source solution is QtiPlot (http://soft.proindependent.com/qtiplot.html).

6. Assaying the ATP Hydrolysis Activity of the Degradosome and the Suv3 Protein

The ATPase activity of the yeast mitochondrial degradosome is associated with the Suv3 protein, which belongs to the DExH/D superfamily of RNA helicases (Cordin *et al.*, 2006; Dziembowski *et al.*, 2003; Stepien *et al.*, 1992) and contains the Walker A and B ATPase motifs. The ATPase activity is evident both for the complex of Suv3p with Dss1p and for the Suv3 protein alone, and is in both cases similar ($\sim$3 μmol·min^{-1}·mg^{-1}). The ATPase activity of Suv3p is stimulated by RNA or short single-stranded DNA; this induction is, however, most evident in the entire degradosome complex (approximately sixfold), whereas the significantly higher background activity in the absence of RNA of the Suv3 protein alone reduces the induction ratio to just approximately twofold (Malecki *et al.*, 2007).

6.1. The ATPase reaction and the TLC assay

1. The reactions are performed in a 20-μl reaction volume, in a buffer containing 10 mM Tris-HCl, pH 8.0, 25 mM KCl, 10 mM MgCl$_2$, 1 mM DTT, and 1 mM ATP. Approximately 0.1 to 0.4 μg of protein and 10 fmol of α-^{32}P-ATP are added to each assay. Optionally, 1 μg of total-yeast RNA, short RNA oligonucleotide, or a 20-nt DNA oligonucleotide is added to induce the ATPase activity. Reactions are performed at 30 °C for 30 min and terminated by adding 2 μl of 0.5 M EDTA.

2. One microliter of the reaction mixture is then spotted on the PEI-cellulose F TLC plate (Merck) prepared and developed as described previously for the RNase assay. Unhydrolyzed substrates (α-^{32}P-ATP) and reaction products (α-^{32}P-ADP) separated on the plate are visualized by autoradiography (film or PhosphorImager). Typical results are shown in Fig. 22.3A.

6.2. Measuring reaction kinetics

Assays for estimating the kinetic properties of the ATPase activity of the Suv3 protein alone or in the degradosome complex are performed essentially as described previously, with 1 μg of total yeast RNA added to induce the enzyme. Substrate concentration is adjusted by varying the amounts of unlabeled ATP added to the reaction (typically in the 0.1 mM to 1.5 mM range); the amount of radioactive ATP (10 fmol per reaction) is negligible in the calculations. Samples (4 μl each) are removed from each reaction at 0, 2, 5, and 8 min and immediately stopped by addition of EDTA to 25 mM; 1-μl

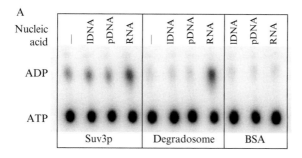

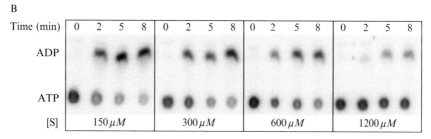

Figure 22.3 ATPase activity of the yeast mitochondrial degradosome. All assays were performed as described in the text and separated on TLC plates. (A) An assay showing that the ATPase activity of the degradosome and the Suv3 protein is stimulated by RNA but not by either long linear dsDNA (lDNA) or plasmid DNA (pDNA). The background activity in the absence of stimulation is visibly higher for Suv3p alone than for the complex, whereas the induced activity is similar. No detectable activity is observed with the BSA negative control (C). (B) Typical time course results obtained with the entire mitochondrial degradosome complex at several substrate (ATP) concentrations. Reaction velocities at different substrate concentrations required for the Michaelis–Menten analysis can be obtained from such assays.

samples from each time point from every reaction are separated by TLC as described previously and products and substrates quantified with Phosphor-Imager and Image Quant (Molecular Dynamics) from spot intensities. Calculations and curve fitting are performed exactly as described previously for the RNase assay. Examples of time-course measurements at several substrate concentrations are shown in Fig. 22.3B.

7. ASSAYING THE HELICASE ACTIVITY OF THE DEGRADOSOME AND THE SUV3 PROTEIN

The Suv3 protein has sequence motifs typical of the DExH/D super-family of RNA helicases. RNA unwinding *in vitro* was demonstrated only for a few members of this protein family (Cordin *et al.*, 2006; Rocak and Linder, 2004). Our results (Malecki *et al.*, 2007) indicate that the helicase activity is apparent only when Suv3p is in complex with Dss1p, and is undetectable for the Suv3 protein alone. The mitochondrial degradosome complex unwinds RNA/RNA, DNA/RNA, and DNA/DNA duplexes, provided that they all have a protruding 3′-end. Blunt-ended substrates, as well as those with a 5′-protruding end, are not efficiently unwound by the degradosome.

7.1. Preparing the substrates

Substrates for the helicase activity assays are prepared by annealing complementary synthetic RNA and/or DNA oligonucleotides. The following nucleotides (RNA sequence shown) were used in our laboratory

T: 5′-CAAACUCUCUCUCUCUCAAC
5W: 5′-AGAGAGAGAGGUUGAGAGAGAGAGAGAGUUUG
3W: 5′-GUUGAGAGAGAGAGAGAGUUUGAGAGAGAGAG
B: 5′-GUUGAGAGAGAGAGAGAGUUUG

The substrate with the protruding 3′-terminus (substrate A) is prepared by annealing oligonucleotides 3W and T; the substrate with the protruding 5′-terminus is prepared by annealing oligonucleotides 3W and T; and the blunt-ended substrate is prepared by annealing oligonucleotides B and T. Only the substrate with the protruding 3′-terminus is efficiently unwound by the mitochondrial degradosome helicase activity. Either the loading strand (e.g., 3W) or the complementary strand (B) can be labeled at the 5′-terminus with ^{32}P with the T4 polynucleotide kinase (New England Biolabs) according to the manufacturer's instructions and cleaned up by ethanol precipitation.

The appropriate unlabeled complementary oligonucleotide is added to the labeled strand in threefold molar excess in the annealing buffer (20 mM Tris-HCl, pH 8.0, 0.5 M NaCl, 1 mM EDTA) in a total volume of 20 μl, heated to 80 °C for 10 min, and then slowly cooled to room temperature (disconnect the heat block and allow it to cool on the bench). Subsequently, loading buffer (10 mM Tris-HCl, pH 7.6, 0.03% bromophenol blue, 0.03% xylene cyanol FF, 60% glycerol, and 60 mM EDTA) is added, and the duplexes are purified by electrophoresis through a 15% nondenaturing polyacrylamide gel, the position of the desired product is located by autoradiography and the band excised from the gel and isolated as described previously (for the RNase substrates).

7.2. Unwinding assay

The reactions are performed in 10 mM Tris-HCl, pH 8.0, 25 mM KCl, 10 mM MgCl$_2$, 1 mM DTT, 1 mM ATP, with 0.1 pmol of labeled duplex RNA and 0.1 μg of protein at 30 °C in a total volume of 20 μl; 1 pmol of cold RNA corresponding to the labeled strand ("trap") is also included in the reaction mixture to avoid reannealing of the reaction products. The reaction is stopped by addition of loading buffer (10 mM Tris-HCl, pH 7.6), 0.03% bromophenol blue, 0.03% xylene cyanol FF, 60% glycerol, and 60 mM EDTA); 10 μg of proteinase K is then added to each tube, and samples are incubated for 15 min at room temperature. This degrades the remaining proteins that would otherwise bind the nucleic acids and prevent them from entering the gel. Samples are run on a native 15% polyacrylamide gel (approximately 4 h at 200 V for a 15-cm gel, at RT) and visualized by autoradiography. Heat-denatured substrate should always be included as a positive control.

When RNA/RNA substrates are used with the degradosome complex containing the wild-type Dss1 protein, the RNase activity will rapidly degrade the displaced strand, and the products of the unwinding reaction will be visible only in the first minutes of the reaction (perform a time course to optimize). RNA/DNA heteroduplexes, with the loading strand made of RNA and the complementary radiolabeled strand of DNA, are also good substrates, and because Dss1p will not hydrolyze DNA, the products will remain detectable for a long time. DNA/DNA duplexes are also unwound, albeit with a lower efficiency. Figure 22.4 shows typical results obtained with a DNA/DNA and a DNA/RNA substrate.

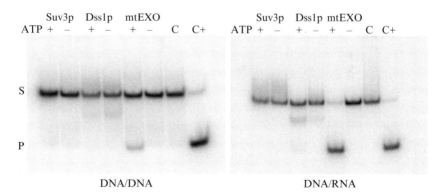

Figure 22.4 Helicase activity of the yeast mitochondrial degradosome. DNA/DNA and DNA/RNA duplex substrates with the protruding 3′-termini were prepared by annealing oligonucleotides 3W and T as described in the text. In the case of the RNA/DNA heteroduplex substrate, the loading strand was RNA. The strand complementary to the loading strand (DNA in both cases) was labeled at the 5′-end with ^{32}P. Suv3p or the reconstituted degradosome complex was incubated with the substrate in the presence or absence of ATP as indicated and analyzed with a native 15% polyacrylamide gel as described in the text. C is the double-stranded substrate; C+ is this substrate denatured by heating to 65 °C.

8. PROTEIN-RNA BINDING ASSAY

Both subunits of the yeast mitochondrial degradosome, as well as the entire complex, are capable of binding RNA with similar affinity, corresponding to a K_d value of approximately 200 nM (short oligonucleotide substrates) to approximately 20 to 30 nM (long polynucleotide substrates) (Malecki *et al.*, 2007). RNA binding properties of the proteins can be estimated with the modified double-filter binding assay (Tanaka and Schwer, 2005; Wong and Lohman, 1993), which involves rapidly passing the preincubated protein-RNA mixture through two filters: nitrocellulose on top and nylon on the bottom. Protein-bound RNA is retained on the nitrocellulose membrane, and the remaining unbound RNA binds to the nylon membrane. To avoid RNA hydrolysis by the RNase activity of the degradosome the buffer contains no divalent metal cations, and ATP is also omitted.

Nitrocellulose filters are presoaked in 0.5 M KOH for 10 min and washed in water until the pH is neutral. Nylon filters are washed once in 0.1 M EDTA (pH 8.0) and three times in 1.0 M KCl for 10 min each time, rinsed once with 0.5 M KOH, and washed with water until the pH is neutral. Both filters are then equilibrated in the binding buffer (5 mM EDTA, 25 mM KCl, 10 mM Tris-HCl, pH 8.0, 1 mM DTT, 10% glycerol) at 4 °C for at least 1 h before use. The filters are then placed in a 96-well dot-blot apparatus connected to a vacuum source (we used the BioRad Dot SF apparatus).

Radiolabeled RNA substrates are prepared exactly as described previously for RNase and helicase substrates. Both single-stranded oligonucleotides labeled at the 5′-end and longer polynucleotides obtained by *in vitro* transcription can be used; oligonucleotide substrates usually give cleaner results.

With the oligonucleotide substrate, binding is performed in 20 μl volume in the binding buffer with 1.75 fmol radiolabeled RNA (87.5 pM concentration) and proteins ranging in concentration from approximately 5 nM to approximately 1200 nM prepared as a series of twofold dilutions. BSA (or a similar neutral, non-RNA binding protein) should be used as the negative control. The proteins and RNA are incubated for 20 min at room temperature. Reactions that use polynucleotide substrates are prepared in a similar manner; the molar concentration of the labeled substrate will be, however, much lower for a comparable activity; unlabeled substrate can be added to bring up the RNA concentration to the desired value.

Each well of the dot-blot apparatus is washed with 100 μl of the binding buffer; the binding reaction is applied and promptly washed again with 100 μl of the binding buffer. It is important to perform the wash and application steps quickly in succession with vacuum to pass the solutions through filters.

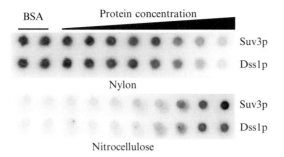

Figure 22.5 An example of the RNA binding assay with Suv3p and Dss1p. An oligonucleotide labeled at the 5'-end was used in a double-filter assay at the concentration of 87.5 pM (1.75 fmol per 20-μl reaction) with a series of twofold dilutions of each protein starting with ~8 nM and ending with ~1024 nM (the exact values differed slightly for the two proteins) as described in the text. Protein-bound RNA is retained on the nitrocellulose membrane, whereas the remaining unbound RNA binds to the nylon membrane. The two leftmost spots correspond to the negative control (BSA), which does not exhibit RNA binding.

Filters are then dried and visualized and quantified with PhosphorImager and Image Quant (Molecular Dynamics). Figure 22.5 shows results of a typical binding assay with Suv3 and Dss1 proteins.

Bound RNA fraction for each protein concentration is estimated as the product of the signal on the nitrocellulose filter and the total signal on both filters. K_d can be estimated by a nonlinear regression fit of the data to the formula $B = [P]/(K_d + [P])$, where B is the bound RNA fraction and $[P]$ is the protein concentration. It is important to ensure that both ends of the binding curve are included in the range of concentrations used; the bound RNA fraction for the highest protein concentrations should approach 95 to 100% and should not increase with further protein concentration increase. On the other end, the bound RNA fraction for the lowest protein concentrations should approach that of a negative control (e.g., BSA). When the experiment is performed properly, this should not exceed 1%. K_d can also be rapidly estimated as the protein concentration in the reaction giving ~50% bound RNA (equal signal on the two filters).

9. Concluding Remarks

Techniques described previously provide a basic foundation for working with the yeast mitochondrial degradosome, both *in vivo* with mutant strains and *in vitro* for determining the biochemical properties of the enzyme and its subunits purified after expression in *E. coli*. Most of them should be readily adaptable to other systems of similar functions. General principles of working with degradosome-deficient mutants will apply to other yeast strains bearing mutations in genes encoding proteins important for the expression of mitochondrial genes. Degradosome-deficient strains also provide a system of selection for hypomorphic mutants with decreased mitochondrial transcription, which can be useful for studying this process. Biochemical *in vitro* techniques described for the yeast mitochondrial degradosome and its subunits should be readily applicable to orthologous proteins from other organisms and to analogous enzymatic complexes involved in the metabolism of RNA.

ACKNOWLEDGMENTS

This work was supported by the Ministry of Science and Higher Education of Poland through the Faculty of Biology, Warsaw University Intramural Grants BW#1720/46 and BW#1680/40, by the Intramural Research Program of the National Cancer Institute, the CoE BioExploratorium project: WKP_1/1.4.3/1/2004/44/44/115/2005, and by grants 2P04A 002 29, 2P04A 054 26, and N N301 2386 33 from the Ministry of Science and Higher Education of Poland. We are grateful to Prof. Ewa Bartnik for critical reading of the manuscript.

REFERENCES

Amiott, E. A., and Jaehning, J. A. (2006a). Mitochondrial transcription is regulated via an ATP "sensing" mechanism that couples RNA abundance to respiration. *Mol. Cell* **22,** 329–338.

Amiott, E. A., and Jaehning, J. A. (2006b). Sensitivity of the yeast mitochondrial RNA polymerase to +1 and +2 initiating nucleotides. *J. Biol. Chem.* **281,** 34982–34988.

Bendtsen, J. D., Nielsen, H., von Heijne, G., and Brunak, S. (2004). Improved prediction of signal peptides: SignalP 3.0. *J. Mol. Biol.* **340,** 783–795.

Binder, S., and Brennicke, A. (2003). Gene expression in plant mitochondria: Transcriptional and post-transcriptional control. *Philos. Trans. R. Soc. Lond. B Biol. Sci.* **358,** 181–188; discussion 188–189.

Boldogh, I. R., and Pon, L. A. (2007) Purification and subfractionation of mitochondria from the yeast Saccharomyces cerevisiae. *Methods Cell. Biol.* **80,** 45–64.

Brachmann, C. B., Davies, A., Cost, G. J., Caputo, E., Li, J., Hieter, P., and Boeke, J. D. (1998). Designer deletion strains derived from *Saccharomyces cerevisiae* S288C: A useful set of strains and plasmids for PCR-mediated gene disruption and other applications. *Yeast* **14,** 115–132.

Carpousis, A. J. (2002). The *Escherichia coli* RNA degradosome: Structure, function and relationship in other ribonucleolytic multienzyme complexes. *Biochem. Soc. Trans.* **30,** 150–155.

Coburn, G. A., Miao, X., Briant, D. J., and Mackie, G. A. (1999). Reconstitution of a minimal RNA degradosome demonstrates functional coordination between a 3′-exonuclease and a DEAD-box RNA helicase. *Genes Dev.* **13,** 2594–2603.

Cordin, O., Banroques, J., Tanner, N. K., and Linder, P. (2006). The DEAD-box protein family of RNA helicases. *Gene* **367,** 17–37.

Dmochowska, A., Golik, P., and Stepien, P. P. (1995). The novel nuclear gene DSS-1 of *Saccharomyces cerevisiae* is necessary for mitochondrial biogenesis. *Curr. Genet.* **28,** 108–112.

Dziembowski, A., Malewicz, M., Minczuk, M., Golik, P., Dmochowska, A., and Stepien, P. P. (1998). The yeast nuclear gene DSS1, which codes for a putative RNase II, is necessary for the function of the mitochondrial degradosome in processing and turnover of RNA. *Mol. Gen. Genet.* **260**, 108–114.

Dziembowski, A., Piwowarski, J., Hoser, R., Minczuk, M., Dmochowska, A., Siep, M., van der Spek, H., Grivell, L., and Stepien, P. P. (2003). The yeast mitochondrial degradosome. Its composition, interplay between RNA helicase and RNase activities and the role in mitochondrial RNA metabolism. *J. Biol. Chem.* **278**, 1603–1611.

Dziembowski, A., and Stepien, P. P. (2001). Genetic and biochemical approaches for analysis of mitochondrial degradosome from *Saccharomyces cerevisiae*. *Methods Enzymol.* **342**, 367–378.

Fernandez-Silva, P., Enriquez, J. A., and Montoya, J. (2003). Replication and transcription of mammalian mitochondrial DNA. *Exp. Physiol.* **88**, 41–56.

Fox, T. D., Folley, L. S., Mulero, J. J., McMullin, T. W., Thorsness, P. E., Hedin, L. O., and Costanzo, M. C. (1991). Analysis and manipulation of yeast mitochondrial genes. *Methods Enzymol.* **194**, 149–165.

Gagliardi, D., Stepien, P. P., Temperley, R. J., Lightowlers, R. N., and Chrzanowska-Lightowlers, Z. M. (2004). Messenger RNA stability in mitochondria: Different means to an end. *Trends Genet.* **20**, 260–267.

Gaisne, M., Becam, A. M., Verdiere, J., and Herbert, C. J. (1999). A "natural" mutation in *Saccharomyces cerevisiae* strains derived from S288c affects the complex regulatory gene HAP1 (CYP1). *Curr. Genet.* **36**, 195–200.

Giaever, G., Chu, A. M., Ni, L., Connelly, C., Riles, L., Veronneau, S., Dow, S., Lucau-Danila, A., Anderson, K., Andre, B., Arkin, A. P., Astromoff, A., *et al.* (2002). Functional profiling of the *Saccharomyces cerevisiae* genome. *Nature* **418**, 387–391.

Gietz, R. D., and Woods, R. A. (2002). Transformation of yeast by lithium acetate/single-stranded carrier DNA/polyethylene glycol method. *Methods Enzymol.* **350**, 87–96.

Golik, P., Szczepanek, T., Bartnik, E., Stepien, P. P., and Lazowska, J. (1995). The *S. cerevisiae* nuclear gene SUV3 encoding a putative RNA helicase is necessary for the stability of mitochondrial transcripts containing multiple introns. *Curr. Genet.* **28**, 217–224.

Golik, P., Zwolinska, U., Stepien, P. P., and Lazowska, J. (2004). The SUV3 gene from *Saccharomyces douglasii* is a functional equivalent of its *Saccharomyces cerevisiae* ortholog and is essential for respiratory growth. *FEMS Yeast Res.* **4**, 477–485.

Kapust, R. B., Tozser, J., Fox, J. D., Anderson, D. E., Cherry, S., Copeland, T. D., and Waugh, D. S. (2001). Tobacco etch virus protease: Mechanism of autolysis and rational design of stable mutants with wild-type catalytic proficiency. *Protein Eng.* **14**, 993–1000.

Krause, K., and Dieckmann, C. L. (2004). Analysis of transcription asymmetries along the tRNAE-COB operon: Evidence for transcription attenuation and rapid RNA degradation between coding sequences. *Nucleic Acids Res.* **32**, 6276–6283.

Malecki, M., Jedrzejczak, R., Stepien, P. P., and Golik, P. (2007). *In vitro* reconstitution and characterization of the yeast mitochondrial degradosome complex unravels tight functional interdependence. *J. Mol. Biol.* **372**, 23–36.

Margossian, S. P., Li, H., Zassenhaus, H. P., and Butow, R. A. (1996). The DExH box protein Suv3p is a component of a yeast mitochondrial 3′ to 5′-exoribonuclease that suppresses group I intron toxicity. *Cell* **84**, 199–209.

Meyer, S., Temme, C., and Wahle, E. (2004). Messenger RNA turnover in eukaryotes: Pathways and enzymes. *Crit. Rev. Biochem. Mol. Biol.* **39**, 197–216.

Min, J., Heuertz, R. M., and Zassenhaus, H. P. (1993). Isolation and characterization of an NTP-dependent 3′-exoribonuclease from mitochondria of *Saccharomyces cerevisiae*. *J. Biol. Chem.* **268**, 7350–7357.

Mitchell, P., Petfalski, E., Shevchenko, A., Mann, M., and Tollervey, D. (1997). The exosome: A conserved eukaryotic RNA processing complex containing multiple 3′ to 5′-exoribonucleases. *Cell* **91**, 457–466.

Mitchell, P., and Tollervey, D. (2000). mRNA stability in eukaryotes. *Curr. Opin. Genet. Dev.* **10**, 193–198.

Mitchell, P., and Tollervey, D. (2001). mRNA turnover. *Curr. Opin. Cell Biol.* **13**, 320–325.

Newbury, S. F. (2006). Control of mRNA stability in eukaryotes. *Biochem. Soc. Trans.* **34**, 30–34.

Peattie, D. A. (1979). Direct chemical method for sequencing RNA. *Proc. Natl. Acad. Sci. USA* **76**, 1760–1764.

Py, B., Higgins, C. F., Krisch, H. M., and Carpousis, A. J. (1996). A DEAD-box RNA helicase in the *Escherichia coli* RNA degradosome. *Nature* **381**, 169–172.

Rocak, S., and Linder, P. (2004). DEAD-box proteins: The driving forces behind RNA metabolism. *Nat. Rev. Mol. Cell Biol.* **5**, 232–241.

Rogowska, A. T., Puchta, O., Czarnecka, A. M., Kaniak, A., Stepien, P. P., and Golik, P. (2006). Balance between transcription and RNA degradation is vital for *Saccharomyces cerevisiae* mitochondria: Reduced transcription rescues the phenotype of deficient RNA degradation. *Mol. Biol. Cell* **17**, 1184–1193.

Saint-Georges, Y., Bonnefoy, N., di Rago, J. P., Chiron, S., and Dujardin, G. (2002). A pathogenic cytochrome b mutation reveals new interactions between subunits of the mitochondrial bc1 complex. *J. Biol. Chem.* **277**, 49397–49402.

Schafer, B. (2005). RNA maturation in mitochondria of *S. cerevisiae* and *S. pombe*. *Gene* **354**, 80–85.

Seraphin, B., Boulet, A., Simon, M., and Faye, G. (1987). Construction of a yeast strain devoid of mitochondrial introns and its use to screen nuclear genes involved in mitochondrial splicing. *Proc. Natl. Acad. Sci. USA* **84**, 6810–6814.

Sherman, F. (1963). Respiration-deficient mutants of yeast. I. Genetics. *Genetics* **48**, 375–385.

Sherman, F. (2002). Getting started with yeast. *Methods Enzymol.* **350**, 3–41.

Stepien, P. P., Kokot, L., Leski, T., and Bartnik, E. (1995). The suv3 nuclear gene product is required for the *in vivo* processing of the yeast mitochondrial 21s rRNA transcripts containing the r1 intron. *Curr. Genet.* **27**, 234–238.

Stepien, P. P., Margossian, S. P., Landsman, D., and Butow, R. A. (1992). The yeast nuclear gene suv3 affecting mitochondrial post-transcriptional processes encodes a putative ATP-dependent RNA helicase. *Proc. Natl. Acad. Sci. USA* **89**, 6813–6817.

Tanaka, N., and Schwer, B. (2005). Characterization of the NTPase, RNA-binding, and RNA helicase activities of the DEAH-box splicing factor Prp22. *Biochemistry* **44**, 9795–9803.

Thomas, B. J., and Rothstein, R. (1989). Elevated recombination rates in transcriptionally active DNA. *Cell* **56**, 619–630.

Tropea, J. E., Cherry, S., Nallamsetty, S., Bignon, C., and Waugh, D. S. (2007). A generic method for the production of recombinant proteins in *Escherichia coli* using a dual hexahistidine-maltose-binding protein affinity tag. *Methods Mol. Biol.* **363**, 1–19.

Winzeler, E. A., Shoemaker, D. D., Astromoff, A., Liang, H., Anderson, K., Andre, B., Bangham, R., Benito, R., Boeke, J. D., Bussey, H., Chu, A. M., Connelly, C., *et al.* (1999). Functional characterization of the *S. cerevisiae* genome by gene deletion and parallel analysis. *Science* **285**, 901–906.

Wong, I., and Lohman, T. M. (1993). A double-filter method for nitrocellulose-filter binding: Application to protein-nucleic acid interactions. *Proc. Natl. Acad. Sci. USA* **90**, 5428–5432.

Zuo, Y., and Deutscher, M. P. (2001). Exoribonuclease superfamilies: Structural analysis and phylogenetic distribution. *Nucleic Acids Res.* **29**, 1017–1026.

CHAPTER TWENTY-THREE

MEASURING mRNA DECAY IN HUMAN MITOCHONDRIA

Asuteka Nagao, Narumi Hino-Shigi, *and* Tsutomu Suzuki

Contents

Abstract

Human mitochondria contain a genome encoding 13 proteins, all of which are components of respiratory chain complexes. Mutations in human mitochondrial DNA often have pathological consequences. Although 12 of the mitochondrial mRNAs are generated from the same polycistronic transcript, the steady-state level of each mRNA differs. The stability of each mitochondrial mRNA is post-transcriptionally controlled by polyadenylation and deadenylation. However, the molecular mechanism by which each mRNA attains a unique stability is not fully understood. In this report, we describe a practical method for measuring the half-lives of human mitochondrial mRNAs using quantitative real-time reverse transcription PCR.

Department of Chemistry and Biotechnology, Graduate School of Engineering, University of Tokyo, Tokyo, Japan

Methods in Enzymology, Volume 447

ISSN 0076-6879, DOI: 10.1016/S0076-6879(08)02223-4

1. Introduction

The mitochondrion is a cellular organelle required for energy production through oxidative phosphorylation. Human mitochondrial (mt)DNA is a circular double-stranded DNA that encodes 13 essential proteins required for oxidative phosphorylation (OXPHOS). Transcription from the heavy (H)-strand of mtDNA generates a large polycistronic transcript containing 12 protein-encoding genes; only one protein, ND6, is transcribed from the light (L)-strand. The primary transcript of the H-strand is processed to produce 12 mature mRNAs. In principle, these mRNAs are generated at the same time during transcription. However, the steady-state level of each mRNA is different (Duborjal *et al.*, 2002), indicating that the stability and half-life of mitochondrial mRNA is posttranscriptionally controlled. The molecular mechanism by which the copy number and stability of each mRNA is controlled is not fully understood.

Polyadenylation of mRNAs plays a pivotal role in regulating mRNA stability. In eukaryotes, polyadenylation confers mRNA stability, promotes translation initiation, and plays a role in the transport of processed mRNAs from the nucleus to the cytoplasm (Beelman and Parker, 1995; Chekanova and Belostotsky, 2003; Sachs *et al.*, 1997; Shatkin and Manley, 2000). In prokaryotes and chloroplasts, in contrast, polyadenylation constitutes the signal for RNA degradation (Carpousis *et al.*, 1999; Hajnsdorf *et al.*, 1995; Kudla *et al.*, 1996; O'Hara *et al.*, 1995; Py *et al.*, 1996). Human mitochondrial mRNAs have a short poly(A) tail that is generated by a human mitochondrial (hmt) poly(A) polymerase (PAP) (Nagaike *et al.*, 2005; Tomecki *et al.*, 2004). In human mitochondria, polyadenylation by hmtPAP is essential to create UAA stop codons for seven of the mitochondrial mRNAs that lack genetically-encoded stop codons.

A 3′-end microdeletion ($\mu\triangle 9205$) of ATP6, which is associated with transient lactic acidosis, shortens the poly(A) tail and decreases the steady-state level of the ATP6 mRNA (Temperley *et al.*, 2003). Previously, we reported that silencing of hmtPAP expression by RNA interference results in shortened poly(A) tails and decreased steady-state levels of CO1, CO2, CO3, and ATP6 mRNAs (Nagaike *et al.*, 2005). These results suggest that polyadenylation stabilizes a certain subset of mitochondrial mRNAs. Precise measurement of the half-lives of mitochondrial mRNAs is essential to investigate the molecular mechanism regulating mRNA stability in mammalian mitochondria.

Here, we describe a detailed method to measure the half-lives of human mitochondrial mRNAs using a quantitative real-time reverse transcription (RT)-PCR method. In one application of this method, we quantified changes in the steady-state levels of mitochondrial mRNAs following siRNA-mediated knocking down of hmtPAP.

2. QUANTITATIVE RT-PCR METHOD FOR MEASURING mRNA DECAY

Quantitative RT-PCR (qRT-PCR) has become a powerful tool to measure steady-state amounts of individual mRNAs in small samples of total-cell RNA (Lutfalla and Uze, 2006). Initially, mRNAs are converted to cDNAs by reverse transcription using various primers, such as an oligo(dT) primer, a random primer, or a specific primer for a gene of interest. The cDNAs are used as templates and specifically amplified by a pair of primers for the transcript of interest, using an instrument for real-time PCR. As the PCR product increases, a fluorescent signal generated by double-stranded DNA binding dyes in the reaction solution increases and can be quantified in real time. Finally, a crossing-point (CP) value that corresponds to the amount of the target mRNA is calculated from the amplification curve. The relative amount of mRNA is quantified by comparison to a reference mRNA, such as glyceraldehyde-3-phosphate dehydrogenase (GAPDH) mRNA. In this study, we quantified 11 of the 13 mitochondrial transcripts in total RNA samples from HeLa cells.

3. PRIMER DESIGN

We designed specific oligonucleotide primers for human mitochondrial genes, as shown in Table 23.1, on the basis of the following parameters. First, a pair of primers for each gene was designed to amplify approximately 140 bp of the expected PCR product. Second, pairs of primers were designed to have a melting temperature of approximately 58–62 °C. Third, the primers were designed to be 20 nucleotides.

Some mtDNA sequences are present in nuclear genome (Pakendorf and Stoneking, 2005; Ruiz-Pesini et al., 2007). To specifically measure the half-lives of mitochondrially transcribed mRNAs, it is important to avoid amplification of mitochondrial sequences encoded in the nuclear genome. To verify that the primers used in this study do not amplify nuclear mitochondrial sequences, we used total RNA from rho0 cells, which were depleted of mtDNA (Hayashi et al., 1990), as a template for RT-PCR using the primer pairs designed in the preceding, and found that no amplification products were detected (data not shown).

As a reference mRNA, we used GAPDH mRNA, with the following primers: 5′-GTCTTCACCACCATGGAGAAGG-3′ (forward) and 5′-ATGATCTTGAGGCTGTTGTCAT-3′ (reverse).

Table 23.1   List of primers used in qRT-PCR of human mt mRNAs

mRNAs	Forward primers	Reverse primers
ND1	GAGCAGTAGCCCAAA CAATCTC	GGGTCATGATGGCAGG AGTAAT
ND2	TACGCCTAATCTACT CCACCTC	GGAGATAGGTAGGAGTAG CGTG
ND3	CCACCCCTTACGAGTGCGGCTT	TTGTAGGGCTCATGGTAGGGGT
ND4/4L	ACCTACTGGGAGAA CTCTCTGT	GGTGAGTGAGCCCCATTGTGTT
ND5	ACCGCACAATCCCCTATCTAGG	TTGGGTTGAGGTGATGATGGAG
ND6	TGGGGTTAGCGATGGGAGGTAGG	AATAGGATCCTCCCGAATCAAC
CO1	GGAGCAGGAACAGGTTGAACAG	GTTGTGATGAAATTGATGGC
CO2	CCCTTACCATCAAATCAATTGGCC	ATTGTCAACGTCAAGG AGTCGC
CO3	TCCTCACTATCTGCTTCATCCG	CCCTCATCAATAGATGGA GACA
ATP6/8	CACAACACTAAAGGACGAACCT	GGGATGGCCATGCTAGGTTTA
CytB	CTGATCCTCCAAATCACCACAG	GCGCCATTGGCGTGAAGGTA

4. ETBR TREATMENT

To measure half-life of mitochondrial mRNA, mitochondrial transcription must be turned off. Ethidium bromide (EtBr) selectively inhibits transcription of circular DNAs like mitochondrial DNA by altering the tertiary structure of the molecule (Hayashi *et al.*, 1990; Yasukawa *et al.*, 2000; Zylber *et al.*, 1969). To optimize the EtBr concentration, we examined the effect of different EtBr concentrations (0, 50, 100, 250, 500, and 1000 ng/ml) in the medium on the degradation rate of mitochondrial mRNAs and the growth rate of HeLa cells. In the range from 0 to 500 ng/ml EtBr, the degradation rate of each mitochondrial transcript increased in proportion to the EtBr concentration. In the presence of 1000 ng/ml EtBr, the degradation rate was similar to that observed in 500 ng/ml EtBr, which suggests that 500 ng/ml EtBr is a minimum saturating concentration for inhibiting mitochondrial transcription. The growth rate of cells treated with 500 ng/ml EtBr was similar to the growth rate of untreated cells, and the growth rate of cells treated with 1000 ng/ml EtBr was significantly decreased (data not shown). Therefore, we used 500 ng/ml of EtBr to disrupt mitochondrial transcription and enable us to measure the half-lives of the mitochondrial mRNAs.

5. CELL CULTIVATION AND RNA EXTRACTION

Semiconfluent cultured HeLa cells grown in DMEM containing 10% fetal bovine serum (FBS) were trypsinized and collected, and equal volumes of the cell suspension were plated in fresh dishes. When the cells reached approximately 80% confluency, the cells in each dish were washed with PBS and the medium was replaced with DMEM containing 10% FBS and 500 ng/ml EtBr (EtBr treatment) to start measuring mRNA degradation (t = 0 h). At various times (t = 0, 0.5, 1, 1.5, 2, 2.5, 3, 4, 5, and 6 h), the medium in each dish was discarded, and the cells were immediately solubilized by adding 1 ml of the TRI Reagent (SIGMA). The lysates were collected with a rubber scraper, transferred to a microcentrifuge tube, and vortexed. Then, 200 μl of chloroform was added to each tube and vortexed vigorously. The tubes were centrifuged at 20,000g for 15 min at 4 °C. The supernatant (600 μl), containing total RNA from HeLa cells, was carefully transferred to a new tube. An equal volume (600 μl) of 2-propanol was added to each supernatant and mixed thoroughly to precipitate the RNA, which was then centrifuged at 20,000g for 15 min at 4 °C . We used the RNeasy Mini Kit (QIAGEN), according to the manufacturer's instructions, to further purify the RNA samples. The amount of total RNA was quantified by measuring UV absorbance using a Gene Spec spectrometer (HITACHI).

6. RT-PCR

Total-cell RNA was treated with RNase-free DNase I (PROMEGA) according to the manufacturer's instructions. This step is required to avoid the amplification of mtDNA. Total-cell RNA samples (100 ng) were used as templates in each reverse transcription (RT) reaction with 5 (or 6) reverse primers (Table 23.1) specific for the genes of interest, including the GAPDH gene as the control in each tube. Reverse transcription (RT) was performed at 50 °C for 1 h using the PrimeScript 1st strand cDNA Synthesis Kit (TAKARA), according to the manufacturer's instructions.

We employed a LightCycler480 instrument (Roche Diagnostics) to perform real-time PCR. PCR reactions were carried out in a 20-μl reaction volume containing 10 μl of 2× SYBR *Premix ExTaq* II Buffer (TAKARA), 1 μl of cDNA as the template, and 250 nM of each primer pair (Table 23.1). Cycling conditions were as follows: 20 sec initial denaturation at 95 °C, followed by 50 cycles of 6 sec at 95 °C for denaturation, and 30 sec at 60 °C for annealing and extension. Fluorescence was detected at the end of each extension phase. After amplification, a melting curve analysis of the amplified DNA was performed between 65 and 95 °C. Data were analyzed and CP values were calculated using the LightCycler480 software.

7. HALF-LIVES OF MITOCHONDRIAL mRNAs

The CP values of each mitochondrial mRNA were normalized by subtracting the CP value of GAPDH mRNA. The percentage of each mitochondrial mRNA remaining over time was plotted and presented as a semilogarithmic plot. As shown in Fig. 23.1, ND1 mRNA decays rapidly compared to the ND4/4L transcript. Since the slope ($-\lambda$) of mRNA decay is correlated to the mRNA stability, the half-life ($t_{1/2}$) of each mRNA can be obtained from the formula, $t_{1/2} = \ln2/\lambda$. In Table 23.2, we summarize the $t_{1/2}$ values of all mitochondrial transcripts in the EtBr-treated HeLa cells. The HeLa cell mitochondrial mRNAs have half-lives ranging from 68 min (ND6) to 231 min (CO3), approximately a 3-fold difference in stability. The average of the half-lives was 125 min.

8. POLYADENYLATION STABILIZES MITOCHONDRIAL mRNAs

We previously reported that, in human mitochondria, polyadenylation stabilizes a certain subset of mitochondrial mRNAs (Nagaike *et al.*, 2005). To investigate whether the stability of each mitochondrial mRNA is

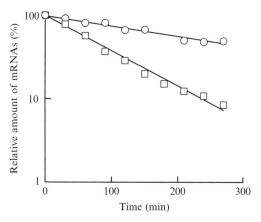

Figure 23.1 Decay of mitochondrial mRNAs in HeLa cells treated with EtBr. The fraction of the ND4/4L (circles) and ND1 (squares) mitochondrial mRNAs remaining at each time.

largely controlled by polyadenylation, we performed an siRNA-mediated knockdown of hmtPAP in HeLa cells, as described previously (Nagaike *et al.*, 2005), and measured the steady-state levels of mitochondrial transcripts by qRT-PCR. siRNAs for hmtPAP (sense strand, 5′-CUCUGAGA UGCAAAAUGAA-3′) and luciferase (sense strand, 5′-CGUACGCG-GAAUACUUCGA-3′) were designed using the siRNA design algorithm "siExplorer" (Katoh and Suzuki, 2007), and synthesized as described previously (Katoh *et al.*, 2003). HeLa cells (7.5×10^5 cells) were plated in 100-mm tissue culture dishes and cultured for 24 h. The medium was then changed to OPTI-MEM I (GIBCO) (10 mL per dish). The cells were transfected with 50 pmol of the siRNA for hmtPAP or luciferase (final concentration, 5 n*M*) using 20 μl of Oligofectamine (Invitrogen) for 5 h, and then added with 5 ml of DMEM containing 30% fetal bovine serum. At 96 h posttransfection, total RNA was extracted from the cells, as described previously. To estimate the efficiency of the knockdown, qRT-PCR was carried out to measure the level of hmtPAP mRNA using the following primers: 5′-AACCTCAGCGCTCACAAGAT-3′ (forward) and 5′-GGAACTTGTCAAGGCAATCC-3′ (reverse). We confirmed that the level of hmtPAP mRNA in the siRNA-treated cells decreased to less than 20% of the control cells treated with the luciferase siRNA (data not shown). We then measured the steady-state level of mitochondrial mRNAs from both cells. The relative amounts of mitochondrial mRNAs are shown in Fig. 23.2. The data revealed that the steady-state levels of CO1, CO2, CO3, and ATP6/8 mRNAs were significantly decreased, and levels of ND1, ND2, ND3, ND5, and CYTB mRNAs were unchanged.

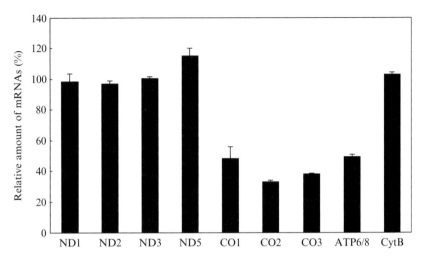

Figure 23.2 Relative steady-state level of mitochondrial mRNAs in HeLa cells treated with an siRNA targeting hmtPAP. The steady-state level of mitochondrial mRNAs from siRNA-treated cells was measured by qRT-PCR. The CP value of each mRNA from cells treated with the hmtPAP-siRNA was compared to the CP value of the same mRNA from cells treated with a luciferase-siRNA to derive the relative steady-state level of each mRNA (%).

9. Discussion

We have demonstrated that the steady-state level and half-lives of mitochondrial mRNAs can be measured by qRT-PCR. In this method, we used 500 ng/ml EtBr to inhibit mitochondrial transcription, to observe mitochondrial mRNA decay. Other inhibitors, such as actinomycin D, could also be used to halt mitochondrial transcription. However, actinomycin D is a strong inhibitor for nuclear transcription. EtBr is an ideal inhibitor for mitochondrial transcription. The obtained values for mRNA half-lives may vary in different cell lines or when measured using different methods. Importantly, our method enabled us to detect not only the intrinsic stability of the mitochondrial mRNAs but also the relative changes in mRNA stability in cells cultured under distinct conditions. Although 10 transcripts are generated as a single primary mRNA from the H-strand, the half-life of each of these mitochondrial mRNAs varied, ranging from 74 to 231 min. This observation indicates that there are protein factors involved in regulating the stability of each mitochondrial mRNA posttranscriptionally. Using Northern blotting analysis for mitochondrial mRNAs, we previously observed that the steady-state levels of mRNAs for CO1, CO2, CO3, and ATP6 decreased significantly when hmtPAP was knocked down (Nagaike *et al.*, 2005). However, the level of ND3 mRNA was unchanged in these experiments (Nagaike *et al.*, 2005), indicating that polyadenylation stabilizes a certain subset of mRNAs but not all mRNAs. Here, we used qRT-PCR to precisely measure the steady-state level of nine transcripts when hmtPAP was knocked down. We found that the mRNAs for CO1, CO2, CO3, and ATP6/8 were significantly destabilized, and the stability of the mRNAs for ND1, ND2, ND3, ND5, and CYTB was unchanged (Fig. 23.2). These results are consistent with our previous observations (Nagaike *et al.*, 2005). According to the half-life of each mRNA determined in this study (Table 23.2), mRNAs for CO1, CO2, CO3, and ATP6/8 have longer half-lives (138–231 min) than the average (125 min), whereas mRNAs for ND1, ND2, ND3, ND5, and CYTB (74–94 min) have shorter-than-average half-lives. Piechota and colleagues (Piechota *et al.*, 2006) showed that inhibition of hmtPAP caused accumulation of ND3 mRNAs with short oligo(A) tails but did not affect the overall stability of ND3 mRNA. This result is also consistent with our observations in this study. Thus, we conclude that long-lived mitochondrial mRNAs are largely stabilized by polyadenylation.

Polyadenylation of mitochondrial mRNAs is required not only for creating UAA stop codons but also for stabilizing mitochondrial mRNAs. A putative poly(A)-binding protein in mitochondria may be involved in this stabilization mechanism (Temperley *et al.*, 2003). In yeast mitochondria, all mRNAs lack a poly(A) tail, and mitochondrial mRNAs are stabilized by mRNA-specific factors, such as *NCA2*, *NCA3*, *NAM1*, and *AEP3* which stabilize the ATP6/8 mRNA (Camougrand *et al.*, 1995; Ellis *et al.*, 2004; Groudinsky *et al.*, 1993; Pelissier *et al.*, 1995). Although no mammalian homologues of these proteins have been identified, it is possible that

Table 23.2   Half-lives of mitochondrial mRNAs in HeLa cells

Transcripts	Half-life (min)
ND1	74 ± 2
ND2	80 ± 6
ND3	91 ± 4
ND4/4L	183 ± 5
ND5	77 ± 7
ND6	68 ± 10
CO1	166 ± 29
CO2	231 ± 18
CO3	138 ± 10
ATP6/8	175 ± 13
CytB	94 ± 24
Average	125

mRNA-specific factors may be involved in the stabilization of mammalian mitochondrial mRNAs. In that case, inhibiting nuclear transcription by actinomycin D indirectly influences stability of mitochondrial mRNAs.

ACKNOWLEDGMENTS

We are grateful to Drs. T. Nagaike and T. Katoh for technical support and advice. We are also thankful to J. Hayashi (Tsukuba Univ) for providing us rho0 cells. This work was supported by the Global COE Program for Chemistry Innovation (to A.N.) and grants-in-aid for scientific research on priority areas from the Ministry of Education, Science, Sports, and Culture of Japan (to T.S.).

REFERENCES

Beelman, C. A., and Parker, R. (1995). Degradation of mRNA in eukaryotes. *Cell* **81,** 179–183.

Camougrand, N., Pelissier, P., Velours, G., and Guerin, M. (1995). NCA2, a second nuclear gene required for the control of mitochondrial synthesis of subunits 6 and 8 of ATP synthase in *Saccharomyces cerevisiae. J. Mol. Biol.* **247,** 588–596.

Carpousis, A. J., Vanzo, N. F., and Raynal, L. C. (1999). mRNA degradation: A tale of poly(A) and multiprotein machines. *Trends Genet.* **15,** 24–28.

Chekanova, J. A., and Belostotsky, D. A. (2003). Evidence that poly(A) binding protein has an evolutionarily conserved function in facilitating mRNA biogenesis and export. *RNA* **9,** 1476–1490.

Duborjal, H., Beugnot, R., De Camaret, B. M., and Issartel, J. P. (2002). Large functional range of steady-state levels of nuclear and mitochondrial transcripts coding for the subunits of the human mitochondrial OXPHOS system. *Genome Res.* **12,** 1901–1909.

Ellis, T. P., Helfenbein, K. G., Tzagoloff, A., and Dieckmann, C. L. (2004). Aep3p stabilizes the mitochondrial bicistronic mRNA encoding subunits 6 and 8 of the H+-translocating ATP synthase of *Saccharomyces cerevisiae. J. Biol. Chem.* **279,** 15728–15733.

Groudinsky, O., Bousquet, I., Wallis, M. G., Slonimski, P. P., and Dujardin, G. (1993). The NAM1/MTF2 nuclear gene product is selectively required for the stability and/or processing of mitochondrial transcripts of the atp6 and of the mosaic, cox1 and cytb genes in *Saccharomyces cerevisiae. Mol. Gen. Genet.* **240,** 419–427.

Hajnsdorf, E., Braun, F., Haugel-Nielsen, J., and Regnier, P. (1995). Polyadenylylation destabilizes the rpsO mRNA of *Escherichia coli. Proc. Natl. Acad. Sci. USA* **92,** 3973–3977.

Hayashi, J., Tanaka, M., Sato, W., Ozawa, T., Yonekawa, H., Kagawa, Y., and Ohta, S. (1990). Effects of ethidium bromide treatment of mouse cells on expression and assembly of nuclear-coded subunits of complexes involved in the oxidative phosphorylation. *Biochem. Biophys. Res. Commun.* **167,** 216–221.

Katoh, T., Susa, M., Suzuki, T., Umeda, N., Watanabe, K., and Suzuki, T. (2003). Simple and rapid synthesis of siRNA derived from *in vitro* transcribed shRNA. *Nucleic Acids Res. Suppl.* 249–250.

Katoh, T., and Suzuki, T. (2007). Specific residues at every third position of siRNA shape its efficient RNAi activity. *Nucleic Acids Res.* **35,** e27.

Kudla, J., Hayes, R., and Gruissem, W. (1996). Polyadenylation accelerates degradation of chloroplast mRNA. *EMBO J.* **15,** 7137–7146.

Lutfalla, G., and Uze, G. (2006). Performing quantitative reverse-transcribed polymerase chain reaction experiments. *Methods Enzymol.* **410**, 386–400.

Nagaike, T., Suzuki, T., Katoh, T., and Ueda, T. (2005). Human mitochondrial mRNAs are stabilized with polyadenylation regulated by mitochondria-specific poly(A) polymerase and polynucleotide phosphorylase. *J. Biol. Chem.* **280**, 19721–19727.

O'Hara, E. B., Chekanova, J. A., Ingle, C. A., Kushner, Z. R., Peters, E., and Kushner, S. R. (1995). Polyadenylylation helps regulate mRNA decay in *Escherichia coli*. *Proc. Natl. Acad. Sci. USA* **92**, 1807–1811.

Pakendorf, B., and Stoneking, M. (2005). Mitochondrial DNA and human evolution. *Annu. Rev. Genomics Hum. Genet.* **6**, 165–183.

Pelissier, P., Camougrand, N., Velours, G., and Guerin, M. (1995). NCA3, a nuclear gene involved in the mitochondrial expression of subunits 6 and 8 of the Fo-F1 ATP synthase of *S. cerevisiae*. *Curr. Genet.* **27**, 409–416.

Piechota, J., Tomecki, R., Gewartowski, K., Szczesny, R., Dmochowska, A., Kudla, M., Dybczynska, L., Stepien, P. P., and Bartnik, E. (2006). Differential stability of mitochondrial mRNA in HeLa cells. *Acta Biochim. Pol.* **53**, 157–168.

Py, B., Higgins, C. F., Krisch, H. M., and Carpousis, A. J. (1996). A DEAD-box RNA helicase in the *Escherichia coli* RNA degradosome. *Nature* **381**, 169–172.

Ruiz-Pesini, E., Lott, M. T., Procaccio, V., Poole, J. C., Brandon, M. C., Mishmar, D., Yi, C., Kreuziger, J., Baldi, P., and Wallace, D. C. (2007). An enhanced MITOMAP with a global mtDNA mutational phylogeny. *Nucleic Acids Res.* **35**, D823–D828.

Sachs, A. B., Sarnow, P., and Hentze, M. W. (1997). Starting at the beginning, middle, and end: Translation initiation in eukaryotes. *Cell* **89**, 831–838.

Shatkin, A. J., and Manley, J. L. (2000). The ends of the affair: Capping and polyadenylation. *Nat. Struct. Biol.* **7**, 838–842.

Temperley, R. J., Seneca, S. H., Tonska, K., Bartnik, E., Bindoff, L. A., Lightowlers, R. N., and Chrzanowska-Lightowlers, Z. M. (2003). Investigation of a pathogenic mtDNA microdeletion reveals a translation-dependent deadenylation decay pathway in human mitochondria. *Hum. Mol. Genet.* **12**, 2341–2348.

Tomecki, R., Dmochowska, A., Gewartowski, K., Dziembowski, A., and Stepien, P. P. (2004). Identification of a novel human nuclear-encoded mitochondrial poly(A) polymerase. *Nucleic Acids Res.* **32**, 6001–6014.

Yasukawa, T., Suzuki, T., Suzuki, T., Ueda, T., Ohta, S., and Watanabe, K. (2000). Modification defect at anticodon wobble nucleotide of mitochondrial tRNAs(Leu) (UUR) with pathogenic mutations of mitochondrial myopathy, encephalopathy, lactic acidosis, and stroke-like episodes. *J. Biol. Chem.* **275**, 4251–4257.

Zylber, E., Vesco, C., and Penman, S. (1969). Selective inhibition of the synthesis of mitochondria-associated RNA by ethidium bromide. *J. Mol. Biol.* **44**, 195–204.

CHAPTER TWENTY-FOUR

DETECTION AND CHARACTERIZATION OF POLYADENYLATED RNA IN EUKARYA, BACTERIA, ARCHAEA, AND ORGANELLES

Shimyn Slomovic, Victoria Portnoy, *and* Gadi Schuster

Contents

Abstract

The posttranscriptional addition of poly(A) extensions to RNA is a phenomenon common to almost all organisms. In eukaryotes, a stable poly(A) tail is added to the 3′-end of most nucleus-encoded mRNAs, as well as to mitochondrion-encoded transcripts in animal cells. In prokaryotes and organelles, RNA molecules are polyadenylated as part of a polyadenylation-stimulated RNA degradation pathway. In addition, polyadenylation of nucleus-encoded transcripts in yeast and human cells was recently reported to promote RNA degradation. Not only

Department of Biology Technion, Israel Institute of Technology, Haifa, Israel

Methods in Enzymology, Volume 447
ISSN 0076-6879, DOI: 10.1016/S0076-6879(08)02224-6

homopolymeric poly(A) tails, composed exclusively of adenosines, but also heteropolymeric poly(A)-rich extensions, which include the other three nucleotides
as well, have been observed in bacteria, archaea, chloroplasts, and human cells. In
most instances, the detection of nonabundant truncated transcripts with post-
transcriptionally added poly(A) or poly(A)-rich extensions serves as a telltale sign
of the presence of a polyadenylation-stimulated RNA degradation pathway. In this
chapter, we describe several methods found to be efficient in detecting and
characterizing polyadenylated transcripts in bacteria, archaea, organelles, and
nucleus-encoded RNAs. Detailed protocols for the oligo(dT)- and circularized
reverse transcription (cRT) PCR methods, as well as the ribonuclease digestion
method, are outlined, along with examples of results obtained with these
techniques.

1. INTRODUCTION

Polyadenylated RNA molecules are present in most organisms. In eukaryotes, a stable poly(A) tail is added to the 3′-end of most nucleus-encoded
mRNAs. This process is important for mRNA stability, exit from the nucleus,
and translation initiation. Stable poly(A) tails are also present at the 3′-ends of
animal mitochondrial transcripts and are essential to establish functional translational stop codons for transcripts lacking such, as well as other postulated
functions (Gagliardi *et al.*, 2004). In prokaryotes and organelles, RNA molecules are polyadenylated as part of a polyadenylation-stimulated RNA degradation pathway (Slomovic *et al.*, 2006b). This process consists of three
sequential stages, initiating with endonucleolytic cleavage, followed by the
addition of degradation-stimulating poly(A) or poly(A)-rich sequences to these
cleavage products, and ending in exonucleolytic degradation. In addition,
polyadenylation of nucleus-encoded transcripts in yeast and human cells was
recently reported to promote RNA degradation, indicating that a form of
polyadenylation-stimulated RNA degradation is present in the nucleus
of eukaryotic cells as well, probably as part of a quality control mechanism
(Doma and Parker, 2007; Houseley *et al.*, 2006; Isken and Maquat, 2007;
Vanacova and Stef, 2007). In most instances, the detection of truncated
transcripts harboring poly(A) extensions serves as initial evidence hinting at
the presence of a polyadenylation-stimulated degradation RNA degradation
pathway (Slomovic *et al.*, 2008).

Not only homopolymeric poly(A) tails, composed exclusively of adenosines, but also heteropolymeric poly(A)-rich extensions, which include the
other three nucleotides as well, have been observed in bacteria, archaea,
chloroplasts, and human cells (Slomovic *et al.*, 2006b, 2008). Polynucleotide
phosphorylase (PNPase) and the archaeal exosome, which bear strong similarities to one another, both functionally and structurally, were found to poly-
merize the heteropolymeric tails in bacteria, spinach chloroplasts, and archaea.

Rapid degradation is the fate of the truncated RNA molecules, once polyadenylated, and, therefore, an efficient amplification and selection procedure is required to detect these molecules. The amplification power of the PCR method is essential. However, the combination of powerful amplification and the very low abundance of the intermediate degradation products results in nonspecific amplification of nonrelated sequences. To filter out these artifacts, the methods described in this chapter were developed and found to be highly efficient in detecting and/or isolating polyadenylated degradation intermediates derived from bacterial, archaeal, organellar, and nucleus-encoded RNAs (Lisitsky et al., 1996; Portnoy et al., 2005; Portnoy and Schuster, 2006; Rott et al., 2003; Slomovic et al., 2005, 2006a; see also chapters in RNA Turnover, Part B, ed. Maquat and Kiledjian).

Because the first method described here, oligo(dT)-primed RT-PCR, although highly efficient, is selective for relatively long polyadenylated tails, the picture obtained from this form of analysis does not always represent the entire mRNA population of a studied gene in terms of the ratio of adenylated versus nonadenylated transcripts and the span of tail lengths that is present. To obtain a more representative picture of the RNA population and tail (homo-hetero) consistency, a second, more quantitative, method should be used. Such a method is circularized reverse transcription (cRT)—RT-PCR. Actually this technique is usually applied to analyze stable poly(A) tails located at the mature $3'$-ends of animal mitochondrial transcripts and nucleus-encoded mRNAs (Slomovic and Schuster, 2008). In many cases, applying this method to detect RNA degradation intermediates in various systems results in the isolation of molecules either lacking any adenosine tails or containing short extensions of only several adenosines (Slomovic et al., 2006a; West et al., 2006). This is most likely due to the rapid pace of degradation once the endo-cleavage products are tagged by polyadenylation but also shows that, in some systems, most of these posttranscriptional extensions are oligo(A) tails.

The third method described here is the $3'$-end labeling and ribonuclease digestion analysis of poly(A) tails. This method is relatively simple and provides a global picture of the polyadenylation of all the gene transcripts in the tested organism at once. Likewise, organisms in which only unstable or no polyadenylation occurs, such as prokaryotes/organelles or certain archaea, respectively, can be analyzed and categorized with this method (Lisitsky et al., 1996; Mohanty et al., 2004; Portnoy et al., 2005). In eukaryotes containing nucleus-encoded RNA with stable poly(A) tails, this method can disclose the lengths of these extensions (Bonisch et al., 2007).

Additional methods for the detection and analysis of $3'$-end extensions have been developed, including the oligonucleotide ligation RT-PCR technique, which provides results similar to those achieved with cRT-RT-PCR.

This method has the advantage that only sequences attached to the $3'$-end are observed (Aravin and Tuschl, 2005; Elbashir *et al.*, 2001; Pfeffer *et al.*, 2003). However, additional uses of cRT-PCR, such as the analysis of $5'$-processing, cannot be achieved with oligo ligation. See also protocols at http://web. wimit.edu/bartel/pub/protocols_reagents.htm and http://banjo.dartmouth edu/lab/MicroRNAs/mir.html.

The hybrid-selection RT-PCR method, which includes an additional selective step before circularization in the cRT-PCR, enables the detection of very low abundant and gene-specific extensions, for example, tails added to the pre-mRNA from the human β-globin gene during pre-mRNA processing in the nucleus (Rissland *et al.*, 2007; West *et al.*, 2006). The smart RT-PCR protocol enables the detection of a population of adenylated RNA without the need to use a gene-specific oligonucleotide (Holec *et al.*, 2006). Once isolated and sequenced, the transcript is subsequently identified with a BLAST algorithm. This method is mostly useful for the identification of unknown polyadenylated transcripts found in prokaryotes and organelles.

2. Oligo(dT) RT-PCR Detection of Polyadenylated Degradation Intermediates

The detection of nonabundant, truncated, polyadenylated RNA molecules corresponding to a studied gene sequence is considered to be a telltale sign of the presence of a polyadenylation-stimulated RNA degradation pathway, as witnessed in prokaryotes and organelles. These molecules, believed to be degradation intermediates between the sequential stages of endonucleolytic cleavage and polyadenylation, are rapidly $3'$ to $5'$-exonucleolytically digested once polyadenylated. Their consequent low abundance makes them difficult to detect and isolate relative to the full-length transcript and the stable poly(A)-tails characterizing most eukaryotic mRNAs and animal mitochondria transcripts. Therefore, a powerful method composed of a number of stages of amplification and specification is necessary to isolate such molecules.

Oligo(dT)-adapter primed RT-PCR is a technique that allows the isolation and sequence identification of the degradation intermediates described previously. Therefore, the sites within the transcript sequence at which polyadenylation occurred can be detected. Furthermore, the nature of the posttranscriptionally added extensions, in terms of nucleoyide composition, can be determined as well. This, in turn, can serve as the first step in identifying the enzyme responsible for its polymerization (Mohanty and Kushner, 2000; Rott *et al.*, 2003; Slomovic *et al.*, 2008).

2.1. Steps 1 and 2: RNA purification and cDNA synthesis

Figure 24.1 shows a schematic description of the nine steps that comprise this method. When studying an organellar system, although organelles can be isolated before RNA purification, total-cellular RNA purification is usually adequate (step 1). RNA can be generated free of contaminating DNA by incubation with DNase. However, this is usually not necessary. Total-cell RNA is reverse transcribed, primed with an adapter oligo that includes a (dT) track at its 3′-end (step 2). The adapter sequence should be ~20 nt excluding the oligo(dT). The length of the (dT) stretch should be adequate to allow annealing at the temperature at which the RT reaction is performed; T_9 has been applied efficiently but longer oligos of up to T_{18} can be used as well (Lisitsky et al., 1996; Portnoy et al., 2005; Rott et al., 2003; Slomovic et al., 2005).

2.2. Steps 3 to 6: PCR amplification and gel extraction

During the RT reaction, the oligo(dT)-adapter can anneal at any point along the poly(A)-tails and, therefore, it is difficult to confirm its original length. Only transcripts with poly(A) extensions undergo reverse transcription, and the resulting first-strand cDNA contains the adapter sequence at its 5′-end (Fig. 24.1, step 3). An RT reaction with 5 μg of total-cell RNA should finally be diluted to 50 μl from which 1 μl will be used to template each PCR reaction in the next step. In this step, the cDNA is PCR-amplified with a forward primer, termed F1, specific to the chosen gene. The reverse primer is the adapter oligo lacking the oligo(dT) track (step 3). It is recommended that one apply multiple PCR reactions (up to 4 or a total volume of 80 μl) to increase the total amount of product. A Taq polymerase with A-tailing activity (in which an adenosine overhang is added to each end of the PCR product) is essential for the T/A cloning step (step 7) described at a later stage in the protocol.

In step 4, the PCR products can either be purified with an appropriate kit or electrophoresed in and eluted from a 1% agarose gel. In both cases, duplicate or triplicate reactions should be purified together. In all stages of this method, PCR templates should not be used as templates taken directly from the previous PCR step, because remnants of the F1 primer will have a negative effect on the specificity of subsequent stages. Indeed, the preferred method is to run the PCR products through an agarose gel, as shown in Fig. 24.1 (step 4), because this allows the exclusion of nonspecific bands visible on ethidium bromide staining. Because the PCR products are a mixture of molecules with 3′-ends terminating at different sites relative to the full-length mRNA, a smear, rather than a discrete band, is expected after electrophoresis (Fig. 24.1, step 4). Moreover, any discrete bands should be avoided, because they are most likely due to either nonspecific amplification

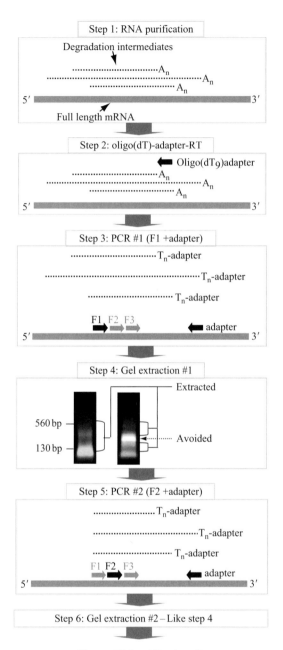

Figure 24.1 (*Continued*)

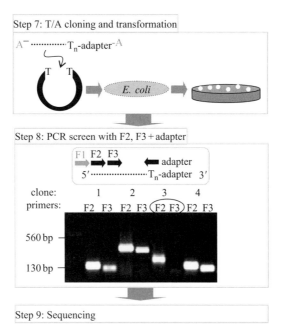

Figure 24.1 Schematic presentation of the oligo(dT) RT-PCR protocol developed for the detection and isolation of nonabundant, truncated, polyadenylated RNA molecules that are intermediates of the polyadenylation-stimulated RNA degradation pathway.

of undesired genes or even annealing of either the F1 primer or adapter in opposing directions (step 4). Such a product could dominate the T/A cloning stage described in the following. Likewise, during all stages of PCR, it is useful to include negative control reactions that use each of the primers alone and a reaction templated by the products from an RT reaction in which no reverse transcriptase has been added. These negative controls allow easy identification of unwanted or irrelevant gel bands that are due to nonspecific amplification and amplification of DNA contaminants.

The region between 100 and 600 bp should be excised from the gel and eluted with an appropriate kit at a final volume of 10 to 15 μl. If the minimal elution volume exceeds this, the purified product volume can be decreased by use of a speed-vac. In step 5, PCR #2 is performed with the adapter oligo paired with F2, a second gene-specific primer, nested immediately downstream of the F1 primer. In PCR #2, 1 μl of gel-purified product from PCR #1 should be used. This second PCR, which uses a nested gene-specific forward primer before the T/A cloning stage, is essential for product specificity as it excludes almost all unwanted products that were amplified during PCR #1 due to chance sequence similarity between the 3'-end of the F1 primer and unrelated cDNA. This simple step increases specificity as much as 500%. PCR #2 products should be purified as described for PCR #1 (step 6).

2.3. Steps 7 to 9: Cloning of the PCR products, selection, and sequencing analysis

In step 7, the entire volume of purified PCR #2 products is inserted into a T/A vector. The ligation reaction is precipitated and resuspended in 8 μl H$_2$O, of which 1 μl is introduced into competent cells via electroporation. LB plates containing the appropriate selective antibiotic(s), β-galactosidase, and IPTG are then used to select colonies that potentially harbor a T/A vector with an insert (step 7). Ideally, each insert is expected to have been derived from a different truncated, polyadenylated RNA molecule corresponding to the studied gene.

In the eighth step, white, antibiotic-resistant colonies are screened with two consecutive PCR reactions after transferring each screened colony to a second LB plate. The first PCR reaction is identical to PCR #2, in which the F2 primer is paired with the adapter (step 8). The second PCR reaction, applied to the same colony, uses a third gene-specific forward primer, F3, which is nested immediately downstream of the F2 primer in the gene sequence, along with the adapter. The products of these two PCR reactions are electrophoresed side by side in a 1% agarose gel (step 8). In each pair, the first lane will ideally display one product band of any size within the range of 100 to 600 bps, according to the region excised from the gel in step 4. If the insert, indeed, corresponds to the studied gene, the second lane of the pair will display a single product band which is ~20 bp smaller than that seen in the first lane, assuming that the F3 primer is nested immediately downstream of the F2 oligo and is of 20 nt in length. In Fig. 24.1, step 8, four PCR pairs from four screened colonies are shown. All but the third pair are positive, because the second lane of this pair shows no product.

The final step in the oligo(dT)-adapter primed RT-PCR method is the sequencing of each positive clone after plasmid purification (step 9). Although strict specificity measures are taken in all steps leading up to this stage, sequencing is essential to fully identify the plasmid insert. When aligned with the full-length mRNA sequence, sites at which the truncated molecules are polyadenylated can be determined. However, one must be aware that as few as six adenosines encoded in the gene sequence are sufficient to cause false annealing of the oligo(dT) during RT. Therefore, cases in which the polyadenylation sites of sequenced clones coincide with A-rich regions within the encoded mRNA sequence should be treated with suspicion. As stated previously, not only the polyadenylation sites but the nature of the extensions, in terms of homo/heterogeneity of nucleotide content, can be examined and used to identify the enzyme responsible for this posttranscriptional activity. Although homopolymeric poly(A) tails are produced by poly(A) polymerases, heteropolymeric tails, consisting of all four nucleotides, are produced by polynucleotide-phosphorylase (PNPase) or the archaeal exosome (Mohanty and Kushner, 2000; Portnoy and Schuster, 2006; Portnoy *et al.*, 2005; Rott *et al.*, 2003; Yehudai-Resheff *et al.*, 2001). An example of poly(A) tails of heteropolymeric and homopolymeric nucleotide content, identified with the oligo(dT) RT-PCR method described previously, is presented in Fig. 24.2.

3. Circularized Reverse Transcription (cRT)-PCR-Sequencing/Labeling Method for the Characterization of Polyadenylated RNA

As described previously, oligo(dT) RT-PCR allows efficient isolation of nonabundant, truncated, polyadenylated RNA molecules. However, the nature by which this method selectively amplifies polyadenylated RNA is the very reason that it cannot be applied if the goal is to not only isolate such molecules, but also assess the unaltered length of the poly(A) extensions, because the oligo(dT) can anneal at any point along the tail during the RT reaction. Also, the oligo(dT) RT-PCR is a "fishing" method for isolating

Cyanobacteria; *Synechocystis: rbcL*
1. $A_2GA_2GA_4G_2A_3G_2A_3GAGAUA_{12}GA_9$
2. $GA_9TA_4UAAG_2A_8GUA_4GU_2A_2UAGA_6GUGCAGA_4CA_8GATA_2CA_2GACA_2GUAUA_3CGA$
   $UA_{12}UA_5CAGAUA_4UA_9U_3ACG_3CACAGA_2UAAGCA_3GAGAUGA_3UAGGACA_6$

Archaea; *Sulfolobus solfataricus:16S rRNA*
1. $A_2GA_2GA_4G_2A_3G_2A_3GAGAUA_{12}GA_9$
2. $AGAUA_3CUGA_2GACAGA_7G_2A_4GA_2UA_4GAUAGAGAUA_4UAGUAGAG_3AUGA_3GACU$
   $A_{12}G_2AUA_{17}$

Human Mitochondria: *CoxI*
1. A_{77}
2. A_{45}
3. A_{23}

Human nucleus: *28S rRNA*
1. $AGAUA_3CUGA_2GACAGA_7G_2A_4GA_2UA_4GAUAGAGAUA_4UAGUAGAG_3AUGA_3GACU$
   $A_{12}G_2AUA_{17}$
2. $A_4G_2A_3GA_3GAGA_4GA_{16}$
3. A_{70}
4. A_{45}

Figure 24.2 Examples of several homopolymeric and heteropolymeric tails isolated and sequenced with oligo(dT) RT-PCR. References describing the detection of these tails are: Cyanobacteria (Rott *et al.*, 2003), Archaea (Portnoy *et al.*, 2005), human mitochondria (Slomovic *et al.*, 2005), and human nucleus–encoded rRNA (Slomovic *et al.*, 2006a).

nonabundant polyadenylated molecules and, therefore, does not indicate which fraction of the transcript population is adenylated and to what extent. For this purpose, a method that is not based on poly(A) selectivity must be used.

cRT-PCR is often used to assess the length and nature of stable post-transcriptional extensions at the mature 3'-ends of mRNA (Nagaike *et al.*, 2005; Slomovic and Schuster, 2008; Temperley *et al.*, 2003). Such stable poly(A) tails characterize most nucleus-encoded mRNA in eukaryotic cells and also mitochondrial RNA in animal cells.

3.1. Steps 1 and 2: RNA isolation, circularization, and RT

Figure 24.3A presents a schematic description of the application of cRT-PCR for the evaluation of stable poly(A) tails. The same guidelines for RNA purification as described for the oligo(dT) RT-PCR method should be applied here. Once purified, ~5 μg of total RNA is circularized by ligation of the 5'- and 3'-ends by T4 RNA ligase. When studying nucleus-encoded mRNA, tobacco acid pyrophosphatase (TAP) must be applied before this stage, because ligation is hampered by the 5'-cap. TAP hydrolyzes the phosphoric acid anhydride bonds in the triphosphate bridge of the cap structure, releasing the cap nucleoside and generating a 5'-phosphorylated terminus. Alternately, oligonucleotide-directed RNase H cleavage can be applied to remove the 5'-region of the transcript, thereby producing a 5'-phosphorylated terminus ready for the ligation step. When studying mRNA generated from polycistronic RNA from organelles or prokaryotes, this step is not necessary, because such molecules lack a 5'-cap or three phosphates of the transcription initiation. If the ligation reaction included DNase to digest contaminant DNA, a phenol/chloroform purification step should be included to avoid DNase contamination at later stages.

In the next step, the entire resuspension of the ligation reaction, after precipitation, is subjected to reverse transcription (RT) and should, therefore, be resuspended in an appropriate volume. As shown in Fig. 24.3A, the reverse oligo used to prime the RT reaction, termed R1, should be designed to anneal to the mRNA sense strand ~100 nt downstream of the predicted 5'-end (if RNase H cleavage was applied, the "new" 5 end). During RT, the enzyme reverse transcribes, initiating at the R1 primer, crossing the adjoined 5'- and 3'-ends, including any posttranscriptional extensions, until it eventually detaches. Once completed, the RT reaction should be diluted to a final volume of 50 μl from which 1 μl will be taken for each subsequent PCR reaction, as described in the following.

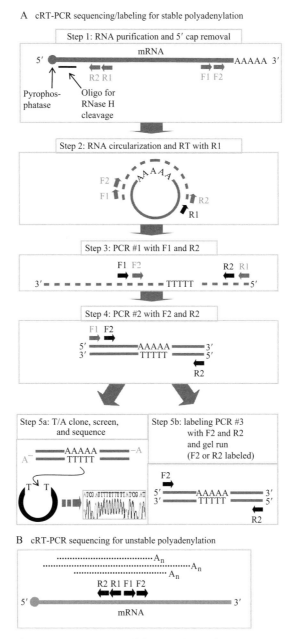

Figure 24.3   Schematic presentation of the circularized (c)RT-PCR protocol for the quantitative analysis of both stable (A) and unstable (B) poly(A)-tails.

3.2. Steps 3 and 4: PCR amplification

In the first PCR reaction, PCR #1, a reverse primer (relative to the sense directionality of the original mRNA sequence) termed R2, which is nested upstream of the R1 primer and ~50 nt downstream of the 5′-end, is used. This oligo is paired with a forward primer, termed F1, which is designed to anneal to the mRNA ~100 nt upstream of the 3′-end (Fig. 24.3A). As described in the previous paragraph for oligo(dT) RT-PCR, duplicate or triplicate reactions should be either run through 1% agarose gel, excised, and eluted or cleaned with an appropriate PCR purification kit. Unlike the oligo(dT) isolation of truncated polyadenylated RNA molecules, when examining mature 3′-end stable polyadenylation with cRT-PCR, the expected product size can be estimated. In the case described here, the expected product size from PCR #1 includes 50 nt from the R2 primer to the 5′-end and 100 nt from the F1 primer to the 3′-end. The product size is increased by any posttranscriptional extensions at the 3′ (or 5′) end. Therefore, when excising the PCR #1 products from the agarose gel, the excised gel fragment should start from the expected product size and include higher molecular weight to accommodate any 3′-tails. These products are then cleaned and eluted to a volume of ~10 μl.

In the next PCR stage, PCR #2, 1 μl from the gel elution is used to template each reaction. The same R2 primer from the PCR #1 step is used, along with a second forward primer, F2, which is nested downstream of the F1 primer and 50 nt upstream of the 3′-end (Fig. 24.3A). The products from the PCR #2 reactions are cleaned by electrophoresis through 1% agarose gel, like the PCR #1 products, but the expected product size is now ~100 bp, plus any increase in size because of the posttranscriptional tail. The elution volume, as before, is ~10 μl.

3.3. Step 5a: Analysis by DNA sequencing

At this point, two different approaches can be chosen: cloning and sequencing or radioactive labeling and detection by gel fractionation and autoradiography. For the sequencing alternative (step 5a), half of the elution volume is used for a cloning reaction with a T/A vector system, as in the case of oligo(dT) RT-PCR, and the other half is reserved for the labeling option, described later. The steps that include T/A cloning, precipitation, and transformation to competent cells are as described previously for oligo(dT) RT-PCR. In contrast to the oligo(dT) RT-PCR colony screening, here, individual resistant white bacterial colonies are PCR screened with a single PCR reaction, identical to PCR #2, with the R2 and F2 primers. Ideally, each positive colony contains a T/A vector with an insert derived from the adjoined 5′- and 3′-ends of the studied mRNA, including any 3′ (or 5′) extensions. The difference in molecular weight of PCR products between different colonies is due only to variations in poly(A)-tail lengths, assuming that the RNA molecules they originated from were properly processed and, therefore, initiated and terminated at the same points (Slomovic and Schuster, 2008).

After plasmid purification from positive colonies and sequencing, the processing of the RNA molecules and the nature and length of the 3'-extensions can be assessed. It must be noted that after sequencing, when applying this cRT-PCR method, it is not possible to determine whether a posttranscriptional extension was originally added to the 5' or 3'-end of the RNA molecule. If, in the given system, it is yet unknown whether 5'-extensions exist and it is, therefore, unclear if the observed tail is 5' or 3', it is possible to apply a similar procedure in which, instead of RNA circularization, an oligonucleotide is ligated to the 3'-end of the transcript and serves as a platform for the annealing of a reverse primer (coupled with a gene-specific forward primer) during RT and subsequent PCR reactions (Elbashir et al., 2001).

3.4. Step 5b: Analysis by radioactive labeling

The second approach that can be applied after the PCR #2 gel excision and elution stage is radioactive labeling (step 5b). Unlike the sequencing approach, which involves analysis of individual clones and, in turn, the assessment of 5'- and 3'-processing and the homo/hetero composition of the poly(A) tail, the labeling approach reveals a more global view of the entire mRNA population of the studied gene at the time of RNA purification. Therefore, trends of lengthening or shortening of the poly(A) tails in the RNA population of the studied gene can be easily detected, whereas with the sequencing approach, a large (statistical) amount of clones need to be sequenced to gain a general picture of the polyadenylation status of a studied gene (Nagaike et al., 2005; Slomovic and Schuster, 2008). The labeling approach is based on the radioactive labeling of the products from PCR #2, during a third PCR reaction, termed PCR #3. To achieve this, either the R2 or F2 oligos must be labeled with $[\gamma^{32}P]$-ATP and polynucleotide kinase (PNK).

We advise first performing two negative control PCR reactions to determine which of the two oligos, R2 or F2, is more suitable to be used in PCR #3. In each of these negative control reactions, 1-μl template from the PCR #2 elution is used, and only one of the primers is present. On examination by ethidium bromide agarose gel staining, the primer to be labeled will be that which does not display any dominant gel bands when reacted with itself, which could interfere with the final analysis. Once either the R2 or F2 primer is chosen, it is labeled in a standard labeling reaction with PNK and $[\gamma^{32}P]$-ATP. 100 ng of the oligo is recommended to be used in the labeling reaction, which, after precipitation, is resuspended in ~ 10 μl H_2O. For the PCR #3 labeling reaction, 1-μl template from the PCR #2 elution is applied per total 20-μl

reaction; 5 ng of the unlabeled oligo and 5 ng of the labeled oligo (either R2 or F2) are added to the reaction. In the case of the labeled primer, it is possible to simply add ~0.5 μl from the 10 μl of labeled oligo, because this is approximately 5 ng (50 nM in the total PCR volume of 20 μl). Between 6 and 10 cycles are sufficient to achieve labeled PCR products.

In the final stage, the [^{32}P]-labeled PCR #3 products are fractionated through denaturing 10% acryl amide gel; 5 to 10 μl of the 20 μl are sufficient and should be denatured in a formamide sample buffer for 2 min at 90 °C before loading. The gel run should be long enough to achieve high resolution and then exposed in a radiography cassette. A molecular weight marker can be achieved by producing an [α^{32}P]-UTP–labeled synthetic RNA (transcribed with T7, SP6, or T3 RNA polymerases and a transcription plasmid) or by labeling a DNA nucleotide sequence of known length with PNK and [γ^{32}P]-ATP. These markers supply a point of reference from which band length (in nucleotides) can be counted. Because nucleotide-nucleotide resolution can be reached with 10% acryl amide gel, in the PCR #3 products lane, every possible length is usually represented by a gel band, because of natural detachment of the Taq polymerase during PCR. If a clear nucleotide-nucleotide representation is not apparent in the PCR product lane, such a ladder can be achieved by producing a second [α^{32}P]-UTP–labeled synthetic RNA and fragmenting it by alkaline hydrolysis, by incubating it at 90 °C in buffer carbonate (pH = 11) for various time points. From the point of reference supplied by the [^{32}P]-DNA marker or the [α^{32}P]-UTP–labeled synthetic RNA, one can identify the ~100 nt region, which is the expected size of the PCR product lacking any additional extensions (poly(A) tail), by counting the bands either in the PCR product lane or the [^{32}P]-labeled RNA fragmented by alkaline hydrolysis. From this "zero" point, one can climb the gel bands in the PCR product lane like a ladder while counting, until reaching a point at which the gel bands drastically intensify. The distance in nucleotide units (gel bands) from the "zero" point (3' in Fig. 24.4) to the intense gel bands is the length of the poly(A) extensions. Because the PCR #3 products originated from a population of mRNA molecules of the studied gene with poly(A) tails of varying lengths, a discrete band is not expected to be observed, rather, a number of bands comprising an average length (Fig. 24.4).

Figure 24.4 shows an example of two gel runs of the human mitochondrial mRNAs, COX1 (A) and ND3 (B) (Slomovic and Schuster, 2008). In the COX1 gel (A), the first lane displays a [^{32}P]-DNA marker, the second lane shows a [^{32}P]-labeled RNA, fragmented by alkaline hydrolysis, and the third lane is the PCR #3 products lane. The "zero" point is marked as (3'), and a polyadenylated fraction with tails ranging from ~35 to 55 adenosines in length is apparent. In the ND3 gel (B), two fractions of the studied mRNA population can be observed in each of the two lanes; an oligoadenylated fraction, with tails of up to 15 nucleotides in length, and a

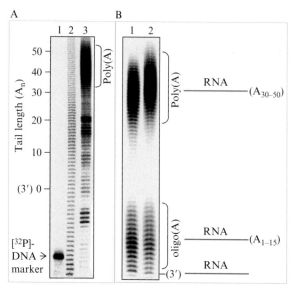

Figure 24.4 Example of results obtained with cRT-PCR labeling. (A) The stable poly (A)-tails located at the 3'-end of the human mitochondrial COX1 transcript were analyzed as described in the text and Fig. 24.3. Lane 1, [^{32}P]-DNA marker; Lane 2, *in vitro*-transcribed [^{32}P]-labeled RNA ladder after alkaline hydrolysis; Lane 3, total RNA from HeLa cells. The DNA marker is used, along with the alkaline hydrolysis ladder, to identify the band, in Lane 3, that represents the COX1 mRNA lacking any 3'-extensions (position marked as 3'-0). From this "zero" point, the rungs of the ladder are counted until reaching the polyadenylated fraction (here, between 35 and 55 adenosines). (B) Human mitochondrion-encoded ND3 mRNA was analyzed by cRT-PCR. Total-RNA purified from HeLa cells (Lane 1) and from a HeLa cell line with stable RNAi-mediated silencing of polynucleotide phosphorylase (PNPase) (Lane 2) was subjected to the protocol described in Fig. 24.3 (Slomovic and Schuster, 2008). Aside from two distinct RNA fractions observed in both lanes, one oligoadenylated and the other polyadenylated, a slight increase in the lengths of the poly(A)-tails of the polyadenylated RNA fraction in lane 2 is apparent. To the right of the gel picture, the examples of RNA molecules are schematically shown: a bare mRNA, oligoadenylated mRNA, and polyadenylated mRNA.

polyadenylated fraction with extensions of ∼30 to 50 adenosines in length. In lane 2, in which the process was applied to RNA isolated from a cell line with constitutive RNAi silencing of the enzyme, polynucleotide phosphorylase (PNPase), an increase in the average length of the polyadenylated fraction is apparent compared with the poly(A) fraction of the nonsilenced cell line in lane 1 (Slomovic and Schuster, 2008). Although the difference is clear, it is nonetheless slight and, therefore, would be difficult to detect on sequencing of even a large number of clones, without applying the labeling assay (Nagaike *et al.*, 2005; Slomovic and Schuster, 2008).

4. cRT-PCR for the Detection of Truncated Polyadenylated RNA Degradation Intermediates

cRT-PCR assays can be easily applied for the analysis of stable 3′-polyadenylation, because the PCR product is of a predetermined size, depending on the locations along the gene sequence at which the R2 and F2 primers are designed to anneal, relative to the 5′- and 3′ of the studied transcript. However, when the goal is to isolate and analyze truncated polyadenylated degradation intermediates with cRT-PCR, the product size cannot be anticipated, because there is no knowledge of where the truncated molecules begin or end. Therefore, the design of the F1, F2, R1, and R2 primers is different than described previously. In this case, the four primers should be grouped as close as possible to one another, as shown in Fig. 24.3B. The closer the primers are to one another, the higher the chance that a truncated molecule will span the entire region covered by the primers. Otherwise, its isolation would not be possible with this primer set.

When applying cRT-PCR to truncated molecules, the labeling approach is much less conclusive than its application during the analysis of stable 3′-polyadenylation. However, the cRT-PCR sequencing assay can be applied and yield useful results. Other than the primer design described previously, there are no major differences in the protocol for isolating truncated molecules. One disadvantage of this technique is that a high percentage of isolated molecules, although truncated, lack poly(A) tails (Slomovic and Schuster, 2008; Slomovic et al., 2006a). Such molecules could be between the stages of endonucleolytic cleavage and polyadenylation or, alternately, may have been polyadenylated but already partially exonucleolytically digested from 3′ to 5′. This is a consequence of the low abundance and rapid digestion of these degradation intermediates and the fact that this assay was especially designed to lack a bias toward polyadenylated RNA to avoid altering the natural poly(A) tail length or report a false ratio between polyadenylated and nonadenylated RNA, as occurs when applying oligo(dT) RT-PCR.

4.1. Polyadenylation analysis by 3′-end labeling and ribonuclease digestion

The method of 3′-end labeling, followed by ribonucleolytic digestion, is an efficient way to assess the general polyadenylation status of a particular organism or isolated organelle (Bonisch et al., 2007; Hajnsdorf et al., 1995; Lisitsky et al., 1996; Mohanty et al., 2004; Portnoy and Schuster, 2006; Portnoy et al., 2005; Rott et al., 2003). The principle of this method is presented in Fig. 24.5. First, total RNA is isolated and [^{32}P]-labeled at

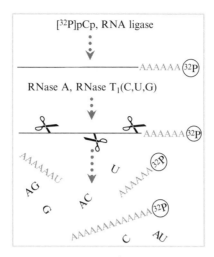

Figure 24.5 Schematic presentation of the 3′-end-labeling and ribonuclease digestion method used for the general assessment of poly(A)-tail length in a given organism or organelle. Total-cell RNA is purified and [^{32}P]-labeled at the 3′-ends, followed by complete digestion with RNase A and RNase T$_1$. Adenosine tracks remain intact, and those from the 3′-end are labeled and, therefore, can be visualized when resolving the digestion products with denaturing polyacrylamide gel electrophoresis and autoradiography.

the 3′-end. This is usually performed by use of [^{32}P]-pCp and T4 RNA ligase. Special care should be taken when choosing the T4 RNA ligase, because some of the vendors' preparations are contaminated with RNA fragments that can interfere when analyzing RNA preparations. Such contamination can cause an organism lacking polyadenylation to appear as if polyadenylation actually exists. Therefore, we advise performing a negative control reaction, in which no RNA is added, before analyzing the tested RNA. From our experience, the T4 RNA ligase of New England Biolabs was contaminated with a small amount of RNA molecules, whereas that obtained from Ambion was found to be clean of any residual polyadenylated transcripts. An alternative to [^{32}P]-pCp labeling is [^{32}P]-labeling with [γ-^{32}P] 3′-dATP by the enzyme, poly(A)-polymerase (Wahle, 1991).

The next step is the complete digestion of the RNA by simultaneously applying RNase A and RNase T1. These enzymes efficiently cleave ribonucleic acids after the G, U, and C residues but not after A. Therefore, only adenosine stretches remain intact after digestion, and only those situated at the 3′-end are radioactively labeled. In a typical reaction, digestion of 20 μg RNA with 25 μg of RNase A and 300 units of RNase T1 (both obtained from Sigma) for 1 h at 37 °C is sufficient. After digestion, the RNA is purified by phenol extraction and ethanol precipitation.

Next, the RNA is fractionated on denaturing polyacrylamide gel. Usually, a long (40 cm) 14% acrylamide/1.5% bisacrylamide gel is used and of adequate acrylamide (AA) percentage to observe tails as short as 4 to 5 adenosines (Mohanty *et al.*, 2004; Portnoy *et al.*, 2005). If shorter tails, of 2 to 3 adenosines, are to be detected, a 20% acrylamide gel can be used. Likewise, if significantly longer tails are of interest, gels of lower percentage should be used. Oligonucleotides of known length, labeled with $[\gamma^{32}P]$-ATP and polynucleotide kinase (PNK), are used as size markers to count the number of adenosines in the tails, similar to the description of markers used in the cRT-PCR labeling procedure. When such a marker is clearly observed in the autoradiogram, the number of adenosines can easily be counted along the ladder of labeled tails that remain intact after the RNase digestion (Fig. 24.6). Alternately, RNA molecules of known length that are synthesized *in vitro* with bacteriophage RNA polymerase and $[^{32}P]$-labeled can be applied.

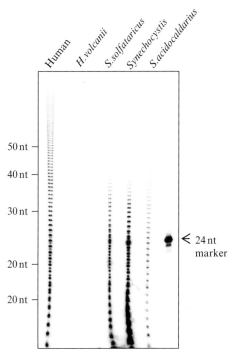

Figure 24.6 Detection of poly(A)-tails with the 3′-end labeling and ribonuclease digestion method. RNA, purified from several resources, as shown at the top of the figure, was $[^{32}P]$-labeled at the 3′-ends, digested with RNase A and RNase T₁, and resolved by denaturing polyacrylamide gel electrophoresis and autoradiography. The lengths of the tails were determined with a 24-nt oligonucleotide size marker as a point of reference. Long poly(A)-tails were observed in the human RNA preparation, reflecting the stable polyadenylation of the mRNAs. Shorter tails were observed in the cases of hyperthermophilic Archaea and Cyanobacteria, reflecting the unstable poly(A)-tails associated with a polyadenylation-stimulated RNA degradation pathway. No adenosine extensions were detected in the RNA from the halophilic archaea, *H. volcanii*, indicating that RNA does not undergo any form of polyadenylation in this organism (Portnoy *et al.*, 2005). (Reproduced with permission from Portnoy *et al.* [2005].)

In systems in which RNA undergoes polyadenylation, a ladder of poly(A) that is almost equally distributed between all the lengths (from a certain minimum to a certain maximum length) is usually observed, indicating that approximately the same amount of each tail length is present at any given time. The reason for this is not completely clear, but it indicates the level of synchronization between the adenylation and degradation rates.

Not only can this technique be efficiently applied to detect changes in the lengths of the stable poly(A) tail population in eukaryotic cells, but it can also be used to analyze unstable poly(A) extensions involved in polyadenylation-stimulated RNA degradation pathways in prokaryotes and organelles. In addition, this method has been used to reveal the few currently known organisms that metabolize RNA without polyadenylation (Fig. 24.6) (Portnoy and Schuster, 2006; Portnoy et al., 2005; Slomovic et al., 2008).

REFERENCES

Aravin, A., and Tuschl, T. (2005). Identification and characterization of small RNAs involved in RNA silencing. *FEBS Lett.* **579,** 5830–5840.

Bonisch, C., Temme, C., Moritz, B., and Wahle, E. (2007). Degradation of hsp70 and other mRNAs in *Drosophila* via the 5′ to 3′-pathway and its regulation by heat shock. *J. Biol. Chem.* **282,** 21818–21828.

Doma, M. K., and Parker, R. (2007). RNA quality control in eukaryotes. *Cell* **131,** 660–668.

Elbashir, S. M., Lendeckel, W., and Tuschl, T. (2001). RNA interference is mediated by 21- and 22-nucleotide RNAs. *Genes Dev.* **15,** 188–200.

Gagliardi, D., Stepien, P. P., Temperley, R. J., Lightowlers, R. N., and Chrzanowska-Lightowlers, Z. M. (2004). Messenger RNA stability in mitochondria: Different means to an end. *Trends Genet.* **20,** 260–267.

Hajnsdorf, E., Braun, F., Haugel-Nielsen, J., and Regnier, P. (1995). Polyadenylation destabilizes the rpsO mRNA of *Escherchia coli*. *Proc. Natl. Acad. Sci. USA* **92,** 3973–3977.

Holec, S., Lange, H., Kuhn, K., Alioua, M., Borner, T., and Gagliardi, D. (2006). Relaxed transcription in *Arabidopsis* mitochondria is counterbalanced by RNA stability control mediated by polyadenylation and polynucleotide phosphorylase. *Mol. Cell. Biol.* **26,** 2869–2876.

Houseley, J., LaCava, J., and Tollervey, D. (2006). RNA-quality control by the exosome. *Nat. Rev. Mol. Cell. Biol.* **7,** 529–539.

Isken, O., and Maquat, L. E. (2007). Quality control of eukaryotic mRNA: Safeguarding cells from abnormal mRNA function. *Genes Dev.* **21,** 1833–1856.

Lisitsky, I., Klaff, P., and Schuster, G. (1996). Addition of poly(A)-rich sequences to endonucleolytic cleavage sites in the degradation of spinach chloroplast mRNA. *Proc. Natl. Acad. Sci. USA* **93,** 13398–13403.

Mohanty, B. K., and Kushner, S. R. (2000). Polynucleotide phosphorylase functions both as a 3′ to 5′-exonuclease and a poly(A) polymerase in *Escherichia coli*. *Proc. Natl. Acad. Sci. USA* **97,** 11966–11971.

Mohanty, B. K., Maples, V. F., and Kushner, S. R. (2004). The Sm-like protein Hfq regulates polyadenylation dependent mRNA decay in *Escherichia coli*. *Mol. Microbiol.* **54,** 905–920.

Nagaike, T., Suzuki, T., Katoh, T., and Ueda, T. (2005). Human mitochondrial mRNAs are stabilized with polyadenylation regulated by mitochondria-specific poly(A) polymerase and polynucleotide phosphorylase. *J. Biol. Chem.* **280,** 19721–19727.

Pfeffer, S., Lagos-Quintana, M., and Tuschl, T. (2003). Cloning of small RNA molecules. *In* "Current Protocols in Molecular Biology" (F. M. Ausubel, R. Brent, R. E. Kingston, D. D. Moore, J. G. Seidmann, J. A. Smith, and K. Struhl, eds.), pp. 26.24.16–26.24.21. Wiley, New York.

Portnoy, V., Evguenieva-Hackenberg, E., Klein, F., Walter, P., Lorentzen, E., Klug, G., and Schuster, G. (2005). RNA polyadenylation in Archaea: Not observed in *Haloferax* while the exosome polyadenylates RNA in *Sulfolobus*. *EMBO Rep.* **6,** 1188–1193.

Portnoy, V., and Schuster, G. (2006). RNA polyadenylation and degradation in different Archaea; roles of the exosome and RNase R. *Nucleic Acids Res.* **34,** 5923–5931.

Rissland, O. S., Mikulasova, A., and Norbury, C. J. (2007). Efficient RNA polyuridylation by noncanonical poly(A) polymerases. *Mol. Cell. Biol.* **27,** 3612–3624.

Rott, R., Zipor, G., Portnoy, V., Liveanu, V., and Schuster, G. (2003). RNA polyadenylation and degradation in cyanobacteria are similar to the chloroplast but different from *Escherichia coli*. *J. Biol. Chem.* **278,** 15771–15777.

Slomovic, S., Laufer, D., Geiger, D., and Schuster, G. (2005). Polyadenylation and degradation of human mitochondrial RNA: The prokaryotic past leaves its mark. *Mol. Cell. Biol.* **25,** 6427–6435.

Slomovic, S., Laufer, D., Geiger, D., and Schuster, G. (2006a). Polyadenylation of ribosomal RNA in human cells. *Nucleic Acids Res.* **34,** 2966–2975.

Slomovic, S., Portnoy, V., Yehudai-Resheff, S., Bronshtein, E., and Schuster, G. (2008). Polynucleotide phosphorylase and the archaeal exosome as poly(A)-polymerases. *Biochem. Biophys. Acta* **1779,** 247–255.

Slomovic, S., Portnoy, V., Liveanu, V., and Schuster, G. (2006b). RNA polyadenylation in prokaryotes and organelles; different tails tell different tales. *Crit. Rev. Plant Sci.* **25,** 65–77.

Slomovic, S., and Schuster, G. (2008). Stable PNPase RNAi silencing; its effect on the processing and adenylation of human mitochondrial RNA. *RNA* **14,** 310–323.

Temperley, R. J., Seneca, S. H., Tonska, K., Bartnik, E., Bindoff, L. A., Lightowlers, R. N., and Chrzanowska-Lightowlers, Z. M. (2003). Investigation of a pathogenic mtDNA microdeletion reveals a translation-dependent deadenylation decay pathway in human mitochondria. *Hum. Mol. Genet.* **12,** 2341–2348.

Vanacova, S., and Stef, R. (2007). The exosome and RNA quality control in the nucleus. *EMBO Rep.* **8,** 651–657.

Wahle, E. (1991). Purification and characterization of a mammalian polyadenylate polymerase involved in the 3′-end processing of messenger RNA precursors. *J. Biol. Chem.* **266,** 3131–3139.

West, S., Gromak, N., Norbury, C. J., and Proudfoot, N. J. (2006). Adenylation and exosome-mediated degradation of cotranscriptionally cleaved pre-messenger RNA in human cells. *Mol. Cell* **21,** 437–443.

Yehudai-Resheff, S., Hirsh, M., and Schuster, G. (2001). Polynucleotide phosphorylase functions as both an exonuclease and a poly(A) polymerase in spinach chloroplasts. *Mol. Cell. Biol.* **21,** 5408–5416.

CHAPTER TWENTY-FIVE

RNA DECAY BY MESSENGER RNA INTERFERASES

Mikkel Christensen-Dalsgaard,* Martin Overgaard,[†]
Kristoffer Skovbo Winther,* *and* Kenn Gerdes*

Contents

Abstract

Two abundant toxin-antitoxin (TA) gene families, *relBE* and *mazEF*, encode mRNA cleaving enzymes whose ectopic overexpression abruptly inhibits translation and thereby induces a bacteriostatic condition. Here we describe and discuss protocols for the overproduction, purification, and analysis of mRNA cleaving enzymes such as RelE of *Escherichia coli* and the corresponding antitoxin RelB. In particular, we describe a set of plasmid vectors useful for the detailed analysis of cleavage sites in model mRNAs.

* Institute for Cell and Molecular Biosciences, Medical School, University of Newcastle, Newcastle, United Kingdom
† Department of Biochemistry and Molecular Biology, University of Southern Denmark, Odense, Denmark

Methods in Enzymology, Volume 447
ISSN 0076-6879, DOI: 10.1016/S0076-6879(08)02225-8

1. INTRODUCTION

Prokaryotic toxin-antitoxin (TA) loci code for a metabolically stable toxin and a metabolically unstable antitoxin. The antitoxin combines with and neutralizes the toxin by direct protein-protein contact. As of 2007, eight different prokaryotic TA gene families had been identified (Gerdes *et al.*, 2005; Makarova *et al.*, 2006). Two large, evolutionarily unrelated TA gene families, *relE* and *mazF*, encode mRNA Interferases (mIs) whose ectopic overexpression inhibits cell growth by mediating mRNA degradation. In their natural contexts, mIs are always expressed together with their cognate antitoxin, and cell death is not usually a consequence of the presence of TA gene loci in a cell. Rather, mIs encoded by TA gene loci can be viewed as regulators that help the cell adjust the degree of global translation during environmental stresses. In this vein, the term "toxin" can be viewed as an anachronism that has adhered since TA gene loci were discovered in the early 1980s. This view is consistent with the important finding that ectopic overexpression is bacteriostatic rather than bacteriocidal (Pedersen *et al.*, 2002). However, ectopic overexpression of mIs is truly toxic in the sense that cell growth is instantaneously blocked (Gotfredsen and Gerdes, 1998; Zhang *et al.*, 2004, 2005).

Toxin-antitoxin loci were discovered on plasmids as a result of their ability to increase plasmid maintenance (Bravo *et al.*, 1987; Ogura and Hiraga, 1983). Thus, the *ccd* and *parD* loci of F and R1, respectively, are TA gene loci that increase plasmid maintenance by inhibiting the growth of plasmid-free segregants (Hiraga *et al.*, 1986; Jaffe *et al.*, 1985; Jensen *et al.*, 1995). Both gene loci encode stable toxins (CcdB and Kid/PemK) and unstable antitoxins (CcdA and Kis/PemI) (Tsuchimoto *et al.*, 1992; Van Melderen *et al.*, 1994). The differential stabilities of the TA-encoded components explain the seemingly paradoxical inhibition of the growth of plasmid-free cells: because the antitoxin is unstable, plasmid-free cells experience activation of the toxin. Consequently, their growth is inhibited and, at the level of the cell culture, this leads to phenotypic stabilization of the plasmid. Careful-flow cytometric analyses were consistent with the proposal that CcdB and Kid/PemK are bacteriostatic rather than bacteriocidal (Jensen *et al.*, 1995).

Early on, homologues of plasmid-encoded TA gene loci were discovered on bacterial chromosomes (Gotfredsen and Gerdes, 1998; Masuda *et al.*, 1993). The later enormous expansion of the prokaryotic DNA databases made feasible the in-depth search for TA gene loci in more than 200 prokaryotic chromosomes (Pandey and Gerdes, 2005). This analysis revealed ≈1200 TA gene loci distributed in a surprising pattern: almost all obligatory intracellular bacteria are devoid of TA gene loci, whereas all

Archaea and almost all free-living bacteria have TA gene loci in their chromosomes. In particular, slowly growing bacteria such as *Mycobacterium tuberculosis* and *Nitrosomonas europaea* have more than 60 and 50 TA gene loci, respectively. Many of these ≈1200 loci (e.g., *relBE*, *higBA*, and *mazEF*) encode mIs that inhibit translation by degradation of mRNA (Christensen-Dalsgaard and Gerdes, 2006; Pedersen *et al.*, 2003; Zhang *et al.*, 2003). Here we describe the methods we have used to characterize mRNA-cleaving enzymes encoded by TA loci.

2. *In Vitro* Analysis of RelE mRNA Interferase

2.1. Expression and Purification of RelE

2.1.1. General considerations

Expression and purification of mIs like RelE require special attention as these enzymes are very efficient inhibitors of cell growth; that is, without their cognate antitoxins, they inhibit their own synthesis. Although strains resistant to the gyrase inhibitor CcdB have been isolated (Bernard and Couturier, 1992), no strains resistant to mIs have yet been reported. We have nevertheless found that *E. coli* C41 (DE3) (Miroux and Walker, 1996), a derivative of strain BL21 (DE3) selected for expression of toxic recombinant proteins, gave higher induction levels and yielded higher cell densities when coexpressing RelB and His_6-RelE than did other common expression strains used. Specifically, we coexpressed RelB and His_6-RelE with sufficient yield the RelB and RelE from a single expression vector, as described in the following.

2.1.2. Plasmid construction

After several trials, we found that coexpression of RelE and RelB yielded a useful path to obtain sufficient amounts of the proteins. As the vector plasmid, we used pMG25, a derivative of pUHE24-2 that contains the efficient phage T7-derived P_{A1} promoter, two *lac* operators O4/O3, and *lacI* (to reduce expression in the absence of the inducer, IPTG) (Lanzer and Bujard, 1988). A map of pMG25 is shown in Fig. 25.1A. Genes SD::*his6*:: *relE* and SD::*relB* were constructed using PCR, primer pairs RelB1 + RelB2 and RelE1 + RelE2 (Table 25.1) and, as template, chromosomal DNA of strain MG1655 that had been digested with appropriate restriction enzymes and inserted into pMG25, yielding pSC2524HE. This plasmid efficiently coexpressed His_6-RelE and untagged RelB when IPTG was added to growing cells.

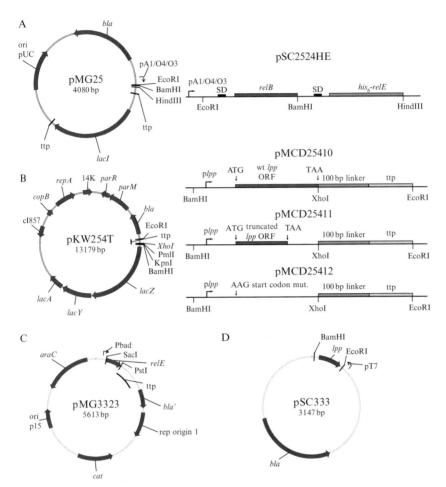

Figure 25.1 Plasmids used. (A) The expression-vector pMG25 (*left*) was derived from pUHE24-2, a pUC19 derivative carrying the strong, *LacI*-regulated T7-derived PA1-O4/O3 promoter (5 × p*lac*). pMG25 was constructed as follows: A fragment carrying the *lacI* gene was PCR-amplified using primers lacINheI and lacI2BsmI and DNA from strain of MG1655 as template. The resulting PCR fragment was restricted using NheI and BsmI and ligated with pUHE24-2 that had been restricted with the same enzymes. Plasmid pMG25 is identical to pUHE24-2 except it contains *lacI* inserted into the *cat* gene of pUHE24-2. pMG25 was subsequently used to construct pSC2524HE. A physical and genetic map of pSC2524HE is shown to the right. The plasmid contains the wild-type *relB* gene upstream of the *his_6*-*relE* gene. On IPTG addition to growing cells, both genes are transcribed into a single mRNA. The resulting mRNA contains efficient ribosome-binding sites for each open reading frame. (B) pKW254T (*left*) is a derivative of the low-copy-number plasmid pOU254 (Gerdes lab), which contains a transcriptional terminator including a 100-bp upstream sequence originating from pMG25. A schematic presentation of the genetic content of plasmids used to express wild-type and mutant forms of *lpp* mRNA is shown in the right side of the figure. The translational start and stop codons are indicated with arrows. The mRNAs are transcribed from the native *lpp* promoter but have been fused to a 100-bp linker-RNA in the

Table 25.1 DNA primers (shown in the 5' to 3'-orientation)

Name	Sequence
TRANSTERM#CCW	CCCCGAATTCGATTCAACCCCTTC TTCGATC
term-BamHI-KpnI-SacI-PmlI-XhoI#CCW	CCCCCGGATCC GGTACCGAGCTCC ACGTGCTCGAGGCCAAGCTTAATTAG CTGAGC
lpp-BamH1-cw	CCCCCGGATCCGAGCTCGGAAGCATC CTGTTTTCTCTC
lpp-XhoI-ccw	CCCCCCACGTGCTCGAGGTACTATTACT TGCGGTATTTAG
lpp-1AAG-cw	GAGGGTATTAATAAAGAAAGCTACTAAAC
lpp-1AAG-ccw	GTTTAGTAGCTTTCTTTATTAATACCCTC
lpp-6ACT-cw	ATGAAAGCTACTAAAACTGGTACTGGGC GCGG
lpp-6ACT-ccw	CCGCGCCCAGTACCAGTTTTAGTAGC TTTCAT
lpp 26	CAGCTGGTCAACTTTAGCGTTCAGAG
T7-hok	TGTAATACGACTCACTATAGGCGC TTGAGGC
T7-3N	AAGGCGGGCCTGCGCCCGCCTCCAGG
lacINheI	CCCCGCTAGCGGAAAAGCATCTT CCGGCGC
lacI2BsmI	CCCCAGCATTCGAGTGAGCTAACTCAC ATTAATTGCG

2.1.3. His$_6$-RelE expression and purification

Plasmid pSC2524HE used to transform C41 (DE3) cells (Miroux and Walker, 1996), and an overnight culture was diluted into 2 L of 2× YT medium containing 100 mg/ml ampicillin (Amp) and cultured at 37 °C. At OD$_{450}$ ~0.5, expression was induced by adding 2 mM of IPTG. After 4 h, the culture was harvested by centrifugation, and cells were resuspended in ice-cold buffer A (50 mM NaH$_2$PO$_4$, 0.3 M NaCl, 10 mM imidazole,

---

3'-end. The plasmid pMCD25410 contains the wild-type *lpp* gene, whereas pMCD25412 contains a mutant form of *lpp*, in which an adenine base has been inserted in the sixth codon, thus creating a frameshift mutation in the gene. Plasmid pMCD25411 contains the *lpp* gene, in which the start codon has been changed from ATG to AAG. (C) pMG3323, which derives from pBAD33 (Guzman *et al.*, 1995), contains a copy of the *relE* gene of *E. coli* inserted downstream of the arabinose-inducible pBAD promoter (Christensen and Gerdes, 2003). (D) The *lpp* gene of *E. coli* was inserted into pGEM3 to generate pSC333 (Christensen and Gerdes, 2003). After linearization with BamHI, the plasmid is used for *in vitro* transcription of an RNA probe that is complementary to the *lpp* mRNA.

5 mM β-mercaptoethanol, pH 8, supplemented with two tablets of Complete EDTA-free Protease Inhibitor Cocktail [Roche Molecular Biochemicals]). Cells were disrupted by passing them three times through a French press, and the soluble fraction was isolated by centrifugation twice for 30 min at 48,400g. The cleared lysate was then incubated with Ni–NTA agarose resin (Qiagen) in batches for 60 min according to the manufacturer's instructions and subsequently loaded onto a gravity flow column at 4 °C. The Ni–agarose resin was washed extensively in buffer B (50 mM NaH$_2$PO$_4$, 0.3 M NaCl, 35 mM imidazole, 1 mM β-mercaptoethanol, pH 8). RelB · His$_6$-RelE was then bound to the column material.

RelB was separated from His$_6$-RelE at room temperature by first eluting in two column volumes of buffer C (100 mM NaH$_2$PO$_4$, 10 mM Tris-HCl, 9.8 M urea, 1 mM β-mercaptoethanol, pH 8) to remove bulk RelB and then incubating the Ni–NTA agarose resin in two column volumes of buffer C in a 50-ml tube overnight with rotation to remove remaining bound RelB. Finally, the resin was applied to a column, washed by another column volume of buffer C, and eluted in buffer D (100 mM NaH$_2$PO$_4$, 10 mM Tris-HCl, 9.8 M urea, 0.5 M imidazole, 1 mM β-mercaptoethanol, pH 8). A fraction (<10%) of His$_6$-RelE may still be bound by RelB. At this point, His$_6$-RelE can be diluted to <0.2 mg/ml and refolded directly by successive steps of dialysis against (a) 1× PBS, pH 7.4, 5 mM DTT, 0.1% Triton X-100, (b) 1× PBS, pH 7.4, 5 mM DTT, and (c) 1× PBS, pH 7.4, 5 mM DTT, 20% glycerol. Alternatively, His$_6$-RelE was diluted to <0.2 mg/ml and dialyzed against buffer E (50 mM MES pH 6, 20 mM NaCl), subjected to cation-exchange chromatography and eluted using a linear gradient of buffer E + 1 M NaCl. His$_6$-RelE was finally dialyzed against storage buffer (1× PBS, pH 7.4, 5 mM DTT, 20% glycerol) and stored in aliquots frozen at −80 °C.

2.1.4. RelB purification

The eluted fractions of RelB in buffer C from above were concentrated, dialyzed against buffer F (20 mM Tris-HCl, pH 8, 20 mM NaCl), subjected to anion-exchange chromatography, and eluted using a linear gradient of buffer F + 1 M NaCl. Finally, RelB was dialyzed against storage buffer (25 mM Tris, pH 8, 100 mM KCl, 1 mM EDTA, 1 mM DTT, 10% glycerol) and stored in aliquots frozen at −80 °C.

2.2. Analysis of RelE-mediated inhibition of *in vitro* translation using *E. coli* S30 extracts

2.2.1. Synthesis of a test mRNA

The preparation and purification of uniformly [³H]-labeled RNA molecules were performed essentially as described (Franch and Gerdes, 1996). Briefly, RNAs were synthesized using T7 RNA polymerase and templates generated by PCR. In all PCR reactions, pBR322 carrying *hok/sok* of R1 (pPR633) was used as the template (Gerdes *et al.*, 1986). The T7 promoter region and wild-type mRNA 5'-end was specified by the upstream oligonucleotide T7-*hok*. The downstream oligonucleotide specifying the 3'-end of the RNA run-off transcript was T7-3N.

2.2.2. In vitro *translation reactions*

Purified RelE was analyzed for inhibition of translation in an *E. coli*–coupled transcription/translation system (Zubay, 1973) purchased from Promega. Each reaction mix contained the following solutions: 6 μl of S30 premix, 4.5 μl of S30 extract, and 1.5 μl of amino acid mix without methionine (1 mM). The assay was started by adding 1 μl of purified RelE (diluted in polymix buffer, 20% glycerol) followed by incubation for 5 min at 37 °C. After this preincubation, 1μl of 3 μM (3 pmol) test mRNA (*hok* mRNA from plasmid R1), 1 μl of [^{35}S]-methionine, and 1 μl of polymix buffer or 1 μl of purified RelB in polymix buffer were added. After an additional 25 min at 37 °C, the reactions were precipitated with acetone and analyzed using SDS–PAGE.

2.3. RelE cleavage of mRNA A-site codons *in vitro*

The *in vitro* analysis of RelE cleavage of mRNA codons positioned at the ribosomal A-site was performed in the Ehrenberg Laboratory and described previously (Freistroffer *et al.*, 1997; Pedersen *et al.*, 2003) and will not be presented here.

3. *In Vivo* Analysis of mRNA Interferase

We have analyzed mI activity using Northern blotting and primer extension analyses of specific RNAs. Whereas Northern blotting provides a quantitative measure of the abundance (and thus stability) of a given RNA species, with or without induction of the mIs of interest, primer extension analysis is suitable for detecting specific mIs-mediated cleavage sites in RNAs. When radiolabeled nucleotides are used to label the DNA or RNA probes, these techniques are extremely sensitive and can be used to study RNAs (including cleavage products) that are present in very low concentrations in the cell.

3.1. Development of a general mRNA probe vector plasmid

Both Northern blotting and primer extension can easily be applied to investigate whether specific endogenously expressed RNAs are degraded by certain mIs. However, when studying the detailed cleavage properties of mIs, it is often useful to generate plasmid vectors that express the model RNA of interest. This has two advantages. First, the RNA can be expressed in higher concentrations than, for example, a chromosome-encoded RNA, thereby increasing the resolution of the analyses. Second, it is possible to introduce specific mutations in the RNAs and observe how they affect RNA cleavage and turnover. It is important to note that when analyzing RNAs expressed from plasmids, the chromosomal gene encoding the RNA usually must be deleted. To study how translation affects mI-mediated cleavage of RNAs, our lab has developed several plasmids that express either wild-type or mutated RNAs. Previously, such plasmids expressing *lpp* mRNA or tmRNA with or without functional start or resume codons have been used to show that RelE and related mIs require translation to degrade RNA (Christensen *et al.*, 2003; Christensen and Gerdes, 2003; Christensen-Dalsgaard and Gerdes, 2006). In this section, we describe how a more recent series of plasmids encoding mRNA probes were generated.

Plasmid pKW254T (Fig. 25.1B) is a derivative of the low copy-number plasmid pOU254 (Gotfredsen and Gerdes, 1998), in which a transcriptional terminator from the plasmid pMG25 (Fig. 25.1A), including an upstream sequence of ~100 bp, has been inserted downstream from a multiple cloning region. The terminator region from pMG25 was amplified using purified pMG25 and primers TRANSTERM#CCW and term-BamHI-KpnI-SacI-PmlI-XhoI#CCW. The PCR product was digested with EcoRI and BamHI and inserted into pOU254. Thus, fragments inserted into the multiple cloning region of pKW254T will be transcribed into fusion mRNAs containing a ~100-bp linker region downstream from the translational stop codon. By annealing a linker region primer to fusion mRNAs, it is possible to analyze full-length reading frames by primer extension. We found that it was often difficult to map 3′-ends of natural mRNAs because of the extensive folding of the 3′-end terminators of wild-type mRNAs.

Plasmids, pMCD25410, pMCD25411, and pMCD25412 are derivatives of pKW254T and carry inserts of either wild-type or mutated forms of the *lpp* gene. The genetic content of each plasmid is schematically illustrated in Fig. 25.1B. As seen, the plasmids were constructed to express three different variants of the *lpp* mRNA fused to the linker DNA as described previously. The composite genes are all transcribed from the native *lpp* promoter. Plasmid pMCD25410 contains the wild-type *lpp* ORF, whereas pMCD25412 or pMCD24411 contain the *lpp* ORF harboring, respectively, a single A insertion within the sixth codon or a point mutation that converts the ATG start codon to AAG (Fig. 25.1B). Briefly, the wild-type *lpp* fragment was amplified using PCR, chromosomal DNA, and primers lpp-BamHI-cw and lpp-XhoI-ccw (see Table 25.1 for primer sequences). The two mutant *lpp* fragments were constructed by site-directed mutagenesis using two rounds of PCR. To introduce the mutation in the start codon of *lpp*, two fragments of the gene were amplified by PCR primers lpp-BamHI-cw and lpp-1AAG-ccw in addition to lpp-1AAG-cw and lpp-xhoI-ccw. The two overlapping PCR products were used as template in a second round of PCR, with the primers lpp-BamH1-cw and lpp-xhoI-ccw. Similarly, the frameshift mutation in *lpp* was constructed by creating two overlapping PCR products using the primers lpp-BamHI-cw and lpp-6ACT-ccw in addition to lpp-6ACT-cw and lpp-xhoI-ccw. The last round of PCR was carried out using the primers lpp-BamHI-cw and lpp-xhoI-ccw with the two overlapping PCR fragments as templates. The three resulting PCR fragments containing the wild-type or mutant *lpp* gene were digested with BamHI and XhoI and inserted into pKW254T.

3.2. Cell growth and RNA purification

Plasmids pMCD25410, pMCD25411, and pMCD25412 were transformed into *E. coli* MG1655Δ*lpp* cells. Subsequently, pMG3323 (pBAD::*relE*), a derivative of the low-copy-number plasmid pBAD33 (Guzman *et al.*, 1995), in which the *relE* gene of *E. coli* has been inserted downstream from the arabinose-inducible pBAD promoter (Fig. 25.1C), was transformed into these strains. The plasmid expresses RelE on addition of arabinose (0.2%) to growing cells.

Cells were grown exponentially in LB at 37 °C. At an OD_{450} of 0.5, the cultures were diluted 10 times and grown to an OD_{450} of 0.5, at which point the transcription of the *relE* gene was induced by addition of 0.2% arabinose, defining time zero. As controls in the RNA analyses, cells of MG1655Δ*lpp* carrying plasmids pMCD25410, pMCD25411, and pMCD25412 were exposed to chloramphenicol (50 µg/ml) at time zero (like *relE* induction, chloramphenicol inhibits translation). Samples of 10 ml were taken at time 2, 10, 30, and 60 min, and harvested at 6000 rpm for RNA purification.

Total RNA was purified from each sample using standard phenol/chloroform extraction. Briefly, cell pellets were resuspended in 200 µl of 0.3 *M* sucrose, 0.01 *M* of Na-acetate, pH 4.5, and mixed with 200 µl of 2% SDS, 0.01 *M* of Na-acetate, pH 4.5, to lyse the cells. The lysis solution was then mixed with 400 µl of phenol (pH 4.5), heated at 65 °C for three minutes, quickly frozen in liquid nitrogen, and spun at 14,000 rpm for 5 min. The aqueous phase was recovered and extraction with 400 µl of phenol was repeated. To gain further purity of the RNA, the aqueous phase was mixed with 200 µl of phenol (pH 4.5) and 200 µl of chloroform and spun at 14,000 rpm for 5 min. Finally, the aqueous phase was mixed with 80 µl of K-acetate (1.5 *M*) and 900 µl of ice-cold 96% ethanol. The RNA was precipitated at −80 °C and the RNA pelleted by spinning the samples at 14,000 rpm for 20 min. The RNA was washed once in 70% ethanol and resuspended in TE-buffer. The concentrations of the RNA samples were then measured and the quality tested in an agarose gel.

3.3. Northern analysis of *lpp* mRNA

Ten μg of total RNA from each sample was fractionated by PAGE (6% low-bis acrylamide) and blotted to a Zeta-probe nylon membrane overnight. The membrane was then prehybridized in ULTRAhyb (Ambion) hybridization buffer for 1 h at 68 °C, followed by the addition of the single-stranded [^{32}P]-labeled riboprobe and left overnight at 68 °C. The hybridization buffer was discarded and the membrane washed in 2 × 5 min in 2 × SSC, 0.1% SDS at 68 °C followed by wash at 2 × 15 min in 0.1× SSC, 0.1% SDS at 68 °C. The membrane was then exposed overnight to a PhosphorImager screen (GE Healthcare), which was scanned on GE Healthcare's Typhoon scanner.

Generally, we use uniformly labeled [^{32}P] RNA probes for Northern analysis that are significantly more efficient than end-labeled DNA probes. A radiolabeled riboprobe, complementary to the *lpp* mRNA, was generated by *in vitro* transcription using T7 RNA polymerase. The DNA template for the transcription reaction was prepared by digesting the plasmid pSC333 (Fig. 25.1D) with BamHI, thus generating a linear DNA fragment containing the reverse complement of the *lpp* gene downstream of the T7 promoter (Christensen and Gerdes, 2003). Alternatively, a DNA template for *in vitro* transcription can be constructed by recombinant PCR using *lpp*-specific oligonucleotide primers where the antisense primer contains the T7 promoter (reverse complement). The following reagents were mixed and incubated at 37 °C for 1 h to generate the RNA probe by *in vitro* transcription: 4 μl of 5× transcription buffer (Fermentas), 2 μl of DTT (0.1 M), 0.8 μl of RNA guard (Amblicon), 4 μl of 3 NTP mix (2.5 mM of ATP, GTP, and UTP), 2.4 μl of 100 μM CTP, 1 μl of linearized template plasmid (0.5 μg/μl), 5 μl of α-[^{32}P]-CTP (50 μCi) (>400 Ci/mmol), and 1 μl of T7 RNA polymerase (Fermentas). The DNA template was removed by adding 1.0 μl of DNase I (NEB) to the Eppendorf tube followed by 15 min of incubation at 37 °C.

A representative example of a Northern blot prepared using the protocol described previously is shown in Fig. 25.2. Here, *relE* transcription is induced at time zero in MG1655Δ*lpp* cells carrying the plasmids pMCD25410 and pMCD25411, which express mRNAs containing the wild-type *lpp* ORF and a nontranslatable *lpp*, respectively. As is shown in Fig. 25.2, the quantity of *lpp* mRNA decreases rapidly on RelE induction with translation of the *lpp* mRNA (Fig. 25.2, lanes 1–4), as opposed to the quantity of nontranslatable *lpp* mRNA, which accumulated after RelE induction (Fig. 25.2, lanes 5–8). The figure clearly shows that RelE cleavage depends on translation of the mRNA (Christensen and Gerdes, 2003).

3.4. Primer extension analysis of *lpp* mRNA

The protocol for primer extension can essentially be divided into five separate sections: (1) labeling of the primer with [γ^{32}P] ATP, (2) Sanger DNA sequencing, (3) hybridization of radiolabeled primer to RNA, (4) reverse transcription, and (5) separation of the synthesized cDNA by PAGE.

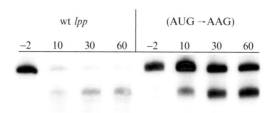

Figure 25.2 Northern blot analysis of wild-type and mutated *lpp* mRNAs. Cells of MG1655Δ*lpp*/MG3323 (pBAD::*relE*) containing either pMCD25410 (wt *lpp*) or pMCD25411 (ATG start codon of *lpp* changed to AAG) were grown exponentially in LB medium at 37 °C. To induce transcription of *relE*, arabinose (0.2%) was added at time zero. Total-cell RNA samples were fractionated using PAGE, and *lpp* mRNA was visualized using Northern blotting. Numbers represent times in minutes after the addition of arabinose.

3.4.1. Labeling of the primer

Primer lpp 26 was used to map the 5′-end of the *lpp* mRNAs expressed from pMCD25410, pMCD25411, and pMCD25412. The following reagents were mixed in a microcentrifuge tube and incubated at 37 °C for 30 min to phosphorylate the primer with $[\gamma^{32}P]$ ATP: 3 μl of primer (5 pmol/μl), 3 μl of polynucleotide kinase buffer, 3 μl of $[\gamma^{32}P]$ ATP (6000 Ci/mmol), 20 μl of H_2O, and 1 μl of polynucleotide kinase (PNK). The PNK enzyme was then inactivated by incubating the reaction mix at 75 °C for 15 min.

3.4.2. PCR sequencing

The DNA sequence of the *lpp* gene was prepared using standard Sanger sequencing using a PCR product of the *lpp* gene as the template and the same primer as for the reverse transcriptase reactions. Six μl of PCR stock (0.06 pmol/μl of radiolabeled primer, 4 fmol/μl of PCR template, 1.25× Taq PCR buffer, 6.25 mM of DTT, and 0.08 U/μl of Taq DNA polymerase (Roche) were added to PCR tubes containing a termination mix for each nucleotide. The four termination stocks contain dideoxynucleotides (ddNTPs) in the respective concentrations: 80 μM of ddGTP, 0.6 mM of ddATP, 0.9 mM of ddTTP, and 0.3 mM of ddCTP. The three remaining dNTPs were added in a concentration of 10 μM to the respective tubes. In addition, 22.5 μM of 7-Deaza-dGTP was added to each sequencing stock. Twenty-five cycles of a standard Taq PCR program were then carried out and 10 μl of formamide loading buffer added to each reaction.

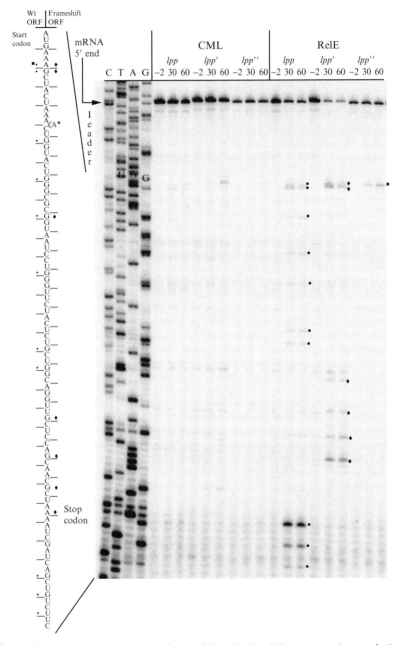

Figure 25.3 Primer extension analysis of *lpp* mRNA. Primer extension analysis of wild-type and mutated *lpp* mRNAs. Strains of MG1655δ*lpp*/pMCD25410 (wt *lpp*), MG1655δ*lpp*/pMCD25412 (*lpp'*—contains a frameshift mutation in the *lpp* gene) and MG1655δ*lpp*/pMCD25411 (*lpp''*—the start codon has been changed from ATG to AAG) were co-transformed with pMG3323 (pBAD::*relE*) and grown exponentially. Arabinose

3.4.3. Primer hybridization and reverse transcription

Ten μg of total RNA from each sample was dried down in a Speedy Vac for approximately 10 min. The RNA samples were then resuspended in 10 μl of hybridization buffer ($1\times$ AMV reverse transcriptase buffer, 0.02 pmol/μl labeled primer, 0.11 mM dNTP mix, 10 mM DTT) and incubated at 80 °C for 2 min on a heating block. The thermostat was then set to 45 °C and the heating block allowed to cool down for approximately 20 min until the temperature was reached. Immediately thereafter, each sample was spun briefly and 1 μl of enzyme mix ($1\times$ AMV reverse transcriptase buffer, 10 mM DTT, 1 U/μl AMV reverse transcriptase) added to each reaction. The samples where then incubated at 45 °C for 30 min, during which time reverse transcription of the RNA occurred. Finally, 10 μl of formamide loading buffer was added to each sample. Samples (1.5 μl), including the sequencing reactions, were then fractionated by PAGE (6% low-bis acrylamide) in a sequencing gel (45 W for 75 min). The gel was dried down and exposed on a PhosphorImager screen (GE Healthcare) overnight, which was scanned on GE Healthcare's Typhoon scanner.

Figure 25.3 shows a primer extension analysis of wild-type and mutated *lpp* mRNAs after the addition of chloramphenicol or arabinose to induce transcription of *relE*. When translation was inhibited by the addition of chloramphenicol, no significant cleavage bands occurred in the mRNA (Fig. 25.3, lanes 5–13), which shows that a general inhibition of translation did not induce mRNA cleavage. However, when translation was inhibited by the ectopic expression of RelE, several specific cleavage bands were observed in the gel (Fig. 25.3, lanes 14–22), which shows that RelE is capable of inducing mRNA cleavage. Only one weak RelE-mediated cleavage band was present in the nontranslatable *lpp* mRNA (*lpp''*, Fig. 25.3, lanes 20–22) as opposed to several significant RelE-mediated cleavage sites in the wild-type and frameshifted *lpp* mRNAs (*lpp* and *lpp'*, Fig. 25.3, lanes 14–19). This observation supports the notion that RelE cleavage depends on translation. Furthermore, no cleavage sites were observed in the nontranslated leader region of the mRNA, again supporting the idea that the presence of ribosomes is essential for proper RelE function. Finally, the two very distinct RelE-mediated cleavage patterns in the wild-type *lpp* mRNA and in the *lpp* mRNA with a frameshift mutation showed that the RelE cleavage pattern depended on the reading frame rather than

---

was added at time zero to induce transcription of the *relE*. The 5′-end of *lpp* mRNA was mapped using the primer lpp26. Numbers represent times in minutes after the addition of arabinose, where −2 is 2 min before addition. Significant cleavage sites in wild-type *lpp*, *lpp'*, and *lpp''* mRNAs are marked with circles, diamonds, or squares, respectively. The sequences of the wild-type and frameshifted *lpp* ORFs are represented to the left, with lines separating the codons. An adenine (marked A*) has been inserted in the sixth codon of the frameshifted *lpp'* mRNA.

on recognition of a specific sequence in the RNA. In support of previous results from Christensen and Gerdes (2003; Pedersen *et al.*, 2003), RelE preferably cleaves between the second and the third base of a codon, with a preference for cleaving on the 5'-end of Gs.

REFERENCES

Bernard, P., and Couturier, M. (1992). Cell killing by the F-plasmid Ccdb protein involves poisoning of DNA-topoisomerase-ii complexes. *J. Mol. Biol.* **226,** 735–745.

Bravo, A., de, T. G., and Diaz, R. (1987). Identification of components of a new stability system of plasmid R1, ParD, that is close to the origin of replication of this plasmid. *Mol. Gen. Genet.* **210,** 101–110.

Christensen, S. K., and Gerdes, K. (2003). RelE toxins from bacteria and Archaea cleave mRNAs on translating ribosomes, which are rescued by tmRNA. *Mol. Microbiol.* **48,** 1389–1400.

Christensen, S. K., Pedersen, K., Hansen, F. G., and Gerdes, K. (2003). Toxin-antitoxin loci as stress-response-elements: ChpAK/MazF and ChpBK cleave translated RNAs and are counteracted by tmRNA. *J. Mol Biol.* **332,** 809–819.

Christensen-Dalsgaard, M., and Gerdes, K. (2006). Two higBA loci in the *Vibrio cholerae* superintegron encode mRNA cleaving enzymes and can stabilize plasmids. *Mol. Microbiol.* **62,** 397–411.

Franch, T., and Gerdes, K. (1996). Programmed cell death in bacteria: Translational repression by mRNA end-pairing. *Mol. Microbiol.* **21,** 1049–1060.

Freistroffer, D. V., Pavlov, M. Y., MacDougall, J., Buckingham, R. H., and Ehrenberg, M. (1997). Release factor RF3 in *E. coli* accelerates the dissociation of release factors RF1 and RF2 from the ribosome in a GTP-dependent manner. *EMBO J.* **16,** 4126–4133.

Gerdes, K., Christensen, S. K., and Lobner-Olesen, A. (2005). Prokaryotic toxin-antitoxin stress response loci. *Nat. Rev. Microbiol.* **3,** 371–382.

Gerdes, K., Rasmussen, P. B., and Molin, S. (1986). Unique type of plasmid maintenance function: Postsegregational killing of plasmid-free cells. *Proc. Natl. Acad. Sci. USA* **83,** 3116–3120.

Gotfredsen, M., and Gerdes, K. (1998). The *Escherichia coli* relBE genes belong to a new toxin-antitoxin gene family. *Mol. Microbiol.* **29,** 1065–1076.

Guzman, L. M., Belin, D., Carson, M. J., and Beckwith, J. (1995). Tight regulation, modulation, and high-level expression by vectors containing the arabinose PBAD promoter. *J. Bacteriol.* **177,** 4121–4130.

Hiraga, S., Jaffe, A., Ogura, T., Mori, H., and Takahashi, H. (1986). F plasmid ccd mechanism in *Escherichia coli*. *J. Bacteriol.* **166,** 100–104.

Jaffe, A., Ogura, T., and Hiraga, S. (1985). Effects of the ccd function of the F plasmid on bacterial growth. *J. Bacteriol.* **163,** 841–849.

Jensen, R. B., Grohmann, E., Schwab, H., az-Orejas, R., and Gerdes, K. (1995). Comparison of ccd of F, parDE of RP4, and parD of R1 using a novel conditional replication control system of plasmid R1. *Mol. Microbiol.* **17,** 211–220.

Lanzer, M., and Bujard, H. (1988). Promoters largely determine the efficiency of repressor action. *Proc. Natl. Acad. Sci. USA* **85,** 8973–8977.

Makarova, K. S., Grishin, N. V., and Koonin, E. V. (2006). The HicAB cassette, a putative novel, RNA-targeting toxin-antitoxin system in archaea and bacteria. *Bioinformatics* **22,** 2581–2584.

Masuda, Y., Miyakawa, K., Nishimura, Y., and Ohtsubo, E. (1993). chpA and chpB, *Escherichia coli* chromosomal homologs of the pem locus responsible for stable maintenance of plasmid R100. *J. Bacteriol.* **175,** 6850–6856.

Miroux, B., and Walker, J. E. (1996). Over-production of proteins in *Escherichia coli:* mutant hosts that allow synthesis of some membrane proteins and globular proteins at high levels. *J. Mol. Biol.* **260,** 289–298.

Ogura, T., and Hiraga, S. (1983). Mini-F plasmid genes that couple host cell division to plasmid proliferation. *Proc. Natl. Acad. Sci. USA* **80,** 4784–4788.

Pandey, D. P., and Gerdes, K. (2005). Toxin-antitoxin loci are highly abundant in free-living but lost from host-associated prokaryotes. *Nucleic Acids Res.* **33,** 966–976.

Pedersen, K., Christensen, S. K., and Gerdes, K. (2002). Rapid induction and reversal of a bacteriostatic condition by controlled expression of toxins and antitoxins. *Mol. Microbiol.* **45,** 501–510.

Pedersen, K., Zavialov, A. V., Pavlov, M. Y., Elf, J., Gerdes, K., and Ehrenberg, M. (2003). The bacterial toxin RelE displays codon-specific cleavage of mRNAs in the ribosomal A site. *Cell* **112,** 131–140.

Tsuchimoto, S., Nishimura, Y., and Ohtsubo, E. (1992). The stable maintenance system pem of plasmid R100: Degradation of PemI protein may allow PemK protein to inhibit cell growth. *J. Bacteriol.* **174,** 4205–4211.

van Melderen, L., Bernard, P., and Couturier, M. (1994). Lon-dependent proteolysis of CcdA is the key control for activation of CcdB in plasmid-free segregant bacteria. *Mol. Microbiol.* **11,** 1151–1157.

Zhang, J., Zhang, Y., Zhu, L., Suzuki, M., and Inouye, M. (2004). Interference of mRNA function by sequence-specific endoribonuclease PemK. *J. Biol. Chem.* **279,** 20678–20684.

Zhang, Y., Zhang, J., Hoeflich, K. P., Ikura, M., Qing, G., and Inouye, M. (2003). MazF cleaves cellular mRNAs specifically at ACA to block protein synthesis in *Escherichia coli. Mol. Cell* **12,** 913–923.

Zhang, Y., Zhu, L., Zhang, J., and Inouye, M. (2005). Characterization of ChpBK, an mRNA interferase from *Escherichia coli. J. Biol. Chem.* **280,** 26080–26088.

Zubay, G. (1973). In vitro synthesis of protein in microbial systems. *Annu. Rev. Genet.* **7,** 267–287.

Author Index

Subject Index

A

Actinomycin D, transcription inhibition in
  Archaea, 384–385
aIF-5A
  purification of *Halobacterium salinarum* protein
    endogenous protein, 394–396
    recombinant proteins
      Escherichia coli, 396–397
      Halobacterium salinarum, 397–399
    ribonuclease activity characterization, 398,
      400–401
Archaea RNA degradation, *see* RNA decay;
  RNA exosome, *Sulfolobus solfataricus*
Aspartate semialdehyde dehydrogenase,
  ribonuclease activity in *Sulfolobus
  solfataricus*, 394

B

Bacillus subtilis messenger RNA decay,
  see RNA decay

C

Circular reverse transcriptase–polymerase chain
  reaction, *see* Polymerase chain reaction
Coimmunoprecipitation
  RNA degradosome co-immunopurification
    from *Escherichia coli*
    immunoglobulin G cross-linking to
      protein A–Sepharose beads, 72–73
    incubation conditions and analysis, 73–75
    materials, 71–72
    overview, 68–71
    polyclonal antibody small-scale affinity
      purification, 75–76
  RNA exosome from *Sulfolobus solfataricus*,
    404–406

D

Dead-box RNA helicases
  activities, 184
  RhlB
    ATPase assay, 193–194
    helicase activity, 185
    messenger RNA degradation
      assay, 195
      pathway, 188–193

ribonuclease E interactions, 187–188, 193
RNA unwinding assay, 194–195
RNA-dependent ATPase activity,
  185–187
types in *Escherichia coli*, 184
DNA microarray, RNA decay analysis in
  Escherichia coli
  cell harvesting, 51–52
  control spots for normalization, 50–51
  data verification, 62
  experimental design, 49–50
  half-life calculation, 59–62
  hybridization, 57–58
  probe preparation, 53–55
  RNA extraction and quantification, 52–53
  scanning, 58–59
  slide preparation, 55–57
  washing, 58
Dss1, *see* RNA degradosome,
  yeast mitochondria

E

Electrophoretic mobility shift assay, ribonuclease
  II–RNA interaction analysis, 147–150
EMSA, *see* Electrophoretic mobility shift assay
EndoA, *Bacillus subtilis*
  assay, 282–284, 303–304
  function, 281
  purification, 281, 283, 304–306
Exosome, *see* RNA exosome, *Sulfolobus solfataricus*

G

Glucose phosphate stress, *see* SgrS
Glycerol gradient centrifugation, RNA exosome
  from *Sulfolobus solfataricus*, 403–404

H

Hfq
  polyadenylation regulation studies in
    Escherichia coli
    global analysis of transcripts, 163–164
    Hfq purification, 374
    rapid amplification of complementary DNA
      ends
      anchors and primers, 166–167
      bacterial strain and growth conditions,
        167–168

555

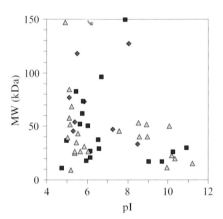

Pierluigi Mauri and Gianni Dehò, Figure 6.4 A virtual 2D map (theoretical pI vs MW) obtained by MAPROMA software for the set of proteins listed in Table 6.1 plus the others with at least one identified peptide. A color/shape code is assigned to each virtual protein spot according the confidence of identification (score value obtained by SEQUEST data handling): yellow/triangles, <20, blue/squares, from >20 to <40, and red/diamonds, >40.